FOOD PROCESSING ENZYMES

FOOD PROCESSING ENZYMES

Recent Developments

Nicholas D. Pintauro

NOYES DATA CORPORATION

Park Ridge, New Jersey, U.S.A.

1979

Library of Congress Catalog Card Number: 79-84426
ISBN: 0-8155-0748-8
Printed in the United States

Published in the United States of America by
Noyes Data Corporation
Noyes Building, Park Ridge, New Jersey 07656

FOREWORD

The detailed, descriptive information in this book is based on U.S. patents, issued since February 1970, that deal with food processing enzymes. This title contains mostly new developments since our previous title *Enzymes in Food Processing and Products* published in 1972.

This book serves a double purpose in that it supplies detailed technical information and can be used as a guide to the U.S. patent literature in this field. By indicating all the information that is significant, and eliminating legal jargon and juristic phraseology, this book presents an advanced, commercially oriented review of recent developments in food processing enzymes.

The U.S. patent literature is the largest and most comprehensive collection of technical information in the world. There is more practical, commercial, timely process information assembled here than is available from any other source. The technical information obtained from a patent is extremely reliable and comprehensive; sufficient information must be included to avoid rejection for "insufficient disclosure." These patents include practically all of those issued on the subject in the United States during the period under review; there has been no bias in the selection of patents for inclusion.

The patent literature covers a substantial amount of information not available in the journal literature. The patent literature is a prime source of basic commercially useful information. This information is overlooked by those who rely primarily on the periodical journal literature. It is realized that there is a lag between a patent application on a new process development and the granting of a patent, but it is felt that this may roughly parallel or even anticipate the lag in putting that development into commercial practice.

Many of these patents are being utilized commercially. Whether used or not, they offer opportunities for technological transfer. Also, a major purpose of this book is to describe the number of technical possibilities available, which may open up profitable areas of research and development. The information contained in this book will allow you to establish a sound background before launching into research in this field.

Advanced composition and production methods developed by Noyes Data are employed to bring these durably bound books to you in a minimum of time. Special techniques are used to close the gap between "manuscript" and "completed book." Industrial technology is progressing so rapidly that time-honored, conventional typesetting, binding and shipping methods are no longer suitable. We have by-passed the delays in the conventional book publishing cycle and provide the user with an effective and convenient means of reviewing up-to-date information in depth.

The Table of Contents is organized in such a way as to serve as a subject index. Other indexes by company, inventor and patent number help in providing easy access to the information contained in this book.

15 Reasons Why the U.S. Patent Office Literature Is Important to You –

1. The U.S. patent literature is the largest and most comprehensive collection of technical information in the world. There is more practical commercial process information assembled here than is available from any other source.

2. The technical information obtained from the patent literature is extremely comprehensive; sufficient information must be included to avoid rejection for "insufficient disclosure."

3. The patent literature is a prime source of basic commercially utilizable information. This information is overlooked by those who rely primarily on the periodical journal literature.

4. An important feature of the patent literature is that it can serve to avoid duplication of research and development.

5. Patents, unlike periodical literature, are bound by definition to contain new information, data and ideas.

6. It can serve as a source of new ideas in a different but related field, and may be outside the patent protection offered the original invention.

7. Since claims are narrowly defined, much valuable information is included that may be outside the legal protection afforded by the claims.

8. Patents discuss the difficulties associated with previous research, development or production techniques, and offer a specific method of overcoming problems. This gives clues to current process information that has not been published in periodicals or books.

9. Can aid in process design by providing a selection of alternate techniques. A powerful research and engineering tool.

10. Obtain licenses — many U.S. chemical patents have not been developed commercially.

11. Patents provide an excellent starting point for the next investigator.

12. Frequently, innovations derived from research are first disclosed in the patent literature, prior to coverage in the periodical literature.

13. Patents offer a most valuable method of keeping abreast of latest technologies, serving an individual's own "current awareness" program.

14. Copies of U.S. patents are easily obtained from the U.S. Patent Office at 50¢ a copy.

15. It is a creative source of ideas for those with imagination.

CONTENTS AND SUBJECT INDEX

INTRODUCTION

Enzymes can be defined as a specialized class of proteins which catalyze chemical reactions. Over a thousand enzymes have been identified and classified according to the substrate on which the enzyme exerts its catalytic action. Enzyme technology is based on protein chemistry applied to production, isolation and purification of enzymes for food, medicinal and industrial purposes. Most enzymes are hydrolytic in action; for example, there are numerous proteases, carbohydrases and lipases which hydrolyze proteins, carbohydrates and lipids from their natural complex structures to simpler, lower molecular weight species which function more effectively. Other catalytic reactions are possible, for example, oxidation-reduction, isomerization, addition to double bonds, and formation of bonds with ATP cleavage.

During the period immediately following World War II, commercial enzymes were used in the food industry mainly for fruit juice clarification, in beer and wine making, in the starch industry, and in dairy and cheese operations. The presence of active enzymes in foods has always been recognized and procedures were developed for enzyme inactivation by heat and chemical methods in order to preserve and stabilize the quality of food during storage.

In the 1960s, the detergent industry became involved in a vast development program in the use of enzymes to clean and remove stains and soil from clothing during regular washing procedures in the home. There was a rapid expansion in the production of proteases and amylases, and in 1969 enzyme production accounted for 3% (a $64 million segment) of the U.S. fermentation industry, of which 50% was produced for the detergent industry and 30% for foods This vast commercialization of enzymes sparked a renewed interest in enzyme chemistry.

Probably the most significant highlight of the enzyme-detergent era was the effort made by the fermentation industry to produce enzymes at higher yields and lower costs to promote wider usage. The peak period for research and development in enzyme chemistry and technology was in the period 1971-1976 when the National Science Foundation funded an enzyme program in its Research Application to National Needs (RANN) program. In 1972 RANN alloted $23 million for 21 projects directly identified as research in enzyme technology.

Developments in enzyme technology, as applied to foods and beverages, are presented in this publication according to food categories, for example, dairy, cheese, meats, starch, fruits and vegetables, etc. The subject matter is organized to show the approaches taken to solve a problem in food manufacture, to improve the specific quality characteristics of food products, to improve storage stability, or to provide greater convenience in recipe preparation.

The researcher has the choice of preparing the enzyme material or obtaining supplies from an enzyme producer or special firm. The cultures that are used to generate enzymes in fermentation processes are designated in this publication by the reference number assigned by the worldwide depositories or centers listed below:

ATCC - American Type Culture Collection
12301 Parklawn Drive
Rockville, Md 20852

CBS - Centraalbureau Voor Schimmelcultures
Julianalaan 67A
Delft (or Baarn), The Netherlands

FRI - Fermentation Research Institute
Agency of Industrial Science & Technology of Japan
Inage, Chiba City, Japan

IAM - Institute of Applied Microbiology
University of Tokyo
Tokyo, Japan

NRRL - Northern Regional Laboratories
U.S. Department of Agriculture
Peoria, Ill 61604

Because enzymes are generally soluble and difficult to recover or purify after complexing with substrates, it was the common practice, up to about ten years ago, to use enzymes without provisions for reuse. Therefore, use of enzymes was limited because of cost and availability. In the early 1970s there was a proliferation of new developments based on the principle of "enzyme immobilization," that is, the binding of enzymes, physically and chemically to inert or insoluble carriers (cellulose, glass beads, resins, collagen, etc.). These new methods made it possible to use column, bed or batch methods without loss of the enzyme because it was possible to recover the carrier and enzyme for reuse.

It can be safely stated that enzyme technology is still in its infancy. Chemical elucidation of protein structure and binding mechanisms of enzymes with substrates are beginning to unfold as researchers continue to probe for answers to problems of enzyme reactions.

LACTASES AND MILK CLOTTING ENZYMES (Cheese)

THERMOSTABLE LACTASES

Lactase from Thermophilic Fungi

Lactose is an abundant disaccharide which is present in milk and in certain dairy products such as yogurt and whey. However, its nutritional value is limited because of the fact that a large proportion of the human population lacks the intestinal enzyme lactase. As a result, the lactose content of milk and other dairy products is not utilized or is only partially utilized by many persons.

Lactase extracted from natural sources may be used to hydrolyze lactose to glucose and galactose, which are more easily assimilated. One of the best known sources of lactase is the yeast of *Kluyveromyces fragilis.* Other known sources are other yeasts, and certain bacteria and fungi. However, lactase as prepared from such sources has exhibited low activity or it is heat labile and rapidly loses its activity at elevated temperatures such as 50° to 60°C. Another problem encountered in providing sources of lactase is that some organisms do not undergo cell growth at an adequately rapid rate. Inasmuch as lactase production is proportional to cell production, yields of lactase from such sources are low.

E.V. Crisan and S.G. Sorensen; U.S. Patent 4,007,283; Feburary 8, 1977; assigned to The Regents of the University of California describe a method for production of lactase from thermophilic fungi at a commercially acceptable rate and in a thermostable and active form. The lactase can be used in low pH dairy products such as buttermilk, yogurt and sour cream in which the lactose component is converted to glucose and galactose for a sweeter product without extra calories.

Not every species, nor every genus of this category (thermophilic fungi) is capable of producing highly active thermostable lactase. Within a given species, there may be variations from one strain or variety to another, some being better than others with respect to cell yield and/or enzyme activity and/or thermal stability. Further, it may be possible, with a strain or species which does not meet these criteria in adequate degree, to induce the desired result by proper selection or manipulation of the culture medium. Thus, given a species and/or given a strain or variety of thermophilic fungi of this category of organisms which does not thrive initially upon a culture medium in which lactose is the sole carbon source, it may be possible to induce such ability by growing the organism on a medium containing both lactose and another more readily assimilable carbon source such as glucose. The organisms are transferred from one culture medium to another, the successive media having diminished proportions of glucose and larger proportions of lactose until organisms result which will grow upon a medium containing lactose as the sole source of carbon.

After practical completion of cell growth (i.e. growth to a point such that further cell growth and/or further lactase production is uneconomical), the mycelium will be harvested. It is not necessary to use drastic separation procedures; and mild procedures may be used which leave the lactase attached in some manner to the ruptured cells. In any event the mycelium is separated from the spent broth or culture medium; it is broken down mechanically to the desired degree and large particles of mycelium are removed.

Well-known separation techniques such as filtration, centrifugation, precipitation with polyacrylic acid, column chromatography using carboxymethylcellulose, gel exclusion chromatography using various Sephadex products, etc. may be used to separate and concentrate the lactase fraction or component of the mycelium. It is not necessary to produce a pure enzyme.

The lactase has a high degree of activity and it retains its activity at temperatures of 60°C or higher. This enables it to be used efficiently in industrial processing of dairy products at elevated temperatures. Among its advantages are the following: At such temperatures the hydrolysis of lactose proceeds more rapidly than at lower temperatures. Where the hydrolysis step is carried out at 60°C or higher, the growth of unwanted organisms such as bacterial or fungal species that commonly induce spoilage of the product is inhibited.

A further advantage of the lactase arises from the fact that much if not all of this lactase exhibits maximum activity at a low pH, e.g. between 4 and 5. Certain fermentation processes such as the production of buttermilk, yogurt and sour cream proceed at low (acid) pH. Further, a mixture of conventional lactase, e.g., from *K. fragilis* (which has its optimum activity at about pH 7), with lactase of this process may be used to advantage, e.g., mixtures of 20 to 80 parts by weight of the process and 80 to 20 parts of conventional lactase.

This mixture of enzymes may be added at a point where the temperature is low (e.g. 40°C) and/or the pH is high (e.g. close to 7) at which point the conventional lactase is most effective. Then during a subsequent phase of the process where the temperature is higher (e.g. 50° to 60°C or higher) and/or the pH is low (e.g. 4 to 5), the process lactase is most effective.

Referring to Figure 1.1, raw milk enters the system at **10** and may contain about 3½ percent butterfat and 8.6 percent of non-fat solids. The milk passes to a storage unit where it is held at a suitable low temperature, for example less than 40°F (4.4°C). The raw milk passes out of storage and may be warmed if desired in a heating unit (not shown) to, for example, 100°F. It then passes to a centrifuge **13** for separation, clarification and/or standardization.

Figure 1.1: Use of Lactase from Thermophilic Fungi

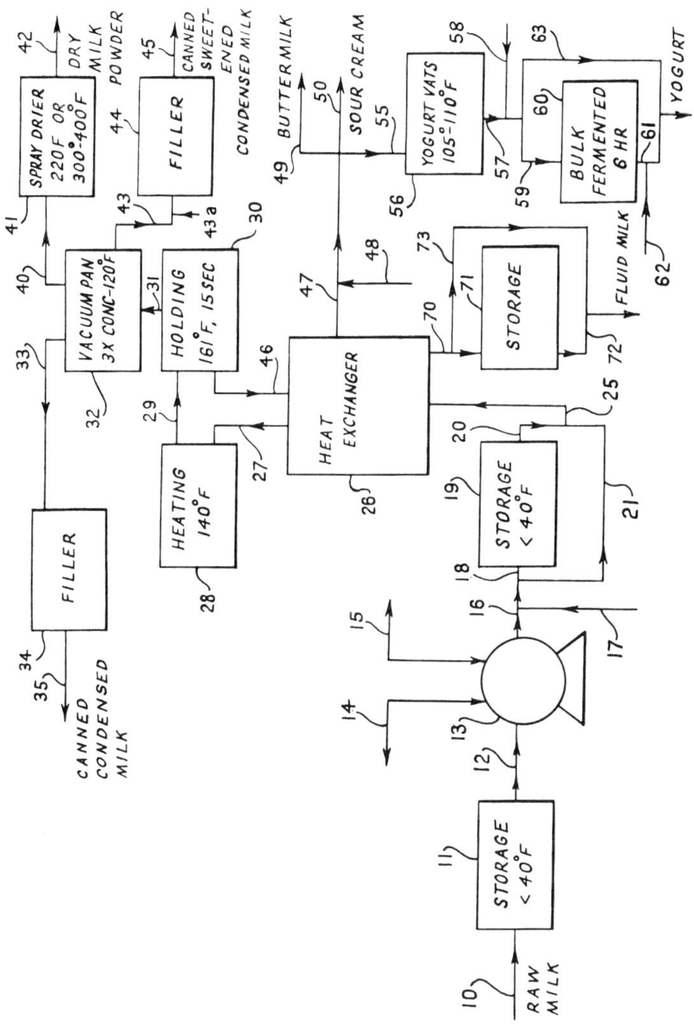

Source: U.S. Patent 4,007,283

Double headed arrows **14** and **15** indicate that in this step butterfat may be removed or may be added and non-fat milk solids may be removed and/or added. The lactase is added through line **17,** for example in an amount equal to about 0.03% of the weight of raw milk. The milk may then pass to a storage unit **19** where it is held at a low temperature, for example below 40°F. Alternatively, the milk may pass directly through line **21.**

In either case the milk passes through a heat regeneration unit or heat exchanger **26** which serves the purpose of preheating the incoming milk in part by heat exchange with outgoing treated milk that has been heated in a subsequent stage of the process. If desired during its passage through the heat exchanger, the milk may be subjected to a homogenizing step (not shown).

In the high temperature short time (HTST) procedure most often used in the dairy products industry, the temperature in unit **30** will be typically about 161°F and the holding time 15 seconds. This accomplishes pasteurization. The preheating unit **28** is used to raise the temperature close to that in the next unit and at the temperature prevailing in unit **28** the lactase is stable and accomplishes hydrolysis of lactose. The process lastase is thermostable in comparison to conventional lactase such as that derived from yeast.

The pasteurized product passes out of unit **30** to a vacuum pan dehydration unit **32.** Assuming that condensed milk is the end product, this dehydration will be partial only and the partially evaporated or condensed milk passes out of the unit **32** to a filler **34** where it is filled into containers. Under the conditions described and illustrated, most or all of the lactose will have been hydrolyzed to glucose and galactose.

In another application whey is treated. Whey is a waste product in the manufacture of cheese and its disposition is a severe problem. It may be treated with the lactase and held at a temperature of about 60°C for a time sufficient to convert the lactose to glucose and galactose and may then be used as a food ingredient for human or animal consumption.

An added advantage is that by hydrolyzing the lactose content of whole milk, whey or other dairy products to glucose and galactose, a sweeter product having no greater calorie content is produced. Yet, another advantage is that by removal of lactose its tendency to crystallize in the manufacture of ice cream and other concentrated milk products is eliminated.

Cultures of 54 strains of thermophilic filamentous fungi, representing over 15 species of 12 genera, were tested for their ability to produce lactase. All strains were capable of exhibiting some growth at 45°C on a culture medium containing lactose as a sole carbon source. 25 strains exhibiting the best growth on the lactose medium were recultured in shaker culture.

Enzymatic activities for selected strains from the genera *Chaetomium, Humicola, Malbranchea, Mucor, Sporotricum* and *Torula* are shown in the table. Freeze-dried preparations from the mycelium of the fungi tested hydrolyzed between 0.03 to 0.95 μmol lactose/mg (dry wt) of cells after 1 hr at pH 6.6 and 37°C. Mycelial dry weight was determined after oven drying at 70°C for 36 hr. Four strains of *Mucor pusillus* exhibited the greatest enzymatic activity, 0.81 to 0.95

μmol/mg, and were studied to determine the pH and temperature optima of their thermophilic lactase. At the optimum pH of 4.4 these strains exhibited activities of 1.15 to 1.36 μmol/mg that were 45% greater than those at pH 6.6. The *Mucor* enzyme preparations exhibited a broader range of activity with respect to pH, with more than 75% of the maximum activity being maintained through a range of pH 4.1 to 5.5. At pH 4.4, the greatest catalytic activity, 1.58 to 2.58 μmol/mg, occurred when lactose hydrolysis was carried out at 61°C, and at least 75% of the maximum activity was retained between 53° and 63°C.

Preparations from fresh mycelium were more active than preparations from freeze-dried cells and hydrolyzed 4.15 to 7.21 μmol of lactose/mg and 1.56 to 2.61 μmol/mg, respectively. By using the optimum pH and temperature determined here, the enzymatic activities of crude preparations from fresh mycelium were 4 to 6.7 times higher than those observed during the preliminary assays conducted at pH 6.6 and 37°C using preparations from freeze-dried cells.

A comparison of the thermostability of the *Mucor pusillus* (3-4-1) enzyme with that of *K. fragilis* showed that although the yeast produced approximately three times more activity per hour, 96% of this activity was lost within 10 min at 60°C. The thermostable *Mucor* enzyme lost only 27% of its activity after 1 hr at 60°C and retained more than 60% of its activity after 2 hr at this temperature. This difference in thermostability can also be noted in terms of the half-life of enzymatic activity.

The yeast enzyme had a half-life of less than 1 min at 60°C while the thermophile enzyme had a half-life of 180 min at this temperature. The thermostability of the *Mucor* enzyme was significantly greater at other temperatures as well.

Lactase Activity of Thermophilic Filamentous Fungi

Organism	Strains Screened	Strains Assayed*	Activity, μmol/mg
Aspergillus sp.	2	–	–
Chaetomium sp.	3	–	–
C. thermophile var. *coprophile*	3	1	0.12
Humicola grisea var. *thermoidea*	2	2	0.03-0.12
H. isolens	1	–	–
H. lanuginosa	2	1	0.03
Malbranchea pulchella var. *sulfurea*	3	2	0.07-0.14
Mucor miehei	3	2	0.15-0.29
M. pusillus	15	11	0.13-0.95
Mucor sp.	7	2	0.28-0.89
Myriococcum albomyces	1	–	–
Papulospora thermophila	2	–	–
Penicillium sp.	1	–	–
Rhizopus sp.	1	–	–
Sporotrichum sp.	1	1	0.65
S. thermophile	2	2	0.18-0.33
Thermoascus aurantiacus	2	–	–
Torula sp.	1	–	–
T. thermophila	1	1	0.58

*Only strains exhibiting good growth on lactose-containing media were assayed quantitatively.

Streptomyces Lactase

Lactase is an enzyme which is not used commerically to any great extent, although these are some possible applications of lactase: (1) increased sweetness in fluid milk without addition of sugars or sweeteners; (2) softening of cast center candies containing lactose using lactase; and (3) improved fermentation in flour doughs if lactose is converted to the more readily fermented monosaccharides.

Lactase has been found in almonds, the intestinal mucosa of young animals, and in various microorganisms including yeasts, bacteria and fungi. Most products which have been used as lactases commercially have been derived from lactose-fermenting yeasts, usually *Saccharomyces fragilis, Zygosaccharomyces lactis, Saccharomyces lactis,* or *Candida pseudotropicalis (Torula cremoris).* Fungal preparations which have been found to be active as lactases, under certain conditions, are derived from *Aspergillus luchuensis, A. oryzae* and *A. niger.* The classical source of bacterial lactase is *Escherichia coli.*

A major limitation to using lactase in processing milk products has been the relatively low heat stability of the available lactases, typically yeast derived. Use of lactases required either long incubation (several days) at low temperature, or a relatively short (2 to 4 hours) incubation at 35° to 40°C. Any longer incubation at 35° to 40°C promoted very significant bacterial growth. The recent observation that a heat stable lactase can be prepared from *Streptomyces* will alleviate the bacterial spoilage problem.

It is quite clear that a lactase of high heat stability would have a distinct advantage over any lactase preparation previously available or described. It could be used under conditions where the substrate would probably remain sterile, or be rendered essentially sterile during the incubation.

The availability of a heat stable lactase prepared from *Streptomyces* is reported by *A.E. Collinge, C.E. Neubeck and J.R. Udinsky; U.S. Patent 3,816,259; June 11, 1974; assigned to Rohm and Haas Co.* Experimental work indicates that *Streptomyces* spp. can produce a unique lactase, in that activity is exhibited at a temperature (70°C) which is above the usual milk pasteurization temperature. All previously described lactases show optimal activity below the pasteurization range.

Properties of β-Galactosidase (Lactase) of Streptomyces coelicolor: This enzyme is produced by culturing an inoculum organism of the genus *Streptomyces* in the temperature range of 20° to 50°C, on a nutrient medium containing lactose under aerobic conditions. The medium also requires incorporation of a proteinaceous material, typically chosen from one of autolyzed yeast (or yeast extract), liver and gland meal, liver meal, and blood meal, constituting up to three weight percent. An example of growing nutrient for *S. coelicolor* and process for preparation of a lactase enzyme concentrate is given in the table below.

Composition of Exemplary Growing Nutrient for *S. coelicolor* to Produce Heat Stable Lactase

	Percent
Lactose	2

(continued)

	Percent
Liver and Gland Meal	1
K_2HPO_4	0.6
$MgSO_4 \cdot 7H_2O$	0.1
$CoCl_2$	0.03

The lactase enzyme concentrate was prepared as follows. A concentrate is prepared from crude culture by adding four volumes of cold isopropanol to a crude culture grown on lactose medium. Initial precipitate is allowed to settle, and washed two times with fresh isopropanol. The final precipitate was dried in vacuo; yield is 46.4 g from 3,150 g of culture.

The β-galactosidase activity of *S. coelicolor* grown on lactose medium at 40°C was determined at pH 4, 5, 6, and 7, using o-nitrophenyl-β-galactoside (ONPG). The results were as follows:

pH	Activity (μmol hydrolyzed/min/mg of E)
4.0	0
5.0	0.00195
6.0	0.00515-0.00595
7.0	0.0032

The pH optimum was thus settled on as 6.0, though a range of slightly acid to neutral pH is operable. Similar preparations of *S. coelicolor* variant grown on a medium free of lactose produced significantly less activity (0.00016 μmol/mg/-min), at the optimum pH. The medium requires a carbohydrate component at a concentration of from 0.25 to 2 percent by weight.

The preparation made from *S. coelicolor* grown on lactose medium, was then tested for β-galactosidase activity on ONPG at 50°, 60°, 65°, 70°, and 80°C, with the following results obtained at pH 6.0.

Temperature (°C)	Activity (μmol hydrolyzed/min/mg of E)
40	0.006
50	0.0116
60	0.0154
65	0.0174
70	0.0149
80	0

The optimum temperature was thus settled on as 65°C, though 60° to 70°C is an operable range. The temperature range is sufficient to cause lactose hydrolysis at a rate of 0.02 to 0.1 percent per hour. Coincidentally, pasteurization (low temperature process) is usually carried out for 30 minutes at 61.5° to 63°C (142.5° to 145°F). At 65°C the enzyme activity is about three times greater than at 40°C, which is presently only about one-tenth the level exhibited by known fungal and yeast concentrates, necessarily prepared and used at 40°C. Yet the fungal and yeast enzyme concentrates have no activity at 65°C. Published work with glucose isomerase derived from species of *Streptomyces* had indicated that this enzyme has a marked sensitivity to certain metallic ions, among them Cu^{++}, Fe^{+++}, and Ca^{++}. Since all of these are found in milk, it was of interest to determine the sensitivity of the present β-galactosidase from *S. coelicolor* to these ions, using the ONPG method at pH 6.0 and 65°C. The β-galactosidase was tested by the ONPG method in the presence of 0.001 and 0.0001 molar concentrations of copper acetate, ferric chloride, and calcium acetate, then compared with a control free of added salts. The results were as follows:

Molar Concentration of Metallic Salt	β-Galactosidase Activity, Percent of Control Free of Metallic Salt
Copper acetate	
0.0001	38
0.001	7
Ferric chloride	
0.0001	100
0.001	11
Calcium acetate	
0.0001	104
0.001	93

Copper appears to have a very harmful effect on this β-galactosidase activity. The effect of iron is somewhat harmful, but only at relatively high levels. The effect of calcium was not significant.

The effect of the *Streptomyces* derived β-galactosidase was also checked on a milk substrate prepared from dried skim milk, diluted with water to the normal solids concentration present in milk (10% ≈ 4.5% lactose). The milk was incubated at 65°C both at pH 6.55 (ambient pH), and pH 6.0 for 30, 60, and 120 minutes, using levels of enzyme equivalent to 1 and 10% of the lactose concentration. After incubation, the milk samples were deproteinized with sulfuric acid, and the clear filtrates were tested for the presence of glucose and fructose to measure the extent of hydrolysis of the lactose and its isomer lactulose in the milk.

There was no significant production of fructose under the incubation conditions used. The production of glucose from the lactose in the milk was about as expected on the basis of the ONPG activity. At pH 6.0, the activity was 0.0175 μmol lactose hydrolyzed per mg E per minute and at pH 6.5 the activity was 0.0145 μmol lactose hydrolyzed per mg E per minute.

The *Streptomyces* used for the production of lactase, has been designated as *Streptomyces coelicolor.* A subculture of this variety can be obtained from the permanent collection of ATCC 21666.

LACTASE AND FILTERED FRACTIONS

An apparatus and process was developed by *R.A.N. Edwards, P.M. Cantrell, and J.J. Miller; U.S. Patent 4,091,116; May 23, 1978* for special hydrolysis of lactose in milk. The process consists of filtering fractions and recombining of fractions with addition of enzyme to selected fractions. Two filters are used in the process.

If the pore size of the two filters is the same, or if the pore size of the second filter is larger than the pore size of the first filter, the final composition of the liquid will be the same as its initial composition, with the exception of the enzymatic conversion of the substrate compound. This process can thus be used to effect the economic conversion of specific substances in a heterogeneous solution or suspension provided that the substance can be enzymatically converted by an enzyme having a larger molecular size than that of the substance. The process is of particular use where it is desired to remove from the solution a compound, other than the substrate compound, which would also be converted by the enzyme or which would deactivate the enzyme. It has been found that the removal of the larger particles or molecules from the solution prior to enzymatic conversion of the substrate compound increases the activity shown by the enzyme.

If desired, an enzyme may be grafted onto a soluble substrate in order to increase its molecular size and hence the rate at which the reaction solution may be filtered. Apparatus for the adjustment of the lactose content of skim milk using β-galactosidase includes a feed tank **10**, ultrafilters **11** and **12** and an enzyme reactor **13** (Figure 1.2).

Figure 1.2: Apparatus for Adjusting Lactose Content of Skim Milk

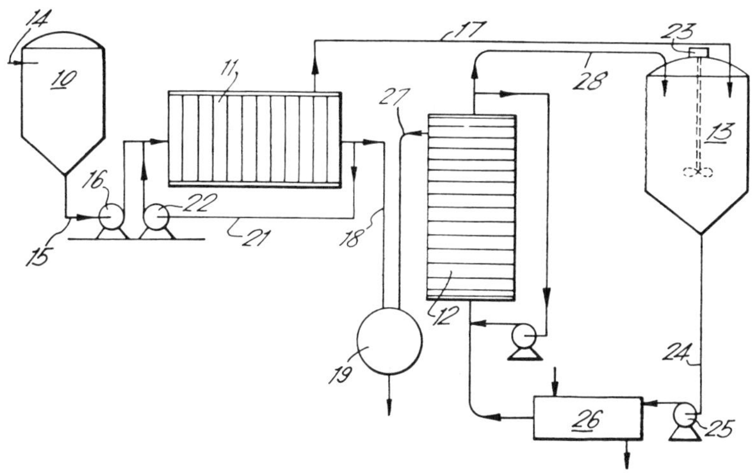

Source: U.S. Patent 4,091,116

Skim milk is fed intermittently into feed tank **10** through line **14** and is fed continuously out along line **15** through pump **16** to the first ultrafilter **11**. Milk passing through ultrafilter **11** is separated into a filtrate consisting of a solution of molecules of a molecular size smaller than that of the enzyme and a concentrate containing larger molecular size molecules. In practice it was found convenient to operate the ultrafilter under conditions wherein 90% of the skim milk is in the filtrate and 10% is separated off into the concentrate.

The filtrate is passed along line **17** to the enzyme reactor **13** while the concentrate is passed directly through line **18** to an in-line mixer **19** for recombination with enzyme treated filtrate. In order to facilitate the starting up of the apparatus a recycle line **21** and pump **22** are provided to recycle milk concentrate through the ultrafilter **11**. Line **17** carries the filtrate from ultrafilter **11** to the enzyme reactor **13** which is fitted with an agitator **23**. The enzyme β-galactosidase is added to the filtrate and converts the lactose in the milk filtrate to glucose and galactose. A conversion rate of 90% has been found to be obtainable with short residence times in the reactor.

The reacted milk is carried by line **24** through pump **25** and heat exchanger **26** to the second ultrafilter **12**. The heat exchanger **26** is used to maintain the temperature in the enzyme reactor **13** at the temperature which corresponds to the maximum activity of the enzyme present in the reactor.

Ultrafilter **12** separates the reacted milk into a filtrate containing the reacted milk constituents and a concentrate containing the enzyme. The filtrate is carried by line **27** to the in-line mixer **19** for recombination with the milk concentrate produced in the first ultrafilter **11**. The enzyme concentrate from the second ultrafilter **12** is returned to the enzyme **13** through line **28**.

The lactose-adjusted milk contained 19% of its initial lactose content. The residual lactose content can be adjusted in the process by varying the ratio of separation achieved by the first filter, or by adjusting the degree of hydrolysis of the filtrate in the enzyme reactor.

Milk having an adjusted lactose can be fed to persons who are unable to metabolize lactose, a condition common in large areas of Asia and the Middle East. The adjusted milk has considerable use in food processing, e.g., lactose tends to precipitate out from frozen dairy products such as ice cream; the use of the treated milk alleviates this problem as well as reducing the amount of sucrose which must be added to the product due to the greater sweetening power of glucose and galactose as compared with the original lactose. Baked goods such as buns and cakes have better browning properties when made with lactose-adjusted milk. The apparatus described above can be used to carry out other processes as shown in the following examples.

Example 1: This example describes the adjustment of the glucose content of egg albumen. Egg albumen normally contains about 0.5% of glucose. When egg albumen is dried to a low moisture content, the glucose will react with amino groups supplied by the major constituent of albumen, the proteins, to form dark colored polymers which detract from the appearance, solubility and functional characteristics of albumen when used for cooking purposes.

It is customary to remove the glucose from albumen prior to drying. This may be achieved by using glucose-oxidase as the enzyme, and filters that are of such a pore size so as to retain glucose-oxidase. In this case it is necessary to dilute the albumen with water prior to feeding to the first filter, as albumen viscosity is high and hampers flow through the system.

Example 2: This example describes the removal of glucose from blood serum prior to drying. Blood is fed to the first filter where red blood cells and molecules larger than glucose-oxidase are removed. The residual serum passes through the filter and the glucose contained in it is converted to gluconic acid by the glucose-oxidase enzyme. The reasons for removal of glucose from blood are similar to those given in Example 1, i.e. if dried blood is to be used as a feed, or as a microbiological media, it undergoes browning reaction if glucose and amino groups are allowed to interact to form colored polymers.

Example 3: This example describes the removal of the bitter principle from orange juice. Orange juice that is bitter contains the material limonin which may be broken down and debittered by the enzyme limoninase.

Orange juice is fed to the first filter where insoluble solids and soluble matter of molecular size greater than limoninase are removed. The enzyme limoninase operates on limonin between the two filters. It is not clear what constitutes the breakdown products of limonin; however it has been found that they are smaller than limonin and are non-bitter.

IMMOBILIZED LACTASE SYSTEMS

Metal Oxide Carrier

Attaching enzymes to insoluble supporting materials has been developed and can now be applied to practical, commercial processes such as producing glucose from starch, isomerizing glucose to the sweeter sugar fructose, hydrolyzing proteins and sugars, clarifying fruit juices and beer, carrying out reactions for producing antibiotic pharmaceutical reagents, treating human blood for promoting desired chemical reactions such as decomposition of urea, and many other useful and potentially useful purposes.

A major advantage of using enzymes in such insoluble form bound to a solid supporting material is that the catalytically active enzyme may be physically retained in the reaction vessel and contacted there with a continuously flowing liquid process stream. Before it was possible to bond enzymes to such insoluble supports, the enzymes would remain in the liquid process stream or could be separated only with difficulty with the result that the enzyme could be reused only with difficulty or not at all.

Now that it is possible to bond enzymes to insoluble supports, it becomes possible to use them in much the same way as ordinary heterogeneous catalysts on inert, insoluble supporting carriers that are well known in the chemical process industry. Such use now permits the continuous, convenient reuse of the same insolubilized enzyme catalyst for contacting a continuously flowing liquid reaction process stream thereby catalyzing a desired chemical reaction within the stream but without the necessity of separating the enzyme from the reaction products or the possible disadvantage of losing the enzyme entirely and not being able to reuse it.

However, some major problems have been encountered in the practical application of such insolubilized enzymes. They have usually been bonded to natural and synthetic high polymers and then used in fixed bed reactors; in such systems disadvantageously high pressure drop and plugging have been encountered and this behavior can be attributed to the small particle size and to the deformable, gel-like properties of the polymeric supporting material. More recent attempts to circumvent these kinds of problems have involved using inorganic supporting materials such as apatite and glass. While the latter materials are less deformable than the organic polymeric supports, columns packed with these materials still are susceptible to plugging and cause high pressure drop when small particle sizes are used in fixed, packed beds. Moreover such materials are often either not available in large quantities of controlled properties or are very expensive. When the particles are made sufficiently large to avoid such adverse plugging and pressure drop behavior, their very size causes large mass transfer resistances both within the liquid film surrounding the support particles and within the support particles themselves if they are porous.

Such mass transfer resistance can prevent the reactants from reaching the immobilized enzyme as fast as they can react and thereby can result in inefficient utilization of the enzyme bound to the solid support. To minimize mass transfer resistances, both within the solid supporting particles themselves and in the liquid film surrounding them, the particles should be as small as possible. However, as has already been stated, fixed packed beds of small particles are markedly susceptible to plugging of columns.

R.W. Coughlin and M. Charles; U.S. Patent 4,016,293; April 5, 1977 developed a method to prepare immobilized enzymes for expanded bed operations. The term expanded bed refers to a process wherein the particles of the bed are suspended and agitated by the liquid but do not mix or circulate within the bed to any appreciable extent. The term fluidized bed refers to a process wherein the suspended and agitated particles do circulate and mix within the bed, and the term suspended bed is generic to the terms fluidized bed and expanded bed and used to refer to either one or both of them.

The carrier or support for the enzymes can be particles of metallic nickel or nickel oxide sinter, e.g., Nickel Oxide Sinter 75 and Nickel Oxide Sinter 90. The density of this material is about 8 g/cm^3 and an appropriate particle-size range is 10 μm to 0.25 inch for fluidization by water. Another useful support for fluidized bed operation is fine, particulate stainless steel such as Glidden 316-L-Si stainless steel powder (of size range 149 μm to 105 μm). Still another suitable support is the porous alumina particles of size range 250 to 150 μm, specific surface area is 4 m^2/g and mean pore diameter 0.44 μm (catalyst support SAEHS-33 supplied by Carbarundum Co.).

A particularly effective, simple and low-cost technique of enzyme binding is to contact the carrier particles with a solution of the enzyme to be bound causing the enzyme to sorb on the particles. After sorption the enzyme-bearing particles are treated with a bifunctional reagent such as glutaraldehyde which is believed to react with free amino groups along the polypeptide chain of the enzyme; this treatment crosslinks sorbed enzyme molecules thereby fixing them in place. Such crosslinking can stabilize enzyme activity over wider ranges of pH as compared to the free enzyme.

Also it was found that particles of relatively low surface area can be coated with a porous coating of a metal oxide such as titanium oxide by simply adding $TiCl_4$ to an aqueous suspension of the particles; the $TiCl_4$ hydrolyzes producing titanium oxide which is deposited in a thin porous layer of high surface area on the particles. In this way coated stainless steel particles with titanium oxide coatings are formed that are readily visible under the electron microscope. Particles so treated show a greatly enhanced ability to sorb enzyme and the enzyme appears to be so strongly sorbed within the porous coating that it does not wash off.

Preparation of enzyme catalysts immobilized by bonding to dense particles permits their use in expanded bed or fluidized bed reactors in which the particles are maintained in fluidized suspension by the flow of a stream of liquid (which contains the substrate or reactant) through the bed. Such liquid fluidization provides the advantages of approximate plug flow of the liquid through the bed in contrast to a stirred, slurry-type reactor and also relative freedom from plugging and from high pressure drop in contrast to fixed bed operation.

In the usual type of fluidized operation the fluidized particles will be well mixed within the bed but it is possible to avoid the mixing of the particles from one part of the bed to another by packing the bed with larger particles which do not move or mix themselves and thereby prevent the small enzyme-bearing particles from mixing within the bed, but at the same time permit the fluidization of the small particles within the interstices of the packed large particles.

Example 1: 100 g of Sinter 90 (nickel oxide particles; 35 to 80 mesh sieve size) in 300 ml H_2O were placed in a flask and chilled in an ice bath. To this mixture 65 ml of $TiCl_4$ was slowly added. After addition of $TiCl_4$ the mixture was heated at 80°C for about 30 min, following which the Sinter 90 particles were separated and washed thoroughly to remove non-adhering fine particles of precipitated titanium oxide. It has been found that such treatment or activation of carrier particles with water and a hydrolyzable metal salt (such as $TiCl_4$) can increase their specific surface area by about ten times, thereby rendering the particles much better sorbents for enzymes.

The treated Sinter 90 particles were then contacted with a buffered solution of lactase [β-D-galactosidase produced from *Aspergillus niger*] in the following proportions: 1 g treated Sinter 90, 3 ml citrate-phosphate buffer (pH is 4.0) and 24 mg of the lactase.

The resulting mixture was maintained at 5°C for about 12 hours and then the Sinter 90 particles were separated and washed with buffer solution; these particles then displayed a specific enzymatic activity of about 100 to 300 lactase units (LU) per g. Higher specific enzymatic activity could be obtained by various means such as simply increasing the proportion of enzyme in the contacting phase of this procedure. Similar results were obtained by applying essentially the same procedure to coat stainless steel particles (149 to 105 μm particle size) and then contacting with enzyme in the same way.

It has also been found that subsequent treatment of these particles with glutaraldehyde does not affect specific enzymic activity but such treatment can improve the stability of the bound enzyme. Enzymatic activities of lactase preparations were assayed by measuring the conversion of lactose to glucose using the Glucostat analysis kit.

The Sinter 90 particles as described above were packed in a 1-inch diameter column to a bed height of about 7.6 cm and the resulting device used to treat a solution of lactose (~20 mg/ml) in citrate-phosphate buffer (pH 3.6) at a temperature of 30°C and at flow rates ranging from about 0 to 4 ml/sec-cm^2 of column cross-sectional area (i.e. at superficial velocity of about 0 to 4 cm/sec). Various conversions of lactose to glucose were obtained depending on flow rate or space time in the reactor and the results are plotted in Figure 1.3 as percentage conversion vs (activity units/flow rate), the abscissa being proportional to space time or residence time in the reactor. Conversions were computed from glucose concentrations of the feed and effluent of the column; glucose concentration was measured by the glucose oxidase method using the Glucostat reagent kit.

Figure 1.3: Conversions of Lactose to Glucose

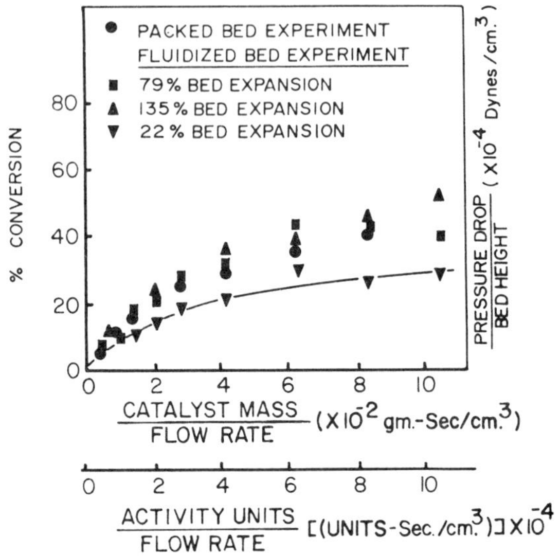

CATALYST SUPPORT - NiO
CATALYST DENSITY - 8.6 gm./cm.^{3}
PARTICLE SIZE - 35-80 MESH
ORIGINAL CATALYST ACTIVITY - 100 UNITS/gm.
CATALYST AGE - 12 DAYS
BONDING TECHNIQUE - EXAMPLE I
BED DIAMETER - 2.54 cm.
PACKED BED HEIGHT - 7.6 cm.
LACTOSE CONC. - 20 mg./ml.
CITRATE - PHOSPHATE BUFFER
pH - 3.6
TEMPERATURE - 30°C

Source: U.S. Patent 4,016,293

The column containing particles of immobilized lactase was operated in the fixed bed mode using downflow with results indicated in Figure 1.3 by circular data points and in the fluidized bed mode using upflow with corresponding results indicated on Figure 1.3 by square data points (for 79% bed expansion) and triangular data points (apex of triangle pointing upwards for 135% bed expansion and apex of triangle pointing downwards for 22% bed expansion).

The superiority of fluidized bed operation with large bed expansion is evident from Figure 1.3. Operation in the fluidized bed mode produces far lower pressure drop (square data points) than operation in the fixed bed mode (circular data points).

Example 2: Glidden stainless steel powder (105 to 149 μm) was contacted at room temperature with solutions of Wallerstein lactase LP enzyme in citrate-phosphate buffer (pH is 3.5) for about ½ hour with gentle agitation. The steel powder was then separated from the enzyme solution and then contacted at room temperature with a 5% solution of glutaraldehyde in citrate-phosphate buffer (pH is 5) for about 80 min with gentle agitation.

After removal from the glutaraldehyde solution the steel powder was recontacted for 30 min with the original enzyme solution at room temperature with gentle agitation. The concentrations and proportions of solutions used for two typical preparations were as follows:

Preparation	..... Enzyme Solution ...... Concentration (mg/ml)	Proportion (ml/g steel)	Glutaraldehyde Solution Proportion (ml/g steel)	Resulting Activity (LU/g)
A	8	2	2	95
B	16	1.2	2	237

It was found, as suggested by the activities of the two preparations detailed above, that the specific activity of the resulting immobilized enzyme bound to steel could be increased by increasing the enzyme concentration of the contacting solution.

Porous Glass Carrier

An important application of lactase in the conversion of lactose-containing whey waste liquor to useful sweeteners is described by *H.H. Weetall and S. Yaverbaum; U.S. Patent 3,852,496; December 3, 1974; assigned to Corning Glass Works.* The process uses porous glass particles as carriers for the lactase enzyme because such particles have a large surface area per g (greater than 100 m^2/g). The flow sheet of the process is shown in Figure 1.4 for recovery of galactose and glucose, and for enzyme conversion of glucose to fructose.

The carriers in porous form, may be used to covalently bond (through silane coupling agents) the enzymes β-galactosidase (lactase) and glucose isomerase. The inorganic carriers should be porous and must have a high surface area (e.g., greater than about 5 m^2/g) to assure sufficient loading of the enzymes. The average pore diameter is 550A±10 percent. Also, the surface of the inorganic carriers must have available hydroxyl or oxide groups to bond with the silicon portion of the silane coupling agents.

Particle size is also important and it should be between about 20 and 80 mesh, U.S. Standard Sieve, so that the enzymes can readily diffuse into the porous network for bonding and so that the respective substrates can readily diffuse into the carrier to the bonded enzymes.

The whey is subjected to the steps of ultrafiltration and reverse osmosis to yield valuable milk protein and an ultrafiltrate of waste products containing lactose which can be further treated to yield useful sweeteners. The ultrafiltrates (waste products) are treated by electrodialysis or fractional crystallization and ion exchange (to remove such undesirable ions as calcium) to remove most salts.

Figure 1.4: Immobilized Lactase and Glucose Isomerase

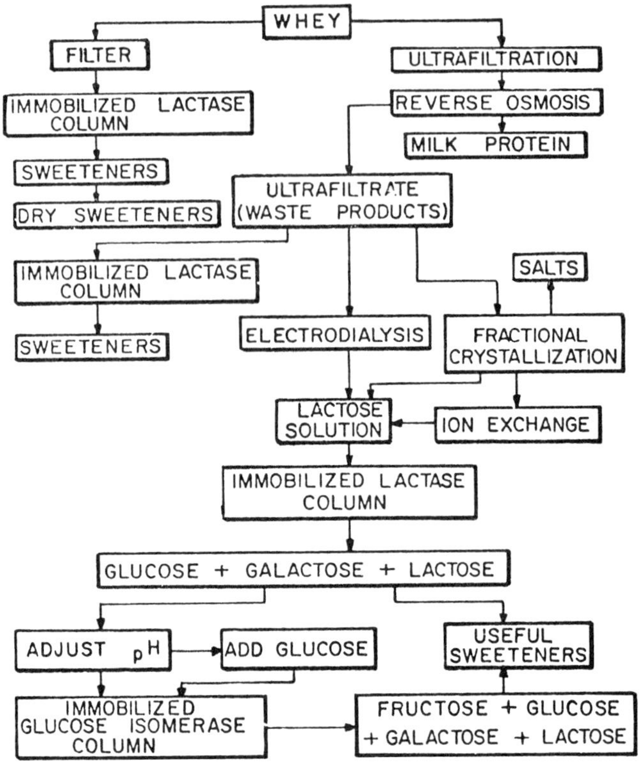

Source: U.S. Patent 3,852,496

The steps of ion exchange and fractional crystallization are to remove ions or salts which may interfere with the product taste or with the subsequent treatments with immobilized enzymes. For example, it was found that the presence of calcium ions in the whey ultrafiltrate tended to interfere with the immobilized glucose isomerase. Hence, it is desirable that such ions be removed by either of the above methods prior to enzymatic treatment.

After removal of the majority of salts and/or undesirable ions, the waste product consists mainly of lactose and small quantities of riboflavin and lactic acid. If desired, the riboflavin may be removed by passing the product through, for example, a column containing activated charcoal. The product then consists mainly of lactose in solution. By passing the lactose-containing solution through a plug flow-through column reactor, containing the immobilized lactase, the lactose hydrolyzes to the sugars glucose and galactose with some lactose remaining.

This solution of sugars can be used directly (e.g., by a dairy) as a useful sweetener, or, treated further by passing the solution through another column reactor containing immobilized glucose isomerase. Since glucose isomerase isomerizes glucose to fructose, the solution, after passing through the column, will consist of fructose, glucose, galactose and some lactose.

The final product may be used as a sweetener in, for example, the manufacture of ice cream. It is important to note that prior to the passage of hydrolysate through the column containing the immobilized glucose isomerase, the pH must be adjusted to a range in which glucose isomerase has an optimum pH.

Preparation of Immobilized β-Galactosidase (Lactase): The reagents used were all reagent grade or better unless otherwise stated. Fungal lactase (lactase-M) was supplied by Miles Laboratories. Yeast lactase (lactase-Y) was supplied by Kyowa Hakko Kogyo Co. The glucose isomerase was from USDA Northern Regional Laboratory. The whey and whey ultrafiltrate came from a local diary.

The lactase enzymes were immobilized on zirconiacoated porous 96% silica glass particles having an average pore diameter of 550 A ±10% (Corning Code MZO-3900). The zirconia-coated porous glass particles were silanized with γ-aminopropyltriethoxysilane by an aqueous method as described in an article by H.H. Weetall and N.B. Havewala, *Biotechnol. Bioeng. Symp.* No. 3, 241 (1972). The enzymes were covalently attached to the alkylamine glass particles, using glutaraldehyde, by the procedure generally described in the above article.

Prior to their use in the treatment of whey and whey ultrafiltrates, the immobilized lactase enzymes were assayed as were free common soluble specimens of the same enzymes. This was done to determine whether or not there would be any effect of the immobilization procedure on the pH optima of the immobilized enzymes.

The soluble lactase-Y was assayed in 20% lactose dissolved in 0.025 M sodium phosphate buffer, pH 6.5, containing 5 x 10^{-4} M manganese and cobalt chlorides, respectively. The lactase-M was assayed in the same buffer without the metal salts. All assays were for 30 min at 60°C. Percent hydrolysis was determined by glucose appearance, since one mol of lactose produces one mol of glucose and one mol of galactose. Glucose measurements were performed using the Glucostat reagent produced by Worthington Biochemical Corporation. One unit of activity was defined as the production of 1 micromol of glucose/min at 60°C at optimum pH. All assays were carried out in a 5 ml total volume at optimum pH.

The immobilized enzyme derivatives were assayed in a total substrate volume of 10 ml in 25 ml flasks. The flasks were shaken to assure thorough mixing during the assay. Sample size ranged from 5.0 mg to 50.0 mg immobilized enzyme derivatives. Specific activity of the immobilized enzymes was determined as units/g on a dry weight basis.

The pH profiles were determined for both the soluble and immobilized lactase enzymes and are quite different. In the case of lactase-M the soluble enzyme shows an optimum at pH 4.0. Immobilization on zirconia-coated CPG causes a shift in pH optimum from 4.0 to pH 3.5. However, at 4.0 the immobilized enzyme still shows 96% of maximum activity. The lactase-Y results were quite unusual. Immobilization caused a pH shift of 3.25 pH units from pH 6.25 to pH 3.0, such that the immobilized enzyme had only 40% of maximum activity at 6.25. The differences in the behavior of the two lactases is not fully understood.

In the experiments below, both the ultrafiltrate of whey and cheddar cheese whey were initially exposed to immobilized enzymes without any adjustments in their general pH range of 4.3 to 4.6.

Example 1: This example describes the treatment of a large batch of whey ultrafiltrate for use in ice cream as a sucrose replacement. Approximately four gallons of whey ultrafiltrate were continuously passed through a lactase-Y plugged flow-through column containing 22,000 units of activity over a seven day period. The starting material was approximately 15% solids of which 70% was lactose. The remainder included calcium, lactic acid, riboflavin, and some residual protein.

The hydrolyzed sample was collected over seven days of operation, pooled, filtered through activated charcoal to remove the riboflavin, and neutralized with a 50% solution of NaOH for treatment with the immobilized glucose isomerase. On neutralization, a large white precipitate formed. This was removed and found to consist mostly of calcium phosphate. The remaining clear neutralized hydrolysate was then tested with immobilized glucose isomerase for conversion to fructose.

Glucose isomerase was prepared in a manner similar to that used for the preparation of the immobilized lactase. Glutaraldehyde was used for the coupling procedure. 1 g of immobilized glucose isomerase containing 525 units (one unit is equivalent to the production of one micromol of fructose at 60°C per minute) was added to 50 ml of whey ultrafiltrate previously adjusted to pH 7.5 and buffered to 0.005 M in magnesium ions by adding $MgCl_2$. The test was carried out for 1 hr at 60°C in a shaking water bath. No conversion was observed.

It became apparent that the glucose isomerase had been inactivated by the presence of calcium ions found in whey. The calcium ions were removed by ionic change with a cation exchange resin. The assay was repeated again and with success. The quantity of glucose in the assay solution was decreased by 50% indicating the maximum isomerization possible.

The whey was continuously passed through a similar column containing the immobilized glucose isomerase. To insure that the final product would be at least equivalent to sucrose on a sweetening basis, an additional 5% by weight glucose was added to the remainder of the four gallons of hydrolyzed whey ultrafiltrate before isomerization. The isomerization was carried out in a column containing 50 g of the immobilized glucose isomerase described above.

The flow rate was maintained at 200 ml/hr. The product was submitted to a dairy for use in preparing an ice cream sample, the results of which are described below.

The final product, after hydrolysis with lactase-Y, the addition of glucose and magnesium chloride, and after isomerization, was found to have the following constituents by weight of solids: 1 percent lactose, 45 percent galactose, 20 percent fructose, and approximately 25 percent glucose. The total solids were initially 15 percent while the final solids content was approximately 10 percent.

Example 2: This example describes the hydrolysis of whole cheddar cheese whey with immobilized β-galactosidase. Whole cheddar cheese whey was filtered through diatomaceous earth (Celite) to remove insoluble particles. The clarified solution was then passed through a column containing lactase chemically coupled to zirconia-coated porous glass particles, as described above. The column was a plugged flow-through glass column reactor (1 cm by 10 cm in length) and contained 8 g of the immobilized lactase having 550 units of enzyme activity per g. The clarified whey was passed through the column at a flow rate necessary to permit 50% hydrolysis of the lactose present in the whey (about 100 ml per hr).

A two liter sample of the hydrolyzed product was evaporated to 40% solids and used for the preparation of an ice cream sample at the maximum acceptable level for the addition of whey as permitted by law (25% wt/wt). From this example, it was found that whole cheddar cheese whey also can be successfully hydrolyzed to useful sweeteners with an immobilized enzyme composite consisting of lactase coupled chemically to zirconia-coated porous glass particles.

Phenol-Formaldehyde-Glutaraldehyde Carrier

Insolubilized but active enzymes are prepared from enzymes which are in a normal or native (soluble) state by reacting them with a phenol-formaldehyde resin and glutaraldehyde. The reaction is generally conducted in an aqueous medium, and the resin and glutaraldehyde are sequentially reacted with the starting enzyme. For the sake of brevity, the insolubilized (immobilized) enzyme products are referred to by the term enzyme-RG, and specifically as lactase-RG, amylase-RG, etc.

The details of the process and specific applications are given by *A.C. Olson and W.L. Stanley; U.S. Patent 3,767,531; October 23, 1973; assigned to the U.S. Secretary of Agriculture.* The most important use of the insoluble (immobilized) enzyme is with lactases for hydrolysis of lactose in whey for recovery of galactose and glucose.

The phenol-formaldehyde resins used are commercially available resins produced by the condensation polymerization of phenol, or a substituted phenol, with formaldehyde under acidic or basic conditions. The resin prepared from phenol and formaldehyde has been shown to contain repeating units of the structure below. Resins of this type are available as Bakelite and Duolite.

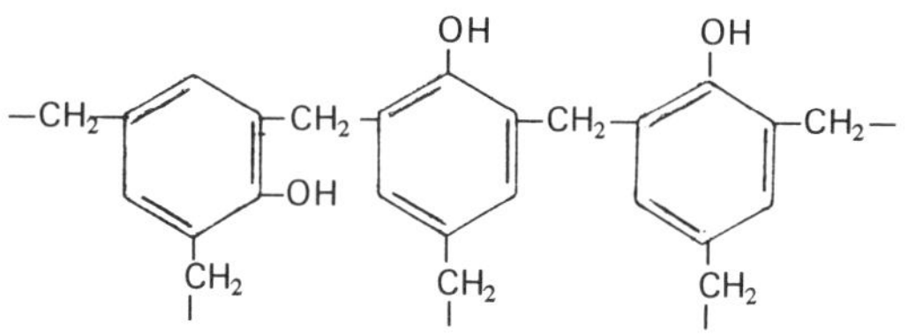

The enzyme to be insolubilized is dissolved in distilled water. Where necessary, the pH of the water is adjusted to a level at which the enzyme is soluble. Next, the aqueous dispersion of the starting enzyme is added to the resin granules. Generally, about 20 to 100 parts of resin per part of enzyme are used. The mixture is gently agitated by conventional means such as shaking or stirring. Having adsorbed the enzyme on the resin, an aqueous solution of glutaraldehyde is added to the above suspension. The amount of glutaraldehyde is not critical. Usually, a large excess, e.g., 10 to 50 parts per part of enzyme, is used; the unreacted residue is removed in a subsequent washing step. The resulting mixture is held for a period of time to ensure formation of the enzyme-RG product. Usually, the holding is for a period of about 12 to 24 hours in a cold room at about 10°C. However, a somewhat shorter holding period can be realized if the holding is conducted at temperatures between 10° to 25°C. The product is then collected by filtration and rinsed several times with distilled water to remove excess reagents and other impurities. The so prepared enzyme-RG is then ready for use.

In an alternative process the following procedure is used. The resin granules, after sieving and removal of impurities, are packed into a column to which is then added an aqueous solution of the enzyme at the proper pH. It should be obvious that the column must be cooled in order to conduct the reaction at temperatures below room temperature. Such cooling can be accomplished by conventional means, such as cold room, jacketed column, and so forth. Good contact between the resin particles and the enzyme is afforded by recirculating the solution through the column. Generally, the flow of solution through the column is contained for a period up to 5 hours; again, the higher the temperature, the shorter the time of contact required.

Following the adsorption of the enzyme by the resin, an aqueous solution of glutaraldehyde, in an amount sufficient to fix the enzyme to the resin, is pumped slowly onto the column. Prior to the addition of the glutaraldehyde, it may be desired to remove excess enzymic material by passing a small quantity of distilled water through the column. The flow rate of the glutaraldehyde solution should be such that the materials are contacted with resin for 12 to 24 hours. Alternately, the glutaraldehyde solution can be applied rapidly, and the column contents held for 12 to 24 hours. Afterwards, the column is flushed with distilled water to remove excess reagents and impurities, and any excess liquid is allowed to drain from the column. Thus prepared, the column is ready for use.

One particular application of the process concerns the disposition of whey, the watery part of milk left over from cheese manufacture. Currently, whey is discarded into rivers, lakes, and other bodies of water with detrimental effects on the environment. The biological oxygen demand (referred to in the art as BOD) of the water is raised, thus producing an imbalance in nature.

Protein can be separated from whey, leaving a material composed essentially of lactose. Although the protein is beneficial as a food supplement, the lactose cannot be utilized. Thus, recovery of protein from whey is uneconomical. However, lactose can be efficiently converted into galactose and glucose. These sugars are useful as fermentation media, e.g., in the culturing of yeast.

Example 1: Preparation of Lactase-RG – The starting enzyme was a commercial product the lactase content of which was estimated to be about 10%. The resin was a commercial phenol-formaldehyde resin.

The resin was sieved to obtain particles of 30 to 40 mesh. After washing the resin with distilled water, it was soaked overnight in 0.1 M (0.58%) sodium chloride and rewashed with distilled water. A solution of 228 mg of the lactase preparation in 20 ml of distilled water was added to 10 g of wet, drained resin at 25°C. After stirring gently for 15 min, 3 ml of a 25% aqueous glutaraldehyde solution was added. The reaction mixture was left in the cold (5°C) for 16 hr and then rinsed several times with distilled water. The immobilized enzyme product exhibited 60% of the activity of the original enzyme.

Example 2: Lactase-RG Hydrolysis of Lactose – Lactase-RG prepared as described in Example 1 was packed into a 1.2 x 15 cm jacketed column over a bed of sand. The column was then washed several times with pH 4.5 (0.02 M) phosphate buffer.

An aqueous lactose solution (3.0% in 0.02 M phosphate buffer) was then pumped through the column at varying flow rates. The column temperature was regulated by passing warm water through the column jacket. The extent of hydrolysis was determined by analyzing for glucose in the effluent. The results are summarized below:

	Temperature											
	30°C				45°C					55°C		
Flow, ml/min	0.2	0.5	1	2	0.2	0.5	1	2	5	0.5	1	2
Hydrolysis, %	90	76	58	43	100	93	81	64	44	100	91	75

Example 3: Preparation of Invertase-RG – The starting enzyme was a commercial product, the invertase content of which was estimated to be about 0.5 to 1.0%. The resin was a commercial phenol-formaldehyde resin.

The resin was sieved to obtain particles of 20 to 30 mesh. After washing the resin with distilled water, it was soaked overnight in 0.1 M (0.58%) aqueous sodium chloride and then rewashed with distilled water. An aqueous solution of 50 ml of the invertase preparation was added to 20 g of wet, drained resin at 25°C. The mixture was stirred gently for 15 minutes and then left at 5°C overnight, after which 2 ml of a 25% aqueous glutaraldehyde solution was added. The reaction mixture was left in the cold (5°C) for 16 hr and then rinsed thoroughly with distilled water. The invertase-RG exhibited 9% of the activity of the original enzyme.

Example 4: Invertase-RG Hydrolysis of Sucrose – Invertase-RG prepared as described in Example 3 was packed into a 1.2 x 15 cm jacketed column over a bed of sand. The column was washed with water. Aqueous sucrose solutions (10 and 30% in 0.02 N sodium acetate buffer, pH 4.5) were then fed through the column at varying flow rates. The column temperature was regulated at 37°C by passing warm water through the column jacket. The extent of hydrolysis was determined by analyzing for glucose in the effluent. The results are summarized below:

Sucrose concentration in solution, %	10	10	10	10	30	30
Flow rate, ml/min	0.2	0.5	1.0	2.0	0.2	0.5
Hydrolysis, %	100	100	100	>95	100	90

ACID-ACTIVE LACTASE

Lactose-hydrolyzing lactase enzymes are known to be produced by various yeasts, bacteria and fungi. Among the organisms useful for this purpose are yeasts such as *Saccharomyces fragilis, Torula cremoris* and *Torula utilis,* bacteria such as *Escherichia coli* and *Lactobacillus bulgaricus,* fungi such as *Aspergillus oryzae, Aspergillus flavus* and *Aspergillus niger.* The lactase enzyme preparations produced by these organisms generally have pH optimums on the alkaline side or in the weakly acid pH range of about 5 to 7. Yeasts, which are the primary source of commercial lactases, are known to produce lactases having pH optimums of about 7. Most of these conventional lactase enzyme preparations contain other enzymes, for example, proteases and amylases, which are the predominant components in the mixture.

In the stomach the gastric fluid provides strongly acidic conditions in the pH range of about 1 to 3. Therefore, the activity of lactase enzyme preparations which are generally highly effective in alkaline or weakly acidic media are for the most part destroyed or inactivated when contacted with the gastric fluid.

T. Cayle; U.S. Patent 3,629,073; December 21, 1971; assigned to Baxter Laboratories, Inc. identified and developed a procedure for preparation of a lactase enzyme which is acid-active and acid-stable and also suitable for oral ingestion. The lactase enzyme preparation is provided by growing a culture of *Aspergillus niger* and separating a lactase enzyme preparation which exhibits at least about 90% of its activity at pH 2.5 to 5.0 and contains at least about 50,000 lactase units (LU) per g of enzyme preparation. The term (LU) is defined as that quantity of enzyme which will produce 10^{-8} mol of o-nitrophenol (ONP) per min at 37°C, pH 4.4, at a concentration of o-nitrophenyl-β-D-galactoside (ONPG) of 0.0005 M.

The quantity of lactase enzyme preparation used in oral dosage form can vary within wide limits, depending in part upon the lactase activity of the particular enzyme preparation, the magnitude of the lactase deficiency or lactose intolerance in the particular subject requiring the dietary supplement of lactase and the particular dietary characteristics of these subjects.

In general, the amount of lactase enzyme preparation in the unit oral dosage form is sufficient to provide that amount of lactase enzyme needed to hydrolyze the lactose normally present or normally produced in the subject requiring the lactase supplement.

The lactase enzyme preparation can also be produced in powdered or granular form for direct mixture with the normal food and feed products used by the subjects requiring the dietary supplement of lactase. For example, in the case of an infant requiring a dietary supplement of lactase, a suitable amount of the lactase enzyme preparation in a powdered or granular form can be added directly to the milk or other food regimen of that infant in any suitable amount.

Example 1: A lactase enzyme preparation having an activity of at least 50,000 LU per g is prepared as follows: To 100 parts of wheat bran was added 60 parts of 0.2 N hydrochloric acid containing 0.62 ppm of $ZnSO_4$, 0.62 ppm of $FeSO_4$ and 0.88 ppm $CuSO_4$. The mixture was sterilized with steam and, after cooling, was inoculated with a sporulated inoculum of *Aspergillus niger.* The inoculated bran was maintained at a temperature of 30°C by passing moist air through the mixture until testing indicated the presence of substantial quantities of lactase after growth for four days.

An aqueous extract of the growth product was prepared by washing the above mixture with four volumes of water. The extract was concentrated by evaporation to a specific gravity of 1.1. A suitable quantity of the above aqueous extract was treated with calcium hydroxide and the pH brought to 6.5 to 6.8. The mixture was filtered after stirring for approximately 30 minutes.

A 10% aqueous bentonite slurry was added to the above filtrate in an amount of about 2.5 to 3% by weight of the filtrate. The pH of the mixture was adjusted to 3.9 to 4.1 and stirred for 30 minutes, after which time it was filtered. The filter cake was sparged with a 45% aqueous acetone solution. The filter cake was then dispersed in a quantity of water equal to approximately twice the cake weight and ammonium hydroxide was added until the pH of the mixture was in the range of 6.9 to 7.2. After stirring for 30 minutes, the mixture was filtered. The filtrate was then evaporated to a dry product by spray drying.

The *A. niger* lactase enzyme preparation produced has an activity of at least 50,000 LU/g, and enzyme preparations can be obtained with activities of about 500,000 to 600,000 LU/g by this method, depending upon the activity and purity of the starting inoculum of *Aspergillus niger.*

Example 2: The activity and stability of the lactase enzyme preparation of example 1 is shown as follows: The activity of the *A. niger* lactase as a function of pH was studied with lactose as substrate. Activity was measured at 37°C and the results are shown below.

Effect of pH on Relative Activity of *A. niger*

	. . . . Relative Activity, %	
pH	37°C	55°C
1.9	32	–
2.1	76	–
2.5	92	–
2.8	98	95
3.2	100	98
3.75	100	100
4.5	96	99
5.5	81	89
6.25	46	63
6.8	24	39
7.2	14	21
7.6	9	13

At either temperature at least 50% of the maximum activity is exhibited between pH 2 and 6, or at least 90% between pH 2.5 and 5.0. The relative activity of the *A. niger* lactase was determined at pH 3.5 on a lactose substrate at 30°, 37° and 55°C. These data are summarized below.

Effect of Temperature on Relative Activity of *A. niger*

Temperature, °C	Relative Activity, %
30	50
37	100
55	100

MILK CLOTTING PROCESSES

Preliminary Starter Culture

Preliminary to addition of milk clotting enzymes such as bacterial rennet in cheese making, it is necessary to lower the pH of milk to insolubilize milk protein. The pH value may be lowered by adding an acid or acid forming materials, but mostly it is effected by the action of a starter. A starter is a culture of one (pure culture) or more (mixed culture) strains of bacteria which are capable of reacting milk sugar into lactic acid.

Such a starter is added in a small quantity, in the order of 1% of the volume of the milk, to the cheese milk and the starter bacteria will begin to multiply. This process promotes the desired conditions for curding of the milk and/or for the subsequent exudation of the whey from the curd during its treatment, both actions taking place under the influence of the lowering of the pH value as a result of the production of lactic acid, and the desired degree of ripeness of the cheese as the result of the action of the produced enzymes.

According to one of the usual methods, a starter is prepared by inoculation of a small quantity of the desired bacteria into a proper medium and incubating it. In general, the medium is pasteurized skim milk and the incubating temperature is approximately 20°C. As already stated, the pH value of the medium is lowered following the production of lactic acid. Moreover, other changes take place in the medium (other reaction products are formed). This altogether causes a decrease of the growth rate. Preferably, the starter is used immediately after its preparation or it must be cooled and stored at a low temperature, as otherwise the activity is reduced, which means that the rate of acid production by the bacteria during the preparation of the cheese is reduced.

G. Hup and J.J. Stadhouders; U.S. Patent 4,053,642; October 11, 1977; assigned to Stichting Bedrijven Van Het Nederlands Instituut voor Zuivelonderzoek, Netherlands found that the disadvantages of known methods can be eliminated by using a culture medium prepared by cultivating a culture of at least one strain of milk sugar-fermenting bacteria in a culture medium including: (1) whey, such as diluted whey, whey having a low lactose content, and aqueous dispersions of whey powder; (2) an aqueous dispersion of non-whey protein selected from the group consisting of soya protein and milk protein in such an amount that the culture medium contains about 0.13 to 0.26 weight percent of non-whey protein, and (3) a buffer mixture. The whey, the aqueous dispersion of non-whey protein and the buffer mixture are each pasteurized separately.

This method makes it possible to obtain a concentrated culture of lactic acid bacteria which, just as in the usual bulk starters, can be introduced directly into the cheese milk without previous cultivation. To prevent a flocculation of protein, which may reduce the starter activity, the components of the culture medium may not be pasteurized after having been brought together for inoculation with the bacteria.

The pasteurization is usually carried out by heating at a temperature of 100°C for a period of 30 minutes (steaming). If use is made of cheese whey, heated to 75°C to eliminate rennet and starter activity, and cooled and centrifuged, and if this liquid is steamed, contingent upon the previous addition of a phosphate mixture or skim milk (or another protein dispersion) or both, then a coagulation of protein occurs, which is thus no longer available for assimilation. This results in a lower starter activity after the starter bacteria have been cultivated therein. If the same process is carried out after having previously digested the whey proteins partially by means of trypsin, then, indeed, after steaming no protein sediment will be present; but this method, again, is expensive.

However, a satisfactory result is obtained by first steaming and subsequently centrifuging the whey. After cooling to a lower temperature, the treated whey is mixed with the buffer mixture and the already separately steamed skim milk. If the buffer mixture is used as an aqueous solution it should be pasteurized as well. The mixing occurs in such a manner that the mixture obtained has a temperature not exceeding 65°C. Otherwise, undesirable precipitates might be formed, though this is not always the case.

Using this separate pasteurization, no co-precipitate of whey and milk protein will be formed, and the activity of the starter is normal when 5% skim milk is used. Steaming of rennet whey is not absolutely necessary. If rennet whey is acidified with hydrochloric acid to a pH of 6.0 then heated for 10 minutes at a temperature of 85°C, and subsequently centrifuged before phosphates and skim milk are added, then the starter, even after addition of as little as 5% of skim milk, has a substantially normal activity.

If the bacteria after inoculation into the culture medium are incubated sufficiently long, they can then be separated from the aqueous phase, for example, by means of freeze drying or by means of a bactofuge, and such that, whether or not the pH of the medium has initially been increased to a more neutral value, the small, but still present, quantity of milk protein can be prevented from being thrown out together with the bacteria, by dissolving the protein by means of sodium citrate or a protein dissolving agent. If the protein is not dissolved previously, a bactofuge may be stopped up very quickly.

Therefore, if skim milk is used as a culture medium it is always neutralized and, subsequently, its casein is dissolved (clarified) by means of the addition of 1% sodium citrate. On centrifuging a culture in skim milk and a process culture there is obtained in both cases a quantity of sediment of the same order of magnitude if both are first neutralized to a pH of 6.6 and then clarified with a solution of sodium citrate. With the aid of a bactofuge it is possible to concentrate more than 90% of the bacteria of a culture into a volume that is 2% of that of the original culture, and thereby, the concentrate is obtained in a continuous stream and can be processed without any previous resuspension being necessary.

The bacteria concentrate can be brought into a form suitable for storage and transport, for example, by rapidly freezing it to a temperature of -40°C, after the addition of lactose, and further storing it at a temperature of preferably not above -10°C. The concentrated bacteria culture may also be lyophilized. The bacteria concentrate can simply be introduced into the cheese-making milk, and exhibit a good activity. Thus, cheese can be prepared from milk which has been coagulated by the addition of rennin and bacteria concentrate as the starter, the coagulated milk cut into small pieces, the resulting mixture gently stirred and, if necessary, heated to promote exudation of whey, and whey drained off, and the curd collected and shaped in a mold.

Pepsin-Resin Carrier

Pepsin preparations for clotting milk are bound to a polymer carrier which is a reaction product of ethylene-maleic anhydride copolymer (EMA) and p,p'-diaminodiphenylmethane (EMA-MDA resin), 2,6-diaminohexane (EMA-HMD resin) and hydrazine (EMA-hydrazide).

The process is given by *L. Goldstein and E. Katzir; U.S. Patent 4,013,511; March 22, 1977; assigned to Yeda Research & Development Co. Ltd., Israel* for an application of pepsin enzyme to clot milk. The process gives details for preparation of EMA-MDA and EMA-HMD resins.

Synthesis of EMA-MDA and EMA-hydrazide resins

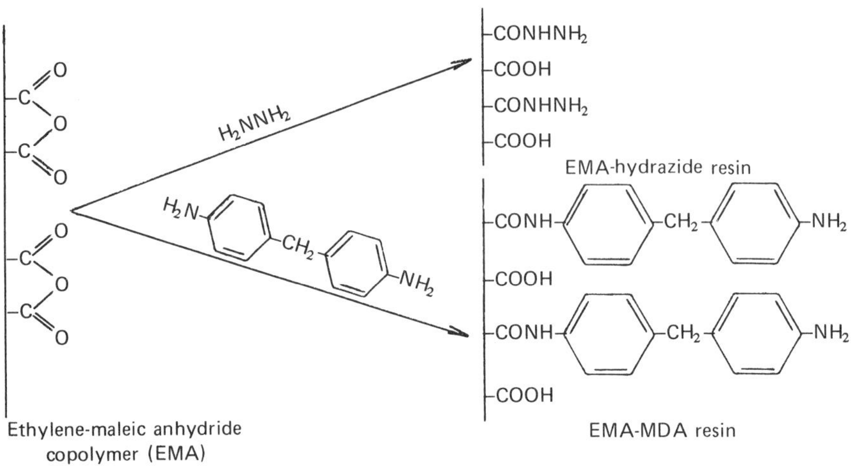

Coupling of anionic EMA-hydrazide resins to proteins: A cold freshly prepared solution of an enzyme in 0.2 M borate buffer pH 8.4 (10 to 40 mg in 2 ml) was slowly added to the magnetically stirred suspension of EMA-hydrazide (100 mg in 5 ml of the same buffer). The reaction mixture was left stirring overnight at 4°C. The water-insoluble EMA-hydrazide enzyme derivative was separated by centrifugation and washed with water, 1 M KCl, again with water and then suspended in water.

Coupling of anionic EMA-MDA resins to proteins: A cold solution of an enzyme in 0.2 M phosphate buffer pH 7.8 (10 to 40 mg in 2 ml) was slowly added to a magnetically stirred suspension of diazotized EMA-MDA (100 mg) in the same buffer (4 ml). The reaction mixture was left stirring overnight at 4°C. The water-insoluble EMA-MDA enzyme derivative was separated by centrifugation or filtration, washed with water, 1 M KCl and again with water and resuspended in water.

Coupling of pepsin to EMA-HMD resins: EMA-HMD resin (100 mg) was suspended in 2 ml 1 N HCl and stirred at room temperature for about 10 minutes. The swollen resin was separated on a suction filter, washed on the filter with 100 ml water, brought to pH 4, resuspended in 2 ml distilled water, and the suspension brought to pH 4.0. A freshly prepared solution of 10 mg pepsin dissolved in 1 ml distilled water was added to the magnetically stirred suspension and the pH readjusted to 4.0. 20 mg EDC [1-ethyl-3-(3-dimethylaminopropyl)carbodiimide] in 1 ml water was added and the pH maintained at 4.0 with 0.1 N HCl. After 30 to 60 minutes, when no measurable changes in pH could be observed, the reaction mixture was transferred to the cold room and left stirring overnight. The immobilized enzyme was separated by centrifugation, washed with 1 M KCl, acidified to pH 4.0 (100 ml) and then with 0.01 N HCl; it was then suspended in 5 ml 0.01 N HCl and stored at 4°C.

Representative data on the material protein binding capacities of the various resins are summarized in the table below. The capacity of EMA-hydrazide resins is considerable higher than that of EMA-MDA resins. This could be due to the higher hydrophilicity of the EMA-hydrazide network, leading to a more open structure, and thus to a higher capacity for proteins.

Protein Binding Capacities of Anionic EMA-MDA and EMA-Hydrazide Resins for Various Enzymes

	. . . . EMA-MDA Resin				. . .EMA-Hydrazide Resin. . .			
	Protein Content		Active Bound Protein		Protein Content		Active Bound Protein	
	(A)	(B)			(A)	(B)		
Enzymes	(mg/100 mg)		(mg)	(%)	(mg/100 mg)		(mg)	(%)
Trypsin	10	9.7	3.4	34	30	29	12.0	40
Chymotrypsin	10	0.8	2.0	20	25	24.1	3.2	13
Subtilisin Novo	20	18	5.4	27	10	9.7	1.0	10
Subtilisin Carlsberg	20	24	2.0	10	10	9.8	1.0	10
Papain	30	31	9.4	31.4	45	44.2	27	60

Note: (A) = calculated from binding curves
(B) = calculated from amino acid analyses

The EMA-MDA enzyme derivatives had particulate form and were easily filtered. The immobilized enzyme samples derived from EMA-hydrazide and EMA-HMD were more swollen, of softer texture, and filtered slowly.

The recovery of enzymic activity in insoluble derivatives prepared under standard conditions (10 mg enzyme per 100 mg resins, in the coupling mixture) was higher when the support was EMA-hydrazide. The total activity recovered in immobilized form was found to depend on the charge characteristics of the support; the anionic resins yielded in general higher total immobilized activity.

Aqueous suspensions of anionic EMA-MDA derivatives of chymotrypsin, subtilisin Carlsberg and subtilisin Novo could be stored at 4°C for 3 to 4 months without significant loss of activity. Suspensions of EMA-MDA papain and EMA-MDA trypsin could be stored under the same conditions for periods of up to 8 months with no decrease of activity. The corresponding cationic EMA-MDA derivatives lost 20 to 30% of their activity under the same conditions.

The EMA-hydrazide derivatives (both anionic and cationic) of trypsin, chymotrypsin, subtilisin Novo, subtilisin Carlsberg and papain exhibited extremely high storage stabilities. No loss of activity was recorded after storing suspensions at 4°C, for about 1 year, in the presence of a bactericide. The EMA-HMD derivatives of pepsin could be stored at 4°C for 4 months without significant loss of activity.

On lyophilization the anionic EMA-MDA derivatives retained a considerable amount of activity. The EMA-hydrazide derivatives retained most of their activity on lyophilization. On lyophilization EMA-HMD pepsin retained 70 to 80% of its activity. EMA-HMD trypsin retained 25% of its activity. Storage of the lyophilized powders in a desiccator at 25°C for about 1 year was accompanied by no loss of activity.

Relative to the corresponding native enzymes all anionic EMA-hydrazide, anionic EMA-MDA derivatives and the EMA-HMD derivatives exhibited higher stabilities in the alkaline pH-range; pH optima were also displaced towards more alkaline values. Enhanced stabilities in the acid pH-range were found for the cationic EMA-hydrazide and EMA-MDA derivatives.

The temperature stabilities of the EMA-hydrazide, EMA-MDA and EMA-HMD derivatives were higher than those of the respective native enzymes. With the enzyme derivatives of the cationic EMA-MDA and EMA-hydrazide resins, decrease in temperature stability relative to the anionic analogies was found in all cases.

DIRECT ACID METHODS

Initial Controlled Acidification

I. Loter, H.G. Dissly and R.E. Schafer; U.S. Patent 3,882,250; May 6, 1975; assigned to Diamond Shamrock Corporation uses a preliminary direct acid addition to acidify milk without precipitation of casein, then enzymes are added to form the curd which is used to make various kinds of cheeses.

One of the major advantages of this process is that the temperature of the milk remains approximately constant from the time the acidifying agent is added until the curd is formed, thus eliminating the need for complex temperature adjustment and special heating or cooling equipment. A primary factor in successful acidification is the avoidance of casein precipitation (acid coagulation) caused by excessive local acidity. Depending upon the conditions under which the milk is to be processed, various methods of acidification can be used.

With proper equipment it is possible to have a continuous introduction of acid until the milk reaches the desired pH. It has been found however, that a preferred acidification is to lower the pH to about 5.15 to 5.20 (at a temperature of about 20°C to about 30°C) and then reduce the rate of introduction of acid until the pH has dropped to about 5.10. The reduction of the rate of flow of acid allows careful monitoring of the pH and it must be noted that milk at the temperatures used in the process is sensitive to even minor adjustments of pH.

There is no specific time limit for the acidification of the milk since the milk will not coagulate upon standing without the addition of an acidogen and a proteolytic enzyme, provided that the temperature is not raised. Obviously, the time for introduction of the acid can be minimized by holding the milk at the lowest temperature and by increasing the rate of flow or agitation of the milk to which the acid is being added. The proteolytic enzyme can be commercial rennin, a diluted rennin extract, a pepsin-rennin mixture, a vegetable derived enzyme or clotting agent. Other enzymes such as pepsin, papain and ficin may be used alone or in combination with rennin. The rennin extract is usually a solution of the enzyme rennin, stabilized and with preservatives added, and standardized in strength to a certain coagulating or clotting power.

The cheese curd is cut and cooked in a conventional manner. The cooking rate is usually faster than the cooking rate in the cultured process. Draining and washing of the cooked curd can be carried out using the standard cheese making procedures.

To make cottage cheese, the acid cheese curd is processed with the additional steps of: Cutting the curd into cubes, expressing the whey from the curd, cooking and stirring the curd for from about 30 min to 90 min at a temperature of about 38° to 65°C and washing and draining the curd.

To make baker's cheese, the acid cheese curd is processed with the additional steps of: Breaking up the curd, setting the curd and expressing the whey and hanging and draining the curd for about 1 hr to help express further whey.

To make Neufchatel cheese or cream cheese, the acid cheese curd is processed with the additional steps of: Cutting the curd, cooking and stirring the curd until the temperature reaches about 55°C and then continuing cooking for about 15 additional minutes. Cooling the curd to about 15°C, draining the curd overnight at about 3° to 4°C to obtain a dry, soft curd, mixing the dry, soft curd with sufficient lactic acid to reduce the pH of the curd to 4.60 and blending the curd into a smooth homogenous cheese mass.

In making cream cheese, the original milk should contain enough cream so as to have a fat content of about 10.5 to 11.5%. In making Neufchatel cheese, the original milk should contain enough cream so as to have a fat content of about 5 to 6%.

Example 1: 20 gal (76 liters) of fortified skim milk were prepared from 10 gal (38 liters) of skim milk to which was added 8.5 lbs (3.9 kg) instant non-fat milk solids dissolved in 77.5 lbs (35.2 kg) of water. The milk was warmed to 76°F (24.5°C) by circulating warm water through the jacket of the cheese vat in which the fortified milk was prepared. The pH was 6.7. The milk was recirculated through a centrifugal pump with a tee (Bernoulli) connection placed at the suction side of the pump so that acid could be injected into the milk. The speed of the pump caused acid to be sucked in at the impeller of the pump. The milk was then discharged to the rear of the cheese vat and recirculated. The acid was 85% phosphoric acid diluted with 3 volumes of water.

The acid was introduced into the milk flow in four steps as follows. In the first step 400 ml of acid was introduced during a 6 min 8 sec period, reducing the pH to 5.60. In the second step 325 ml of acid was introduced during a 12 min 54 sec period further reducing the pH to 5.11. In the third step 50 ml of acid was introduced during a 4 min 20 sec period reducing the pH to 5.10. In the fourth step 36 ml of acid was introduced during an 8 min 4 sec period reducing the pH to 5.03, its lowest value. The milk at this stage showed no signs of precipitation. 15 g of $CaCl_2 \cdot H_2O$ dissolved in 30 ml of water, followed by 228 g of D-glucono-delta-lactone and 24 ml of rennin coagulator dissolved in 500 ml of water, were added to the fortified skim milk. The milk was then allowed to set in a quiescent state for 1 hr and 20 min while maintained at the original temperature of 76°F (24.5°C). A firm curd resulted which was cut in the conventional manner by cheese knives. The pH of the whey was 4.78. A mixture of 11 ml phosphoric acid in 1 pint (0.75 liter) of water was added to the whey to reduce the pH to 4.60. The curd was cooked to 120°F (49°C) for 1 hr, drained, washed and drained to yield 25.25 lb (11.5 kg) of curd.

Example 2: Cheese was prepared using commercial dairy equipment set up in the following manner. A 150 gal (570 liter) storage vat was connected by piping to a centrifugal pump which was connected by further piping to a 200 gal (760 liter) cheese vat. Means for injecting acid into the pipe was attached between the pump and the cheese vat (i.e., the receiving vat) and a pH meter was introduced into the piping between the point of introduction of the acid and the cheese vat.

Firstly, the equipment was calibrated by pumping 10 gal (38 liters) of water from one tank to the other. This took 15 sec which is equivalent to 40 gal (152 liters) per minute. This was calculated to be the equivalent of 37 to 38 gal/min (140.6 to 144.4 l/min) of milk. Accordingly, it was further calculated that the addition of 1 gal (3.8 liters) of 85% phosphoric acid diluted with 3 parts water per 100 gal (380 liters) or 1,440 ml of dilute acid per minute would be required. As a safety factor, only 95% of this quantity or about 1,360 ml per min was added.

The acid injection means was set to deliver 1,360 ml per min dilute acid and it was calculated that this volume of acid to 38 gal (188.4 liters) of milk should lower the pH to 5.20.

The milk used had a Quevenne Lactometer (milk specific gravity tester) reading of 37.5 at 60°F (11.5°C), which is equivalent to a specific gravity of 1.0375. Non-fat milk solids were about 9.4%. 150 gal (570 liters) of milk were poured into the storage vat and warmed to 72°F (22°C). The valves were opened and the centrifugal pump was started. The pH meter was read at the discharge point into the vat and indicated 4.9 with the milk precipitating. The rate of introduction of the dilute acid was slowly decreased until the pH of the milk at the discharge point read about 5.2, and the casein no longer precipitated.

The total time to empty the vat of milk was 4 min 25 sec which meant that the rate was 33.5 gal/min. The total volume of acid injected was 4,820 ml. The pH of the milk equilibrated to 5.25 without any visible precipitation.

While maintaining the milk at the same temperature of 72°F (22°C), it was set with 6 lbs (2.7 kg) D-glucono-delta-lactone dissolved in 5 gal (19 liters) water to which was then added 7½ oz (221.8 ml) rennin coagulator. The milk was set for 1.5 hr while maintained quiescent and at the same temperature, and then cut. The coagulum was very good and was easily cut with half inch knives. The whey pH was 4.90. 150 ml of 85% phosphoric acid in 10 gal (38 liters) of water at 90°F (32°C) was poured into the vat gently. After gentle stirring, the pH of the whey was 4.60. The curd was then cooked out to 102°F (39°C) in 1 hr. It was then drained and washed in the usual manner.

Example 3: 1 qt (0.95 liter) of skim milk containing 1% added non-fat milk solids was warmed to 80°F (26.5°C) in a warm water bath and held at that temperature. With vigorous agitation, a solution of 1 part of 85% phosphoric acid and 3 parts of water was slowly added dropwise in increments with the following results:

Total Volume of Acid, ml	pH After Acid Addition	pH After Stirring 5 Minutes	Precipitate
0.0	6.7	–	none
2.0	6.1	–	none
4.0	5.65	–	none
5.0	5.50	–	none
6.0	5.35	5.40	none
6.5	5.30	–	none
7.0	5.22	5.22	none
7.5	5.19	5.20	none
8.0	5.11	5.12	none
8.5	5.06	5.08	none
9.0	5.00	5.01	trace

The total time for addition of the acid was about 45 min. Acid addition was stopped when a trace of precipitation became discernable. A solution of 3.36 g of D-glucono-delta-lactone and 0.3 ml of rennin coagulator (Hansen's Standard Strength) was added to the acidified milk which was allowed to set for 1 hr. Although a trace of precipitate was visible at the surface, a good curd was cut, which was cooked out in the usual manner. The pH of the whey was 4.9. This indicates that at the given temperature and rate of acid addition, milk could successfully be acidified to a pH of just above 5.01.

Controlled Proteolysis

A highly active proteolytic enzyme is used by *E.E. Rice and O.J. Lantero, Jr.; U.S. Patent 3,886,288; May 27, 1975; assigned to Swift and Company* which is added to milk to produce a stable coagulum and inactivated before excessive hydrolysis occurs. It is possible to reduce the time to form a coagulum in cheese manufacture by a direct acid addition to the rennet milk mixture under certain conditions.

The added acid produces an almost immediate formation of a coagulum which reduces the time normally required as when only a culture and rennet are incorporated. It appears that the added acid increases the solubility of bivalent alkaline earth metal ions in the mixture which precipitate the milk protein to form the coagulum.

It has been observed that, the enzyme rennet alters the milk protein casein to paracasein. Soluble calcium (and magnesium) salts present in the milk then precipitate the paracasein, thereby forming insoluble calcium (and magnesium) salts of the altered casein commonly known as curd. The acid produced by the action of the bacterial starter and the acid added by the cheese making step described above aid in solubilizing the calcium (and magnesium) so that it will readily precipitate the paracasein. Additionally, the continual incubation of the added culture promotes distinct flavors and textures in the cheese curd during later curing and aging processes.

The substitution of certain enzymes (ones that exhibit high proteolytic activity) for rennet in conventional commercial methods of cheese manufacture, has usually been unsatisfactory to produce a high quality cheese curd with high yield and organoleptic scores. It is generally recognized that these highly active proteolytic enzymes exhibit more proteolytic activity than the enzyme rennet, and it has been found that such enzymes, when added to milk in otherwise conventional methods of cheese curd manufacture, produce an unstable coagulum thereby preventing the formation of a strong curd. This is believed to be due to an excessive proteolytic hydrolysis or digestion of the protein casein by the added highly active enzyme.

It was found that a high quality cheese curd can be produced through the use of highly active proteolytic enzymes to alter casein by carefully controlling the enzyme proteolysis. This may be accomplished by inactivating the enzyme after a sufficient period of time has passed to alter the casein but prior to the time excessive hydrolysis begins. Several enzyme inactivating systems have been generally recognized, for example, a heat system or an hydrogen peroxide catalase system, but such systems have not been incorporated into conventional methods of manufacturing high quality cheese for it has been found that the systems produce the disadvantages of altering the final flavor and/or destroying bacterial cultures added to increase flavor and texture development upon aging and curing of high quality cheeses.

This process gives procedures for controlled enzyme proteolysis and also specifies the addition of a bacterial starter that will develop flavor and texture during aging and curing.

It is preferred to alter the milk protein by adding a plant derived proteolytic enzyme, for example, bromelain, papain, or ficin. These plant-derived enzymes have been observed to be highly active and are believed to be generally undesirable for utilization in conventional commercial cheese making processes due to high proteolytic activity. These plant derived enzymes are relatively cheap and are readily available in commercial preparations.

Each of the highly active proteolytic enzymes varies in its ability to alter the protein casein in differing concentrations and in differing lengths of time. The enzyme must be allowed to act on the milk protein casein for a sufficient time to alter the casein to paracasein but not for such a period of time that excessive hydrolysis occurs thereby preventing the formation of a stable coagulum.

It may be desirable to follow the enzyme activity to specifically determine when the casein has been sufficiently altered. Tests have shown that neither the desired alteration of the casein which permits proper coagulation nor the excessive hydrolysis which prevents the formation of a stable coagulum can be observed visually.

However, the extent of protein change can be followed with an appropriate test, for instance, at time zero the enzyme can be added to the milk and at specific time intervals an aliquot of milk can be removed. Immediately upon removal, calcium chloride is added to the aliquot to determine if the milk will clot. When one of the aliquots clots immediately, the casein has been altered sufficiently and the enzyme should be inactivated as soon as possible.

After the highly active proteolytic enzyme has been in contact with the milk for a sufficient time to alter the casein, it is then inactivated and/or destroyed by an enzyme inactivating system to prevent any further proteolytic hydrolysis.

A hydrogen peroxide catalase system destroys the activity of the enzyme and leaves no effect upon flavor or texture development of the cheese curd. The amount of hydrogen peroxide to be added should be an amount sufficient for complete inactivation by oxidizing the enzyme present. It has been determined that a preferred amount of enzyme in the milk mixture can be totally inactivated by the addition of about 10 to 20 ml of 3% hydrogen peroxide solution per liter of milk.

Any excess hydrogen peroxide remaining after an optimum deactivating time of about 2 to 10 min can be removed by incorporating a solution of commercial catalase into the mixture. It is most important to remove all excess hydrogen peroxide from the mixture, for if any remains it will destroy any subsequently added bacterial culture resulting in a loss of flavor and texture development of a high quality finished cheese curd. A commercial catalase preparation, for example, beef liver catalase or fungal catalase, is used.

After it has been determined that the starter culture is growing well, a coagulum is formed by precipitating or coagulating the paracasein by additional acidulation or incorporating additional soluble calcium ions into the mixture. The coagulum will form by the acid produced by the starter culture without the addition of one of these coagulating adjuncts, but the procedure is quite slow and therefore not preferred. It appears the pH of the milk bears a direct relation to the formation of the coagulum, and precipitation occurs almost immediately when the pH of the mixture is lowered to between about pH 5 to pH 6.3. The acidulation can be accomplished by the addition of any food acid such as lactic acid, acetic acid, citric acid, etc.

If desired, coagulation of the enzyme treated milk culture mixture can be accomplished by acidulation from ester hydrolysis. For example, the ester D-glucono-delta-lactone can be used. The ester, when incorporated into the mixture, appears to slowly hydrolyze to produce the acid form of the compound which lowers the pH of the enzyme treated milk culture mixture sufficiently to form a coagulum. Approximately 8 to 12 g of D-glucono-delta-lactone added per liter of milk will form a coagulum in about 10 to 30 min.

It may also be desirable to precipitate the mixture by furnishing additional soluble calcium ions and thus strengthen the curd formed. A firm coagulum is formed when a soluble calcium salt, such as calcium chloride or calcium gluconate, is added to the mixture.

Example 1: A 100 ml quantity of 5% ficin solution was added to 54 gal of milk that had been adjusted to 88°F, the temperature normally used with rennin. The enzyme was inactivated 13 min later by adding 1,970 ml of 3% hydrogen peroxide solution. The excess hydrogen peroxide was destroyed by adding 90 ml of beef liver catalase 10 min after the peroxide addition. The treated milk was held for 35 min to make sure all the residual peroxide was destroyed. Then, a 5 pound quantity of culture was added. It was determined that the culture was growing well 65 min later and a 990 ml quantity of 44% calcium chloride solution was added, causing immediate clotting. The curd was handled in the normal fashion by cutting, cooking, cheddaring, salting and finally pressing.

Example 2: A 100 ml quantity of a 6% papain solution diluted to 2,000 ml with cold tap water was added to 54 gal of milk that had been adjusted to 85°F. The enzyme was inactivated 15 min later by adding 2,000 ml of a 3% hydrogen peroxide solution, then 10 min after the peroxide addition an 80 ml quantity of beef liver catalase solution diluted to 1,000 ml with cold tap water was added, destroying the excess peroxide.

The milk was stirred rather vigorously for 10 min, after which a 4.6 pound quantity of lactic starter culture was added. After 50 min of incubation at 85°F the culture was growing well, at which time the pH of the milk was lowered from 6.6 to 5.5 with a 10% solution of lactic acid which caused immediate coagulation. Precipitation of the milk protein was observed when the pH of the enzyme treated milk was below 6.0. After coagulation the curd was handled in the normal method of preparing cheddar cheese.

Coated Acid Addition

In the manufacture of cheese curd, it is conventional practice to inoculate milk with a bacteria culture which produces lactic acid to acidify the milk. When such milk reaches an acidity at or near its isoelectric point, agglomeration of the casein occurs to form a curd and effect a separation of the whey. Although when properly regulated and controlled cultured cheese curd, such as is produced in the manufacture of cottage cheese, is of high quality, the difficulties and disadvantages encountered are many.

An obvious expedient in circumventing the difficulties encountered by the utilization of bacteria cultures in the manufacture of cheese curd is the direct addition of acid to milk in quantities needed to effect the desired acidity for casein coagulation. However, cheese curd, such as cottage cheese curd, to be marketable must be of a gelled texture that is obtained only by the relatively slow and uniform coagulation which occurs at an acidity which is at or near the isoelectric point of the milk.

When concentrated acids are added directly to milk to increase its acidity, the localized area of the milk where the acid is introduced acquires an excessive acidity before a homogenous mixture of milk and acid can be obtained. Premature casein agglomeration occurs in the localized areas and uneven precipitation occurs generally. The texture of the resultant loose precipitate is unsatisfactory usually being hard, dry, and uneven in size as well as being generally unpalatable.

Casein agglomeration in milk effected by the slow addition of concentrated acids accompanied by agitation to avoid premature precipitation requires prohibitive time and is difficult to control. Agglomeration of casein effected by the addition of acid of a dilution to avoid premature casein precipitation results in a thin, watery, and undesirable product.

It is known that when concentrated acid is added to refrigerated milk the casein does not readily agglomerate. Culture acidification of milk is generally at 70°F or higher since casein does not readily coagulate at lower temperatures.

L.L. Little; U.S. Patent 3,792,171; Febuary 12, 1974 utilizes dry acid with an edible coating which retards or dampens solubility of the acid and hence the rate of pH drop. Basically the process as it is utilized commercially is as follows:

(1) With the milk at a temperature below 50°F (about 40°F), acidify to a pH of about 4.85 to 5.20, using a concentrated acid.

(2) Warm the milk to ambient temperature (about 50° to 80°F).

(3) Add enzyme (about 1 to 100 cc per 1000 lb of milk) and hold the milk quiescent until coagulated firm enough to cut.

(4) Cut the curd. This is usually done at a pH of 4.85 to 5.20 (in the preferred procedure the curd is cut at a pH of 4.95 to 5.05).

(5) After cutting the curd, make a second addition of acid. This should be in an amount sufficient to reduce the pH of the whey at the end of the cooking process to 4.30 to 4.50. Also, it is desired to reduce the pH of the finished cottage cheese to 4.85 or below. However, the absorption of acid from the whey into the cubes of curd is very slight and in actual practice it rarely drops as low as 4.85.

    When the pH of the finished cottage cheese is 4.95 or above, the cheese has a tendency to develop off-flavors during storage. This is a very serious difficulty and it is mandatory that the pH be kept below this point if a satisfactory product is going to be produced. Also, when the pH is 4.85 or below, consistently better yields are attained. At present no means have been found to cause the curd to absorb more of the second addition of acid.

(6) Cook the cubes (while in the whey) at a temperature of from about 100° to 150°F.

(7) Separate cubes (curd) from the whey.

(8) Wash the cubes with water.

The above steps can be modified by adding acidogens such as GDL (glucono-δ-lactone) at Step No. 3 to assure acid coagulation but more importantly to improve the pH or acid characteristics of the resultant curd (i.e., to 4.85 or below).

One of the disadvantages of the above direct acidification process is that it is not possible to reduce the pH of the milk as low as is desired without encountering premature coagulation. It is well known that a cheese curd that exhibits a pH of about 4.95 or greater has poor keeping qualities or a tendency to develop off-flavors during storage. Acidogens added to Step 3 above provide at least some acid development after the milk has set or curd has formed thus assuring a lower pH in the developed curd and a product with good keeping qualities having no tendency to develop off-flavor flavors during storage.

The addition of acidogens such as GDL causes no immediate reduction in the pH of milk because it is neutral and only forms acid as it hydrolyzes slowly in solution to gluconic acid. There is only a very slight amount of hydrolysis of acidogens such as GDL during the setting period but the rate of hydrolysis increases rapidly as the temperature is raised during cooking. Consequently there is a highly desired acid development within the curd at this critical point of the process. A typical profile of the pH of the milk and curd during processing is as follows:

(a) With the milk at about 40°F, acidify with 75% phosphoric acid at the rate of 3.75 cc per 100 g solids in the milk.

(b) Warm the milk to about 70°F. At this point the pH will be approximately 5.05.

(c) Add GDL at the rate of 2 g per 100 g solids in the milk. Then add an enzyme solution (cottage cheese coagulator) at the rate of 330 cc per 100 gal of milk and agitate thoroughly. At this point, the pH is normally 5.00 to 5.05.

(d) Allow the milk to set in a quiescent state for 90 min, then cut the coagulated milk into cubes using ½ inch curd knives. At this point, the pH would normally be 4.90 to 4.95.

(e) Allow the cheese to set for 15 min, then add 75% phosphoric acid at the rate of 0.50 cc per 100 g of solids in the milk.

(f) Start heating the cheese and after about 10 min stir very carefully. Continue heating and stirring the cheese until it reaches a temperature of 125°F.

(g) Stop the heating and drain the whey from the curd. The pH of the whey at this point should be about 4.40 to 4.50.

(h) Rinse the cheese by filling the vat about ¾ full of chilled water that has been acidified to a pH of 6.00 or below and chlorinated with 10 parts per million of available chlorine. Allow the water to remain on the curd about 15 min; then drain the water from the curd. Wash the curd a second time by repeating the process just outlined.

(i) Drain off the last rinse water, then cream and package the dry curd. The pH of the dry curd at this point should be 4.75 or below.

The drawbacks encountered by using acidogens such as GDL in the manner described above continue to relate to costs since the amount of acidogen required remains relatively high denying the process a true competitive position relative to conventional culturing. Acidogens hydrolyze slower than is desired causing a lag in acid development. Hydrolysis is only about 85% complete so about 15% of the acidogen does not function and is lost.

The gluconic acid produced has a high equivalent molecular weight compared to other organic acids. For example, gluconic acid's equivalent molecular weight of 196 compares to 58 for fumaric acid. The cost of GDL is about two to three times the price of fumaric acid. Thus, on a use basis the cost of using GDL would be about ten times the cost of using fumaric acid. Additionally because of the slow rate of GDL hydrolysis, it is difficult to accurately control the pH profile of the milk from the time the curd is cut until the cooking process is completed.

It was found that solid acid particles or crystals can be coated with a wide variety of materials that will dampen or retard their dissolution so that their rate of acidification of the liquid media is gradual and similar to that of the acidogens. By using this process it is possible to control the rate of acidification and thus eliminate the disadvantages of the slower than desired dissolution properties of the acidogens.

The use of coated or dampered solid acids as a replacement for acidogens in finishing off acidification with other acids is also significantly advantageous. The reason for this, of course, is that liquid mineral acids such as hydrochloric or phosphoric acid are more readily available at substantially lower costs than the organic acids. Thus, the combined use of mineral acids and the solid coated acids of this process are significant advantages over the prior art culturing or acidogen procedures. Coated acid particles can also be used as a substitute for acidogens as additions to milk prior to coagulation effected by other means (i.e., acid-enzyme action) so as to be entrapped in the curd and effect subsequent acidification. The preferred acids can be classified into two distinct groups, as follows:

	Grams of Acid per 100 Grams of Water	
	77°F	212°F
High solubility		
Tartaric acid	147	343
Citric acid	184	525
Malic acid	139	*
Low solubility		
Fumaric acid	0.6	10.0
Adipic acid	1.9	83.0
Succinic acid	7.1	21.0

*Extremely soluble.

It was discovered that a great many materials may be used to coat the solid acid particles, thereby giving greater flexibility to the rate and amount of damping that can be applied to a particular acid. Also, it was discovered that any coating can be easily modified to lessen the damping effect by blending the coated acid with a small amount of inactive powdered diluent.

Particular success has been found with animal or plant oils or fats, monoglycerides of fatty acids and fatty acids. Such materials do not dissolve in an aqueous media such as milk but do disperse at a relatively controllable rate. Additionally, such materials have a tendency to soften or become less viscous upon heating so as to more rapidly disperse. Thus, the rate of solubility or damping effect may be controlled by both the materials and the heating steps under which processing is effected.

Buffered acid particles or crystals may be entrapped in cheese curd avoiding the necessity of refrigeration. In this procedure milk is acidified to a pH of about 5.10 to 5.30 at ambient temperatures without coagulation, i.e., the formation of flaking or precipitation of protein. Dampered acids are then added in sufficient quantity to develop the desired pH profile without any immediate signicant reduction in the pH. Enzymes are added to the milk after the addition of the dampered acids in quantities to effect casein coagulation. The milk is allowed to set in a quiescent state after the enzyme addition until coagulation occurs.

This procedure provides some significant advantages in that many dairies do not have sufficient refrigeration capacity or plate cooling capacity to cool the milk to 40°F. Cooling the milk to 40°F and subsequently warming to 70°F after acidification is in itself a significant expense. Warming milk in the cheese vat from 40° to 70°F is very slow, and generally requires approximately from one to two hours time. Time saving is one important advantage of direct acidification process. Plate heat exchangers can cut the warming time to a minimum of fifteen to twenty minutes, but this requires a very substantial investment in heat exchange equipment and storage vats.

Where direct acid additions are made prior to the addition of the dampered acid crystals or granules such as where the dampered acids are used to finish off acidification such direct acids may consist of any food grade acids. For example, the direct acids may consist of hydrochloric acid or any of the dampered acids recited above (uncoated).

The following procedure exemplifies the process of the subsequent acidification of cheese curd.

(1) Skim milk is acidified at about 40°F with 75% phosphoric acid at the rate of about 3.75 cc per 100 g solids in the milk.

(2) The milk is then warmed to about 70°F. At this point the pH is approximately 5.05.

(3) In conventional finishing off processes acidogens such as GDL are added at this point for the development of acid in the curd subsequent to coagulation. At this point the pH is normally 5.00 to 5.05 and is not immediately affected by the GDL addition. In the example below dampered acid is substituted for acidogen. Also, an enzyme solution (cottage cheese coagulation) is added at this point at a rate of about 330 cc per 100 gal of milk.

(4) The milk is then allowed to set in a quiescent state for about 30 to 90 min while the curd forms. At this point the pH is normally (using GDL in place of dampered acids) about 4.90 to 4.95 (pH when set). The curd is then washed with water and cut into ½ inch cubes.

(5) After setting for 15 min additional 75% phosphoric acid is added (to the whey-curd mixture) at the rate of 0.50 cc per 100 gal of solids in the milk.

(6) The whey-curd mixture is then heated with agitation until it reaches a temperature of about 125°F.

(7) The whey is then drained from the curd. At this point the pH (of the curd) will be within the pH range of 4.40 to 4.50.

(8) The curd is rinsed with cold water at a pH of 6.00 or below, chlorinated with 10 parts per million of available chlorine. The water is allowed to remain on the curd for about 15 min and then drained. Such washing is repeated a second time.

(9) After draining the last rinse water the curd is creamed and packaged. The pH of the dry curd should be 4.75 or below.

Example: The dampered acid formula used was 90% Monsanto fine crystal fumaric acid, and 10% corn oil.

pH profile at 2.0 g/100 g solids	
When set	5.10
After 30 minutes	4.92
After 2 hours	4.92
pH after heating to 100°F	4.25
Comments: Smooth coagulation, no whey	

CONTINUOUS CURD PROCESSES

Pulsating Curd Pump Method

The production of curds for making cheese traditionally involves a succession of batch operations such as heating the milk to the renneting temperature, renneting, coagulation, slabbing the coagulum, more or less pronounced syneresis of the curd and molding. This traditional method is attended by various disadvantages, including dispersion of the physiochemical and bacteriological characteristics of the various batches of curd obtained, and the difficulty of rationalizing the various stages of production.

Continuous processes also come up against a serious difficulty due to the fact that the coagulum sticks to the walls of the coagulation chamber, generally tubular in shape, in which the product circulates. This is because the formation of aggregates adhering to the wall of the coagulation chamber is an obstacle to the normal circulation of the product and can result in more or less pronounced blockage of the coagulation unit. In addition, when these aggregates drop off the wall of the coagulation chamber, they form lumps of curd whose physiochemical characteristics differ from those of the rest of the product and impair the homogeneity of the curd obtained.

It is possible to obtain uniform curd in a continous production operation without blockage of the coagulation unit in the process of *P. Stenne; U.S. Patent 3,899,595; August 12, 1975; assigned to Claudel SA, France.* The renneted milk is introduced into the coagulation chamber under a pulsating effect, which produces appreciable variations in the rate of flow of the milk, by means of a suitable device such as a variable-output pump, for example a piston pump, or a valve which is alternately open and closed and whose throughflow cross section is periodically varied.

The coagulation chamber is formed by a pipe-line in which the milk circulates during coagulation. The geometry of this pipe and the material of which its wells are made are selected in such a way that this wall is deformed under the influence of the pulsations applied to the milk. These pulsations produce elastic expansions and contractions in the wall which are reflected in particular in slight localized variations in the throughflow cross section of this conduit. These variations in the throughflow cross section counteract the phenomena which cause the coagulum to adhere to the wall, and prevent blockage of the coagulation unit. The throughflow cross section, the length and the general layout of the pipe-line (winding, bends, determining its loss of pressure) are of course selected in such a way that the flow pattern of the product, more particularly its linear rate of flow at all points, is compatible with the cohesion requirements of the coagulum formed, and in such a way that the residence time of the product in the conduit is sufficient to enable the coagulum to be formed.

Additional deformation can be produced in the coagulation conduit by applying mechanical forces to it, for example by using a helically wound conduit and by locally modifying the winding diameter through the application of radial pressures to the conduit.

In the process described above, the coagulum formed is generally fragmented on leaving the coagulation conduit under the effect of its own weight for example when it is collected on a draining screen. A number of differently sized particles or lumps is formed under these conditions, their distribution being fairly heterogeneous so that they offer a fairly large syneresis surface. A coagulum of this kind is thus particularly suitable for rapid draining and, after draining, forms a curd whose mineral substance, especially calcium, content is relatively high.

However, it can be of advantage, for example in the manufacture of such cheeses as Camembert or Brie, to obtain a curd whose mineral substance content, and especially its calcium content, is considerably lower. A curd of this kind, known as a lactic curd, is generally obtained by slow draining which leaves the calcium enough time to react with the lactic acid to form lactates which are eliminated during draining.

It has been found that the cohesion characteristics of the curd formed in the coagulation conduit, i.e., the ability of this curd to withstand uncontrolled fragmentation, occurring for example spontaneously at the outlet end of the coagulation conduit under the weight of the curd itself, is greatly influenced by the coagulable protein content of the treated milk. This resistance to uncontrolled fragmentation increases when the coagulable protein content of the treated milk increases. Accordingly, it is possible, by adjusting the coagulable protein content of the milk to a suitable value, to collect a coagulum with improved cohesion, which can be more or less pronounced according to requirements, at the outlet end of the coagulation conduit.

If necessary, a coagulum of this kind can be subsequently fragmented, for example by slabbing, into pieces of suitable size offering a syneresis surface such that draining takes place at the required rate to give a curd with a suitable mineral substance content after draining.

If, on the other hand, the coagulable protein content of the milk is adjusted to high values, corresponding for example to total concentrations of non-fat solids of the order of 25 to 30%, it is possible to collect at the outlet end of the coagulation conduit a coagulum of extremely firm consistency which can be cut into unitary slabs corresponding in volume to the required volume of coagulum to obtain, after draining, a unitary cheese.

This modification has proved to be of particular interest in the manufacture of cheese intended to be sold by the piece, because it enables unitary slabs of curd to be obtained whose weights show greatly reduced dispersion. The coagulable protein content of the milk can be increased either by adding a suitable quantity of coagulable proteins to the milk or by treating the milk by a selective filtration process, such as ultrafiltration. Coagulable proteins can be added to the milk in the form of an aqueous solution of these proteins in their native or non-degraded state in which the concentration of these proteins is higher than in the milk. An aqueous solution of this kind can be obtained by subjecting milk to ultrafiltration or by filtering the milk on a gel and collecting the first elution fraction. It is also possible to add to the milk a dried product, for example, in powder form obtained from an aqueous solution of these proteins by drying under such conditions that the coagulable proteins do not undergo any appreciable changes modifying their behavior with respect to the coagulating agents.

If the coagulable protein concentration of the milk to be treated is increased by subjecting the milk to ultrafiltration, the serum protein content of the milk is increased correlatively in proportions governed by the characteristics of the semi-permeable membranes used. Such an increase in the serum protein content can prove to be of advantage because it enables a curd to be obtained whose protein content is greater than that of a curd obtained from an equivalent quantity of milk by a conventional process. As shown in Figure 1.5a, the installation has the following components connected in series, in the order listed, by a pipeline **11**:

Figure 1.5: Continuous Production of Curds

a.

b.

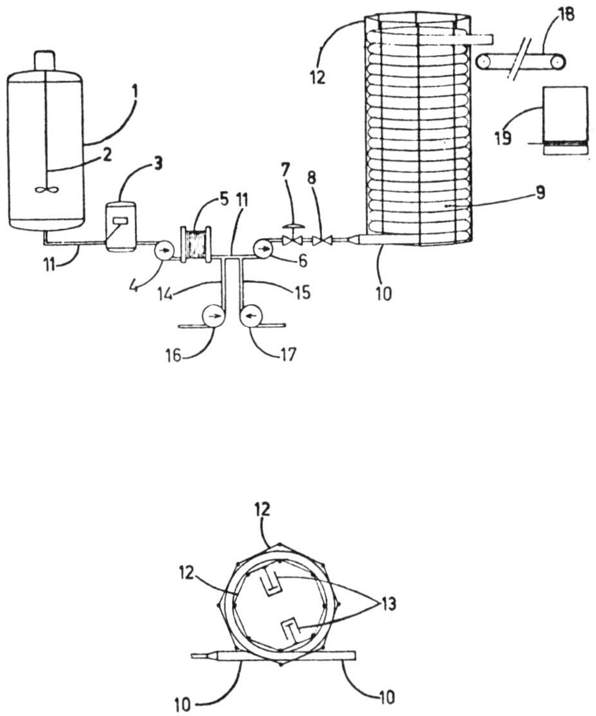

Source: U.S. Patent 3,899,595

A storage tank **1** equipped with a stirrer **2** and a device for keeping it at a constant temperature (not shown), a constant level chamber **3**, a centrifugal pump **4**, a plate heater **5**, a centrifugal pump **6**, a pneumatically controlled valve **7**, a variable opening valve **8**, and a coagulation unit **9** in the form of a helically wound tube **10** of polyvinyl chloride. This tube **10**, which has a diameter of 100 mm, a thickness of 2 mm and a length of 20 m, is arranged in an octagonal frame **12**. This frame is subjected to the action of pneumatic jacks **13** (shown in Figure 1.5b) and is capable of deforming to such an extent that the radius of curvature of the helical tube **10** is modified by about 30%.

The installation also has two pipelines **14** and **15** connected in parallel to that section of the pipeline **11** situated between the heater **5** and the pump **6**. By means of the pipelines **14** and **15** equipped with metering pumps **16** and **17**, respectively, it is possible to introduce the quantities of lactic ferment and rennet required to obtain the type of cheese desired. These additives are intimately mixed with the milk by the centrifugal pump **6**.

Finally, the installation has a drainage screen **18** in the form of a moving belt which is permeable to liquids. A perforated hopper **19**, whose cross section is several times greater than that of the required cheese is arranged beneath the screen **18** to collect the drained curd.

Example: A milk containing 34 g/l of fats, stored at 10°C in the tank **1**, is heated to 35°C through circulation in the plate heater **5**. 1% by volume of a ferment consisting of an association of *Streptococcus lactis, Streptococcus cremoris, Streptococcus citrovorus* and *Streptococcus diacetilactis*, and 0.03% by volume of a rennet with a strength of 1/10,000, are then introduced into the milk circulating by means of the metering pumps **14** and **15**. The lactic ferment and the rennet are mixed with the milk by the centrifugal pump **6** whose output is at least twice the output of the installation.

The pneumatically controlled valve **7** is operated by an automatic device which closes it for 1 sec and then opens it for 1 sec in repetitive fashion. This operation of the valve produces a pulsating effect upon the flow of milk which in turn produces deformation of the coagulation conduit **10** reflected in local variations in the diameter of the conduit of the order of 0.15%. The opening of the valve **8** is regulated to adjust the rate of flow to 940 l/hr, which corresponds to an average linear rate of flow of the milk during coagulation in the tube **10** of the order of 2 m/min, and to a residence time of the product in the order of 10 min.

A pulsating effect is produced by opening and closing the valve **7**, which is reflected in a periodic variation in the output of product at the outlet end of the coagulation unit. On the other hand, pressures are periodically applied to the frame **12** by means of the jacks **13**, thus modifying the radius of curvature of the helical conduit by about 30%.

The curd collected at the outlet end of the conduit **10** is distributed onto the drainage screen **18** whose drainage rate is regulated to 2 m/min. The drained curd then drops into the hopper **19** where it is available for molding and the subsequent operations involved in the manufacture of cheese.

Linear Coagulation Method

Attempts have been made for partial mechanization and improvements in the technology of certain phases of making telemea cheese. There are known processes for crushing the curd in a cheese vat followed by drawing it out by means of large curd scoops or pails and pouring it into a curd draining table, pressing it in molds, and coagulating it directly on a curd draining table lined with polyethylene sheet. In fact, all these processes do not truly mechanize the cheese making process which remains a manual process in other stages, but they represent only the improvement of technological operations or phases, leading in the end to discontinuous stages.

The continuous processes, whose coagulation phase takes place in ascendent vertical direction, are not suitable for telemea cheese production because of the fact that they cannot carry out the transition to the treatment and pressing phases which take place in an horizontal and descendent vertical direction because it is not possible to maintain the integrity of the curd when the direction changes produce the breaking of the curd.

As a result, the bedding of the milk cannot be maintained unchanged from coagulating phase up to the obtaining of the finished product. All the attempts to modify telemea cheese making processes by passing the curd through a curd crushing phase produced a cheese having pressing caverns, a low conservation index and an unsatisfactory quality.

V. Nikolic; U.S. Patent 3,518,094; June 30, 1970; assigned to Ministerul Industriei Alimentare; Romania developed a process which uses linear coagulation and pressing to permit continuous operations and at the same time maintain curd integrity. The unit for this process is shown in Figure 1.6. The unit consists of a centrifugal milk pump **1**, two metering pumps **2** for the coagulation solution, a tank **3** for the coagulation solution, a coagulation unit **4** inclined at 25° to 35° connected with a unit **5** consisting of two lateral walls **6**.

Figure 1.6: Linear Coagulation Method

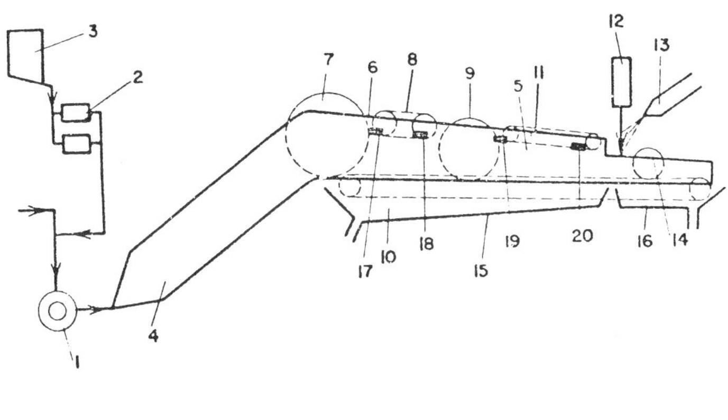

Source: U.S. Patent 3,518,094

Between the lateral walls are provided a set of knive-revolving disks **7**, a short pressing belt **8**, a second set of knife-revolving disks **9**, a conveyor belt **10**, a belt for final pressing **11**, a guillotine cutter **12** with hydrulic drive for transverse cutting, a set of nozzles **13** for cool water spraying, another set of disk knives **14**, a trough **15** for collecting whey and a trough **16** for collecting cooling water. The lateral walls **6** are provided with orifices **17**, **18**, **19**, and **20** for eliminating the whey. The belts and the knives are synchronistically driven but it is possible to vary the treatment rhythm within ±20%.

Milk is prepared for coagulation by pasteurization, addition of correction agents (calcium chloride, starting culture, and possible also alkaline nitrates), the acidity is brought (in terms of density degrees in the Thömer system) to 20° to 22° T by means of diluted lactic or hydrochloric acids and the temperature is established at 28.5° to 30.5°C. The coagulation solution is simultaneously prepared by dissolving rennet in water. The concentration of the coagulation solution is determined by means of the relation:

$$c = \frac{100\,k \times D \times Pr}{d}$$

where c is the concentration of the coagulation solution, as percent; k is a correction factor of the passage from coagulation in static state to coagulation in flow state (a non-dimensional coefficient); D is the flow rate of the milk pump, in l/hr; Pr is the relative strength of rennet expressed as conventional units (kg/l); and d is the flow rate of the coagulation solution pumps, in l/hr. The correction factor determined by experiment for this method for telemea cheese making is k = 1.4 to 1.8. The notion of relative coagulation strength is necessary in order to express the strength of rennet which is related to a certain quality of the milk to be treated. The quantity of the coagulation solution is determined by the relation:

$$Q = \frac{1.1 \times L \times d}{D}$$

where Q is the quantity of the coagulation solution in liters; L is the quantity of milk in liters which is to be treated after adding the correction agents; D and d have the same meaning as above. The required quantity of rennet is calculated from the values of c and Q.

When milk and the coagulation solution are prepared, the pumps **1** and **2** are operated. Milk enters the lower part of the coagulation body **4** by flowing upwards in the form of a laminar, linear, and uniform flow over a period of 30 to 35 min. When it arrives in front of the knives **7**, the curd advances and is cut vertically by the blades of the knife disks and crosswise by the wires, which join the borders of the outside disk within 10 cm. The cut curd is taken up by the conveyor belt **10** and the treatment devices. The belt **8** exerts a light gradual pressure and the whey is separated and eliminated in the upper part of curd, by draining it through the orifices **17** and **18**. The fused curd is cut again by the set of knife disks **9** which are closer together than the former set.

The eliminated whey is drained by decantation through the lateral orifice **19**. The curd is then pressed by the belt **11** and the whey separated is eliminated through the orifice **20**. A strip of curd is obtained whose width depends on the transversal sizes of the unit and which must be a multiple of the block sizes.

Delayed Rennet Action

Another process for continous production of curd in cheese making is given by *N.J. Berridge; U.S. Patent 3,520,697; July 14, 1970; assigned to National Research Development Corporation, England.* A difficulty which is usually encountered in processes for the continuous production of curd is that due to the continuous nature of these processes the milk is usually moving during curd formation, and this movement causes a reduction in the amount of fat retained in the curd.

It is known that if rennet is added to milk at low temperature, rennet action occurs without formation of curd, and that if the cold renneted milk is then heated coagulation takes place rapidly. Processes have been devised using this effect to enable coagulation to take place quickly thus reducing the effect of disturbance during coagulation. These processes have involved passing the cold renneted milk over a heated surface in the form of a thin film. A disadvantage which has been encountered is that the curd tends to stick to the heated surface.

The improved process for continous production of curd required the steps of renneting milk at a temperature such that rennet action can occur without curd formation, holding the milk at that temperature at least until rennet action is nearing completion or is complete and contacting the renneted milk with one surface of a semi-permeable membrane, the other surface of which is in contact with water at elevated temperature so that the milk is heated to a temperature at which curd formation can occur.

The milk should be contacted with the membrane in such a form that the milk is rapidly heated throughout its bulk to the required temperature. For example, the membrane can be in the form of a thin tube through which the milk is caused to flow. Alternatively, two sheets of membrane placed close together to form effectively a flat tube can be used or the milk can be flowed as a thin layer over a sloping sheet membrane. Regenerated cellulose (Visking dialysis tubing) and collodion (cellulose nitrate) are suitable materials for the membrane.

The milk can first be inoculated with a suitable starter culture such as *Streptococcus thermophilus* and its pH adjusted, for example, between 6.1 and 6.3. For cheddar cheese the preferred pH is between 6.2 and 6.3. The milk is then cooled, to a temperature between 0°C and 10°C, and treated with rennet, and is held at this temperature at least until the specific rennet action is nearing completion or is complete as shown by coagulation when a drop of milk is allowed to fall from a height of 2 mm into water at 50° to 60°C. If desired the milk can be held at the holding temperature for times considerably in excess of the minimum time needed for the completion of rennet action. A long holding time may modify the properties of the curd.

The milk is then brought into contact with the membrane by flowing it over one surface of the membrane, while hot water is flowed over the other surface of the membrane. During this stage the milk should be heated to a temperature sufficient to cause clotting but not so high as to kill the starter culture or to cause protein to be deposited from the whey, e.g. to a temperature between 30° to 50°C. In some circumstances, e.g. when making soft cheeses, the starter can be added at a later stage in the process and then the milk can be heated to temperatures above those at which cultures are killed. The required temperature of the water on the other side of the membrane will depend on the length of time that the milk is in contact with the heated membrane, on the initial temperature of the milk, the nature and thickness of the membrane, and rate at which the water flows over the membrane. The membrane should not be heated at the point where the milk first comes into contact with it, so that the milk flows over the surface of the membrane for a short distance before coming in contact with the heated part of the membrane where curd formation occurs and the tube or ducts leading to the membrane do not become heated. This ensures that curd formation does not occur in the tube or ducts leading to the membrane.

The amount of curd sticking to the membrane surface is decreased if the water is acidified with phosphoric acid or hydrochloric acid. Even when acidified water is used a small amount of protein may stick to the membrane, and it may be necessary periodically (say every 8 hours) to interrupt the milk flow and flush the membrane with water followed by an acidic or alkaline cleansing solution. If any further particles of protein remain stuck to the membrane, treatment with pepsin solution may be required. The hot water may be recirculated, but should be changed from time to time. A vertical tube of semipermeable membrane **1** (Figure 1.7) is joined at its upper and lower ends to glass tubes **2** and **3** respectively. The tube **2** has U-shaped portion **4.**

Figure 1.7: Continuous Production of Curd

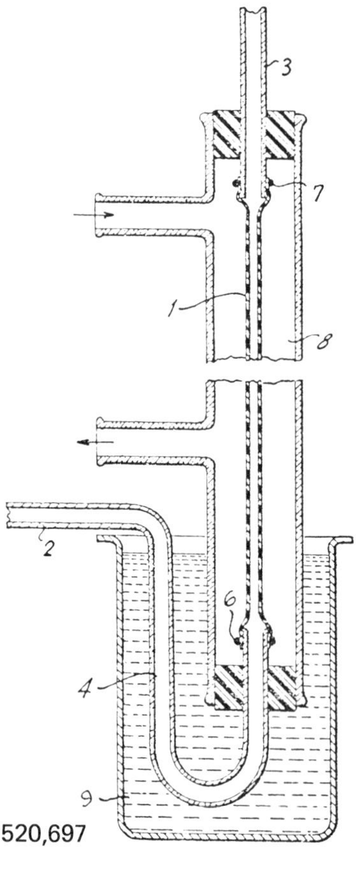

Source: U.S. Patent 3,520,697

The ends of the tube **1** are slipped over the ends of the tubes **2** and **3** and held in place by rubber rings **6** and **7**. The internal diameter of tube **1** is slightly smaller than the internal diameter of tube **3** so that the curd is not forced against the edge of the tube **3**. The tube **1** and a short length of each of the tubes **2** and **3** are enclosed in a water jacket **8**. The U-shaped portion **4** of the tube **2** and part of the water jacket **8** are immersed in a cooling bath **9**. The water jacket **8** is immersed to a depth such that the end of the tube **1** is below the surface of the cooling fluid in the cooling bath **9**. The tube **1** is 3 feet long and is made of ¼ inch diameter Visking tubing.

In operation, the bath **9** is kept at 0 to 10°C, and tap water containing 0.105% (w/v) phosphoric acid is circulated through the jacket **8** at a temperature of 60°C, and at a rate of 1,260 ml/min. The flow pattern of the water in the jacket **8** is such that the hot water does not flow to the bottom of the jacket, and thus does not heat the tube **2**. Milk which has been treated with a suitable starter culture and then renneted at pH 5.9 to 6.1 and 0°C to 10°C and held for about 1 hr, is pumped into the tube **2** at the rate of 80 to 90 ml/min.

Curd forms in the heated portion of the tube **1**, becomes detached from the wall and moves upwards carried by the flow of whey, which is exuded from the curd. The curd leaving the top of the tube **1** has a temperature of 44°C. The curd, on emerging from tube **3**, can be received into whey where it is held at scalding temperature or otherwise handled, for example, according to the processes normally used according to the type of cheese required.

DRAINED CHEESE MILK PROCESS

J.-L.J. Maubois, G.P.C.G. Mocquot, and L.J. Vassal; U.S. Patent 3,914,435; October 21, 1975; assigned to Institut National de la Recherche Agronomique, France uses milk concentrate as a starting material for cheese production. This concentrate is prepared by ultrafiltration and has essentially the same composition as cheese, but in a liquid form.

It is known that milk, and especially cow's milk, is more difficult to coagulate with rennet when heated above a certain temperature. This is the case with milk boiled at about 100°C. It is the same thing with milk which had been heated to autoclave sterilization temperatures (110° to 115°C for 10 to 20 min), as well as milk which has undergone known ultra-high temperature (UHT) heating, during which the milk is heated to a temperature in the range of 110° to 150°C, and maintained, usually about 4 to 10 sec, and then cooled.

When, after having been subjected to such known heating, milk was brought to 30°C and rennet was added, it was observed that either coagulation did not occur, or a very soft coagulum was formed slowly, which did not harden, did not exude whey by syneresis, did not drain, or drained very badly. Such a coagulum is therefore not very suitable, or is even totally unsuitable for conversion into cheese.

On the other hand, when milk, having been subjected to UHT heating between 110° and 150°C, for 2 to 8 sec is then subjected to ultrafiltration until its protein concentration is twice that of the protein concentration of the starting milk, has its temperature adjusted to 30°C and the rennet added, a consistent, firm coagulum is observed to form, which is suitable under conventional operations for converting milk into cheese. Instead of starting with a heat-treated milk, such as a UHT milk, it is possible, to heat the retentate obtained from raw or pasteurized milk to a temperature higher than 110°C and then use the retentage for the manufacture of cheese. The consistency and firmness of the coagulum obtained after the action of rennet on the retentate subjected to such a heat treatment were found to be substantially the same as those of the coagulum obtained by the action of rennet on a retentate obtained from milk subjected to heat treatment at a temperature higher than 110°C.

A certain desired weight of raw milk is used as a starting product and subjected to ultrafiltration. An aliquot fraction of the retentate is taken at regular intervals of time. As ultrafiltration continues, a series of samples with increasing protein concentrations is thus obtained; the first sample corresponding to the first aliquot fraction of the retentate has, for instance, a protein concentration equal to 1.5 times that of the starting milk, the second to twice, the third to three times the protein concentration of the starting milk, etc. The last sample taken at the end of ultrafiltration has, for example, a protein content 6 times that of the starting milk.

If the same amount of rennet (for example, 0.01 ml of commercial liquid rennet, of a strength of 10,000) is added to a predetermined weight (20 g for example) of each of the samples of retentate having a temperature of 30°C, the duration of coagulation is observed to be substantially the same, whatever the protein content of the various samples of retentate examined. In other words, when the protein content of the retentate is, for instance, six times that of the starting milk (milk as it is obtained from the cow), the amount of rennet, based on 1 g of protein present in the sample of retentate in question, necessary to add to obtain coagulation by a time t, is one-sixth, also based on 1 g of protein present, that required to coagulate the starting milk in the same time t.

The ultrafiltration of milk having undergone heat treatment, such as UHT milk, or the heat treatment, for example UHT heating of the retentate or previously ultrafiltrated milk, has many advantages. Firstly, in all cases a starting material is obtained which is perfectly suited for the manufacture of cheese as it is well coagulated by rennet under normal conditions.

Secondly, it is possible to have a starting material from which all undesirable microorganisms, such as fermentation microorganisms which have a detrimental effect on cheese, have been eliminated. The process thus makes use of the preliminary sterilization of the milk (or the protein-enriched milk which constitutes the retentate). The process provides a starting material in which the spore forming bacteria belonging to the genus *Clostridium* are destroyed. It has long been known that bacteria of the genus *Clostridium* are responsible for the butyric fermentation (late fermentation) of cooked cheese, notably Gruyere cheese.

A moderate (110°C) UHT treatment of milk, associated with ultrafiltration, makes it possible to obtain a milk which coagulates well, and is therefore suitable for making cooked cheeses, but without the risk of their undergoing butyric fermentation. This opens up new prospects for milk-producing regions where cooked cheese is made, notable Gruyere, as these regions can thus envisage feeding silage to cows, which is much easier to use than normal fodder (hay) whereas feeding silage to the cows in Gruyere-making areas is presently forbidden owing to the well known risk arising from *Clostridia* which are often present in large numbers in silage.

Example: The manufacture of a soft cheese (Camembert type) is described in this example. Using filtration techniques 10 kg of milk (32 g/kg fat content), having undergone ultra-high temperature treatment (150°C for 2 sec), circulated across a membrane of an ultrafiltration apparatus. The temperature of the milk was 10°C.

When the weight of the filtrate recovered reached 5.45 kg, the retentate was collected. It had the following composition: Nitrogen substances, 6.5 g %; total dry extract 18.5 g %; fat content 6.6 g %; and ash 0.8 g %. The pH was 6.7. 3 kg retentate was heated to 30°C and 1.6 g of a suspension of frozen, concentrated starter *(Streptococcus lactis)* was added. The mixture was maintained at this temperature. When the pH reached a value of approximately 6.2, 0.4 ml of a commercial rennet solution per kg of retentate was added. Setting (formation of coagulum) occurred 7 min after the addition of the rennet.

The coagulum, which was soft at the start, became firm in about 15 min. One hour after the addition of rennet, the cheese was molded according to the technique normally used in the manufacture of Camembert. The last part of the coagulum draining process occurred under conditions comparable to normal conditions. Sixteen hours after the addition of rennet, the cheeses were removed from the molds and salted according to the conventional technique.

CHEESE AND CHEESE FLAVORS

ANIMAL RENNET PROCESSES

Bovine Rennet Preservation

Calf rennet remains the industry standard against which other coagulants are measured. However, declining calf slaughter has resulted in a worldwide shortage of stomach raw materials for rennet production. This is reflected in the growing use of adult bovine stomachs with their lower rennet content, and processes for increasing the recoverable yield of rennet from these tissues. Packing in rock salt or sodium chloride crystals is a traditional method for meat preservation and prevention of microbial growth. In the present practice of rennet production, it is customary to wash the calf and adult bovine stomachs with water immediately after slaughter and pack them in solid sodium chloride prior to storage under refrigeration. The rennet is subsequently extracted from the tissues with water or 5 to 10% w/v sodium chloride solution.

R.B. Dardas; U.S. Patent 3,753,860; August 21, 1973; assigned to Pfizer Inc. proposes a more effective method of preservation. It has been found that dipping these stomachs in a 1 to 10% w/v propylene glycol solution of a 1:1 mixture of the methyl and propyl esters of p-hydroxybenzoic acid prior to storage under refrigeration allows for increased yields of extracted rennet of 50% or more than those obtained using solid sodium chloride.

The choice of esters of p-hydroxybenzoic acid (generically referred to as Parabens) is not critical, and they may be used singly or in combination. The preferred solution is a 1:1 mixture of methyl and propyl Parabens in propylene glycol. Calf and adult bovine stomachs, after slaughter, are washed with water, dipped in the propylene glycol-Parabens solution and stored under refrigeration.

Example 1: Calf stomachs are washed with water immediately after slaughter and set aside under refrigeration (4° to 5°C) for approximately 12 hours. Individual stomachs are dipped in 3% w/v and 5% w/v propylene glycol solutions of a 1:1 mixture of methyl and propyl Parabens. A control group of stomachs is packed in solid sodium chloride.

The preservative treated calf stomachs are then stored under refrigeration (4° to 5°C). After a time period of 1 week, 2 weeks and 5 weeks, the individual stomachs are placed in 10% w/v sodium chloride solution at a 3:1 solution to meat ratio and held at 37°C for about 48 hours each for a total of 3 such extractions. The extracts from each stomach are combined and assayed for rennet. The results are shown in the following table with the body of the table expressed in arbitrary rennet units per gram of tissue.

	...... Storage time (weeks)		
Preservative	**1**	**2**	**5**
Sodium chloride	90	55	54
3% Parabens	83	89	88
5% Parabens	104	111	121

Example 2: Adult bovine stomachs are washed with water immediately after slaughter, and individual stomachs are dipped in 1%, 5% and 10% propylene glycol solutions of a 1:1 mixture of methyl and propyl Parabens. A control group of stomachs is packed in solid sodium chloride. After time periods of 1 week, 2 weeks and 5 weeks storage under refrigeration (4° to 5°C), individual stomachs are placed in water at a 3:1 ratio of water to meat and held at 37°C for about 48 hours for a total of 3 such extractions. The extracts from each stomach are combined, and assayed for rennet. The results are shown in the following table with the body of the table expressed in arbitrary rennet units per gram of tissue.

	...... Storage time (weeks)		
Preservative	**1**	**2**	**5**
Sodium chloride	17	10	27
1% Parabens	27	–	32
5% Parabens	51	63	67
10% Parabens	34	40	54

Poultry Rennet

A rennet-like milk curdling enzyme can be obtained from the stomachs of poultry animals in a process described by *M. Horisberger, T. Sozzi and R. Pousaz; U.S. Patent 4,081,330; March 28, 1978; assigned to Societe d'Assistance Technique pour Produits Nestle SA, Switzerland.*

Although generally speaking the stomachs of poultry, especially domestic poultry, yield a milk-curdling enzyme of suitable quality, glandular or proventricular stomach tissue is used because it is in these organs that the maximum concentration of pepsinogen, the precursor of pepsin, is found. The treatment involves immersing tissues in an aqueous saline solution with a pH value of below 4 which may be adjusted by means of a buffer solution or any acidification agent in the form of a mineral or organic acid. The salt used is sodium chloride, essentially in view of the "food-grade" application of the milk-curdling enzyme.

The various stages of the process are shown in thick lines in Figure 2.1 in which optional modifications are also shown, but in thin lines. In a first modification, the ionic force value of the liquid phase of the suspension is readjusted or modified before phase separation, because it is preferred to carry out this separation at values which enable a large quantity of the uninteresting protein fractions extracted by the acid treatment to be precipitated without, at the same time, introducing too much salt into the liquid phase and, hence, interfering with the subsequent reductions in volume. For example, if sodium chloride is used as the salt, the concentration of that salt is kept within a range from approximately 6 to 10%. At the same time, the pH value of the suspension may be readjusted to a value below 4, for example, to the initial value, because it has been found that the pH value varies during the treatment.

Figure 2.1: Production of Poultry Rennet

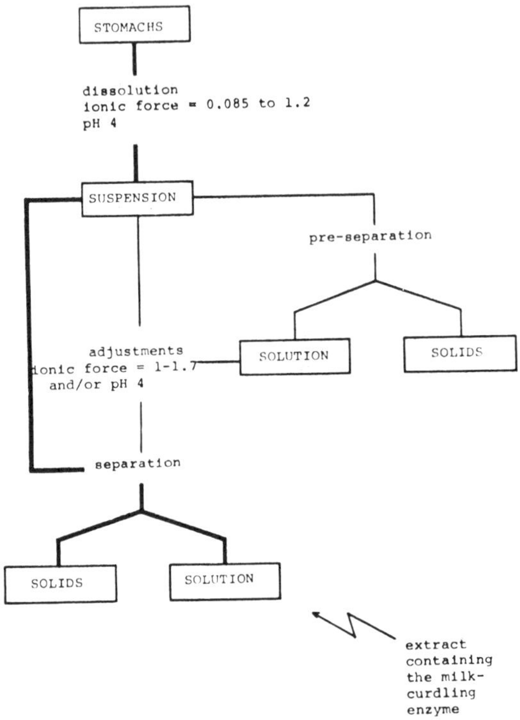

Source: U.S. Patent 4,081,330

In a second modification, which is an improvement in the first modification, the phases of the suspension are separated, for example, by centrifuging, the ionic force and/or pH changes taking place in the solution freed from insoluble materials, for example, the supernatant phase. The solution obtained after separation, or after preseparation and separation, is an extract which contains the milk-curdling enzyme. This extract is in the form of a clouded liquid which may be directly used for coagulating milk or clarified to remove fats and oils from it. In addition, the pH value of the clarified extract is reduced to a value around 2, which is the pH value at which the milk-curdling enzyme develops its greatest activity. This pH value on the one hand and the presence of salt in a significant concentration on the other hand provide this enzyme in solution with excellent keeping qualities.

It is also known that the gastric system of certain birds, especially carnivorous birds, is the center of intense proteolytic activity of the pepsic type. A process is used for extracting a milk-curdling enzyme by means of an alkaline agent or even by pressing stomachs. The enzyme prepared by acid extraction has an activity greater than that of the pepsin obtained from poultry by alkaline extraction.

Example 1: 56 kg of the glandular stomachs of chickens are ground in a Stephan mill fitted with a screen having holes 1.5 mm in diameter, and the product of grinding immersed in 165 liters of deionized water containing 1.45 kg of NaCl. A 12.5% solution of HCl is then added until a pH value of 3 is obtained. The suspension is then vigorously stirred for 15 minutes and subsequently centrifuged at 2,000 rpm in a Heine centrifuge, yielding 55 kg of sediment, which is discarded, and 167.6 kg of supernatant phase to which 11.7 kg of sodium chloride are added, followed by the introduction of 400 g of 12.5% HCl to adjust the pH value to 2.95.

After recentrifuging at 5,000 rpm in a de Laval centrifuge, the supernatant phase is freed from fats by treatment at 28°C in a de Laval centrifugal separator rotating at 7,000 rpm. 131.5 kg of solution are obtained in this way. 60 kg of this solution are removed and acidified to pH 2 by the addition of 500 g of 12.5% HCl. The dosages of this acidified solution containing the milk-curdling enzyme produced the following results (acid extraction). The results obtained with a solution acidified to pH 2, obtained by alkaline extraction under otherwise exactly the same conditions, are also given by way of comparison.

	Acid Extraction	Alkaline Extraction
	. . . .%	. . . .
Dry matter	9.90	7.15
Fat	0.04	0.01
Calcium	0.005	0.05
Sodium chloride	7.95	6.10
Nitrogen (total)	0.30	0.17
Nitrogen (soluble)	0.27	0.16

Example 2: The remaining 71.5 kg of the solution obtained in Example 1 are concentrated after fat extraction by evaporation in vacuo at 42°C to a weight of 21 kg. The concentrated solution is adjusted to pH 2 by the addition of 370 g of 12.5% HCl. The results of the dosages are as follows.

	Acid Extraction	Alkaline Extraction
	. . . .%.	. . . .
Dry matter	31.40	29.13
Fat	0.01	0.031
Calcium	0.017	0.011
Sodium chloride	24.4	24.0
Nitrogen (total)	0.95	0.71
Nitrogen (soluble)	0.75	0.66

Example 3: 80 kg of whole milk pasteurized for 5 seconds at 65°C are inoculated with 1% *S. thermophilus* strains, and the milk thus inoculated is kept at 37°C for 1 hour. The pH value is then 6.20. This milk is then divided into two equal fractions of 40 kg each. The first fraction is then renneted with 20 ml of the concentrated solution of Example 2, an acidified solution containing the milk-curdling enzyme and having a strength of 1:13,000, followed by the addition of 10 g of $CaCl_2 \cdot H_2O$, where strength = 10,000 x 240/t, and t is the time required to coagulate 100 ml of milk at 37°C with 0.1 ml of enzyme solution. At the same time, the second fraction is renneted with 1.5 g of Hansen's rennet of strength 1:150,000, again followed by the addition of 10 g of $CaCl_2 \cdot H_2O$. Each of the milks is coagulated after 10 minutes.

These curds are then treated in the conventional production of mozzarella, which is: working the curds for 6 minutes; placing in molds; turning over after 15 minutes; draining for 4 hours at 37°C; extruding the paste at 60°C; salting to 1% and storing in a cold room in iced water. The mozzarellas obtained are tasted after 24 hours. The mozzarella prepared with the solution containing the milk-curdling enzyme is barely distinguishable from the traditional mozzarella prepared with Hansen's rennet.

Pepsin-Rennet Combination

Pepsin-rennet preparations in a 1:1 ratio have been used to replace calf rennet in cheese making. The amount of proteolytic enzyme needed to coagulate milk protein can be reduced by at least 66% in a two-step process where the temperature and acidity expressed as °SH of the milk are controlled during each step. Thus, hard, soft, and cottage cheeses may be manufactured using only 100,000 Soxhlet activity of proteolytic enzyme (calf rennet or its substitutes) per 100 liters of vat milk.

In the first step, proteolytic enzyme is added to milk in insufficient quantities to cause milk coagulation at an acidity of °SH of 6.0 to 8.5 at temperatures ranging from 0° to 32°C. In the second step, a second proteolytic enzyme is added to the milk and the temperature raised to from 20° to 45°C and the acidity as expressed in °SH is raised to from 7.0 °SH to about 12.5 °SH.

The following examples are given by *M.Szadkowska, J. Pisarek, B. Dzik, E. Staniak, D. Lipniewska and F. Bijok; U.S. Patent 4,082,853; April 4, 1978; assigned to Instytut Przemyslu Mleczarskiego, Poland,* to illustrate the process.

Example 1: The following concerns the manufacture of hard cheese, such as Edam cheese. 100 liters of pasteurized milk at an acidity of °SH 7.0, are standardized to the required fat content, e.g., 45% fat content in dry solids, 2.4% to 2.8% fat content in fluid milk, and brought to a temperature of from 29° to 30°C. 100 mg of pork pepsin of 1:4,000 coagulating activity is dissolved in 200 ml of 0.01 N hydrochloric acid 15 to 30 minutes before use.

To the milk is added about 2% starter and 20 g of anhydrous calcium chloride, and the dissolved pepsin is poured into the milk in a thin stream. The milk is agitated for 10 minutes, and the temperature is maintained at 28° to 30°C for 1 hour. The acidity of the milk as expressed in °SH is raised to 8.0, and the milk is heated to 32°C. 1 g of calf rennet, with 1:100,000 Soxhlet coagulating activity, is added, and the milk is agitated for 3 minutes. Milk movement is stopped, and curd sufficiently firm to cut is obtained after 30 to 40 minutes.

Further treatment of the curd is carried out in the traditional manner for making Edam cheese.

Example 2: The following concerns the manufacture of soft cheese such as Camembert. 100 liters of pasteurized milk at an acidity of °SH 7.8, is brought to a temperature of from 28° to 30°C. From 1.5 to 3.0% of starter for Camembert cheese, 20 g anhydrous calcium chloride, and mold culture are added to the milk. 50 mg of pork pepsin with 1:4,000 coagulating activity are dissolved in 200 ml of 0.01 N hydrochloric acid 15 to 30 minutes prior to use in the process. The pepsin solution is poured in a thin stream into the milk, and the treated milk is agitated for 10 minutes. The treated milk is held at a temperature of from 28° to 30°C for 1 hour, after which time the milk acidity is approximately 8.5 °SH.

The milk is heated to 32°C and spread in cheese vats for coagulation and curd treatment. Calf rennet of coagulating activity 1:100,000 Soxhlet is added in quantities of 0.33 to 0.66 g. This represents one-third of the normally required 1 to 2 g of calf rennet of 1:100,000 coagulating activity to obtain curd after 60 to 90 minutes in 100 liters of milk. After addition of the rennet, the milk is agitated for three minutes, and milk movement is then stopped. Curd sufficiently firm to cut is obtained after 60 to 90 minutes. Further treatment of the curd and mass is in accordance with the traditional methods of manufacture of Camembert cheese.

Example 3: The following concerns the manufacture of an acid-rennet cottage cheese obtained by a centrifugal method. 100 liters of skimmed milk of acidity 7.5 °SH are brought to a temperature of from 25° to 30°C, preferably 27°C. From 0.5 to 1.2%, preferably 0.8%, of starter and 20 g of calcium chloride are added. From 15 to 30 mg, preferably 20 mg, of pork pepsin of 1:4,000 coagulating activity (specified according to the Polish Pharmacopea IV) are dissolved in 200 ml 0.01 N hydrochloric acid 15 to 30 minutes before use.

The pepsin is poured in a thin stream into the milk, and the milk is agitated for 10 minutes. The treated milk is maintained at a temperature of from 25° to 30°C, preferably 27°C, for 2 hours. The milk is then treated with 50 mg of calf rennet and ripening is continued for 12 to 15 hours at a temperature of from 25° to 30°C, preferably 27°C. Further procedure is in accordance with the requirements for acid-rennet cottage cheese manufacture.

Example 4: The following concerns the manufacture of Cheddar cheese from milk stored at 10° to 12°C. 100 liters of pasteurized milk of standardized fat content is brought to a temperature of between 10° to 12°C, and from 0.5 to 0.8% of starter is added. 50 mg of pork pepsin of 1:4,000 coagulating activity are dissolved in 200 ml of 0.01 N hydrochloric acid 15 to 30 minutes before use. The pepsin is poured into the milk in a thin stream, and the milk is agitated for 10 minutes. The milk is maintained at a temperature of from 10° to 12°C for 6 hours. At this time the milk has an acidity of °SH of from 8.2 to 8.4.

The treated milk is heated to 30°C and 30 g of anhydrous calcium chloride and 1 g of calf rennet with coagulating activity of 1:100,000 are added. This amount of calf rennet is one-third of the normally required 3 g of calf rennet of this activity for 100 liters of milk. The milk is agitated for 3 minutes and then milk movement is stopped. Curd sufficiently firm to cut is obtained after 30 minutes.

The curd is then treated in conformance with the traditional process for cheddar cheese manufacture.

BACTERIAL RENNETS

Rennet Equivalent Substitute

In the production of cheese it is necessary to coagulate the milk in order to be able to separate the casein from the whey. It is usual to use an enzyme isolated from calf stomach in this process. In order to obtain a satisfactory yield of cheese and in view of the subsequent ripening of the cheese it is highly important that the proteolytic activity of the rennet enzyme is low, so that as little casein as possible dissolves during the production and ripening of the cheese. Rennet from calf stomach is satisfactory for the above purpose. However, it is to be expected that the demand for rennet cannot be met with enzymes from calf stomach in the future. Several attempts have been made in order to discover other milk-coagulating enzymes having low proteolytic activity.

A long series of microorganisms produce proteinases having milk-coagulating effect, but most of these also possess a considerable proteolytic effect, so that the yield obtained in the production of cheese becomes too low, bitter-tasting decomposition products of the casein (peptides) at the same time being formed during the cheese ripening. This applies to e.g., proteinases from *Bacillus subtilis* and *Aspergillus saitoi.* Enzyme from *Mucor pusillus* appears to possess a more favorable ratio between milk-coagulating and proteolytic activity. However, this microorganism would appear to be most suitable for surface cultivation and provides relatively low yields by submerged cultivation.

A new enzyme was identified by *K. Aunstrup; U.S. Patent 3,988,207; Oct. 26, 1976; assigned to Novo Terapeutisk Laboratorium A/S, Denmark* which has low proteolytic activity and therefore avoids low yields and bitter-tasting decomposition products (peptides) of casein. The enzyme is produced from a submerged cultivation of *Mucor miehei* Cooney et Emerson CBS 370.65 as described in the following sections.

The following describes the pilot plant experiment. In an inoculation tank there was prepared a nutrient medium composed as follows:

Potato starch, kg	2
Soy meal, kg	1.5
Barley flour, kg	5
Bacterial amylase	
Novo 5000 SKB, g	5
$CaCO_3$, g	500
Water, liters	~40

The mixture was heated to about 70°C and kept at that temperature for about 30 minutes. The mixture was then boiled for 90 minutes at 120°C under direct injection of steam, the volume after cooling being about 50 liters. Following cooling to 34°C the medium was inoculated with spores from a Fernbach flask containing YPPSS-agar of the following composition:

Yeast extract, g	4
K_2HPO_4, g	1
$MgSO_4 \cdot 7H_2O$, g	0.5
Soluble starch, g	15
Agar, g	20
Water, ml	1,000

The Fernbach flask had been inoculated with a lyophilized culture of *Mucor miehei* Cooney et Emerson CBS 370.65 and incubated at 40°C for sporulation. After inoculation, stirring (240 rpm) and aeration (60 liters per minute) were started. Fermentation was continued for 48 hours until a good growth in the tank was observed. The contents of the tank were then transferred to a main fermentation tank containing:

Potato starch, kg	8
Soy meal, kg	6
Ground barley, kg	20
Bacterial amylase Novo 5000 SKB, g	25
$CaCO_3$, kg	2

The medium was boiled and sterilized in the same manner as the inoculation tank medium, and the final volume was 200 liters. After the inoculation, stirring was started (400 rpm) as well as aeration (0.2 m^3/min), and soy oil was added as an antifoaming agent when necessary. During fermentation the pH value was constant at 6.5. After 129 hours the milk-coagulating enzyme content of the culture liquid was 2,580 Kunitz rennet units (KRU) per liter.

Example: In 500 ml Erlenmeyer flasks 25 g of a mixture of wheat bran and water (1:1) was sterilized. The flasks were sterilized by heating to 120°C for 45 minutes, and after cooling to 40°C they were inoculated with spores of *Mucor miehei* Cooney et Emerson strain CBS 370.65 as described above. The flasks were incubated for 3 days at 40°C. The bran was then extracted with 200 ml of water, and the milk-coagulating activity of the extract was found to be 4,500 KRU per liter.

The milk-coagulating enzyme is compared with rennet from calf stomach in the following: The activity against sodium caseinate at pH 6.3 and 37°C is compared with that of rennin. The same amount of KRU of the two enzymes was used, and the optical density (OD) was determined at 280 mμ after precipitation with trichloroacetic acid. After 180 minutes the OD for the rennin-treated caseinate had not altered (constant at 0.137) while the OD for *Mucor miehei* enzyme had increased from 0.136 to 0.265, an increase which is rather slight in practical cheese making.

High Clotting System

Another microbial rennin obtained from the culture *Mucor miehei* NRRL 3420 is described by *R.L. Charles, D.P. Gertzman and N. Mclachouris; U.S. Patent 3,549,390; December 22, 1970; assigned to Miles Laboratories, Inc.* This enzyme has a high ratio of clotting activity to proteolytic activity and can be used under usual calf rennet milk-clotting conditions to produce cheese of good flavor and having desirable body and texture.

The organism was isolated from dried sediments from a sewage disposal plant and is classified as *Mucor miehei*.. The particular strain of *Mucor miehei* useful in the production of the milk-clotting enzyme has been given the identification number NRRL 3420.

The organism is grown under submerged fermentation conditions for about 2 to about 7 days at a temperature of from about 30° to about 50°C. At temperatures below about 30°C the yield of desired enzyme is too low to be economically practical, while at temperatures above about 50°C the desired enzyme becomes inactivated. The preferred growth temperature is from about 35° to about 40°C. The desired milk-clotting enzyme is present in the beer outside the mycelium and can be recovered in liquid form by simply filtering off the mycelium from the fermentation beer and retaining the filtrate. The resulting liquid enzyme has storage stability comparable to calf rennet. The enzyme can be recovered in a powdered form.

The filtrate can be frozen and dried from the frozen state using known freeze-drying techniques. The fermentation beer filtrate could also be treated with ethanol to precipitate the enzyme. The precipitated enzyme is then washed with ethanol and vacuum dried at room temperature. Acetone or other organic solvents as well as ammonium sulfate can also be used to precipitate the enzyme from an aqueous solution. The milk-clotting enzyme has good activity with respect to clotting milk while also having a low level of proteolytic activity. It thus does not appreciably attack or soften the curd formed during the milk-clotting. This enzyme can be used under usual calf rennet milk-clotting conditions and it is therefore adaptable to current cheese making operations.

It can be used in place of all or part of the calf rennet in standard cheese making. Cheese trials have shown that cheese made from curd produced by the *Mucor miehei* NRRL 3420 milk-clotting enzyme, even after an extended curing period, exhibits desirable flavor and freedom from bitterness.

Example 1: A culture of *Mucor miehei* NRRL 3420 was transferred to several 1 liter Erlenmeyer flasks each containing a sterilized medium consisting of 10 g corn starch, 12.5 g ground corn, 7.5 g soy meal, 1.25 g calcium carbonate and 250 ml water. This medium also contained 100 mg bacterial amylase to liquefy the starchy components of the medium. The inoculated flask contents were then shaken in a reciprocal motion shaker with a 2" stroke at 176 strokes/minute for 6 days while maintained at a temperature of 32.5°C. The resulting fermentation beer was then filtered from the mycelium and the combined filtrate from all the flasks was assayed to contain enzyme activity of 1,233 Soxhlet units/ml.

A portion of the filtrate was frozen and dried from the frozen state. The dried powder was assayed to contain 114,823 Soxhlet units/g and 727 proteolytic units/g. The ratio of Soxhlet units to proteolytic units was 157.9.

An enzyme prepared in the above manner was then used to make cheese. A 240 lb quantity of pasteurized whole milk was placed in a cheese vat and heated to 88°F (31°C). A 2.4 lb quantity of lactic acid starter culture was then added to the milk. After an hour a quantity of the above prepared liquid enzyme filtrate containing a total enzyme potency of about 800,000 Soxhlet units was added to the milk in the cheese vat. The milk clotted, and a satisfactory curd was prepared in 28 minutes.

It was then cut into desired shapes and heated to 102°F (39°C). The whey was drained off and the usual steps of cheddaring, milling, salting and hooping were carried out to obtain green cheese. A control cheese was prepared at the same time using a calf rennet under similar conditions. The cutting quality of the curd and the texture of the green cheese prepared from the microbial enzyme were found to be comparable with those prepared from calf rennet. The cheeses were ripened at 45°F (7°C) and 90% relative humidity for 60 days. The organoleptic evaluations, such as smelling, tasting and breaking indicated that the cheese made from the microbial enzyme had normal body and texture and normal pleasing flavor completely free of any bitterness, rancidity or other off flavors and was substantially the same as that produced from calf rennet.

Example 2: A culture of *Mucor miehei* NRRL 3420 was transferred to several 2,800 ml Fernbach flasks each containing a sterilized medium consisting of 8.0 g dextrose monohydrate, 12.5 g ground corn, 5.0 g Peptone T and 250 ml water. The inoculated flask contents were then shaken in a reciprocal motion shaker with a 3" stroke at 176 strokes/minute for seven days while maintained at a temperature of 32°C. The resulting fermentation beer was filtered from the mycelium and the combined filtrate from all the flasks was assayed to contain 2,400 Soxhlet units/ml. The filtrate was then lyophilized and the resulting powder was assayed to contain 298,804 Soxhlet units/g and 2,321 proteolytic units/g. The ratio of Soxhlet units to proteolytic units was 128.7. The above-produced enzyme can then be used to clot milk in a cheese making process.

The enzyme has a higher ratio of milk-clotting activity to proteolytic activity than any other microbial rennin. Two separate batches of enzyme believed to have been produced from prior *Mucor miehei* strain Ka 3111 had the following properties:

	Clotting activity SU/g	Proteolytic activity PU/g	SU/PU
Strain Ka 3111 Batch			
1	295,200	3,348	88.2
2	304,180	3,126	97.3

An enzyme produced from *Mucor miehei* strain University of California 21 contained 28,846 Soxhlet units/g and 433 proteolytic units/g for a SU/PU ratio of 66.6. This material produced cheese having an undesirably rancid flavor. Since the amount of a given milk-clotting enzyme to be used in a cheese making process is generally directly related to its clotting activity, for a given level of clotting activity these enzymes will supply a significantly higher amount of proteolytic activity than is present in the enzyme from *Mucor miehei* NRRL 3420.

This higher proteolytic activity will tend to form unsatisfactory curd and undesirably bitter flavor in the finished cheese. In contrast to this, the higher activity ratio of the new enzyme enables satisfactory cheese to be made which is comparable to that produced through the use of natural calf rennet.

Inactivated Cellulase Component

An improved microbial enzyme, which is a substitute for calf rennet is produced from *Mucor pusillus* and *Mucor miehei* and has its cellulase component inactivated.

These microbial rennet preparations frequently contain cellulase impurities which sometimes cause the disadvantage of disintegration and/or reducing the durability of cloth used in cheese making.

It was found by *Y. Kokusho, S. Higashi, H. Machida and S. Iwasaki; U.S. Patent 3,950,221; April 13, 1976; assigned to Meito Sangyo Kabushiki Kaisha, Japan,* that by heat-treating microbial rennet in an aqueous medium in the presence of sodium chloride, the cellulase impurities can be inactivated without substantial loss of milk-clotting activity in the microbial rennet.

The heat-treatment of microbial rennet can be accomplished in various forms. It can be in the form of aqueous whole cultures or fermentation beers, or filtrate or extract of fermented koji or concentrate, or in the form of a microbial rennet-containing solution of each step in recovering and purifying. It can also be in the form of dried material or material precipitated with organic solvents or with salts. The rennet is then dissolved in an appropriate aqueous media. The concentration of the microbial rennet in the aqueous medium is not critical, unless the volume of the aqueous medium is so small as to prevent the dissolution of sodium chloride. The time of the addition of sodium chloride is not critical, as long as sodium chloride is present at the time of heat-treatment. The presence of sodium chloride will enhance the thermostability of microbial rennet and will promote the inactivation of cellulase impurities at the same time.

Generally, a higher concentration of sodium chloride is desirable in order to minimize the inactivation of the milk-clotting activity of microbial rennet and to maximize the inactivation of cellulase activity. The amount of sodium chloride is about 15 to 30% by weight. Where the concentration of the microbial rennet is higher, the same effect can be obtained with lower concentration of sodium chloride. The process is carried out under thermostable pH conditions for the microbial rennet. Originally, the thermostable pH range is inherent to each enzyme and is variable according to the properties of the microbial rennet. For example, microbial rennet produced by *Mucor pusillus* is most thermostable at a pH about 5.0.

In order to achieve the maximum removal of cellulase activity, it is desirable to perform the heat-treatment at the highest possible temperatures for the longest possible period of time, unless the milk-clotting activity of the microbial rennet is disadvantageously reduced. However, the milk-clotting activity will be reduced at unreasonably high temperatures, and excessively low temperatures will require too long periods of treating time, resulting in a reduction of the efficiency of operation.

Example: A culture medium consisting of 10 parts of wheat bran, 0.1 part of ammonium sulfate and 7 parts of water was sterilized and was inoculated with a culture of *Mucor pusillus* IAM F-27 (IAM: Institute of Applied Microbiology, University of Tokyo), and cultivated for 72 hours at 35°C. A solution of microbial rennet was prepared by extracting the fermented wheat bran koji with water. One ml portion of this solution contained 1.4 CU cellulase and 25,800 SU milk-clotting enzyme. Sodium chloride was added to aliquot portions of this solution to achieve 5%, 10%, 15% and 20% (w/v basis) concentrations respectively and the pH of the solution was adjusted to 5.0. Milk-clotting activities and cellulase activities which remained in the resulting solution after the heat-treatment at 60°C for 5 minutes are shown in the table below.

The values in the table are expressed as percentage to each enzyme activity in the original solution. A blank test without the addition of sodium chloride was carried out for comparison.

	Comparative Example	Run 1	Run 2	Run 3	Run 4
Concentration of NaCl (%)	0	5	10	15	20
Remaining cellulase activity	85	40	30	20	10
Remaining milk-clotting activity	91	100	100	100	100

Proteolytic Component Removal

The discovery of a number of microorganisms producing a milk-clotting enzyme, called microbial rennet, with a very low proteolytic activity is of very great importance. It is known that rennets have a specific proteolytic activity in respect to the protective colloid, kappa casein, and that by digestion the casein of the milk is clotted. Although the proteolytic activity of microbial rennet is almost negligible, the rennet may be contaminated a little with undesired proteolytic enzymes, which enzymes also decompose the caseins of the curdled milk. There are various causes of these contaminations. Thus, for example, a rennet may be infected with another microorganism excreting undesired unspecific proteolytic enzymes on isolation from a culture of the relative microorganism.

It is also possible that during prolonged fermentation of the organism, for example, to obtain a higher yield, an unspecific proteolytic enzyme is excreted together with microbial rennet. Furthermore it may happen that an interesting microorganism for the production of rennet cannot be used because the percentage of unspecific proteolytic enzymes is too high. Thus it is known that certain species of bacteria produce much rennet but also large quantities of unspecific enzyme.

Nonspecific proteolytic enzymes contaminate microbial rennets, and not only coagulate or clot milk but also digest the caseins of the curd which causes bitter flavor and unsatisfactory consistency during the aging of the cheese. *H.C.T. Moelker and R. Matthijsen; U.S. Patent 3,591,388; July 6, 1971; assigned to Organon Inc.* give a process to purify these rennets and thus improve the flavor and general quality of cheeses.

In a solution of microbial rennet a selective separation could be effected by silicates between two substances that are so related as microbial rennet and unspecific proteolytic enzymes. In this process the unspecific enzymes are for the greater part absorbed, the rennet remaining behind in the solution without having lost its activity. The absorption of the contaminating enzymes can be effected by adding an absorbent to the solution of the rennet and stirring the mixture for some time, followed by removal of the absorbent, for example, by filtration or centrifugation, or by passing the solution to be purified through a column of the absorbent.

Finally an enzyme concentrate or, if desired, a dry enzyme preparation, can be obtained by evaporation of the water or precipitation of the milk-clotting enzyme by any conventional technique such, for example, as salting out or precipitation with an organic solvent. It stands to reason that instead of one absorbent a mixture of two or more absorbents can be used, it also being possible to repeat the treatment with the same or another absorbent. From rennet preparations, in which the ratio between clotting activity and proteolytic activity was about 6,000:11,000, purified rennets were thus obtained, in which this ratio had improved to about 14,000:20,000.

The choice of the pH during the absorption is determined by the stability of the microbial rennet at different pH's and by the absorption properties of the absorbent. The rennet is stable at a pH between about 3 and 9 dependent upon the temperature. On the other side the absorption of proteolytic enzymes occurs within this whole pH range. The absorption properties decrease accordingly as the pH rises. Below pH 4 some microbial rennet, too, is absorbed as a rule, while outside the pH range 3 to 9 the rennet denaturates. The purification is performed at a pH between 4.5 and 6, at which pH the best results are obtained.

As starting material for the purification a powdery rennet can be used which is dissolved in water or a buffer solution at a pH of about 5, but it is also possible to use for this purpose a fermentation liquid containing the rennet. It is even possible to perform the purification during the last phase of the fermentation of a microorganism producing rennet. A submerged culture or an aqueous extract of a surface culture of a microorganism producing rennet is taken as starting material, because in this case the isolation and purification of the rennet can take place simultaneously, for example, by filtration over a silicate.

As examples of suitable absorbents belonging to the groups of natural and synthetic silicates are mentioned: bentonite such as the types BIV 350 bianco, BIV 220 alvorio and Micro BIV, manufactured by Societa AZ Mineraria, Isole Pontine, Rome, Permutit such as Decalso F, supplied by Permutit Company, London; attapulgite and chemically pure colloidal silicon dioxide (Aerosil).

Example 1: 10 g of a solid microbial rennet originating from *Mucor pusillus* was dissolved in 250 ml of distilled water. To the resulting solution was added 10 g of bentonite BIV 350 bianco after which the pH was adjusted to 5.0 by means of 1 N acetic acid. The resulting suspension was stirred for 15 minutes and then filtered over a fluted filter. The clotting and proteolytic activities were determined before and after the absorption step by the methods mentioned in the patent, and included in the following table together with their ratio R:

	Before absorption	After absorption
Clotting activity, u/ml	13,600	13,450
Proteolytic activity, u/ml	2.40	0.96
R	5.67×10^3	14.01×10^3

Example 2: The purified rennet obtained in Example 1 was used in the production of cheese in a comparative experiment with unpurified rennet. To 20 ml of uncooked new milk was added 3 ml of the purified rennet solution obtained in Example 1, 4 g of saltpeter and 1% starter culture. The temperature at which coagulation occurs is 30°C; coagulation time is 31 minutes. In the same manner, 20 liters of milk was coagulated with 3 ml of the unpurified rennet preparation mentioned in Example 1; clotting time is 30.5 minutes.

The curdled milk was cut into slices, during which operations excretion of whey occured. To further this excretion process the mass was regularly kept in motion for 20 minutes with a cheese harp. Then, after removal of 10 liters of whey, 6 liters of water at 45°C was added to the curdled milk, after which the temperature was 37°C. The mass was kept in motion for about 50 minutes with a cheese harp. Then the curdled milk was separated from the whey by means of a pressing plate fitted with perforations. Finally the curdled milk was divided into two portions. These portions were put into warm cheese molds, which were placed under a press. After 6 weeks and 6 months respectively a group of 8 experienced tasters submitted the cheeses to an organoleptic examination by a comparative taste test. The following result was obtained by the group of 8 tasters:

The cheese prepared with the purified rennet is called Cheese A and the cheese prepared with the unpurified starting material is Cheese B.

	Cheese A		Cheese B	
	After 6 weeks	After 6 months	After 6 weeks	After 6 months
Bitter	0	0	1	2
Somewhat bitter	1	0	7	6
Not bitter	7	8	0	0

Example 3: After fermentation in the presence of mycelium 10 g of bentonite Micro BIV was added to 4 liters of a submerse culture of *Rhizopus peka.* The activity of the solution was 120,000 Soxhlet units/liter, and the proteolytic activity 60 units/liter. The pH of the fermentation liquid was adjusted to 5.0 with 1 N hydrochloric acid and the suspension was stirred for another 15 minutes. Then the fermentation was stopped and the solid substances were removed from the culture by filtration. Then the clotting and proteolytic activities were determined in the clear filtrate. A clotting activity was found of 114,000 Soxhlet units/liter and a proteolytic activity of 8 units/liter so that the ratio R between clotting and proteolytic activities had improved from 2,000 to 14,250.

Cheese was prepared with a rennet preparation treated with 0.35 liter of bentonite and with a nontreated rennet preparation. An organoleptic examination by a group of experienced tasters showed that the taste of the cheese prepared with purified rennet had considerably improved as shown in the following table.

	Cheese prepared with purified rennet		Cheese prepared with crude rennet	
	After 6 weeks	After 6 months	After 6 weeks	After 6 months
Bitter	0	0	6	2
Somewhat bitter	1	1	2	6
Not bitter	7	7	0	0

Special Milk-Clotting Specificity

A method for producing milk-coagulating enzyme (bacterial rennet) and use of this enzyme for making cheese is given by *T. Imai, Y. Irie and Y. Kanazawa; U.S. Patent 3,661,594; May 9, 1972; assigned to Godo Shusei Kabushiki Kaisha, Japan.* Milk-coagulating enzyme having high activity is obtained from broth of a selected strain of bacteria belonging to *Bacillus polymyxa* cultivated on several kinds of artificial or natural nutrient media containing carbohydrates, nitrogen sources, growth factors and inorganic salts. As the carbon sources, assimilable carbohydrates such as the flours of corn, soybean or other crops, powdered skimmed milk and various sugars can be used. The inorganic salts may also include phosphate, magnesium, calcium and other salts. The cultivation of the bacterial strain is carried out in a conventional manner in liquid broth at a temperature of 25° to 37°C and at a pH of 5 to 8 for 2 to 7 days. As a result of the cultivation, the milk-coagulating enzyme "bacterial rennet" is produced in the nutrient medium.

The crude product maintains a high milk-coagulating activity at a pH of from 4 to 9 when it is stored at 5°C for 7 days, or at 27°C for 24 hours. However, it becomes unstable at pH value of less than 3 or more than 10. Furthermore, it is observed that the crude enzyme exhibits a strong coagulating activity in acidic solution in the same manner as in the case of the commercial rennet. It decreases in activity in solution as the pH of the solution increases.

In Figure 2.2 the solid line shows the amounts of nitrogen (Kjeldahl-N) transferred to the whey after milk was coagulated by adding the enzyme (obtained from *Bacillus polymyxa)*. The dotted line gives those amounts transferred in the case of commercial rennet under the same conditions. From the figure it is observed that the enzyme exhibits substantially the same efficiency as in the commercial rennet; percentages of nitrogen as low as about 14% of the total amount of nitrogen were constantly transferred to the whey during the incubation periods of from 30 to 150 minutes. This shows that the enzyme can coagulate the majority of casein in milk without much proteolysis. Thus, this enzyme is confirmed to have very high milk-coagulating activity as compared with the commercial one and is, consequently, considered to be useful "bacterial rennet" in cheese manufacture.

Figure 2.2: Special Milk-Clotting Specificity

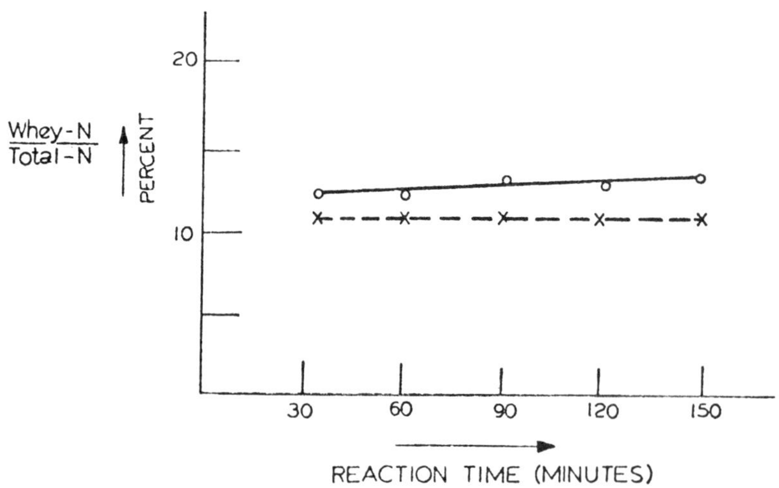

Source: U.S. Patent 3,661,594

Example 1: A nutrient medium containing 10% corn flour, 0.1% powdered yeast extract and 1⁄200 mol of $CaCl_2$ was inoculated with a strain of bacterium belonging to *Bacillus polymyxa* identified as (Ferm-P) 412, and after 4 days' shaking or submerged culture at 30°C, the milk-coagulating activity of the bacterium reached its maximum. The cells were separated by filtration or centrifugation and an enzyme solution was obtained. The milk-coagulating activity of this enzyme solution was 400 Soxhlet units/ml for a 10% solution of powdered skimmed milk containing 1⁄100 mol of $CaCl_2$. By salting out this enzyme solution with ammonium sulfate of 0.3 to 0.8 saturation, substantially all of its activity could be transferred to the precipitated fraction.

Furthermore, the activity of enzyme recovered is on the order of 88%, 80%, 70% or 75%, when precipitated by adding organic solvent in the following amounts: namely 75% methanol, 70% ethanol, 60% acetone or 70% isopropanol. The activity of the crude enzyme precipitated by the 70% methanol was 20 Soxhlet units/mg, while that of the commercial rennet powder of Chr. Hansen's Laboratory, Inc., was 100 Soxhlet units/mg.

To 100 kg of raw whole milk maintained at 30°C, was added 0.1 kg of $CaCl_2 \cdot 2H_2O$ followed by 1 kg of starter. When the acidity of the milk reached 0.18% (as lactic acid), 8 g of bacterial rennet was added and the mixture was stirred. After 25 minutes' standing, the curd was formed for cutting. Cooling was carried out in two consecutive steps at 34° and 38°C and after 80 minutes of cooking at 38°C, casting into molds was performed with compression. The draining of the whey from the curd was almost the same as commercial rennet, and then 8.6 kg of green cheese was salted by soaking in a sodium chloride solution. Test of taste and flavor after 4 months' ripening showed no sign of bitterness and no difference between this cheese and commercial rennet cheese.

Example 2: A nutrient medium containing 7.5% of powdered skimmed milk, 1% of glucose, 0.1% of powdered yeast extract and 1⁄200 mol of $CaCl_2$ was inoculated with a strain of bacterium belonging to *Bacillus polymyxa* designated (FERM-P) 412, and after 4 days' shaking or submerged culture at 30°C, the cells were separated by filtration or centrifugation and an enzyme solution was obtained. The activity of this enzyme solution was 420 Soxhlet units/ml. Thus 1 ml of this enzyme solution was equivalent to 4.2 mg of rennet powder of Chr. Hansen Laboratory, Inc. By means of the method as described in Example 1, the enzyme was precipitated and then it was dried and powdered.

Next, to 80 kg of whole milk mixed with 20 kg of skimmed milk which were maintained at 30°C were added to 0.12 kg of $CaCl_2 \cdot 2H_2O$ followed by 1.2 kg of starter. After a quarter hour 5 g of "bacterial rennet" obtained as mentioned above was added, and the same operations as in Example 1 were performed whereby was obtained 8.8 kg of green cheese (yield 11.0%).

Kappa-Casein Specificity

An enzyme substitute for rennet was developed by *D.R. Farr; U.S. Patent 3,852,478; December 3, 1974; assigned to Societe d'Assistance Technique pour Produits Nestle SA, Switzerland* which has specific activity for the phenylalanine-methionine bond of kappa-casein. This means that the enzyme has a coagulating action which can be considered identical with that of calf rennet.

The process provides a method for producing a milk-clotting enzyme by biosynthesis, in which a microorganism of the species *Physarum polycephalum* producing this enzyme is cultured under aerobic conditions in an aqueous nutrient medium. *Physarum polycephalum* is a known microorganism belonging to the order of the Physarales in the Myxomycetes class. One strain of this microorganism, on deposit at the Centraalbureau voor Schimmelcultures, Baarn, The Netherlands, under the number 491.61, is available to the public.

It is possible, when a pure product is desired, to separate the proteolytic enzyme from the other substances present in the concentrated enzyme solution. For this purpose the protein may be separated from the concentrated solution, for example, by precipitation, and redissolved to provide a second aqueous solution which may be fractionated by a chromatographic technique. Eluted fractions having the desired optical density and proteolytic activity are then isolated to obtain a practically pure product. Examination of the enzyme thus purified has shown that it cleaves specifically the phenylalanine-methionine bond of the kappa-casein peptide chain, which means that it has a coagulating action on milk which may be considered as identical with that of rennin.

Example 1: Culture of a strain of *Physarum polycephalum* is effected under submerged conditions in a 15 liter fermenter containing an aqueous nutrient medium having the following composition:

	Percent
Glucose	1.0
Yeast extract	0.15
Casein hydrolysate (tryptone)	1.0
$CaCl_2 \cdot 2H_2O$	0.06
KH_2PO_4	0.2
$MgSO_4 \cdot 7H_2O$	0.06
$FeCl_2 \cdot 4H_2O$	0.006
$MnCl_2 \cdot 4H_2O$	0.0085
$ZnSO_4 \cdot 7H_2O$	0.0035
Citric acid	0.35
Hematin	0.0005
Tap water, qs 100%	

8 liters of an aqueous solution containing these constituents, with the exception of glucose, are supplied to the fermenter and sterilized by heating under pressure for 15 minutes at 121°C. The glucose is dissolved in water, at a level of 50 g/liter, and sterilized separately. The solution in the fermenter is then cooled to 25°C and the sterile aqueous glucose solution is added. The pH of the mixture is adjusted to 4.5 with a sterile sodium hydroxide solution and the hematin, previously sterilized for 15 minutes in alkaline solution at 121°C, is added.

2 liters of inoculum, obtained by shaked flask culture for 2 days of *Physarum polycephalum* microplasmodia, are then added to the nutrient medium. The nutrient medium is agitated with a stirrer rotating at 150 rpm to assure sufficient oxygenation of the medium, and is maintained at 25°C. Culture of the microorganism is continued for 20 hours under these conditions and a culture medium containing 1% cell dry matter is obtained.

The resulting product having enzyme activity has the following properties:

Influence of pH on proteolytic activity	Examination of the proteolytic activity in a pH range from 3.0 to 7.0 has shown that the activity is at a maximum between pH 4.0 and 5.0 and decreases to 15% of the maximum at pH 7.0
Influence of temperature on activity	The proteolytic activity is unchanged between 0° and 30°C and decreases above 30°C to practically zero at 60°C
Coagulating effect on milk	Mixing of equal volumes, at 37°C, of the supernatant and skimmed milk containing 12% solids leads to coagulation in 10 minutes
	Mixing under identical conditions with skimmed milk containing 24% solids leads to coagulation in 4 minutes

Example 2: The supernatant obtained in Example 1 is freeze-concentrated, by cooling to induce formation of ice crystals. The unfrozen fraction, of which the concentration of proteinaceous substances has increased at the expense of the solid phase, is collected and mixed with an 0.05 M acetate buffer solution, at pH 4.5. The proteins present in the concentrated solution are then precipitated with ammonium sulphate and the precipitate formed is recovered by centrifugation. The precipitate is then dissolved in an 0.05 M acetate buffer solution at pH 4.5 and this solution is dialyzed for 24 hours against the same buffer solution.

The resulting solution is then fractionated chromatographically by passage through a column of diethylaminoethyl cellulose equilibrated with an 0.05 M acetate buffer solution containing sodium chloride at decinormal concentration. Elution is carried out with this buffer solution, at a rate of 50 ml/hr. The optical density of each fraction eluted is measured at a wavelength of 260 nm, and the proteolytic activity of each is also measured. Elution of the enzyme appears as 3 marked peaks of proteolytic activity, and the fractions corresponding to these 3 elution peaks are collected. The activities on kappa-casein of rennin and of each of the eluted fractions are then compared.

Rennin specifically cleaves the phenylalanine-methionine bond of the kappa-casein peptide chain, and if it is reacted with kappa-casein and carboxypeptidase A is added, phenylalanine is liberated and may be estimated. It should be added that addition of carboxypeptidase A by itself does not liberate phenylalanine. It has been found that the reaction with kappa-casein of the eluted fractions corresponding to the 3 peaks of proteolytic activity, in amounts of 1 part by weight of enzyme product per 100 parts by weight of substrate, followed by the action of carboxypeptidase A, leads to the liberation of phenylalanine. This demonstrates that rennin and the enzyme product react with kappa-casein in similar manner.

Low Proteolytic and Lipolytic Type

A substitute for rennet capable of coagulating milk in cheese-making with low proteolytic and lipolytic activity was developed by *T.O. Wiken and G. Bakker; U.S. Patent 3,857,969; December 31, 1974; assigned to Stamicarbon NV, Netherlands.*

This enzyme gives less off-flavor in cheese after curing and aging. The quality of not imparting flavors to the cheeses is especially required in producing cheeses having less pronounced taste of their own, such as the Gouda and Edam type cheese. The enzyme preparations are produced by cultivating fungi of the strain *Mucor lamprosporus* or a natural or synthetic variant or mutant under aerobic conditions on a natural or synthetic culture medium containing carbon, nitrogen and trace nutrients. The cultivation of the fungi is accomplished at a temperature between 15° and 45°C with an incubation period of about 2 to about 12 days.

The fungi can either be cultivated on the surface of the culture medium or submerged in a solution of the culture medium. The enzyme preparation may be recovered by extraction with water from the surface of a culture such as Koji culture. The recovery of the enzyme preparation from a liquid culture medium in which the enzyme is submersed can be accomplished by precipitation of the enzyme by means of a solvent or a salt, or the enzyme preparation can be recovered in the dissolved state by filtration and dialysis.

Example 1: An amount of 1 ml of a spore suspension of *Mucor lamprosporus* was added to 100 g of wheat bran and 70 ml of a 1% aqueous yeast extract contained in a flask. After an incubation of 7 days at 25°C, the medium was extracted with 425 ml of water and the aqueous extract was sterilized by Seitz filtration. In this way, 175 ml of a clean filtrate with a solid matter content of 2.7% by weight was obtained. The coagulating activity of the aqueous enzyme preparation was found to be 4,000 GE; one GE (Gerinnungs-Einheit) is the quantity of coagulant causing visible curdling of 1 ml of milk in 40 minutes at 35°C.

The proteolytic activity of the enzyme preparation was determined by formalin titration. In this titration a quantity of enzyme preparation was added to 125 ml of milk so that the coagulation time equaled 500 seconds. The resulting suspension was titrated with an alkaline solution and held for 36 hours, formalin was added and the suspension titrated with an alkaline solution again. An uncurdled milk was similarly treated as a blank. The proteolytic activity was determined for several curdling agents as well as rennet and the preparation obtained from *Mucor lamprosporus.* The results indicate that a trypsin solution, a *Mucor globosus* preparation and a *Mucor pusillus* preparation all resulted in substantial amounts of alkaline solution required per 10 ml of milk in the proteolytic determination. The *Mucor lamprosporus* preparation was found to result in the same amount of alkaline solution required as for rennet.

Example 2: Cheese of the Gouda variety was prepared using as a coagulant for the milk an enzyme preparation obtained by the process. In all other respects, the Gouda cheese was prepared in the usual manner. After 3 weeks of aging, the average weight of the cheese was 360 g at a moisture content of 33%. Cheese prepared using rennet as the coagulant had an average weight after 3 weeks of 355 g at a moisture content of 35%. An organoleptic examination proved that the cheese made with the coagulant was comparable with the cheese prepared with rennet as regards to consistency, color, smell and taste.

Enzyme-Cellulose Complex

J.P. Bourdeau, J.L. Seris and R. Pornin; U.S. Patent 3,960,666; June 1, 1976; assigned to Societe Nationale des Petroles d'Aquitaine, France claim that enzymes fixed on cellulose materials (immobilized enzymes) have excellent stability, that is, they remain attached to the support and can thus be used repeatedly without dissolving away the enzyme. It was found that celluloses which contain a certain proportion of lignin constitute enzyme supports much better than the pure cellulose. In the past, it was always sought to prepare a cellulose as pure as possible for serving as a support for enzymes, because it was assumed on principle that the presence of lignin and other vegetable substances capable of accompanying the cellulose prevented the fixation of enzymes.

The new complex of enzyme fixed on the cellulose is characterized in that the cellulose support contains 10 to 20% of lignin being dispersed uniformally throughout the cellulose. The cellulose can be derived from any one of the well-known natural sources, for example, wood, of which it is possible to use the waste material, especially chips or sawdust, shavings or paper pulp. One special source of the cellulose as regards the retention of enzymes, is the stalks of cereals and particularly maize cobs.

Just as for pure cellulose, there is no great advantage in activating the lignified cellulose before fixing the enzyme; the appropriate activation leads to extremely stable complexes. The different known activation treatments, such as polydiazotization, aminobenzylation, bromoacetylation, treatment with cyanogen bromide or with glutaraldehyde, etc. can be applied.

One cellulose which leads to particularly advantageous results is the cellulose of maize cobs activated beforehand by the action of thionyl chloride in the presence of an excess of pyridine, so that the final product contains about 3 to 4% of sulfur.

The coupling of the enzyme with the cellulose takes place in an aqueous buffering solution the pH of which is adapted to the particular enzyme. For example, for rennet, a buffering solution of phosphates is used with a pH value from 5.9 to 6.4. The medium is agitated at a moderate temperature, which is generally between ambient temperature and 50°C, for the time which is necessary for fixing the enzyme, this generally being of the order of 24 hours. After separation of the grains thus coupled with the enzyme, it is best to eliminate the enzyme portion which is not fixed by covalence; this can be effected by washing with buffering solutions of suitable pH value and with water.

Example 1: Maize cobs, containing about 80% of cellulose and 15% of lignin, are reduced into grains of 1.5 to 2.5 mm. 50 g of these grains are placed in 18% sodium hydroxide solution and this is stirred for 15 minutes at normal temperature. The grains are then separated and thoroughly washed with cold water, until the filtration has only a slight coloration. The washing is completed with 1 liter of methanol and then with 400 ml of pyridine. The grains are then poured into 800 ml of pyridine, after which 100 ml of $SOCl_2$ are added dropwise to the agitated pyridine during 45 minutes. The temperature of the reaction mixture is raised up to 55°C and the medium suddenly becomes intensely colored.

One hour after completing the introduction of the thionyl chloride, the mixture is poured into ice-containing water, and the grains are then separated and washed with several liters of water and then with an HCl solution of pH 2, until there is complete elimination of the pyridine absorbed on the grains.

The washing is completed with 1 liter of methanol. The grains thus obtained are dried at 40°C. It is found that the treated cellulose contains about 3.5% of sulfur and only traces of chlorine.

Into 160 ml of a buffering solution of phosphates of pH 6.2 are introduced 16 g of the grains prepared as indicated above and 40 ml of rennet. The whole is maintained at 40°C under stirring for 24 hours. The grains are then separated and there is carried out the desorption of the rennet which is not chemically fixed, by washing: first of all with 250 ml of the same buffering solution as above, then with 2 liters of distilled water and finally with 250 ml of acetate buffering solution of pH 4.4. The mother solutions from these treatments are used again. The grains thus obtained of the complex consisting of rennet and maize cob cellulose are kept at 4°C in 250 ml of acetate buffering solution.

Example 2: Grains of the cellulose-rennet complex, prepared following the procedure of Example 1, were used for curdling milk in an experimental production of cheese. The equipment, shown in Figure 2.3 includes a cylindrical vertical vat **12**, containing a coaxial sleeve **13** open at its two ends, within which is lodged a stirrer **15**. Through **14** milk is continuously poured into the coaxial sleeve. The grains **17** of cellulose-rennet complex are placed within the lower part of the vat and sleeve, and they are constantly kept in suspension in the milk, within the sleeve, by the action of the stirrer. While the milk, after contact with the complex, leaves the vat through overflow **16**, the grains tend to settle in the annular space between the walls of **12** and **13**; thus the milk continuously flows out and the grains remain within the vat.

The milk, so pretreated with the enzyme solid complex, is then passed into a usual heated container (nonrepresented in the figure) where it is curdled by heating.

Figure 2.3: Enzyme-Cellulose Complex

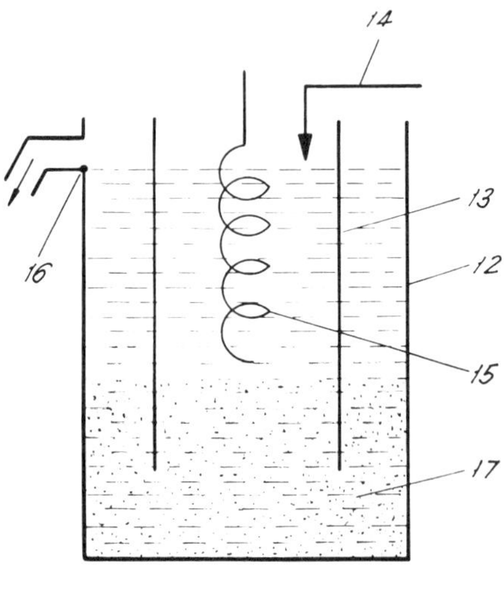

Source: U.S. Patent 3,960,666

8 g of cellulose were used, on which 0.04 g of commercial rennet were fixed as described in Example 1. Milk was passed through the apparatus at a rate of 0.63 liter/hour, and its temperature was 15°C. The region **17** of the suspension of grains extended approximately over one-third of the height of vat **12**; at this level the degree of casein proteolysis reached about 80%, while this degree was about 100% in the milk leaving the vat by the overflow **16**. The milk was then heated to 37°C and rapidly curdled in the usual container into which it flowed from **16**. In this way, 35 liters of milk were treated in 55½ hours with the single charge of 8 g of cellulose complex containing only 0.04 g of rennet.

When the usual method of curdling by directly adding rennet is carried out, the amount of 0.04 g of the enzyme can curdle only 0.4 liter of milk. That means the complex leads to an 87 times greater efficiency of the enzyme, as it permits treating 35 liters instead of 0.4.

CHEDDAR CHEESE PROCESSES

Coagulating Enzyme Complex

A typical Cheddar cheese manufacturing procedure involves the following steps. (1) A batch of pasteurized or unpasteurized whole milk is maintained at 86° to 90°F and is inoculated with one or more species of lactic acid-producing bacteria, such as *Streptococcus lactis, Streptococcus cremoris, Streptococcus acidophilus, Streptococcus thermophilus* and *Lactobacillus acidophilus.* The organisms are allowed to grow within the milk for about 30 to 60 minutes and in so doing produce an increase in acidity. This increase in milk acidity is normally about 0.01 to 0.02% when measured as titratable acidity.

(2) After the desired increase in acidity has been reached a milk-coagulating agent, such as animal rennet, is stirred into the milk and then the system is allowed to stand unagitated for about 30 minutes. Coagulation is usually conducted at a temperature of about 88°F and a gel-like coagulum is formed.

(3) When the coagulum reaches a desired degree of firmness, it is cut into small cubes of about ¼ or ⅜ inch.

(4) The curd and whey are then stirred and heated to a temperature of about 102° to 104°F. This cooking period normally requires about 30 minutes.

(5) The curd and whey are stirred continuously for a further period of about 40 to 50 minutes while being held at the elevated temperature. During this period the effects of elevated temperature, increasing acidity and the coagulating agent all combine to influence the firming and drying of the curd particles and the separation of whey from these particles.

(6) After the curd particles have reached a desired degree of firmness and dryness, and the acidity has increased by about 0.02 to 0.03%, the curd is allowed to settle and the whey is drained from the vat.

(7) The total interval setting time and dipping (draining) time is about 2½ hours. The characteristic final steps of the cheddaring procedure are then followed.

E.D. Murray, V.E. Gruetzner and D.M. Irvine; U.S. Patent 3,542,563; Nov. 24, 1970; assigned to John Labatt Limited, Canada, use an enzyme complex as the rennet substitute. The complex contains a neutral peptidase and an acid peptidase which is derived from the bacterium *Bacillus subtilis.*

When Cheddar cheeses were made according to the conventional process using the peptidase complex as coagulating agent, it was found that the resultant cheese had a moisture content of about 36.5 to 38.5%. Moisture content is a very important index as to quality in Cheddar cheese and is maintained at about 35%. Higher moisture levels tend to produce cheeses which have a weak body, develop discoloration and possess off-flavors.

When the peptidase complex is used in the production of cheese, the neutral peptidase is primarily involved in milk coagulation while the acid peptidase is primarily responsible for firming and drying of the curd. Studies were made on the factors affecting the activities of the neutral and acid peptidases and it was found that the two peptidases possessed maximum enzymic activities at different temperatures. Thus, the neutral peptidase was found to have maximum activity at about 115°F (46°C) while the acid peptidase showed maximum activity at about 129°F (54°C). It was also found that because of differences in the heat activity characteristics of the two peptidases, a temperature increase from 103° to 110°F caused an activation of the acid peptidase (firming enzyme) in excess of 30%, while the neutral peptidase (coagulating enzyme) was activated by less than 5%.

The process also requires, in addition to the milk-coagulating enzyme complex, an extra treatment of the curd to reduce the moisture level of the curd particles before aging and ripening. During application of the flash of additional heat the temperature should be raised from the regular holding temperature as quickly as possible, maintained at the elevated temperature for a short time sufficient to stimulate syneresis and then depending on the starter organisms used, it may be quickly cooled to the normal holding temperature.

Particularly good results have been obtained at an elevated temperature of about 110°F for a period of about 3 minutes. The main point to be realized in applying the flash of additional heat is that the temperature of the batch should be raised to the desired level and returned again to the normal level in order to obtain a desired degree of stimulation of syneresis with the least possible input of heat since this minimizes heat inactivation of the lactic acid-producing bacteria. The flash of additional heat is applied during the first 15 minutes of the holding stage, after the batch reaches the maximum cooking temperature. However, the selection of the actual time during the holding stage for the application of the flash of additional heat can be varied considerably depending on the size of the batch, the size and capabilities of the equipment, etc.

Example: The following concerns the production of cheese on a pilot scale. Whole milk containing about 3.4% fat was pasteurized and cooled. 160 lb were placed into a 200 lb capacity steam-jacketed vat and the temperature of the milk was adjusted to 88°F. 1½% (by weight) of conventional commercial starter culture was added and the milk was allowed to ripen until there was an increase in acidity of about 0.01 to 0.02%, this being obtained in about 30 minutes. After acid development 15.0 ml of a peptidase complex were added to the milk.

The complex of neutral peptidase and acid peptidase derived from the bacterium *Bacillus subtilis* had an activity such that 3 fluid ounces coagulated 1,000 lb of milk. Coagulation resulted and a firm coagulum was produced. When the milk was properly coagulated, it was cut into cubes with regular ¼ inch curd knives. The cut curd was agitated slowly to prevent breaking and matting of the soft particles. The cooking stage was started about 10 minutes after cutting, and the curd and whey were heated to 103°F within a period of about 30 minutes. During the holding stage a temperature of 103°F was maintained until there was an increase in acidity of about 0.01 to 0.015%. This required an additional time of about 25 minutes.

After the increase in acidity, steam was reapplied to the jacket and the temperature of the curd and whey was raised to 108°F within about 5 minutes. The temperature was maintained at 108°F for about 10 minutes, after which cooling water was applied to the jacket and the temperature of the vat was adjusted to 103°F within a period of about 5 minutes. When the curd was firm and dry, acidity had increased by 0.03% and then it was allowed to settle. The whey was then drained off and the normal cheddaring procedure was followed. The resulting cheese possessed a moisture content of about 35%.

Enzyme Complex Production

The enzyme complex of acid and neutral peptidase, described in the previous process, is prepared by *E.D. Murray and M.P. Prince; U.S. Patent 3,507,750; April 21, 1970; assigned to John Labatt Limited, Canada*, using a special mutant of *Bacillus subtilis.*

It has been found that if an asporulant mutant of a bacterium of the genus Bacillus is grown in a nutrient medium containing a carbon source, a nitrogen source and inorganic salts, acid and neutral peptidases are selectively grown to the substantial exclusion of alkaline peptidase, and the yield of acid and neutral peptidases is greatly increased over that obtained with previous methods. Moreover, if the medium is enriched with a source of L-leucine, the acid and neutral peptidases are directly obtained in a ratio which is highly suitable for use in the production of cheese.

L-leucine can be added to the culture medium in the pure form, the racemic form or in a suitable nutrient containing L-leucine. The amount of L-leucine added to the medium can be varied over a wide range but about 0.01% (w/v) L-leucine has been found to be the optimal amount to be added for maximal beneficial adjustment of the ratio of acid peptidase to neutral peptidase. It will, of course, be appreciated that the L-leucine is required only when adjustment of the ratio of acid peptidase to neutral peptidase is necessary.

Example 1: The following concerns the production of a culture filtrate. An inoculum of *Bacillus subtilis* (parent strain) was prepared by transferring organisms from an agar slant to a sterile seed medium of trypticase soy broth (3.0% w/v) fortified with calcium chloride (0.01% w/v) and L-leucine (0.01% w/v), all adjusted to pH 6.2 with 1.0 N HCl. The seed medium was prepared by dispensing 100 ml of medium into 300 ml Erlenmeyer flasks. The inoculated seed preparation was shaken on a rotary shaker for 12 hours at 36°C and after this incubation period the seed was used on 2% v/v basis to inoculate a production medium.

The production medium was prepared by using water-extractable nutrients from wheat bran. To accomplish this, a slurry of wheat bran (3.0% w/v) and water was boiled under pressure at 120°C for 45 minutes. The mixture was cooled and allowed to settle and the clear supernatant was drawn off and fortified with calcium chloride (0.01% w/v) and L-leucine (0.01% w/v). The medium was adjusted with 1.0 N HCl so that after sterilization the final pH was 6.2±0.1.

The inoculated production medium was maintained at 36°C and stirred at 300 rpm in a fully baffled fermentor of 5.0 liters working volume. Sterile air was added at the rate of 2 liters/minute, and normal liquid antifoaming agents were used to suppress foam. The system was allowed to grow until a maximum level of enzymic activity was reached (judged by milk coagulation tests), this growth period being generally 12±1 hours. The culture broth was then clarified by centrifugation and/or filtration.

The final yield of each enzymic component in crude culture filtrate is shown in the table below. For comparative purposes data are included for the non-amino acid-fortified production medium. Activities are expressed as units/ml of culture filtrate. The ratio is of acid peptidase to neutral peptidase.

Medium	Acid Peptidase	Neutral Peptidase	Ratio Acid to Neutral	Alkaline Peptidase
Wheat bran + calcium chloride	2,000	4,740	1:2.37	1,308
Wheat bran + calcium chloride + L-leucine	2,300	4,140	1:1.80	987

Example 2: The following concerns the manufacture of cheeses on a pilot scale. Whole pasteurized milk (16 Imp gal) was heated to 31±1°C. When this temperature was reached, 0.5 to 1.0% cheese starter culture was added and the mixture was stirred thoroughly. The starter organisms were allowed to grow until the acidity of the milk reached 0.17% as lactic acid. At this time 15 ml of concentrated enzyme complex were added. The mixture was stirred thoroughly and allowed to remain, covered and unagitated, in a jacketed vessel for 25 to 30 minutes.

The coagulated milk was then cut with a curd knife and the normal techniques of cheddar cheese manufacture were followed. After pressing, the final cheese was placed in a ripening cabinet at a temperature of 12±2°C with a relative humidity of 70 to 75%. The cheeses were allowed to ripen for periods of 1 to 18 months and evaluation of the cheeses at various time intervals showed them to be of normal cheddar flavor and quality.

Heat-Treated Milk Process

When Cheddar cheese is made from heat-treated milk or subpasteurized milk, certain initial steps and special procedures are necessary to obtain the proper curd and ripening. Often such cheeses lack the flavor characteristics associated with production from raw milk.

M.J. Roberts; U.S. Patent 3,650,768; March 21, 1972; assigned to Kraftco Corporation states that heat treatment destroys needed microorganisms and enzymes, for flavor character. An enzyme process was developed to improve body, flavor and texture of American and Cheddar cheeses from heated milk. The term heat treated milk, refers to a milk which is subjected to heat treatment at least equivalent to heating the milk to 135°F, and cooling it down with no-hold at the temperature. The heat treated milk may be pasteurized, which is generally understood to mean that the milk tests phosphatase negative, or may be sterilized, which is generally understood to mean that the microorganisms and enzymes are substantially or completely destroyed. As used for cheese, heat-treated (differentiated with a hyphen in this specification) milk is milk which has been heated to from 135° to 150°F and cooled with no-hold. Such treatment generally destroys gas-forming microorganisms but is less than pasteurizing conditions.

American cheese is a term which identifies a group of cheeses which includes Cheddar cheese and Colby cheese and further includes Montery and Jack cheeses. Cheddar cheese derives its name from a step in the manufacturing process where the drained cheese curd is cheddared or matted. American cheese generally is a hard cheese, ranging in color from nearly white to a more traditional red-yellow color and is made from whole cow's milk. Many attempts have been made to make American cheese from heat-treated subpasteurized or pasteurized milk, and considerable research has been conducted with respect to American cheese made from heat-treated and pasteurized milk.

When Cheddar cheese in particular is made from heat-treated milk it is usually produced by a so-called, "time schedule" method. Use of the time schedule method makes it possible to control the amount of acid that develops and the rate of acid development during the make procedure.

Proteolysis of the milk protein in the cheese, to provide the desired breakdown of protein, occurs primarily during curing of the cheese curd. The amount of proteolysis is related to the level and type of microorganisms present during curing, as well as the curing conditions. Growth of the micrococcus is slowed after addition of the *Streptococcus lactis* starter culture to the milk due to the formation of lactic acid and the level of the micrococcus is determined primarily by the level of addition to the milk.

It is believed that the micrococcus continues to grow slowly under the acidic conditions present during curing of the cheese. It is believed also that the proteolytic activity of the micrococcus is primarily due to products produced by and during the growth of the micrococcus. It is believed that an enzyme so produced acts to break down the protein to a desired level and does not continue to break down protein to undesired by-products. In other words, the system seems to be self-limiting as to the proteolytic activity.

When the proteolytic micrococcus is used along with heat-treated milk, there is a tendency to produce undesired flavor. However, it has been found that the use of a particular proteolytic micrococcus in combination with a specified lipase produces a desirable American cheese flavor. It is desirable to add a lipase of a type and at a level such that the fat is selectively split into fatty acids and glycerol. The lipase should be added in sufficient amount to contribute to the flavor but should not be added at such a high level as to provide a rancid flavor. At levels of addition of the particular lipase above about 2 g of dry powdered enzyme preparation per 1,000 lb of milk, a rancid flavor may be produced.

At levels of addition below about 0.1 g/1,000 lb of milk there is little flavor contribution from the particular lipase, and an undesired flavor may be produced by the proteolytic micrococcus. The lipase is preferably added at a level of 0.25 g/1,000 lb of milk. The particular lipase for use in the process is that obtained by extraction from the throat tissue of calves. Other suitable lipase may be obtained by extraction from the throat tissue of lambs or kids.

When American cheese or similar type cheese is produced from heat-treated milk, to which a combination of a proteolytic micrococcus and the lipase is added, the cheese has a desirable Cheddar cheese flavor. However, the body and texture of the cheese may be soft or crumbly or otherwise undesirable for some purposes. It has been found that an improved body and texture can be provided when a lactobacillus or a closely related microorganism is also added to the heat-treated milk prior to setting of the milk. Such a lactobacillus may be referred to as a bodying microorganism.

Addition of such a microorganism is optional, but may be used where a particular body or texture is desired. Some of such lactobacillus microorganisms also provide flavor to the cheese. Various homofermentative lactobacilli may be used, such as *Lactobacillus lactis* and *Lactobacillus bulagricus.* The preferred lactobacillus microorganism is a strain of *Lactobacillus lactis.* The lactobacillus microorganism will usually have some proteolytic activity and may contribute some flavor to heat-treated milk.

Example: 25,000 lb of raw milk was heat-treated at a temperature of 162°F for a period of 16 seconds for pasteurization. The milk was pumped to a vat and allowed to cool to 88°F. To the cooled, heat-treated milk was added 50 lb of a milk starter culture including a proteolytic micrococcus. The culture contained 1.75×10^9 cells/g of a proteolytic micrococcus. To the milk was also added 6.25 g of a lipase which had been obtained by extraction from the throat tissue of a calf. After addition of the above culture and the lipase enzyme, 250 lb of a *Streptococcus lactis* starter culture was also added to the milk.

The milk was agitated and allowed to ripen for a period of one hour. 75 oz of a rennet solution were then mixed into the ripened milk, and the stirring was stopped. After about 30 minutes the curd was sufficiently firm to cut and was cut with curd knives into ¼" cubes. Stirring was then commenced and was continued until the whey was drained. Draining was completed in about 30 minutes, and the curd was packed in a layer about 7 or 8" deep on each side. When the curd was firm enough to be turned without breaking, it was cheddared. The curd was then run through a curd mill and was spread evenly over the bottom of a vat and stirred. 3 lb of salt for each 1,000 lb of milk were then mixed in and the curd was piled on each side of the vat while additional whey was drained.

After salting, the curd was transferred to cloth-lined metal hoops, and was pressed, dried and wrapped in a plastic film in accordance with conventional procedures. After wrapping, the cheese was stored in a warehouse at a temperature of 45°F to effect curing. The cheese was cured for a period of 6 months. At intervals of 2 months a sample of the cheese was removed and inspected. The samples of cheese were compared with samples of cheese which had been prepared in a similar manner from the same milk with the exception that the combination of a proteolytic micrococcus and a lipase enzyme was not added to the heat-treated milk prior to setting. The samples of cheese produced by the above method were scored considerably higher in flavor than the control samples. The body and texture of the cheese produced were acceptable.

CHEDDAR AND OTHER CHEESE FLAVORS

Rapid Cheddar Flavor Development

The following review on cheddar cheese flavor and an improved process for rapid flavor development is given by *F.V. Kosikowski; U.S. Patent 3,975,544; August 17, 1976; assigned to Cornell Research Foundation, Inc.*

Rapid inducement of typical flavor in cheddar cheese has received considerable research attention. (A.C. Dahlberg and F.V. Kosikowski, 1947, "The Flavor, Volatiles, Acidity and Soluble Protein of Cheddar and Other Cheeses," *J. Dairy Sci.,* 30:165; D.A. Forss and S. Patton, 1966, "Flavor of Cheddar Cheese," *J. Dairy Sci.,* 49:89; F.V. Kosikowski, 1957 "Flavors of Processed Foods, Cheese Flavors," *Proceedings of Symposium on the Chemistry of Natural Food Flavors, Nat. Acad. Sci.,* Washington, D.C.; L.A. Mabbit, 1961, "The Flavor of Cheddar Cheese," *J. Dairy Res.,* 28:303).

The influence of the lactic streptococci from cheese starters was investigated by Reiter et al., McDonald and Perry (B. Reiter et al., 1967, "The Effect of the Microbial Flora on the Flavor and Free Fatty Acid Composition of Cheddar Cheese," *J. Dairy Res.,* 34:257; I.J. McDonald, 1964, "Characteristics of Proteinases of Three Strains of *Staphylococcus lactis* Isolated From Cheddar Cheese," *J. Dairy Res.,* 31:147; K.D. Perry, 1961, "A Comparison of the Influence of *Streptococcus lactis* and *Str. cremoris* Starters in the Flavor of Cheddar Cheese," *J. Dairy Res.,* 28:221).

Attempts have also been made to impart cheddar cheese flavor to bland curds by adding pure amino acids, fatty acids or synthetic mixtures of identified volatiles to cheese or cheese slurries (G.J. Silverman and F.V. Kosikowski, 1953, "Observations on Cheese Flavor Production by Pure Compounds," *J. Dairy Sci.,* 36:574; and S. Singh et al., 1969, "Factors Affecting Flavor Development in Cheddar Cheese Slurries," *J. Dairy Sci.,* 53:533).

Animal and plant lipase-induced cheese flavor has been reported by Babel and other investigators. (F.B. Babel, 1945, "Fat Degradation in Cheddar Cheese Made from Pasteurized Milk Without and With Added Lipase," *J. Dairy Sci.,* 28:201; J.W. Harper, 1956, "Lipase Systems Used in the Manufacture of Italian Cheese II Selective Hydrolysis," *J. Dairy Sci.,* 40:556; G.H. Richardson et al., 1970, "Gastric Lipase Characterization and Utilization in Cheese Manufacture," *J. Dairy Sci.,* 54:643). Richardson et al. applied gastric lipase preparations in cheese milk for the manufacture of cheddar and provolone cheeses, which were organoleptically preferred. Often, however, the resulting flavors of cheese to which enzymes have been added were accompanied by strong bitterness.

Orla-Jensen suggested the potential importance of microbial enzymes to cheese flavor; but relatively little activity to purposely induce cheese flavor with enzyme preparations has occurred, perhaps because of lack of proper enzyme systems.

(S. Orla-Jensen, 1939, "The Ripening Process of Hard Cheese," *Proc. 3rd Int. Congress of Microbiology*). Kosikowski referred to preliminary experiments dealing with the addition of amino acid, decarboxylase enzyme systems from enterococci cells to cheddar cheese curds in attempts to develop flavor. (F.V. Kosikowski, 1957, "Flavors of Processed Foods, Cheese Flavors," *Proceedings of Symposium on the Chemistry of Natural Food Flavors, Nat. Acad. Sci.,* Washington, D.C.)

In the process described below a mixture consisting of a lipase, protease and salt (salt-enzyme mixture) is used. The lipase may be either an animal lipase or microbial lipase enzyme. The microbial lipase enzyme is derived from a mold, Aspergilli ssp., although other species like Penicillium ssp. or Rhizopus ssp. may be used. Examples of species are *Aspergillus oryzae, Aspergillus niger, Penicillium roqueforti, Penicillium glaucum* and *Rhizopus oryzae.*

The neutral protease enzyme of the enzyme mixture is derived from a mold, Aspergillus ssp., although other species like Mucor ssp., Penicillium ssp., Rhizopus ssp. or Streptomyces ssp. may be used. The peptidase enzyme, which may be included in the enzyme mixture, is also derived from a mold, Aspergilli ssp., Streptomyces ssp., Penicillium ssp., or Rhizopus ssp. may be used.

Example: Ripening the Milk—Milk pasteurized at 72°C for 16 seconds was placed in a cheese vat and 0.5% of active lactic starter was added while mixing. The milk temperature was adjusted to 88°F and annato cheese color was added (15 ml/1,000 lb milk). The starter remained in contact with the milk for between 15 and 30 minutes.

Setting the Milk—Single strength rennet extract (90 ml/1,000 lb milk) diluted with tap water in the ratio of 1 part extract to 40 parts water was added to the milk. After agitating for about 5 minutes the milk was held quiescent for about 25 minutes to allow time for rennet action to form a curd.

Cutting the Curd—When the curd reached the required strength, it was cut into cubes using ¼" wire cheese knives. The cut curd mass was then stirred for about 5 minutes at slow speed.

Cooking the Curds—Cooking the curds was started gradually using steam in the cheese vat jacket until a peak temperature of 100°F was attained over a period of about 30 minutes. At the same time there was steady agitation. The temperature was then maintained for about 45 minutes with intermittent agitation.

Draining the Whey—Agitation was stopped permitting the curds to settle. A strainer was inserted in the exit gate, and while the curds were pushed through the whey to the end of the vat with a curd rake, the greenish colored whey was drained off. The curds were trenched over the length of the vat during the latter part of the curd draining.

Cheddaring the Curd—The trenched curd was allowed to mat for about 15 minutes after removal of the whey and then was cut into blocks which were spaced about 1" apart. After about 15 minutes, the curd blocks were turned over repeatedly at 15 minute intervals, and during the latter part of this period the curd blocks were piled double. The cheddaring process continued until the titratable acidity of the clear whey exiting from the vat reached 0.5 to 0.6%.

Milling the Curd Slabs—The flattened-out curd slabs were fed into a milling machine, and the milled curds were returned to the vat.

Adding the Enzyme-Salt Mixture to the Curds—The milled curds were mechanically forked while spreading a mixture of cheese salt and food grade enzymes over the curds. The cheese was mechanically forked for a period of about 30 minutes. The salt-enzyme mixture contained: sodium chloride, 2.3 lb/100 lb cheese curds; animal lipase enzyme preparation, 2 to 20 g/100 kg cheese curds; and neutral protease enzyme preparation, 5 to 50 g/100 kg cheese curds.

Hooping and Pressing—The cheese curds were placed in metal molds or hoops and maintained in a hydraulic press overnight at a pressure of about 20 to 25 psig.

Drying the Cheese—After removing the whole cheese from each of the hoops, they were placed on shelves in a drying room for 3 to 4 days, the room being maintained at about 60°F and 75% humidity. The cheese was turned daily on the shelves unless packaged in rindless wrappers.

Curing the Cheese—Each cheese was cured for 1 month at 10° to 25°C; this short period was sufficient to produce a high level of flavor with minimum bitterness.

Cheddar Curd Culture Additives

A.B. Storrs; U.S. Patent 3,875,305; April 1, 1975; assigned to Lacto Products Co., Inc. states that if a particular organism, *Streptococcus diacetilactis* is used, the body and flavor of cheddar cheese can be improved provided the culture is added at the proper stage of the process. In addition, the process allows a shorter aging period for optimum body and flavor.

If used in the usual manner of culturing, *S. diacetilactis* normally produces sizable amounts of carbon dioxide which, in the production of cheddar cheese, would result in an unmanageable, floating curd structure. For that reason, *S. diacetilactis* has been rigorously avoided in any phase of the manufacture of cheddar cheese. If the organism is used as provided in this process however, no problems from gas production are encountered.

In the improved process, a small amount of a *S. diacetilactis* culture is mixed into the cheese curd, after the curd has been drained but before being packed into hoops or similar containers. The culture can be active (live, viable) cells, primarily of inactive cells, or a mixture of active and inactive cells. *S. diacetilactis* cultures are well-known and are readily obtainable from many recognized culture collections.

When an active culture is to be used, a freshly grown paste can be added directly to the cheese curd or preserved for future use by freezing in a conventional manner, in which case the paste is usually diluted with a phosphate buffer to a pH of about 6 before freezing. If desired, the diluted paste can be freeze-dried in a conventional manner for storage. When an inactivated culture is to used, a suitable inactivating material, such as acetone, is added to a solution containing viable cells, such as a water-diluted paste or a water suspension containing freeze-dried culture powder, and the precipitated fraction containing the inactivated cells is separated by evaporation or filtering.

The mechanism responsible for the advantageous results achieved by treating a cheese curd with a *Streptococcus diacetilactis* culture is not fully understood at this time. Both active and inactivated cultures are effective in improving the body and flavor characteristics so it would appear that the results are due more to some type of enzymatic process or some other unknown factors than to the metabolic activity of live bacteria.

In order to obtain the best results, the *S. diacetilactis* culture should be thoroughly mixed into and substantially uniformly dispersed throughout the cheese curd. Therefore, before being added to the curd, the culture is mixed with a sufficient amount of water to facilitate this mixing and dispersing. Of course, use of excessive amounts of water which would adversely affect the body texture of the resultant cheese should be avoided. In cheese making processes including a salting step after draining but before hooping, the culture can be added before or after salting and, if desired, can be added along with the salt so long as it is thoroughly mixed into and uniformly dispersed throughout the cheese curd.

The addition of very small quantities of a *S. diacetilactis* culture to the cheese curd is effective for improving the body and flavor characteristics of the cheese products. The specific quantity used depends primarily upon whether or not an active or inactive culture is used and the amount of improvement desired. It has been found that an active culture is somewhat more effective than an inactive culture in improving the body and flavor characteristics of the resultant cheese product by a factor of perhaps 2 to 5 times, especially at higher curing temperatures. Thus, less active culture is required to produce the same results. However, under some circumstances, such as where facilities for handling or storing viable cultures are not available, the cheese maker may find it more advantageous to use inactivated cultures.

Example: The cheddar cheese used was prepared by a conventional, so-called granular process. Two 20 lb samples of cheese curd prepared by this process were used for the test, one being used as a control sample and the other as a test sample. After draining both curd samples were processed in the identical manner except that, after draining and salting but before hooping, an inactivated *S. diacetilactis* culture was added to the test sample. Otherwise the two samples were processed in the identical manner.

A concentrated paste of freshly grown cells was prepared from a culture of *S. diacetilactis* DRC-1. The cells were inactivated by adding an equal volume of acetone to the paste. After about 1 hour, the mixture was subjected to a moderate vacuum to remove substantially all of the acetone and then completely dried under a high vacuum. 100 ppm of the thus prepared inactivated culture (diluted with a small quantity of water to aid in dispersion) was thoroughly mixed into the test sample curd. The control and test samples were placed in separate hoops, pressed, covered and wrapped in the usual manner, and then stored at 50°F for aging.

Sample trier plugs were removed from both samples at intervals during a 12 month storage period and examined by a panel of experienced judges for physical and flavor characteristics. Within the first month of aging, a marked difference between the body characteristics of the two samples was observed. The test sample had begun to develop a body which was "longer" (i.e. more flexible and pliable) and had a smoother, closer texture than that of the control sample. The control sample initially developed an acid flavor and gradually became more mild and mellow during the 12 month period. On the other hand, the test sample did not develop a comparable acid flavor and throughout the 12 month aging period consistently was scored as having a flavor which was smoother and more mellow than that of the control sample.

Flavor Enhancement Enzyme System

A specific enzyme-containing microorganism for enhancement of dairy flavor is given by *N. Mukai and M. Kawai; U.S. Patent 3,858,492; January 7, 1975; assigned to Kyowa Hakko Kogyo Kabushiki Kaisha, Japan.* The culture belongs to the Basidiomycetes group which can be cultivated on suitable mediums containing necessary growth nutrients. The flavor-enhancing enzyme accumulates in the culture medium.

The accumulated enzyme can be separated by the extraction of the solid medium with water or a suitable water-soluble medium. The solution is then concentrated in a suitable manner such as for example, precipitation with organic solvent, salting-out method, concentration in vacuo, cellulose ion exchange resin treatment etc.

Basidiomycetes group which may be used include, for example:

	Deposition No.
Fomitopsis pincola (Fr) Karst	ATCC 20036
Irpex lacteus	IFO 5367
Lenzites saepiaria	ATCC 9419

Basidiomycetes group can advantageously be cultured in liquid medium under aeration-agitation conditions, although solid medium can also be used.

The dairy flavor of green cheese obtained in conventional manner can be enhanced in a simpler way and in a shorter time by adding to the green cheese, prior to ripening, the enzyme-containing substance in an amount of about 0.05 to 1% w/w and allowing it to stand, for example, for about 1 to 5 hours with stirring at about 10° to 20°C until a good dairy flavor is obtained.

Sweet cream butter is conventionally produced by churning a milk cream so as to collect fat globes and working the same, while ripened cream butter is conventionally produced by the use of a ripened nonhuman animal's milk, which has been ripened by adding a butter starter thereto, and treating the ripened milk in an analogous manner to that of the production of sweet cream butter.

Good dairy flavor can be imparted to sweet cream butter in a simpler manner and in a shorter time by adding the enzyme-containing substance either before or after churning.

In the former process, good dairy flavor can be obtained by adding to a cream before churning the enzyme in an amount of 0.1 to 1% w/w together with lactic acid in an amount that would give an acidity of 0.1 to 0.5% calculated as lactic acid and standing the resultant mixture at 15° to 40°C with stirring until a sufficient dairy flavor is obtained. In the latter process which is simpler than the former, even though the enhanced dairy flavor is not stronger than that obtained by the former, a churned cream is added with the enzyme in an amount of 0.1 to 1% w/w and the resultant mixture is then worked at 15° to 25°C until a sufficient dairy flavor can be obtained (for example, for 1 to 10 hours).

Milk products containing lactic acid, such as yogurt, can be given a good dairy flavor in a simpler way and in a shorter time in an analogous manner by the use of the enzyme. It is preferred to use the enzyme-containing substance in an amount of about 0.05 to 1% w/w and lactic acid in an amount that would suffice to give an acidity of 0.5 to 1% calculated as lactic acid.

Nonhuman animal's milk such as whole milk, powdered milk and condensed milk can also be given a good dairy flavor in a simpler way by the use of the enzyme-containing substance in an amount of 0.05 to 1% by weight of solid content. The enzyme-containing substance of the process can be used directly without further purification for enhancing the dairy flavor of various milk food stuffs, such as chocolate, cakes, baby foodstuff, ice cream, margarine, etc. The following nonlimitative examples further illustrate the process.

Example 1: A liquid medium (20 liters) containing saccharide (5% w/w), corn steep liquor (2% w/w), yeast extract (0.03% w/w) and small amounts of inorganic salts was adjusted to a pH of 5.0. After sterilization, the medium was inoculated with a strain of *Fomitopsis pincola* (Fr) Karst (ATCC 20036) and was cultured for 120 hours at 30°C (400 rpm; aeration of 15 liters/minute). The filtration removed microbial bodies from the culture medium. To the filtrate was added ammonium sulfate (70% of the saturating amount) to give precipitates, which were then centrifuged and freeze-dried to yield a substance containing the enzyme for flavor enhancement.

The enzyme-containing substance was then added to a green cheese prepared in conventional manner (0.5% w/w). After working for 2 hours at 15°C there was obtained an agreeable cheese with enhanced dairy flavor similar to that of conventionally ripened cheese.

Example 2: A strain of *Irpex lacteus* (IFO 5367) was cultured in a similar medium under similar conditons to those described in Example 1 and was filtered to remove microbial bodies. The filtrate was treated with ammonium sulfate in a similar manner to that described in Example 1 to give a substance containing the enzyme. A sterilized milk cream containing fat (about 35% w/w) was added with the substance (0.13% w/w) together with lactic acid so as to give an acidity of 0.15% calculated as lactic acid. The mixture was then stirred well at 37°C for about 4 hours, from which a sweet cream butter was prepared in a conventional manner. The butter obtained had an enhanced good dairy flavor similar to that of the conventional ripened cream butter.

Controlled Lipolysis

Spontaneous or uncontrollable lipolysis has plagued the food industry for many years, since undesirable flavors, or harsh rancidity, can result. In recent years, however, the development of controlled lipolysis has resulted in many improved food products. Thus, it is common practice in the dairy industry to treat milk fats with lipolytic enzymes to produce desired flavors.

The lipolytic enzyme acts on the triglycerides to produce free fatty acids which, in turn, can be further converted to other compounds, such as ketones, in some instances. The amount and type of fatty acids produced are dependent on the amount and type of fat in the food product, the amount and type of lipolytic enzyme product and the time and temperature of treatment with lipolytic enzyme.

L.I. Feldman and J.G. Dooley; U.S. Patent 4,065,580; December 27, 1977; assigned to GB Fermentation Industries, Inc. found a highly satisfactory source of fat-flavoring enzyme system derived from the microbial organism *Mucor miehei* (see also processes in U.S. Patents 3,549,390 and 3,950,221).

The enzyme flavoring system is defined in terms of the ratio of esterase activity (EA) to lipase activity (LA), or EA/LA. The esterase activity is that activity of the enzyme on the water-soluble fats, namely those fats having carbon chain lengths of from about 4 to about 10. Conversely, the lipase activity is that activity of the enzyme on the water-insoluble fats, namely those fats having carbon chain lengths of about 12 and higher, particularly 12 to 22.

It has been found that the lipolytic enzyme system produced by certain species of the genus Mucor, and particularly *Mucor miehei*, as opposed to that of other fungi, such as Aspergillus and Rhizopus, has the property of expressing greater esterase activity than lipase activity, in a manner similar to that of commercial pregastric esterases from calf, kid and lamb. That is, it shows an EA/LA ratio of greater than one. By way of comparison, Aspergillus and Rhizopus produced lipolytic systems which were found to show EA/LA ratios of less than one, and consequently are unsuitable for this purpose. The esterase activity is measured by a test on tributyrin and the lipase activity is measured by a test on an olive oil substrate.

The desired enzyme flavoring system is recovered by separation from the fermentation mash. This can be carried out by filtration of the mash at acid pH of about 4 to 5, extraction of the filter cake at alkaline pH of about 10 to 12, acidification of the extract to a pH of about 7, and then recovering by filtration or concentrating the acidified extract by evaporation or ultrafiltration and then drying the concentrate such as by spray drying. For certain uses it is desirable to use a lipolytic enzyme system that is also substantially free from rennet enzyme activity. This can be achieved by using the foregoing recovery procedure.

The recovered lipolytic enzyme flavoring system can be added to milk, milk fat, butter oil, cheese, milk chocolate, margarine oil, coffee whiteners, and other triglyceride fat-bearing foods and food components to produce desirable milk and butter-like flavors. It can be used to accelerate or intensify naturally developed flavors in these foods and food components and can be blended with known flavor developing agents to produce a variety of flavor profiles.

Examples of triglyceride fats which can be teated with the lipolytic enzyme flavoring system to develop desirable flavors are animal fats such as milk fat, lard and tallow, and vegetable fats such as cottonseed oil, soybean oil, corn oil, peanut oil, safflower oil, palm oil, coconut oil and margarine and shortening oil blends. Examples of cheeses where the lipolytic enzyme flavoring system can be used to develop desirable flavors are Provolone, Romano, Feta, Pizza, Mozzarella, Cheddar, Swiss, Blue, Parmesan, Colby, Gouda and Edam.

Example 1: Mucor miehei, NRRL 13042, is transferred from an agar slant under aseptic conditions into a 1 liter Erlenmeyer flask containing 200 ml of the following medium:

	Percent
Soybean flour (Nutrisoy 300°C)	1.5
Dried whey	3.0
Enzymatically degraded cornstarch	12.0
Water	83.5

The flask is incubated on a rotary shaker at 37°C for 114 hours starting with an initial pH of about 6. The desired lipolytic enzyme system is recovered as follows: The fermentation broth is adjusted to pH 5 and then filtered. The filtrate is extracted with dilute NaOH solution at pH 10 to 11 and the extract acidified to pH 7. The acidified extract is then filtered and evaporated to a concentrate having an EA value of 25.8 and an EA/LA ratio of 2.6 in one recovery run (A); and is filtered, evaporated and spray dried to a solid having an EA/LA ratio of 3.5 in another recovery run (B).

Mucor miehei derived lipolytic enzyme produced as above in recovery run A with the stated EA/LA ratio of 2.6 is used to develop a desirable cheese flavor as follows. 1 lb samples of Current A Cheddar cheese (2 to 4 week old cheese, having a bland flavor) are ground with a Sunbeam table model electric grinder and the lipolytic enzyme blended by further mixing until evenly distributed in proportions of 440 mg/lb of cheese. The blended enzyme-cheese combinations are then filled into sterile plastic petri dishes. The dishes are placed in an anaerobic container to inhibit mold growth and several samples are incubated for 4 days at 20°C and other samples are incubated for 11 days at 20°C.

After incubation the cheese was tasted for its organoleptic properties and was found to have a clean flavor profile similar to that produced by commercially available pregastric Esterase C (Dairyland Foods Laboratory) tested in the identical manner. Both cheeses developed a desirable slight to medium rancidity. By way of comparison, an esterase preparation obtained from *Aspergillus flavus* was found to have an EA/LA ratio of 0.08 and produced an undesirably harsh butyric-caproic-capric type profile and a slight metallic after-taste.

Example 2: Mucor miehei lipolytic enzyme is added to cheese milk prior to the addition of the starter culture in an amount of ⅛ to 2 oz/1,000 lb milk for the development of controlled lipolysis to produce flavors ranging from delicate and buttery to peppery and sharp "piccante" in the manufacture of Italian type varieties of cheese, namely Asiago-Mozzarella, Provolone and Romano, respectively, with cheese milk for Romano and other sharp flavored varieties receiving the higher levels of lipolytic enzyme treatment.

Example 3: A Blue cheese is made from raw milk, homogenized at 90°F and bleached with benzoyl peroxide. The milk is heated to 86°F and inoculated with 1% of a *Streptococcus lactis* starter. After a brief holding period, the milk is inoculated with a microbial rennin product from the growth of the strain of *Mucor miehei* designated NRRL A 13042 in an amount of 3 oz/1,000 lb of milk, and ¼ oz of *Mucor miehei* lipolytic enzyme/1,000 lb of milk. The mixture is agitated until a curd of satisfactory firmness is obtained. The curd is then inoculated with spores of *Penicillium roqueforti*, separated from the whey and transferred to hoops. The hooped curd is salted by immersing in brine for 3 days and packaged in an evacuated plastic (Cryovac) bag. Pin holes are made throughout the cheese to admit air and facilitate growth of the mold. The cheese is then ripened by placing in a curing room for 2 months at 50°F and 90% humidity to produce an excellent quality Blue cheese with an improved flavor over that obtained without the added lipolytic enzyme.

Blue Cheese Flavor

F.V. Kosikowski and R.C. Jolly; U.S. Patent 3,973,042; August 3, 1976; assigned to Cornell Research Foundation, Inc. reviews the background on flavor development in Blue cheese by milk lipases and gives a process for Blue cheese flavor improvement.

β-oxidation pathways of free fatty acids' deacylation and decarboxylation to methyl ketones leading to typical flavor have been outlined by Gehrig and Knight (R.F. Gehrig and S.G. Knight, 1963, "Fatty Acid Oxidation by Spores of *P. roqueforti*," *J. of Appl. Micro.* 11:166-170); Hammer and Bryant (B.W. Hammer and H.W. Bryant, 1937, "Flavor Constituent of Blue Cheese," *Iowa State College J. Sci* 11:281); and Katz and Chaikoff (J. Katz and I.L. Chaikoff, 1955, "Synthesis via Kreb's Cycle in the Utilization of Acetate by Rat Liver Slices," *Biochemica et Biophysica Acta* 18:87).

The influence of heat on the inactivation of milk lipases has been reported by Gould and Trout (I.A. Gould and G.M. Trout, 1939, "Lipase Action in Mixture of Raw and Pasteurized Homogenized Milk," *Quart. Bull. Mich. Agric. Exp. Sta.* 22, 101); Harper and Gould (W.J. Harper and I.A. Gould, 1959, "Some Factors Affecting the Heat Inactivation of the Milk Lipase Enzyme System," *Int. Dairy Congress,* 1959; Vol. 1:455); Nilson and Willart (R. Nilson and S. Willart, 1960, "The Influence of Homogenization on the Fat Splitting in Milk," *Rep. Milk & Dairy Research,* Alnarp, Sweden 60); Sjostrom (G. Sjostrom, 1959, "Lipase Problems in Milk and Dairy Products," *Rep. Milk & Dairy Res.* Alnarp, Sweden 58); and Stadhouders and Mulder (J. Stadhouders and H. Mulder, 1959, "Surface Organisms Associated with Fat Hydrolysis in Cheese," *Neth. Milk Dairy J.* 13, 291).

It is generally estimated that Blue cheese from raw or pasteurized, homogenized milk requires approximately 4 to 7 months at 40° to 50°F to obtain peak flavor. Natural materials containing lipase have been added to raw, pasteurized, homogenized milk, but bitterness invariably accompanied the flavor development, Gould and Trout (previously cited); Lassen et al. (P.B. Lassen, G.M. Trout and I.A. Gould, 1941, "Rancidity Studies on Mixtures of Raw and Pasteurized Homogenized Milk," *J. Dairy Science* 24, 771); and Nilson and Willart (R. Nilson and S.W. Willart, 1961, "The Heat Inactivation of the Fat Splitting in Milk," *Rep. Milk & Dairy Research,* Alnarp, Sweden 60).

Possible sources of the lipases in Blue cheese ripening are (1) the milk, (2) natural microflora of the milk and cheese, (3) *P. roqueforti*, *P. glaucum* and related species, and (4) other enzyme extracts from natural materials. The presence of two lipolytic enzyme systems in milk is indicated by Chandan and Shahni (R.C. Chandan and K.M. Shahni, 1963, "Purification and Characterization of Milk Lipase, II, Characterization of the Purified Enzyme," *J. Dairy Sci.* 46:503-509).

Strains of *P. roqueforti* also differ in their lipolytic activity as observed by Proks et al. (J. Proks, J. Dolezalek and Z. Peck, 1959, "A Study on the Coaction of the Genus Torulopsis on the Formation of Methyl Ketones in the Cheese of the Roquefort type," *XV Intern. Dairy Congress* 2(3) 729-735). Lawrence and Hawke (R.C. Lawrence and J.C. Hawke, 1966, "The Incorporation of [1-^{14}C] Acetate into the Methyl Ketones in Steam Distillates of Bovine Milk Fat," *Biochem. J.* 98:25.) demonstrated two lipolytic enzyme systems in *P. roqueforti.* Various attempts have been made to accelerate lipolysis in Blue cheese. Pancreatin, mulberry juice lipase, chicken stomach lipase, erepsin, steapsin, etc. have been tried.

The improved process for Blue cheese requires the combination of a food grade lipase and salt with regular Blue cheese mold spores such as *Penicillium roqueforti*. A laboratory study was conducted to introduce into the curd various microbial and animal lipases together with salt and *P. roqueforti* spores and to observe the speed and quality of typical flavor development in pasteurized, homogenized milk cheese and the rate and type of methyl ketones produced. The following lipases were utilized:

Test Designation	Source
Lipase A	Animal lipase (Lipase L)
Lipase B	Animal lipase (Lipase KL)
Lipase C	Microbial lipase (Lipase 3500)

Blue cheese was prepared from 500 kg lots of (a) raw whole milk, (b) pasteurized skim milk plus homogenized raw cream, and (c) pasteurized whole milk. Commercial lipase preparations were introduced into the curd with salt and *P. roqueforti* spores at 0.5 to 1.5 g/7 kg curds. The curds were subsequently made into Blue cheese. Plug samples from two wheels of each of the several lots prepared were analyzed in triplicate. The moisture and fat content was 45 and 27.5%, respectively, for the control cheese at the end of 90 days.

In various samples of cheese made from control pasteurized milk, homogenized raw cream and pasteurized skim milk, and lipase A, lipase B, and lipase C, respectively, the nitrogen increased from 0.31 to 0.91%. The value for free volatile fatty acids increased from 8.99 to 26.20 and the pH of cheese increased from 5.0 to 6.4 from 0 to end of 90 days respectively. Highest pH increase was noticed at 45th day of ripening for lipase C, 60th day for lipase B and pasteurized milk Blue cheese control and 90th day for lipase A and pasteurized skim milk and raw homogenized cream Blue cheese control.

Blue cheese from raw whole milk developed unclean flavor and was excluded from further analysis. All Blue cheese showed normal body and texture characteristics and uniform mild growth.

Soluble Nitrogen, Free Volatile Fatty Acids and pH of Blue Cheese at 90th Day of Ripening

Cheese Sample	Soluble Nitrogen (percent)	Free Volatile Fatty Acids*	pH
Control-Past. milk (0 days)	0.31	8.99	5.0
Control-Past. milk	0.72	12.27	6.1

(continued)

Cheese Sample	Soluble Nitrogen (percent)	Free Volatile Fatty Acids*	pH
Cream and Skim milk (Control)	0.65	12.32	5.5
Lipase A	0.86	13.49	6.4
Lipase B	0.80	13.94	5.9
Lipase C	0.91	26.20	5.4

*Milliliters 0.1 N NaOH/10 g sample.

The total carbonyls increased with time to maximums at 40°F for lipase B and lipase C and for Blue cheese made from pasteurized milk, pasteurized skim milk and raw homogenized cream and lipase A after 90 days. Odd number methyl ketones with chain length from C_3 to C_{15} were quantified. In lipase C, C_6 methyl ketone was observed after 30 days at 40°F. 2-heptanone and 2-nonanone were major methyl ketones for all the samples over ripening. The pasteurized milk control contained no C_{15}. In lipase C (cheese) the typical pattern of individual methyl ketones from 3 to 15 carbon atoms was obtained. On the 75th day of ripening, the concentration of 2-heptanone was 76.06 μmols per 10 g of dry cheese and the concentration of 2-nonanone was 32.45 μmols per 10 g of dry cheese.

Maximum Concentration of Carbonyls in Blue Cheese

Sample	Days at 40°F	Total Carbonyls	Mono Carbonyls	Methyl Ketones
		......μMols/10 g Dry Cheese......		
Control-Past. milk	0	3.78	3.60	t*
Control-Past. milk	90	165.72	95.25	47.63
Cream and Skim milk (Control)	90	200.07	102.57	58.62
Lipase A	90	200.13	120.90	72.96
Lipase B	75	241.14	181.37	115.23
Lipase C	75	313.41	250.72	150.43

*t=trace

Example: Introducing Spores and Lipase—To 100 kg curds are added 5 to 30 g food grade microbial lipase 4000 (Aspergillus ssp.); 60 to 70 g *Penicillium roqueforti* spores and about 2 kg sodium chloride. The additives are stirred uniformly into the curds for 5 minutes.

Dipping the Curd—Partly salted, mold-inoculated curds are scooped into perforated, circular, stainless steel Blue cheese molds.

Pressing or Molding Curd—The hoops are turned completely over every 15 minutes for the first 2 hours and left on drainage mats overnight at room temperature without any follower or surface weight, but covered with protective clean cheesecloth. The room temperature is kept at about 72°F by proper controls.

Salting—The approximately 5 lb wheel of curd is removed from the mold by encircling it with a spatula. After tapping the metal mold on the table, the cheese will drop out. Then the individual wheels of cheese are placed in a pail or box of plain coarse salt and all the surfaces are liberally covered with the salt. Any salt which remains loose can be brushed off; then the individual wheels are gently laid on their curved sides in a wooden cradle, in a 60°F, 95% humidity room. The cradle prevents the sides of the wheel from splitting. The heavy salting in this manner is repeated for 5 days (4 more times). At the end of this period, the cheese surface is very hard. Alternatively the wheels of cheese may be brined.

Needling Air Holes—After 5 days of salting, about 50 air holes are pierced on each of the cheese's flat sides with a long ice pick, ⅛" in diameter, or with the mechanical head of a needling machine which produces 50 holes/single application. The air holes should penetrate through to the other side.

Mold Growth Period—The perforated cheese wheels are placed on their curved sides, 1" apart, in wooden cradle racks at 50° to 60°F, and 95% relative humidity and are turned a quarter turn and wiped with a clean cloth every 4 days. Mold appears internally by about the 10th day. Turning and cleaning are continued for 20 more days.

Curing and Storing—After 1 month, all cheese surfaces are scraped and cleaned by hand or are buffed with a rotating burlap bag wheel. Temporarily the cheese is wrapped in foil and removed to a cold room, about 40°F. The cheese is held at this cold temperature for 1 to 3 months for full flavor and body development. Then the temporary foil is removed and the surfaces cleaned. It is then wrapped in a clean foil and distributed or stored cold. The cheese held at 35° to 40°F with proper spacing can remain in good condition for several more months.

PROCESSED CHEESE REWORK PROCESS

Many cheese varieties, for example, American, Cheddar, Swiss, Brick, Limburger, etc., are today available in pasteurized process form. In the cooking and preparation of cooked cheese foods, pasteurized process cheese (also referred to as process cheese or pasteurized cheese for the sake of brevity) has many advantages over natural cheese. It melts smoothly and quickly without fat separation or stringiness. Because various types of cheese may be blended together, it is possible to obtain cheese products having balanced flavors and textures. But perhaps the most important characteristic of process cheese is its excellent storage life, that is, stability toward spoilage, rancidity, etc., over a prolonged period of time.

Notwithstanding the fact that cheeses of all types have fairly long storage life, they are, nevertheless, perishable in varying degrees. Some varieties, for example, Parmesan, have a long storage life, while others, for example, Camembert, have only a relatively short storage life. The natural ripening process which yields the distinctive cheese flavors and textures does not cease when the cheeses have reached their prime, but on the contrary proceeds until the cheese is no longer suitable for consumption. An additional loss of cheese yield by deterioration or drying out of the cheese has long been a problem to cheese manufacturers.

Much like the milk from which it is produced, cheese is a complex food product. The final acidity of the cheese and the flavor are mainly influenced by the means used to prepare the cheese curd and the curing conditions of the final curd.

Some desired qualities of commercial cheeses include smooth texture, high resiliency, softness, moistness, pleasing flavor, and the like. Color, a feature of localized consumer preferences, is easily controlled by the addition of pure vegetable coloring matter to the milk. It has been estimated that about 55% of the cheese produced in the United States is of the pasteurized process type, and presumably a large portion of this is ultimately utilized in the preparation of cooked cheese dishes. Natural cheese is less suitable for cooking since the fat tends to separate from the casein when the cheese is subjected to heat. This fat separation is usually prevented in pasteurized process cheese by the use of emulsifying agents, for example, sodium phosphates and sodium citrate.

A process for preparation of a pasteurized processed cheese from a combination of natural cheese and rework process cheese is described by *T.P. Kichline and L.G. Scharpf; U.S. Patent 3,692,630; September 19, 1972; assigned to Monsanto Company*. A proteolytic enzyme is added to the natural and rework cheese mixture containing an emulsifying agent and water. It was found that the optimum product could be obtained after the enzyme had been in contact with the cheese for only a short period of time, thereby minimizing proteolysis and not adversely affecting cheese flavor. While natural cheese is usually prepared by the use of an enzyme such as rennet to curdle the milk, the enzyme activity of the rennet appears to be substantially destroyed during the milk curdling and/or aging procedure due either to a change or breakdown of chemical structure of the rennet or for some other reason.

It is believed that the proteolytic enzymes utilized in the processes cause a chemical reaction to take place in the natural cheese alone or in combination with the rework cheese during the emulsification and pasteurization step and thereby provide a process cheese of improved properties. Also, it is believed that when rework cheese is used such cheese is chemically "broken down" so that it effectively blends together with the natural cheese to provide a homogeneous mixture.

Of the various enzymes which may be used in the processes, those proteolytic enzymes of bacterial or fungal origin, which have a specific activity (that is, protease units/g of enzyme) of about 50,000 to 990,000 are preferred. A protease unit is defined as equal to 0.5 mg of trichloroacetic acid-soluble tryosine liberated from casein on digestion for 10 minutes at 37°C.) The proteolytic fungal type enzyme is commercially available under the trademark Rhozyme P-11 derived from the cultures of the fungus *Aspergillus flavus-oryzae* and has a specific activity of from about 100,000 to about 150,000.

Generally, the amount of enzyme/g of emulsifying agent is an amount to provide less than 12,000 casein units of protease activity, usually, from 200 casein units of protease activity to about 2,500 casein units of protease activity. For example, the enzyme obtained Rhozyme P-11 is used in an amount to provide from about 6 casein units of activity to about 2,500 units of activity per gram of the emulsifier; the enzyme obtained from *Mucor pusillus* Lindt is used in an amount to provide about 2 casein units of activity to about 2,700 casein units of activity/g emulsifying agent.

Example: Separate batches were introduced individually into a preheated (150°F) 1 quart steam-jacketed Baker Perkins mixer. After mixing by means of a single blade mechanical agitator was begun, appropriate quantities of distilled water (to yield the end moisture content) were added followed by addition of an enzyme-emulsifying agent mixture containing approximately 5 g of disodium phosphate duohydrate, 5 g of trisodium phosphate dodecahydrate and appropriate quantities of a proteolytic enzyme commercially available under the trademark Rhozyme P-11, which is an enzyme of fungal origin and which had a specific activity of 110,000 protease units/g of enzyme. The mixing of the ingredients was continued for about 5 minutes to achieve an end (or "cooker out") temperature of about 165° to 168°F. (It took approximately 8 to 10 minutes, respectively, to achieve an end temperature of about 175°F and 186°F.

The cheese temperature was monitored during mixing by means of a thermocouple which was inserted into the side of the mixing chamber and was connected to a temperature recorder. At the end of this pasteurization or heating and mixing period, which was conducted at atmospheric pressure, the separate samples were poured into individual Teflon lined trays, covered with Teflon sheets, rolled out to approximately ⅛" thickness and then cooled down to refrigerator temperature of approximately 10°F. After the pasteurized process cheese melt had sufficiently solidified, it was cut into standard size slices, that is, a 3.5" x 3.5" square.

STARCH, CELLULOSE AND CORN SYRUP SWEETENERS

STARCH CONVERSIONS

Corn Syrups—Basic Process

The basic process for preparing sugars of high purity from starches is illustrated by *K. Sugimoto and M. Hirao; U.S. Patent 3,804,715; April 16, 1974; assigned to Hayashibara Company, Japan*. The process described is specifically designed to prepare syrups of high maltose content by using β-amylase enzyme and α-1,6-glucosidases (isoamylase).

A starch is first made in a homogeneous liquid form. The liquefaction may be effected by (1) a conventional liquefaction process with an acid; (2) a process in which α-amylase, a product obtained from *Bacillus subtilis*, is used; (3) a process in which amylase contained in malt is used; or (4) a process in which starch is stirred under heating for a few minutes at a temperature between 100° to 160°C in the absence of an enzyme to obtain a homogeneous liquid.

If a glucose polymer with an odd number of glucose moieties is formed by liquefaction and is completely decomposed into maltose or disaccharide, in a saccharification process, the formation of glucose or maltotriose cannot be avoided. Thus, the production rate of maltose from short chained glucose polymer decreases, as a result of the formation of glucose or maltotriose. Accordingly, in order to produce highly pure maltose, the dextrose equivalent (DE) of liquefied starch must be lowered as much as possible. In such a case, it is necessary to limit the DE within the range 1 to 2% by employing the high temperature liquefaction (4) or enzyme liquefaction (2).

If starch syrups have a maltose content of about 50 to 80%, purification treatment after the saccharification is easy and retrogradation of dextrin in the course of the saccharification is not feared at a DE in the range of 5 to 30% when using the enzyme liquefaction (2). However, at a DE higher than 10, the final maltose content is gradually reduced as DE increases.

A continuous reaction apparatus as shown in Figure 3.1 is used for the gelatinization and saccharification of starch slurries. In the figure, **1** is a tank into which starch slurries are fed, in which the concentration, pH and amount of α-amylase to be added are controlled. The starch slurry is pumped into a saccharification vessel **2**. During vigorous stirring with the stirrer, steam is introduced and the slurry is heated to a predetermined temperature to be gelatinized quickly.

Figure 3.1: Corn Syrups—Basic Process

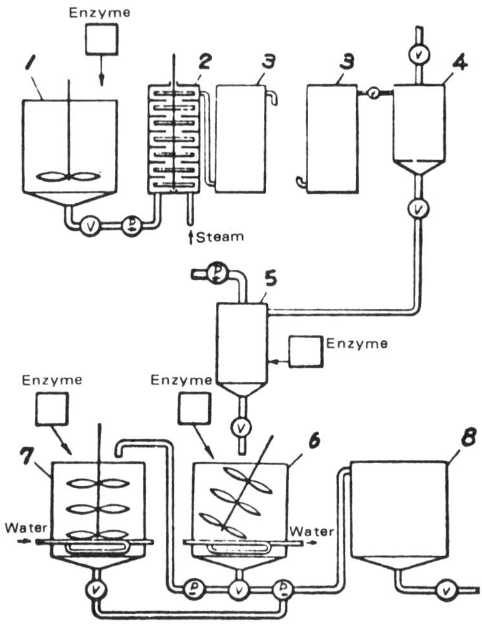

Source: U.S. Patent 3,804,715

The liquefied solution is sprayed in a flash cooler **4** and cooled to a desired temperature. The liquefied solution is taken out with a pump and poured into a first blending tank **6**. A first enzyme is sprayed from the top of a vacuum flash cooler **5** or directly and continuously poured into the first blending tank **6**. In case β-amylase and α-1,6-glucosidase are introduced at the same time, the mixed liquefied solution is held in a saccharification tank **8** to complete the saccharification.

If the saccharification is performed in two stages different from each other in temperature, mixture of enzyme and gelatinized solution is mixed under stirring in the first blending tank **6** for a predetermined time and the mixture is continuously taken out from the bottom into a mixture residence tank **7**, to which a second enzyme is poured continuously to hold the enzyme at a predetermined temperature and then sent into the saccharification tank **8**. Mixing of enzymes and temperature control in the first mixing tank and the mixture residence tank

7 are effected by a strong stirrer and thermoregulator, respectively. Control of pH and temperature is effected separately in each tank.

Liquefaction or Gelatinization of Starch: Starches used as raw materials are corn starch, waxy corn starch and sweet potato starch as representative subastral and subterranean starches. Gelatinization and liquefaction of starch are important since they control maltose content of saccharified starch. Three appropriate methods of gelatinization and liquefaction of starch are shown below.

High Temperature Gelatinization Method – This method involves heating starch slurry quickly with stirring in the apparatus of the figure at a temperature of 100° to 170°C to obtain homogeneous liquid. Starch slurry adjusted to pH 5 to 6 is transferred into a starch liquefier **2** by a pump. While steam is introduced at the bottom of the liquefier **2**, the liquid is heated with stirring to effect liquefaction homogeneously. The DE of starch becomes 0.5 to 5 during heating for 10 to 30 minutes.

The liquefied solution obtained having a low viscosity is in a completely and homogeneously dispersed state, and the degree of polymerization of the molecules is high. Therefore, the method is suitable for the preparation of a product of high maltose content. If the pH of the starch slurry is adjusted to 3.5 to 4 in this method, the decomposition of starch is facilitated to yield DE 5 or higher, but this is unsuitable for the preparation of a product of high maltose content.

Liquefaction with Enzyme – Liquefaction of starch with α-amylase is effected by using 0.2 to 0.5% of commercially available α-amylase product from *Bacillus subtilis* based on starch in the same manner as in the preparation of glucose or maltose syrup. A suitable quality of α-amylase is added into the starch slurry and the pH is adjusted to 6.0. The slurry is added into the starch liquefier **2** at a temperature of 80° to 90°C. Holding time is 1.5 to 2 minutes. The liquid is treated to obtain DE 0.5 to 2. If DE of 5 to 35 is desired, the number of holding tanks **3** is increased and the temperature is adjusted to 80°C. After the completion of the liquefaction, the treated liquid is heated with steam in the final residence tank to 120° to 125°C to inactivate the enzyme, which is then introduced into a cooler.

Any concentration of starch below 40% is available and decomposition and liquefaction are performed homogeneously. According to the treatment proposed in the second saccharification stage, concentrations of liquefied solutions are varied. If hydrolyzate having maltose content of about 50 to 80 is desired, a liquefied solution of a high concentration of 20 to 35% is suitable, and in case the highest maltose content is desired, a concentration of 10 to 20% is suitable. As for DE in the liquefaction of starch, in case a maltose content of about 50 to 80% is desired, DE is controlled to 5 to 30, and in case a high maltose content of higher than 80% is desired, DE is controlled to 0.5 to 3. Those conditions influence the results of the saccharification.

Saccharification of Liquefied Starch Solution – This method is illustrated by the following examples.

Example 1: Into 10% sweet potato starch slurry, 0.2% of α-amylase is added and the slurry is liquefied at pH 6.0 at 90°C until DE 2.7; then there are added

2.5 units, per gram of starch, β-amylase obtained from wheat bran and saccharification is carried out at 55°C for 16 hours (A).

Separately, 10 units of α-1,6-glucosidase obtained from *Lactobacillus* together with 25 units β-amylase are used and the saccharification is carried out in the same manner (B). Comparing the above case (A) with the case (B), it is noted that maltose content when using β-amylase alone is 69.6%, while maltose content when using α-1,6-glucosidase together with β-amylase is 90% (refer to table below).

Case	Maltose (%)	Glucose (%)	Maltotriose (%)	Dextrin (%)
A	69.6	1.1	3.5	25.7
B	90.4	0.4	1.3	7.9

The above results prove that enzyme solutions obtained from the different strains having α-1,6-glucosidase activity can react under the same conditions to yield syrups of high maltose content.

Example 2: A starch slurry of a concentration of 10 to 25% is liquefied in the same manner as in Example 1, to which are then added 20 units of α-1,6-glucosidase and 10 units of β-amylase.

After saccharification at 45°C for 10 hours, the following results are obtained. Appropriate liquefaction rate is about 10%, and an indication is observed that the maltose content decreases as the concentration increases.

Concentration of Starch (%)	DE	. . . . Sugar Component, %. . . . . Glucose	Maltose	Maltotriose	Dextrin
10	56.8	0.2	78.1	7.9	13.8
15	55.5	0.2	78.0	7.9	13.9
20	54.1	0.3	75.0	8.0	16.7

High-Yield Maltose Process

Amylases produced from some strains of *Streptomyces* that give a high yield of maltose from starch and amylose are described by *Y. Koaze, Y. Nakajima, H. Hidaka, T. Niwa, T. Adachi, K. Yoshida, J. Ito, T. Niida, T. Shomura, and M. Ueda; U.S. Patent 3,868,464; February 25, 1975; assigned to Meiji Seika Kaisha, Ltd., Japan.*

The particular enzymatic activities and characteristics of the new amylases of *Streptomyces* are summarized in the table below. For comparison, properties of some known amylases of bacteria, fungi, plants and animals are shown together in the table.

Sources of Amylase	Optimum pH	Limit in Hydrolysis of Starch by Amylase Action (%)	Ratio of Glucose:Maltose Produced (by weight)
Streptomyces albus	4.5-5.0	78	0.058:1
Streptomyces aureofaciens	4.5-5.0	76	0.058:1

(continued)

Sources of Amylase	Optimum pH	Limit in Hydrolysis of Starch by Amylase Action (%)	Ratio of Glucose:Maltose Produced (by weight)
Streptomyces hygroscopicus	4.5-5.0	82	0.055:1
Streptomyces hygroscopicus var. *angustomycetes*	4.5-5.0	80	0.053:1
Streptomyces viridochromogenes	4.5-5.0	79	0.051:1
Streptomyces flavus	4.5-5.0	76	0.058:1
Streptomyces tosaensis nov. sp.	4.5-5.0	79	0.053:1
Bacillus subtilis (bacteria)	5.3-6.8	70*	0.71:1
Aspergillus oryzae (fungi)	5.5-5.9	48*	0.64:1
Germinated barley (plant)	4.7-5.8	80	–
Hog pancreas	6.9	80	0.024:1
Human saliva	6.9	80	–

*Calculated as glucose.

By comparing the properties of the various amylases shown in the table above, it is evident that the amylases of the seven *Streptomyces* species are not coincident with any of the known amylases particularly in view of their optimum pH range, the limit in the hydrolysis of starch and the ratio of glucose:maltose produced. The amylases of this process exhibit a relatively lower optimum pH and a relatively higher value for the limit in the hydrolysis of starch in combination with their lower ratio of glucose to maltose. The process of hydrolysis of starch takes place in such pH and temperature environment that would prevent the contamination of the saccharification mixture by growth of undesired microorganisms. The limit of the enzyme for hydrolysis of starch is no less than 75% of the theoretical maltose.

A starting starch is dispersed in water to a suitable concentration after it is washed well with water at first to remove the foreign matter therefrom. An amylase of a *Streptomyces* may then be immediately reacted with the washed starch in the aqueous dispersion, though it is preferable to effect the step of liquefying the starch by reacting a known liquefying amylase, e.g., bacterial α-amylase prior to the saccharifying treatment of starch with the *Streptomyces* amylase.

For carrying out the step of liquefying the starch, starch may be reacted at pH 5 to 6 and at 70° to 90°C with a known amylase, for example, a bacterial liquefying amylase. It is preferred to use an aqueous dispersion of starch at a starch concentration of 30 to 40% by weight. After this liquefaction is completed, the liquefying amylase is inactivated by heating. Subsequently, the liquefied starch in the aqueous dispersion or solution is subjected to a saccharifying step with the *Streptomyces* amylase.

The *Streptomyces* amylase may be used and added in the form of either the culture filtrate, a crude amylase precipitated from the culture filtrate, or a solution of a partially purified amylase obtained during the purification of the culture filtrate, or a precipitated amylase deposited during the purification of the culture filtrate, or a purified and dried powdery amylase product. It is suitable that the *Streptomyces* amylase is added to the liquefied starch in an amount 200 to 400 saccharifying units per 1 g of starch and that the saccharifying step is carried out at pH 4 to 6 and at 50° to 60°C. The saccharification reaction may proceed until there is formed an aqueous solution of the saccharification products of starch containing maltose at a maltose content of 70% or more by weight on the dry basis, when the saccharification reaction may be deemed as having reached the end point. Then, the enzyme is inactivated by heating the saccharified starch solution. The inactivated reaction mixture, namely, the saccharified starch solution containing at least 70% by weight of maltose may then be used as such as a sweetening material, if desired, but it may be subsequently passed to a filtration process and/or a concentration process and further to a decolorizing process and/or a desalting process, if required. An example of the analysis of the composition of a sweetening material which has been prepared using the amylase of *Streptomyces hygroscopicus* is: maltose, 75.4% by weight; glucose, 3.1% by weight; oligo-sugars (including maltotriose, maltotetraose, etc), 10.1% by weight; and dextrin, 11.4% by weight.

Example 1: A strain of *Streptomyces hygroscopicus* identified as ATCC 21722 was inoculated to 9 liters of a seed medium comprising 2% corn meal, 1% wheat embryo and 0.5% ferma media, adjusted to pH 7.0, and the incubation was made at 28+°C for 24 hours in a jar-fermenter with agitating and aeration to prepare a seed culture. On the other hand, 300 liters of a production medium consisting of 3% corn starch, 1% skimmed milk, 0.2% potassium dihydrogen phosphate, 0.05% magnesium sulfate, 0.01% manganese sulfate and a small proportion of defoaming agent were charged in a 600 liter capacity fermenter, sterilized at 121°C for 30 minutes and cooled.

The seed culture was then inoculated to the sterilized production medium, and production medium was incubated at 28°C for 85 hours with aeration and agitation. The resulting culture broth was filtered to give the culture filtrate which was concentrated below 40°C under reduced pressure to liquid volume of one-fifth times the original volume. To the concentrate was then added a 2-fold volume of cold ethanol, precipitating the amylase which had been produced and accumulated in the culture broth. Drying of the precipitate gave 341 grams of a crude enzyme which showed a potency of 43,800 units/gram.

Example 2: 100 ml portions of medium having the same composition used in Example 1 were placed in shaking flasks of a 500 ml capacity and sterilized. *Streptomyces hygroscopicus* (ATCC 21722) was inoculated and shake-cultured at 28°C for 24 hours to prepare a seed culture. On the other hand, 20 liters of a production medium containing 12% soluble starch, 3% defatted soybean meal and 0.2% potassium dihydrogen phosphate adjusted to pH 7.0 was charged in a jar-fermenter of a 30 liter capacity, sterilized and cooled. To this sterilized production medium was inoculated the seed culture, and the cultivation was made at 35°C for 90 hours. The culture broth obtained was then treated in the same manner as in Example 1 to give 77 grams of crude amylase which showed a potency of 45,000 units/gram.

Example 3: To an aqueous dispersion of 9 kg of potato starch in 12 liters of tap water which was adjusted to an initial pH 6.0 by addition of sodium hydroxide was added 0.1% of a known bacterial liquefying amylase. The resulting mixture was added dropwise into 18 liters of hot water while the temperature was maintained at 70° to 73°C, so that the liquefaction of the starch was completed. The liquefying amylase was inactivated by heating the solution at 100°C for 5 minutes after the completion of the liquefaction phase. When the solution had then cooled to a liquid temperature of 55°C, the crude amylase of the *Streptomyces* as prepared in Example 1 was added in an amount of 400 units/gram of the starch. The mixture was then adjusted to an initial pH 6.0 and held at 55°C for 40 hours to complete the saccharifying reaction. After the completion of the saccharification, the reaction mixture was filtered, routinely decolorized with active carbon and then desalted by means of a known desalting ion exchange resin. The mixture was concentrated by evaporation of water in vacuo to give 10 kg of a clear and colorless syrup of starch hydrolyzate comprising 3.6% glucose, 72.5% maltose, 10.3% maltotriose and 13.6% dextrin (the water content 25% by weight) which was useful as a liquid natural sweetening material mainly composed of maltose.

Maximum Pullulanase Elaboration

Conventional amylases are not completely effective in hydrolyzing 1,6 starch linkages and frequently form undesirable starch hydrolyzate by-products. By using α-1,6-glucosidic hydrolyzing enzymes (e.g., pullulanase) in conjunction with other amylases, the conversion syrup industry would be able to produce a broader spectrum of starch hydrolyzates and conversion syrup products. If pullulanase were available at an economically feasible cost, starch conversion syrups of a superior quality could be produced at a significantly lower price.

In general, pullulanase is produced by inoculating and fermenting a culture media containing assimilable carbon, nitrogen and mineral nutrients with *Aerobacter aerogenes* under conditions conducive to its growth. In some processes, pullulanase production by the *Aerobacter aerogenes* is permitted to occur simultaneously with cell propagation. In other processes, pullulanase elaboration does not occur until late in growth phase. In some processes, pullulanase production is intentionally repressed by using a culture media deficient in saccharide materials which are deemed essential for pullulanase synthesis. Upon achieving sufficient *Aerobacter aerogenes* growth, the cells are then induced to produce pullulanase with a carbohydrate such as maltose, maltotriose or pullulan.

Pullulanase production and its applications in the manufacture of corn syrup products are described by *A.A. Bulich; U.S. Patent 3,963,575; June 15, 1976; assigned to A.E. Staley Manufacturing Company*. The commercial merits of using pullulanase in the production of sweeteners from corn are dependent on whether microbically produced pullulanase can be easily and economically recovered from a production culture media. This process describes the critical conditions for high yields and efficient recoveries of this enzyme.

Due to the high proportion of pullulanase elaboration in extracellular and superficially bound form, the processing ease and cost encountered in recovering high pullulanase yields in a commercially usable form are a particular advantage of the process. The amount of intracellular or internally bound pullulanase (which normally requires cellular disintegration for recovery) can be minimized to less than about 5% of the total pullulanase production. During the enzyme production stage, a high proportion of pullulanase is elaborated by the mutant into the aqueous culture media phase as a water-soluble enzyme.

When the mutants are incubated with the correct aqueous nutrient medium, they are further characterized as being free from mucilaginous polysaccharide materials. Excessive, cell-produced, mucilaginous polysaccharide will adversely affect the yields of recoverable pullulanase. In the production and recovery stages, the absence of this mucilaginous material facilitates the transfer and dispersion of pullulanase into the aqueous phase of the production culture media. Upon completion of the pullulanase production stage, the superficially bound pullulanase is not imbibed or occluded by cell-produced mucilaginous material. By the addition of detergents, the superficially bound cell-bound pullulanase can be easily converted to a soluble form.

The organism used possesses the capacity of yielding (on a commercial production basis) greater than 350 units of pullulanase for each milliliter of production broth. A particular processing advantage indigenous to these organisms is the characteristics of the pullulanase which the organisms produce. The organism's capacity to produce pullulanase substantially in the extracellular and superficially bound form makes it easy to recover at least 80% of the total pullulanase in the final production broth.

In a production culture media consisting essentially of amylopectin as a carbohydrate source, pullulanase yields of superficially bound and extracellular pullulanase of at least 500 units/milliliter can easily be prepared. The process is conducted with a nutrient culture media under incubation conditions such that the pullulanase yields are greater than 750 units/milliliter of culture media and more than 85% (e.g., between 85 to 95% or more) of the total pullulanase produced in the extracellular and superficially bound form.

Pullulanase elaboration by mutants is also significantly increased when the production culture medium contains a small amount of an assimilable iron. Trace amounts of iron, in culture mediums to sustain cell propagation, are generally insufficient for effective elaboration of pullulanase by the mutants. By supplementing the production culture medium with a small amount of excess assimilable iron (i.e., an amount more than required for cell propagation), pullulanase production is significantly enhanced.

Example 1: A production culture yielding 790 units of pullulanase for each milliliter of culture media broth was prepared from *Klebsiella pneumoniae* NRRL-B-5780 as a pullulanase-producing organism. The inoculum nutrient medium consisting of (on weight basis) 2.0% waxy maize starch, 2.0% BactoPeptone, 0.5% ammonium acetate, 0.05% sodium citrate ($Na_3C_6H_5O_7 \cdot 2H_2O$), 0.1% K_2HPO_4, 0.1% KH_2PO_4, 0.05% $MgSO_4 \cdot 7H_2O$, 0.05% KCl, 0.001% $FeSO_4 \cdot 6H_2O$ and water was prepared.

The resultant mixture was heated on a steam bath under agitation to provide a homogeneous culture media with the waxy maize starch being thoroughly pasted. To each of five 500 milliliter bottom baffled DeLong flasks, 100 milliliters of the inoculum media and one drop of an antifoam agent was added. The inoculum media was then sterilized at 121°C for 15 minutes.

The flasks containing the nutrient media were then inoculated with *Klebsiella pneumoniae* NRRL-B-5780 (agar slants). The flasks were then incubated for 16 hours at 28°C at 265 rpm on a rotary shaker and the resultant inoculum was utilized as seed inoculum for the production culture media. A pullulanase production nutrient media was prepared by dissolving 60 parts by weight of sodium citrate ($Na_3C_6H_5O_7 \cdot 2H_2O$) in 1,000 parts (1 liter) by weight tap water (20°C). 350 parts by weight of corn steep liquor (dsb) was slowly added to the sodium citrate solution (with stirring) to provide a sodium citrate-corn steep liquor solution. The resultant solution was adjusted to pH 6.3 by the addition of 50% solution of potassium hydroxide. A pasty waxy maize preparation was then separately prepared by slowly adding, with agitation, 350 parts by weight of the low viscosity waxy maize starch to 5,000 parts by weight (5 liters) water and heating to 96°C. After thoroughly pasting the waxy maize starch, 50 parts by weight potassium chloride and 70 parts by weight ammonium sulfate were added.

The sodium citrate-corn steep liquor solution and the above aqueous starch paste media (at 96°C) were then mixed together with additional hot tap water to provide a 10-liter aqueous production culture media. This 10-liter aqueous production media was then placed in a fermentor and autoclaved at 121°C for 45 minutes. A sterile (autoclaved at 121°C for 15 minutes) 0.036 molar acidified, ferrous sulfate solution dissolved in 0.08 molar hydrochloric acid solution was prepared to provide supplemental ferrous ion. Immediately prior to inoculating the fermentation media with the seed inoculum, 20 ml of the ferrous sulfate solution was dissolved in the fermentor to provide a production nutrient media containing 4.0 mg of ferrous ion (20 mg of ferrous sulfate) per liter of aqueous production media.

The above aqueous production media was then inoculated with the seed inoculum (initial pH 6.3). The inoculated aqueous production media was incubated at 28°C, a stirrer speed of 550 rpm and aeration rate of 6 vvm (volume air per volume fermentation broth per minute) for 23 hours. During the initial 9 hours of fermentation, the pH gradually increased to 7.75. The pH was maintained throughout the production cycle at 7.75 by means of an automatic sensing and metering device, whereby it was neutralized with 14 molar aqueous acetic acid solution (80% acetic acid). After fermenting for 23 hours, the optimum pullulanase yields (790 units of pullulanase per milliliter of fermentation broth) had been achieved.

Example 2: Enhanced pullulanase yields are achieved when the production culture media contains supplemental ferrous ions. Except for different levels of ferrous ion, corn steep liquor and incubation time, pullulanase production under this example was the same as Example 1. The results are tabulated in the table below:

Run No.	Corn Steep Liquor (%, dsb)	MgFe^{++} (mg/l)	Yield (units/ml culture media)	Time (hr)
1	4.0	3.0	744	28.0
2	3.5	4.0	790	26.0
3	3.0	4.4	730	27.0
4	3.0	0.0	656	28.0
5	3.5	4.4	780	27.5
6	3.5	6.4	775	27.5

As evident by comparing Runs 3 and 4 above, use of supplemental ferrous ions (corn steep contained 0.7 mg of ferrous ion) per liter, provided more than a 10% increase in pullulanase yields. On a comparative basis, the 656 units/ml yield for Run 4 still represents a superior yield improvement over those obtained with *Aerobacter aerogenes*. As illustrated by a comparison of the yields in Runs 2, 3, 5 and 6, an increase in ferrous ions above 4.0 mg/l provided higher yields than those achieved without supplemental ferrous ions, but did not enhance yields above that of Run 2.

Malt Diastase and Pullulanase Combination

R.E. Heady and F.C. Armbruster; U.S. Patent 3,565,765; February 23, 1971; assigned to CPC International Inc. illustrate how high maltose syrups can be produced by maltose-producing enzymes in combination with a pullulanase preparation. Maltogenic enzymes are widespread in nature. Good sources for these enzymes include malted grains such as barley, sorghum and wheat. Other sources include microorganisms, e.g., maltogenic enzymes may be obtained by submerged culture fermentation of the strains of the bacterium *Bacillus polymyxa*. The enzyme pullulanase, also designated dextrin α-1,6-glucosidase, is elaborated during the submerged culture fermentation by strains of the bacterium *Aerobacter aerogenes*.

The partially hydrolyzed starch that is used as the starting material is obtained by acid hydrolysis or enzyme hydrolysis of any conventional starch. Suitable starches include cereal starches such as corn, grain, sorghum and wheat; waxy starches such as waxy milo and waxy maize; and root starches such as potato starch and tapioca starch. Crude starch sources may also be used, such as ground cereals, macerated tubers or the partially purified starches therefrom.

Prior to hydrolysis, the starch is solubilized by gelatinization. Gelatinization is carried out by heating the starch to a temperature exceeding about 60°C in the presence of moisture. Acid hydrolysis of starch is carried out in a conventional manner to a DE between about 10 and 20. Enzyme hydrolysis of starch is carried out using suitable liquefying enzymes to attain a DE between about 5 and 20. The debranching enzyme, pullulanase, used in the process is produced by members of the bacterial species *Aerobacter aerogenes* when suitably incubated under conditions of aerobic culture.

Example 1: Production of High Maltose Conversion Products – This example illustrates the simultaneous use of a maltogenic enzyme and pullulanase in the preparation of high maltose starch conversion products from partially enzyme hydrolyzed starch. This example also illustrates varied dosages of pullulanase and malt diastase in combinations to produce the high maltose, high fermentables conversion products.

A 30% by weight suspension of corn starch was enzymatically liquefied at 91°C, pH 5.5, using a bacterial α-amylase preparation at a dosage of 0.05% on a dry basis. The enzyme liquefying action was stopped by heating the liquefied starch to 121°C for 15 minutes when the partially hydrolyzed starch reached a DE between 2 and 5. The partially hydrolyzed starch was cooled to 55°C, adjusted to pH 5.8 and then dosed with the respective levels of pullulanase and malt diastase to yield syrups having the maltose contents indicated below:

Dosage, u/100 g ds							
Pullulanase	Malt Diastase	DE	Dextrose (%, db)	Maltose (%, db)	Maltotriose (%, db)	DP_4 and Higher (%, db)	Fermentables (%, db)
48 Hours							
0	50	47.6	3.3	64.5	11.6	20.6	79.4
0	100	51.1	4.0	67.0	12.5	16.5	83.5

(continued)

Dosage, u/100 g ds Pullulanase	Malt Diastase	DE	Dextrose	Maltose	Malto-triose	DP$_4$ and Higher	Ferment-ables
			(%, db)				
48 Hours							
0	200	51.9	3.8	71.1	11.7	13.5	86.5
0	25	43.5	1.9	61.6	9.8	26.7	73.3
50	25	51.5	4.1	72.0	14.4	9.5	90.1
100	25	52.9	1.5	75.4	15.9	7.2	92.6
200	25	52.8	1.2	76.0	15.9	6.9	93.1

Example 2: Production of High Maltose Conversion Products from Partially Enzyme Hydrolyzed Waxy Milo Starch – This example illustrates the enhancing effect of pullulanase when used with malt diastase in the conversion of waxy starches to high maltose, high fermentable conversion products. Waxy milo starch was enzymatically liquefied to 5 DE as outlined in Example 1, cooled to 50°C, adjusted to pH 5.7 and dosed with pullulanase and malt diastase enzymes simultaneously as shown in the table below:

Dosage, u/100 g ds Pullulanase	Malt Diastase	DE	Dextrose	Maltose	Malto-triose	DP$_4$ and Higher	Ferment-ables
			(%, db)				
46 Hours							
0	20	38.3	2.75	54.5	8.35	35.4	64.6
100	20	47.6	3.4	73.0	16.9	6.4	93.6
200	20	48.9	1.3	73.5	16.3	7.6	92.4
0	80	44.7	4.1	60.5	10.3	25.1	74.9
100	80	51.4	3.2	73.0	16.7	7.1	92.9
200	80	51.9	3.1	75.5	19.3	2.1	97.9

Stabilized β-Amylase

β-amylase is present in various plants such as barley, wheat, malt, sweet potato and soybean as well as *Bacilli* and is obtained by extraction with water or a buffer solution. Since the extraction usually contains a large amount of soluble organic materials such as saccharides and protein decomposition products in addition to β-amylase, it is apt to be putrefied. When the treatment is carried out at a relatively low temperature from about 20° to 30°C, the putrefaction and inactivation can be avoided. But, in such case, the efficiency of the concentration and purification of β-amylase is considerably decreased.

O. Shibata and H. Taniguchi; U.S. Patent 4,024,000; May 17, 1977; assigned to Fuji Oil Company Ltd., Japan found that β-amylase in an aqueous medium can be stabilized even at a high temperature (e.g., about 40° to 55°C) when a divalent or trivalent metal ion is present and the pH is properly adjusted. It has also been found that the presence of the metal ion at the required pH is effective in prevention of the putrefaction of an aqueous extract containing β-amylase.

In order to obtain a β-amylase product of high purity, it is preferred to eliminate previously materials of low solubility in the stable region of β-amylase. Typical examples of such materials are proteins having an isoelectric point at pH 4 to 6, which may be eliminated by conventional procedures such as precipitation with acids, salting out at low saturation degrees and addition of alkaline earth metal salts. To the aqueous β-amylase solution obtained, a divalent or trivalent metal compound which can liberate a divalent or trivalent metal ion in an aqueous medium is added. Metal compound examples are hydroxides of calcium, magnesium, barium, aluminum, etc. Salts of these metals with acids such as hydrochloric acid, phosphoric acid, lactic acid, citric acid and tartaric acid are also utilizable. The metal compound is incorporated in an amount to make a concentration of not less than about 0.1% by weight to not more than 1% by weight in the aqueous β-amylase solution.

The pH of the resulting aqueous β-amylase solution is adjusted to a pH of about 5.0 to 7.0. Within this range, a lower pH is more effective in prevention of putrefaction. The resultant aqueous β-amylase solution incorporated with the metal compound and adjusted to the correct pH is quite stable to heat and can be subjected to chemical and/or physical treatment at a relatively high temperature (e.g., about 45° to 55°C) for a comparatively long period of time without any substantial inactivation and putrefaction.

Among various treatments to which the stabilized aqueous β-amylase solution may be subjected, the most advantageous one is membrane separation. The term membrane separation as used is intended to mean a conventional procedure for separation of water and low molecular weight materials from high molecular weight materials by the utilization of the selective permeability of a membrane so as to enhance the concentration of the high molecular materials with elevation of their purity.

In addition to the stabilization and putrefaction preventing effects as explained above, the presence of the divalent or trivalent metal ion is effective in improvement of the permeation efficiency and enhancing the effect in salting out. The concentration and purification of conventional aqueous β-amylase solutions by membrane separation have been carried out at a relatively low temperature (e.g., below about 25°C) in order to avoid the inactivation and putrefaction. Since the permeation ability of a membrane is generally increased with a higher temperature, low temperature results in the decrease of the efficiency of membrane separation.

For stabilized aqueous β-amylase solution, membrane separation is applicable at a high temperature (e.g., 45° to 55°C) and thus the purpose of the concentration and purification of β-amylase can be achieved efficiently. The concentrated and purified β-amylase solution obtained by the membrane separation may be, if necessary, salted out by addition of an appropriate salt such as ammonium sulfate or sodium chloride to precipitate a fraction containing β-amylase. Collection of the precipitate and drying affords a β-amylase product of high purity.

Example 1: Sweet potato juice after separation of starch was adjusted with hydrochloric acid to pH 5.2, and coagulated proteins and other impurities were removed to give a solution having a β-amylase activity of 300 units/ml. The solution (100 liters) was adjusted to pH 5.5, and a solution of aluminum chloride (200 g) in water (200 g) was added. The resulting mixture of pH 6.8 was concentrated at 50°C by ultrafiltration until the total volume became 10 liters. The β-amylase activity of the concentrated solution was 2,700 units/ml.

Example 2: Crushed soybean was extracted with an acetate buffer of pH 5.4, and insoluble materials were removed to give a solution having a β-amylase activity of 160 units/ml. To the solution (80 liters), there was added calcium hydroxide (200 g), and the pH was adjusted to 6.6. The resulting mixture was concentrated at 50°C by reverse osmosis using a membrane AS-197 until the total volume became 8 liters. The time required for concentration was 5 hours. No putrefaction was observed. The β-amylase activity of the concentrated solution was 1,460 units/ml.

Glucoamylase Preparation

Production of glucoamylase for converting starch to dextrose is explained in the process of *R.G. Dworschack and C.A. Nelson; U.S. Patent 3,660,236; May 2, 1972; assigned to Standard Brands Incorporated.* Glucoamylase is elaborated by many types of microorganisms. Certain strains of fungi belonging to the *Aspergillus* group such as the *Aspergillus niger* group and the *Aspergillus awamori* group, certain strains of the *Rhizopus* species and certain strains of the *Endomyces* species are suitable.

Of particular importance in the use of glucoamylase preparations are the presence and amounts of transglucosidase and α-amylase. Transglucosidase catalyzes the formation, particularly from maltose, of unfermentable dextrose polymers, such as isomaltose and oligosaccharides which contain α-D(1→6) glucosidic linkages. This is, of course, detrimental to the production of dextrose since lower yields of dextrose are obtained. α-Amylase acts on starch to catalyze the formation of saccharides of lower molecular weight such as maltose which may be relatively easily broken down to dextrose by the glucoamylase.

The presence of α-amylase is generally not considered detrimental to the production of dextrose and may actually hasten the formation of dextrose; however, it does produce saccharides of the kind which are susceptible to being polymerized to unfermentable dextrose polymers by transglucosidase when relatively large quantities of α-amylase are present.

In commercial practices, it is recommended to produce glucoamylase in a series of steps. Reasons for growing glucoamylase-producing microorganisms in stages are manifold, but are primarily dependent upon the conditions necessary for the growth of the microorganisms and/or the production of glucoamylase. These conditions include stability of the microorganisms, proper nutrients, pH, osmotic relationships, degree of aeration, temperature and the maintenance of pure culture conditions during fermentation. For instance, to obtain maximum yields of glucoamylase, the conditions of fermentation in the final stage may have to be changed somewhat from those practiced to obtain optimum growth of the microorganism in the culture development stages.

Maintaining the purity of the medium alsc is an extremely important consideration, especially where the fermentation is performed under aerobic conditions. After the fermentation is completed, the cellular material may be removed from the enzyme by filtration, centrifugation, etc. The glucoamylase preparation may be used directly to saccharify starch, or concentrated and then used.

This process is based upon the discovery that ammonia gas or ammonium hydroxide may be incorporated into a fermentation medium during the production of glucoamylase in order to maintain the pH in the desirable range. This discovery provides a number of important advantages. For example, it has been found that by using ammonium hydroxide and/or ammonia gas, it is possible to eliminate from a standard growth medium NaOH, the nitrogen salts and part of the corn steep liquor. This results in a glucoamylase preparation with lesser quantities of soluble solids present. Consequently, these glucoamylase preparations, when used to convert starch to dextrose, contribute very little to the ash content of the hydrolysate. This provides the advantages that in refining the starch hydrolysate the ion exchange resins used for lowering the ash content may be used for longer periods of time; the life of the ion exchange resins increased and the number of regenerations decreased.

Furthermore, recovery of the glucoamylase from the nutrient may be achieved very easily and, in some cases, has been about 70% higher than achieved by conventional processes. Moreover, the glucoamylase preparations are generally higher in glucoamylase activity and lower in transglucosidase activity. This is, of course, extremely important when the preparation is used for converting starch into dextrose since higher yields of dextrose are obtained.

Example 1: This example illustrates the use of ammonia gas for the production of glucoamylase and compares this with older methods in which NaOH is used as the neutralizing agent. A baffled 2,000-gallon fermentor having a sterile air source and impeller agitation means was charged with a presterilized medium containing 800 lb corn starch, 15 lb $(NH_4)_2HPO_4$, 25 gallons corn steep liquor and sufficient water to achieve a volume of about 1,000 gallons. Inoculum of *Aspergillus awamori* NRRL 3112 was inoculated into this fermentor. The temperature of the medium was maintained at 100°F and the aeration rate was 0.5 volume of air/volume of medium/minute. The fermentation was allowed to proceed for about 24 hours.

Inoculum from this stage was inoculated into a 6,000-gallon fermentor having a sterile air source and impeller agitation means. The fermentor was charged with a presterilized medium containing 15% corn starch, 2% corn steep liquor and 1% $(NH_4)_2HPO_4$. The temperature of the medium was maintained at 95°F and aerated at an aeration rate of 0.8 volume of air/volume of medium/minute during the fermentation. The fermentation was allowed to proceed for about 63 hours. During the fermentation, the pH of the fermentation medium was maintained in the range of 3.8 to 4.2 by the addition of a 30% NaOH solution. Four fermentations were carried out in this manner and the glucoamylase activities determined. These activities are shown below.

Another series of fermentations were carried out in exactly the manner described above but no $(NH_4)_2HPO_4$ was used and the pH of the medium was maintained in the range of 5.5±0.2 by the continuous addition of ammonia gas. Three fermentations were carried out in this manner and the glucoamylase activities determined. These activities are shown in the table below:

Constituents of Medium	pH	Neutralizing Agent	Glucoamylase Activity
Corn starch, 15% Corn steep liquor, 2% $(NH_4)_2HPO_4$, 1%	3.8-4.2	NaOH	162
Corn starch, 15% Corn steep liquor, 2% $(NH_4)_2HPO_4$, 1%	3.8-4.2	NaOH	152

(continued)

Constituents of Medium	pH	Neutralizing Agent	Glucoamylase Activity
Corn starch, 15% Corn steep liquor, 2% $(NH_4)_2HPO_4$, 1%	3.8-4.2	NaOH	148
Corn starch, 15% Corn steep liquor, 2% $(NH_4)_2HPO_4$, 1%	3.8-4.2	NaOH	163
Corn starch, 15% Corn steep liquor, 2%	5.5±0.2	NH_3 gas	232
Corn starch, 15% Corn steep liquor, 2%	5.5±0.2	NH_3 gas	270
Corn starch, 15% Corn steep liquor, 2%	5.5±0.2	NH_3 gas	278

From the above table, it is readily apparent that the use of NH_3 gas greatly enhances the production of glucoamylase.

Example 2: This example illustrates the use of glucoamylase preparations to convert starch into dextrose. A suspension of 35% corn starch by weight in water was enzyme liquefied to a dextrose equivalent (DE) of about 15. To the liquefied starch was added an amount of glucoamylase preparation to achieve a level of 90 units of glucoamylase activity per lb of starch dry substance. After about 60 hours at 60°C, the hydrolysate was analyzed for DE. Two series of ten hydrolysates each were prepared in this manner. The glycoamylase preparation used for the first series was prepared by the use of ammonia gas. The other series was prepared with glucoamylase preparations prepared using NaOH as the neutralizing agent. The DE of the hydrolysates are shown below:

Fermentation Medium*	pH	Neutralizing Agent	DE of Hydrolysates**
Corn starch, 15% Corn steep liquor, 2% $(NH_4)_2HPO_4$, 1%	3.8-4.2	NaOH	94.3
Corn starch, 15% Corn steep liquor, 2%	5.5±0.2	NH_3	95.7

*Used to prepare the glucoamylase preparation.
**Average of 10 experiments.

From the above table it is seen that when glucoamylase prepared with NH_3 gas was used to convert starch to dextrose, greater yields of dextrose were obtained than glucoamylase preparations prepared with NaOH. This indicates that the glucoamylase preparations contain lesser quantities of transglucosidase.

CELLULOSE CONVERSIONS

Cellulose-Cellulase Suspensions

Cellulose, a principal constituent of all plant life, is the most abundant polymer on the earth having an estimated total annual production in excess of 10^{11} tons.

Aside from this natural worldwide abundance, cellulose is the major component of much waste material, e.g., agricultural waste in the form of straw, leaves, stalks, hulls, shells, bagasse, etc.; wood waste such as sawdust, chips, bark, etc. and urban waste principally in the form of paper refuse. A nonsoluble polymer, cellulose is made up of repeating units of glucose, a monosaccharide that can be directly assimilated by most organisms and is the chief source of energy for plants and animals. The utilization of cellulose in the production of glucose or other useful by-products is hampered by the fact that the cellulose polymer is extremely resistant to degradation, although some microorganisms have the ability to enzymatically hydrolyze cellulose and some animals, e.g., ruminants, can convert cellulose to animal protein.

M.H. Mandels and J.A. Kostick; U.S. Patent 3,764,475; October 9, 1973; assigned to U.S. Secretary of the Army used cellulose as support and absorbent for cellulase for enzymatic hydrolysis of cellulose to glucose. A strain for the production of cellulase is *Trichoderma viride* mutant QM 9123. This mutant strain has a two-fold increase in cellulase activity over the parent strain. The organism is cultured in a mineral medium containing 1% cellulose in shake flasks or submerged cultures. After 10 to 14 days culture at 28° to 29°C, the culture broth is filtered off to remove the residual solids yielding a culture filtrate containing 0.5 to 0.7 mg of protein per ml. The pH of the culture filtrate is adjusted to 4.8 to 5.0 with a 0.5 molar citrate buffer and Merthiolate at 0.005% is added to prevent microbial growth. This culture filtrate is used directly as the crude enzyme without further processing.

The activity of the cellulase enzyme in the culture filtrate can be expressed or measured by the carboxymethylcellulose hydrolysis method (Cx units/ml) or the Filter Paper Assay Procedures (FP activity) ("Advances in Chemistry," ACS, 95, 391-414). The activity of freshly prepared culture filtrates of the mutant strain ranges from about a FP activity of 2.0 to 4.0 which is a two-fold increase in activity over that produced by the parent (Tv QM 6a) organism.

If a more highly active enzyme solution is required to bring about more rapid hydrolysis of cellulose, the culture filtrate may be concentrated by standard ultrafiltration techniques. Membranes having molecular size cut-off capabilities of from 20,000 to 30,000 have proven effective in concentrating the cellulase enzymes. A five-fold volume concentration brought about by ultrafiltration results in a corresponding three- to four-fold increase in enzyme activity as measured by the Assay Procedures. For purposes of this process, the cellulase enzyme should have a FP activity of at least 2.0 and preferably in excess of 4.0.

Cellulose, the raw material or substrate, is a straight chain polymer of glucose units and if hydrolyzed, is converted to glucose and to other polysaccharides such as cellobiose, cellotriose, cellotetraose and cellopentaose. It is desirable that the cellulose be fairly pure to reduce unwanted soluble impurities in the process. It is important that the resistant nature of native cellulose due to the strong H-bonding across the cellulose chains of crystalline microfibrils be overcome by rupturing the cell fibers thereby reducing the particle size and increasing the reactive surface.

In addition to the reduction in particle size of the cellulose substrate, the ratio of substrate to enzyme has a significant effect on the reaction rate. Dense suspensions of finely ground cellulose where the solids contents of the substrate in the cellulase suspension was 10 to 30% were found to be highly reactive in a system in which the suspension is vigorously agitated in the presence of a highly concentrated (1 to 2 mg protein/ml) enzyme culture filtrate solution. It is desirable, however, in such a system to remove the sugar products being formed from the suspension to prevent product inhibition.

Removal of soluble sugar products from the cellulose-cellulase-sugar system appeared as a significant obstacle to the development of a practical process for the hydrolysis of cellulose. Due to the water solubility of the enzyme material, simple mechanical filtration would result in the loss of or withdrawal of enzyme along with the sugar solution. Separation of the sugar component without removal of enzyme can be effected by means of pressure filtration through a molecular sieve membrane (ultrafiltration membrane). Ultrafiltration is a necessarily slow and relatively costly procedure, but one that was considered to be necessary if the valuable enzyme component was to be retained and reused. It was discovered, however, that cellulose strongly adsorbs cellulase enzyme at a pH of 4.0 to 5.0 and at temperatures within the range of 25° to 50°C, conditions which are also optimum for enzyme hydrolysis. The adsorbed enzyme is released as the cellulose is hydrolyzed and is readsorbed to be reused on fresh or excess cellulose.

A 10% concentration of cellulose is adequate to adsorb 90% of the enzyme component in a concentrated culture filtrate (1.0 mg protein/ml). As hydrolysis proceeds, it is, of course, essential to replenish the system to maintain the desired cellulose concentration and to likewise replenish the water lost with the sugar component with fresh buffered solution. Dilute enzyme (0.1 mg protein/ml) can be added to the replenishing buffer to make up for losses due to incomplete adsorption and/or denaturation or inactivation of enzyme. With the effective immobilization of the cellulase enzyme on the unhydrolyzed cellulose substrate, the sugar solution can be separated from the enzyme-cellulose complex by any simple separation or filtration technique, such as mechanical filtering, decanting, or elution from a packed column but not including techniques such as ultrafiltration.

Example 1: 300 g of milled cellulose was suspended in 3 liters of cellulase (0.7 mg protein/ml) solution in a 5-liter vessel immersed in a water bath at 50°C. The mixture was continuously stirred with a motor driven stainless steel impeller blade at 100 rpm. Buffered water was fed continuously into the mixture at a rate to maintain a 3-liter volume in the reaction vessel. The sugar solution formed by the enzymatic hydrolysis was removed through the top of an L-shaped arm originating at the base of the reaction vessel by means of a pump. Almost all of the cellulose in the unstirred side arm settled back into the reactor, but a coarse sintered glass filter is set in the side arm to exclude any particles from the harvest.

The cellulose used was a purified spruce pulp, Solka-Floc SW 40A, which was milled dry for 24 to 70 hours in a porcelain pot mill. The average particle size of the milled cellulose was 50 microns. The enzyme cellulase was obtained from the growth of the fungus *Trichoderma viride* QM 9123 in a 0.75% cellulose medium for 13 days. The culture was filtered through glass wool and the culture filtrate which contained 0.5 to 0.7 mg protein/ml was adjusted to a pH of 4.8 with 0.05 molar citrate buffer. Milled dry cellulose was fed continuously from time 0 at a rate of from 2.3 g/hr to maintain the cellulose concentration at 10%. At 23 hours, when glucose in the reactor approached 5%, a feed of buffered water was begun and glucose syrup removed at an equal rate from the side arm. The process was terminated after 164 hours. Glucose concentration in the reactor remained between 4.7 and 6.1% from 23 to 164 hours despite removal of 1.4 volumes of effluent with no replenishment of enzyme. Glucose production from 0 to 164 hours equaled 0.8 g/l/hr.

Ethylene Treatment

The following preliminary steps are used by *A.H. Freytag and J.C. Linden; U.S. Patent 4,097,333; June 27, 1978; assigned to The Great Western Sugar Company* to hydrolyze a wide variety of cellulose containing materials to glucose and other by-products: (1) milling of cellulosic material to about 100 to 150 mesh; and (2) if the material contains a significant amount of lignin, subjecting the material to pretreatment with sulfuric acid or sodium hydroxide. A hydrolysis mixture is formulated comprising the cellulosic material, cellulase and water. If cellulase is produced in the mixture by cellulase-producing microorganisms, then the mixture will further comprise a suitable conventional amount of nutrient materials. For best results, a sufficient amount of water is added to the mixture to effectively disperse the cellulosic material throughout the mixture. Additionally, the mixture is treated with a sufficient amount of ethylene to enhance the hydrolysis reaction either prior to hydrolysis, as a pretreatment, and/or during the hydrolysis reaction.

The hydrolysis reaction is then carried out in a conventional manner. For optimum results, the temperature of the hydrolysis mixture is maintained at about 40° to 55°C, and the pH of the hydrolysis mixture is maintained, with the use of a suitable buffer, at or near an optimum level for the particular cellulase used, e.g., at about 4.0 to 5.5 during the hydrolysis reaction.

Example 1: Two aqueous dispersions of cellulosic materials are prepared by grinding for each 0.5 g of Whatman No. 114 filter paper in a Wiley Mill and adding the ground paper to 50 ml of distilled water containing a sufficient amount of NaOH to make the dispersions 2 N in NaOH. The dispersions are then boiled for 1 hour. After cooling to 50°C, 5 ml of 0.1 M acetate buffer (pH = 4.5) and 5 mg of Rhizopus Type III Crude Cellulase are added to each dispersion and the mixtures are incubated at 50°C on a shaking water bath.

During incubation, ethylene gas is sparged through one of the mixtures at the rate of 1 cc/min. On the first, second and third days following initiation, 10 ml samples are taken from each mixture and the samples are filtered, frozen and analyzed for glucose by increase in reducing value. Higher levels of glucose in the supernatant are readily apparent in the ethylene-treated hydrolysis mixture with glucose production being increased over the conventional hydrolysis mixture by about 9% after 24 hours of incubation.

Example 2: The procedure of Example 1 is repeated except that the mixtures are placed into compartmentalized fermentation vessels having the mixtures in a first compartment of the vessel and an equal volume of 0.1 M acetate buffer (pH 4.5) in a second compartment of the vessel, the first compartment being separated from the second compartment by a 0.45 micron Metrical 25 mm filter in a Millipore filter holder to allow glucose to migrate through the filter into the second compartment. The effect of the presence of ethylene gas on hydrolysis with glucose removal by diffusion is shown on the following page.

Ethylene Present During Hydrolysis	Time from Initiation (hr)	Reducing Sugar in Supernatant First Compartment	Second Compartment	Total
		(mg/ml)		
No	24	24.5	3.4	27.9
No	48	17.7	8.2	25.9
No	72	15.7	10.2	25.9
Yes	24	20.0	5.2	25.2
Yes	48	16.2	11.2	27.4
Yes	72	15.0	13.0	28.0

As shown above, the production of reducing sugar is significantly enhanced by sparging ethylene gas through the mixture during hydrolysis.

CYCLODEXTRIN PRODUCTION

A special cyclodextrin glycosyltransferase (CGT) has been isolated by *M. Shiosaka; U.S. Patent 3,988,206; October 26, 1976; assigned to Hayashibara Biochemical Laboratories, Incorporated, Japan* to give higher yields of cyclodextrin from starch and higher transfer ratios of the glucose moiety in starch to the glucose moiety in sucrose. The new CGT is produced by *Bacillus stearothermophilus* and gives better heat stability and higher yields than previous CGTs.

Bacillus macerans CGT was reported to act on starch solution to form at first α-cyclodextrin consisting of six glucose residues and next β-cyclodextrin consisting of seven glucose residues. If the solution contains a glucose acceptor such as sucrose, the CGT transfers glucose from starch to the acceptor.

The enzymes of *Bacillus stearothermophilus* and *Bacillus macerans* differ in heat stability. When both enzymes were allowed to stand at various temperatures for 15 minutes, the residual activity of *Bacillus macerans* CGT was 100% at 50°C and 30% at 60°C, whereas that of *Bacillus stearothermophilus* CGT remained 100% even at a temperature of 65°C. Since a 15°C increase in temperature increases the reaction rate and protects the reaction mixture from contamination by microorganisms, *Bacillus stearothermophilus* CGT is more advantageous in industrial practice.

When the ability of transferring glucose to sucrose (referred to as transfer activity) is expressed as the ratio of hydrolysis activity (dextrinogenic activity) in a sucrose-bearing starch solution to that in a solution containing starch only, the transfer activity of *Bacillus macerans* CGT is 0.58, while that of *Bacillus stearothermophilus* CGT is about 1.62 to 1.84.

The effects of *Bacillus megaterium* and *Bacillus stearothermophilus* CGTs on solutions containing sucrose and starch were compared, and the observed enzyme activities were expressed in dextrinogenic activities necessary to give a 50% or higher ratio of glucose-transferred sucrose to total sucrose. The test showed that *Bacillus stearothermophilus* CGT permits the same transfer ratio to be attained at one-seventh of the dextrinogenic activity required for *Bacillus megaterium* CGT. When *Bacillus macerans* CGT was added to a starch solution in an amount of 10 units of dextrinogenic activity per gram starch and the solution was allowed to stand at 55°C, cyclodextrin was obtained in a 43% yield from starch, whereas *Bacillus stearothermophilus* CGT gave a 56% yield when the amount was one-tenth of that of *Bacillus macerans* CGT.

Enzyme Purification: A culture broth of *Bacillus stearothermophilus* FERM-P No. 2218 was centrifuged to remove cells and $(NH_4)_2SO_4$ was added to the supernatant to 15% saturation. The solution was passed through a starch column. The adsorbed enzyme was eluted from the starch with water, $(NH_4)_2SO_4$ was added to the eluate, and the precipitates formed between saturations of 25 and 50% were recovered and subjected to gel fractionation to purify the enzyme. The results are shown in the table below:

Step	Activity (u/ml)	Volume (ml)	Total Activity (u)	Specific Activity (u/mg protein)	Yield (%)
Supernatant culture broth	10.4	1,180	1.23×10^4	1.27	100
Eluate from starch	40.4	300	1.21×10^4	5.6	99
$(NH_4)_2SO_4$ precipitate	1,020	10.7	1.09×10^4	52.3	89
Gel fractionation by Biogel 150	128	80	1.02×10^4	129	82.3

Assay of the purified CGT by disc gel electrophoresis and isoelectric focusing showed a single absorbance peak. The isoelectric point of the CGT was at pH 4.45.

Enzymatic activities were determined measuring dextrinogenic activity and cyclodextrin-producing activity and the transfer ratio of glucose to sucrose. As correlation was noted between the activities and the absence of α-amylase, the determination methods were found to be quantitatively accurate. Based on these findings, enzymatic activity was determined mainly by dextrinogenic activity measurement.

Transfer Activity to Sucrose: 1 ml of an enzyme solution was added to 3 ml of a solution containing 6.7 g/dl soluble starch, 6.7 g/dl sucrose, 2×10^{-3} mol $CaCl_2$ and 0.03 mol acetate buffer, pH 5.5. After incubation for 10 minutes at 40°C, 0.5 ml of the mixture was poured into 9 ml of 1 N HCl to terminate the reaction. A 0.05 ml sample was poured into 15 ml of 0.02 N H_2SO_4, and after admixing 0.2 ml of a 0.1 N I_2-KI solution, the optical density of the mixture was determined at 660 nm. One unit of activity was designated as the amount of enzyme that caused a 10% reduction in iodine stain of 200 mg soluble starch at 40°C within 10 minutes.

Dextrinogenic Activity: A mixture prepared by adding 0.2 ml of an enzyme solution to 5 ml of a solution containing 0.3 g/dl soluble starch, 0.02 mol acetate buffer, pH 5.5 and 10^{-3} mol $CaCl_2$ was allowed to react at 40°C for 10 minutes. 0.5 ml reaction mixture was collected, poured into 15 ml of 0.02 N H_2SO_4 to terminate the reaction and assayed for optical density at 660 nm after addition of 0.2 ml of 0.1 N I_2-KI solution. One unit of activity was designated as the amount of enzyme that caused disappearance of iodine stain in 15 mg soluble starch at 40°C within 10 minutes.

Example: Bacillus stearothermophilus FERM-P Nos. 2217, 2218, 2219, 2220, 2221, 2222, 2223, 2224 and 2225 were cultured on a medium containing (in g/dl):

Soluble starch	2
NH_4Cl	0.5
K_2HPO_4	0.05
$MgSO_4$	0.025
$CaCO_3$	0.5

50 ml portions of the medium were transferred to 500 ml Erlenmeyer flasks, sterilized at 120°C for 30 minutes and inoculated with respective strains. The resulting mixtures were subjected to shaking culture at 50°C on a rotary shaker for 50 hours. The resultant culture broths were analyzed for their dextrinogenic activity, cyclodextrin-producing or Tilden-Hunson activity and transfer activity to sucrose. The results are shown below:

Activity (u/ml)	*Bacillus stearothermophilus* FERM-P No.								
	2217	2218	2219	2220	2221	2222	2223	2224	2225
Dextrinogenic	0.5	2.7	0.4	0.9	0.9	6.8	3.5	3.0	17.0
Tilden-Hudson*	1.5	6.0	1.0	3.0	2.0	20.0	8.0	10.0	80.0
Transfer to sucrose	0.8	4.2	0.6	1.6	1.1	9.0	5.0	5.0	34.0

*Referred to above as "cyclodextrin-producing activity."

AMYLASE RESISTANT MODIFIED STARCH

It is a common practice to etherify starch with various etherifying agents to form starch ether derivatives which have improved properties, principally as sizes or coatings for paper and textiles. As far as is known, no sugar-type syrups or other sugar-type products have been commercially manufactured from etherified starch. While it is known that etherified starch can be hydrolyzed, very little is known about the character of the resulting hydrolyzates. It has been suggested that ether linkages may survive the hydrolysis even when the hydrolysis is carried to the stage of forming monosaccharides.

C.C. Kesler, P.L. Carey, and O.G. Wilson; U.S. Patent 3,505,110; April 7, 1970; assigned to Penick & Ford, Limited developed and tested a process for the hydrolysis of starch which was etherified with propylene oxide to a predetermined level of hydroxypropyl substitution. The hydroxypropylated starch is then hydrolyzed to produce a syrup having a specified dextrose equivalent (DE) and fermentables (FE). If desired, the syrup can be converted to a dry product by spray drying.

The product either as a syrup or in dry form can be used as a substitute for ordinary sugars. There has long been a need for relatively nonfermentable sugars for use in special foods. There is evidence that amylase-resistant saccharides may be of value in controlling dental caries, and such products can also be used to reduce calorie intake. Since the saccharide products are highly resistant to the action of digestive enzymes, including salivary amylases, pancreatic enzymes, and animal diastases, there will be substantially no enzymatic digestion of the saccharides in the mouth or in the gastrointestinal tract. Furthermore, since the products are nontoxic and highly palatable, they should be more readily acceptable as sugar substitutes than artificial sweetening agents.

Propylene oxide can be reacted with starch by the same procedures as other alkylene oxides such as ethylene oxide. Propylene oxide reacts monofunctionally with starch hydroxyl groups to introduce hydroxypropyl groups into the starch, the hydroxyl groups being attached to the second carbon of the propyl chain rather than to the terminal carbon as is the case with hydroxyethyl-substituted starch. The propylene oxide that does not react with the starch converts to propylene glycol or other nontoxic by-product. Therefore, traces of such by-products in the final product would not interfere with the use of the product in foods or confections. However, the trace by-products can be largely removed by suitable washing and purification.

After etherification, the starch is neutralized and then hydrolyzed in water. The hydrolysis may be accomplished by the regular pressure and acid conversion process used to convert corn starch to corn syrup or the starch may be liquefied with a liquefying enzyme and then treated with a saccharifying enzyme to produce a syrup containing sugar ethers. A combination of acid conversion followed by enzyme saccharification may also be used.

It has been found preferable to use an enzyme-enzyme hydrolysis procedure, first treating the starch with a liquefying enzyme (e.g., bacterial amylase), and then with a saccharifying enzyme (e.g., fungal amylase). The hydrolysis is continued until the starch is converted to the form of a syrup. The hydrolysis may be carried out at a solids concentration at about 25 to 40%. The resulting syrup can be concentrated to produce a final syrup containing from about 60 to 80% solids, such as 70 to 75% solids. The hydrolyzate can be spray or drum dried to produce a dry product. In carrying out the hydrolysis it is important to control the dextrose equivalent and fermentables of the final product.

Example: 34 g of NaOH were dissolved in 2,040 g of isopropyl alcohol and 1,490 g of corn starch (10.1% moisture) were added with stirring. To the slurry were added 210 g of water and then 390 g of propylene oxide. The slurry was stirred at 123° to 125°F for 91 hours. The alkali was then neutralized with glacial acetic acid to a point that an aqueous solution of the starch had a pH of 5.5. The starch was filtered and washed by reslurrying with 3,000 ml of alcohol. The filtered starch was air dried. Analysis of the starch indicated 13.91% hydroxypropyl substitution.

To a 20-gallon stainless steel vessel equipped with heater, agitator and thermometer were charged 42½ lb of water. Over a period of an hour, 20¼ lb of starch (made by the above procedure) and 7 g of Tenase liquefying enzyme were added alternately. The temperature was slowly increased during this period to raise the temperature to 140°F. The temperature was held at 140°F for 1 hour and then increased to 187°F. The solution was cooled to 140°F over a 1 hour period. Then, 7 g of Tenase were added and two of the heating and cooling cycles made, followed by 1 hour at 180° to 186°F. After cooling to 140°F, the pH was adjusted to 4.8 with HCl and 80 g of Diazyme L-30 (amyloglucosidase) added. The solution was placed into a 140°F oven for 86 hours. The syrup was heated with live steam at 210°F for ½ hour. The syrup was filtered through a mixture of Sil-Flo and Celite 110. The syrup was passed through a 2" x 24" column packed with granular decolorizing carbon, filtered to remove the carbon fines and then passed through two series of 2" x 24" cationic and anionic columns. The syrup was concentrated to 76% solids in a vacuum evaporator. The DE was approximately 24%, the fermentables (FE) 27%, and the product analyzed nil maltose. Digestion with salivary amylase for 4 hours at pH 6.8 and 37.5°C did not increase the glucose, maltose, or DE.

MALTULOSE AND MALTOTRIULOSE PROCESS

A chemical and enzyme process for oxidizing maltose and maltotriose after enzyme conversion of starch to the corresponding ketoses is given in detail by *S. Sakai, T. Miyake, and Y. Sato; U.S. Patent 3,691,013; September 12, 1972; assigned to Hayashibara Company, Japan.*

The conversion from maltose to maltulose generally occurs in the presence of a basic catalyst, and has long been known as the Lobry de Bruyn reaction. It is an easy way of isomerizing aldose to ketose and, for that purpose, strong alkali or a strong basic ion exchange resin is used. At a low temperature below 40°C, the decomposition by oxidation of sugar is relatively small, but, with the rise of the temperature, the loss due to the oxidation and decomposition of the sugar is increased. With these results in view, tests were conducted with equipment capable of providing continuous rapid heating and rapid cooling and neutralizing. The results, as shown in the following table, indicate that the sugar loss increased with the rise of the temperature and the extension of the reaction time.

Test Specimen Isomerization Rate, %, and Loss (%)

	After Heating Time of			
Temperature	**5 Minutes**	**10 Minutes**	**15 Minutes**	**3 Days**
35°C	–	–	–	45 (1)
80°C	–	37 (1)	–	–
100°C	30 (1)	–	45 (5)	–

Note: $Ca(OH)_2$ conc. 1.5% of sugar; sugar conc. 40%.

The reaction proceeded satisfactorily with a sugar concentration of 20 to 40%, and, where calcium hydroxide was used, a 0.13% solution was allowed to stand for 3 to 4 days, and the reaction was carried out until no more change was observed in optical rotation. Substantially the same results were obtained with sodium hydroxide and ammonium hydroxide.

The paper chromatogram of a purified solution showed that maltose, maltotriose, and a small amount of tetraose were left in a combined residual amount of at least about 50%. As ketose, there were observed maltulose, maltotriulose, and tetraulose. The solution upon purification with active charcoal and ion exchange resin became a very sweet syrup and was usable as a sweetening agent.

The enzyme systems for conversion to ketose are obtained from bacterial cells from the strains *Pseudomonas graveolens* NRRL 14 and *Pseudomonas fragi* NRRL 25. Other *Pseudomonas* strains which do not involve the hydrolysis of disaccharide may also be used although they extend the reaction time. The bacterial cells obtained by the cultivation above described are added to the 20 to 30% mixed sugar solution, and the mixture is oxidized by aeration and agitation while the drop of the pH value is being prevented in the presence of calcium carbonate and other additives. In this manner, the residual disaccharide and trisaccharide can be oxidized to aldonic acids without any loss of the sugar. The reaction time ranges from 10 to 20 hours, and the degree of oxidation was determined by measuring the amount of calcium salt of the aldonic acid produced.

Maltose as the starting material is prepared by either liquefying corn or potato starches at a high temperature of 160°C or upwards or continuously liquefying the starch with the addition of a liquefying enzyme or α-amylase at a high temperature between 85° and 90°C, thereby forming a low DE liquefied starch having a decomposition rate DE of about 1 to 5%, and then quickly cooling the liquefied starch to 50° to 60°C, and simultaneously adding β-amylase and α-1,6-glucosidase, or reversing the order and modifying the timing for the addition. High purity maltose having a maltose content of over 60% can be produced.

The α-1,6-glucosidase hydrolyzes the α-1,6-glucoside bond of the molecule of amylopectin that interrupts the β-amylolysis, thus permitting full performance of the β-amylose. The glucosidase which may be used is any of the enzymes produced by the strains of *Escherichia intermedia* (ATCC 21073) and *Pseudomonas amyloderamosa* (ATCC 21262). The β-amylase which may be used is β-amylase extracted from wheat bran, β-amylase extracted from soybean, or any other amylase from vegetable origin, or an enzyme produced by a microorganism such as *Polymyxa* (ATCC 8523). The properties of the ketose sweetening agent as a food additive were studied and the following results were obtained.

Sweetness: Sweetness test showed that the maltulose sweetening agent is slightly less sweet than ordinary sugar but sweeter than grape sugar. As for the quality of the sweetness, it is milder than sugar but has more body or depth than grape sugar, owing to the complex structure composed of several saccharides.

In a panel test conducted by 30 examiners, ordinary sugar, grape sugar, maltitol, and the ketose were compared in terms of the intensity of sweetness. By the pair test method, each test material was repeatedly tested 5 or more times. When compared in different sugar concentrations of 35%, 20%, and 10%, the test materials were ranked in the following order of decreasing sweetness:

$$\text{ordinary sugar} > \text{ketose} \cong \text{maltitol} > \text{grape sugar}$$

Thus, ketose is ranked between ordinary sugar and grape sugar and is practically on the same level as maltitol. For the comparison of the strength of sweetness, a 35% solution of ketose was compared with sugar solutions in different concentrations of 10%, 15%, 20%, 25%, and 30%. The results indicated that the 35% solution of the ketose is substantially as sweet as the 25% sugar solution.

Crystallizability and Crystallization-Inhibiting Properties: The maltulose sweetening agent is freely soluble in water, and is only slightly crystallizable in high concentrations at normal temperature.

Moisture- and Flavor-Retaining Properties and Viscosity: Since the sweetener is a ketose mixture consisting of maltulose, maltotriulose, and minor proportions of higher ketoses, it has an excellent moisture-retaining property. It retains flavor and has somewhat greater viscosity than cane sugar because it contains some proportions of tri- and tetra-saccharides. With these features, it is best adapted for applications as a food additive. It proves particularly useful in the making of sponge cakes, and its viscosity and flavor-retaining property combine to impart fresh flavor and unique palatability to canned fruits, fruit juices, etc.

Process for Producing Ketose: Example A-1 – Starch slurry of sweet potato in a concentration of 20% and at pH 6.0 was liquefied at 90°C by the addition of 0.2% of a liquefying enzyme per gram of the starch. A homogeneous liquefied starch with a decomposition rate of 3.5% was obtained. This liquefied starch was rapidly cooled to 50°C, β-amylase and α-1,6-glucosidase (obtained from an *Escherichia* strain ATCC 21073) were added in amounts of 20 and 10 units, respectively, per gram of the starch, and the liquefied starch was saccharified at 45°C and pH 6.0 for 35 hours. After the deactivation of the enzymes, the resultant was decolorized and purified with active carbon, and a sugar solution with a maltose purity of 93% on dry basis resulted. Purification of the maltose was performed by crystallization.

2 kg of this maltose was dissolved in 10 liters of water and, after the addition of 13 g of calcium hydroxide, the mixture was reacted at 35°C for 4 days. The optical rotation of the reacted solution was determined and, when the value was stabilized, the reaction was terminated and the resultant was concentrated in vacuum. It had a maltose content of about 52%.

One-half of this mixed solution was put aside, and 1 liter of a solution, containing 10% of the reaction product as an anhydride basis, was neutralized with CO_2. Into this solution were dissolved 1% corn steep liquor, 0.2% urea, 0.06% potassium dihydrogen phosphate, 0.025% epsom salt, and 2.5 g calcium carbonate. The mixture was sterilized with heat and inoculated with *Pseudomonas graveolens* which was cultured with stirring at 400 rpm and aeration with an equivalent amount of air at 30°C for 50 hours. From the culture fluid, the bacterial cells were centrifugally collected and used as an enzyme source.

Example A-2 – The maltose obtained in Example A-1 was diluted to 2 liters of a solution with a 15% concentration, and the solution was reacted, with the addition of 2.0 g of calcium hydroxide, at 35°C for 3 to 4 days. After the conclusion of the reaction, the reacted solution was neutralized with carbon dioxide gas, decolored with active carbon, and 37 g of calcium carbonate was added to a liter of the solution. With the further addition of 5 g of the bacterial cells obtained in Example A-1, the solution was vigorously agitated with aeration at 30°C for 15 hours.

Thereafter, the reaction was terminated, and the bionic acid produced was precipitated and separated in the form of a basic calcium salt in the manner described in Example A-1, and the filtrate that contained maltulose was decolorized and desalted with an ion exchange resin. From the liquid upon concentration, 150 g of a sweet syrup was obtained. The yield in terms of an anhydride was 47%. The composition contained a few percent of maltotriulose. The maltobionic acid separated as a calcium salt was freed from calcium and decolorized in the same manner as in Example A-1. The yield of maltobionic acid which contained maltotrionic acid was 45%.

Example B-1: Preparation of Carbonated Drink — Of carbonated drinks, aerated cider-like drink was chosen for the object of this example. A condiment mixture of the following composition was prepared in the respective amounts per 10 liters of the drink to be obtained: sweetening agent (maltulose 90% as anhydride), 700 g; saccharin, 12 g; sodium chloride, 2 g; citric acid, etc., 60 g; and flavoring, 50 ml. This blended solution was completely dissolved, filtered and carbonated in the usual manner. Thus, a plain and sweet refreshing drink was obtained as bottled. The maltulose sweetening agent added refined and deep sweetness in good harmony with the sourness and artificial sweetness, and imparted lightness and coolness to the aftertaste of the drink.

Example B-2: Preparation of Concentrated Fruit Juice Drink — Ingredients of the following proportions were used in preparing 1,000 liters of a concentrated juice which contained orange juice. A composition of 79 liters of 5-fold concentrated orange juice, 750 liters of a 60% solution of the maltulose sweetening agent (which contained 87% maltulose), 40 kg of citric acid, 5 liters of orange base, and 2.5 liters of orange essence was diluted with water to a total volume of 1,000 liters.

The maltulose-based sweetening agent harmonized well with the orange and supplied sweetness which made the inherent sourness and flavor of the fruit all the more profound. Since the maltulose sweetener contains small proportions of higher ketoses and is difficult to crystallize, a clear concentrated juice several times as dense as the normal concentration is obtained without the possibility of crystal deposition.

FRUCTOSE ENZYME PROCESSES

DEAE-Bound Glucose Isomerase

There are many processes for producing fructose-containing solutions. These processes may be grouped into three broad categories. In the first category, sucrose is inverted to glucose and fructose by the use of an acid or invertase. In the second category, glucose is converted to fructose by the use of alkaline catalysts.

However, there are a number of distinct disadvantages associated with alkaline isomerization. For instance, due to the nonselectivity of the alkaline catalysts, various objectionable by-products are formed, such as large amounts of colored bodies and acidic materials. To refine alkaline-isomerized liquors to remove the objectionable by-products to produce an acceptable product requires rather complicated and costly procedures.

The third category for producing fructose-containing solutions involves enzymatically-converting glucose in a glucose-containing solution, e.g., corn syrup, to fructose. Various microorganisms are known which produce glucose isomerase. A basic process for enzyme isomerization of glucose to fructose is given by *K.N. Thompson, R.A. Johnson, and N.E. Lloyd; U.S. Patent 3,788,945; January 29, 1974; assigned to Standard Brands Incorporated.*

Because of the economics involved in producing glucose isomerase, it is of the utmost importance to use the isomerase under conditions whereby maximum yields of fructose are produced using minimum quantities of glucose isomerase. Moreover, the conditions for isomerization should be such that minimal quantities of objectionable by-products are produced. Glucose isomerase is produced primarily intracellularly by microorganisms. Thus, the major portion of the glucose isomerase is found within and/or on the cell walls of the microorganisms. When the isomerase is released or extracted from the cells, it is essentially solubilized. It would involve a rather costly and complicated procedure to recover the solubilized isomerase so that it can be used in another isomerization reaction. In this process, glucose isomerase was bound to DEAE-Sephadex and the glucose solution passed through continuously.

There are many methods for the production of glucose-containing solutions. For instance, the methods, which are being practiced commercially, principally involve saccharifying cornstarch into glucose. These methods may be grouped into three categories. The first is an acid process in which a dilute acid solution is used to hydrolyze starch to glucose. The second is an acid-enzyme process in which starch is liquefied by a mild acid treatment and then an enzyme is used to convert the liquefied starch to glucose. The third is an enzyme-enzyme process in which two enzyme treatments are used, the first to liquefy the starch and the second to convert the liquefied starch to glucose.

In this process, it is preferred to use a glucose-containing solution produced by either of the latter two processes since such solutions generally contain greater amounts of dextrose on a dry substance weight basis, lesser amounts of acids, less color and lower amounts of oligosaccharides.

The characteristics of the bed of the bound glucose isomerase are extremely important in respect to the quality of the fructose-glucose solution produced and the commercial utilization of this process. The bed should contain at least 20 IGIU (International Glucose Isomerase Units) per cubic centimeter. If the bed contains less than 3 IGIU of glucose isomerase activity per cubic centimeter, a greater bed volume may be necessary to isomerize equivalent amounts of dextrose. This may present various attendant problems such as greater pressure drops across the bed, longer contact times between the glucose-containing solution and the isomerase to produce the desired fructose and greater capital expense because of the size of the equipment needed to contain the bed of bound isomerase.

Furthermore, as the depth of the bed is increased, there is a greater tendency for compaction of the bed due to the high pressures which must be used to pass the glucose-containing solution. For example, in a relatively shallow bed, the bed can only compact to a small extent while in a bed of greater depth, the compaction can be much greater. When this occurs, the pressure drop across the bed will increase to such an extent that the pressure necessary to pass the glucose solution through the bed may be extraordinarily high, and may be so high that conventionally constructed equipment cannot be used to contain the bed. The stability value of the bound glucose isomerase should be at least 400 hours.

Example: This example illustrates the use of glucose isomerase bound on DEAE cellulose (diethylaminoethyl cellulose) to continuously convert glucose to fructose. *Streptomyces* species ATCC 21175 was grown under submerged aerobic conditions and then separated from the fermentor broth by filtration. 1 kg of filter cake was slurried in 5 liters of deionized water to which was added 50 ml of 0.1 $CoCl_2$ and 8 g of a cationic detergent. The temperature of the slurry was maintained at 60°C, and the pH was maintained at 6.7 to 6.8. After 3.75 hours of stirring, the slurry was filtered under vacuum. The filtrate was concentrated by evaporation under vacuum to 30 IGIU/ml.

500 ml of the concentrated filtrate was diluted to 1,500 ml with deionized water. 5 g of DEAE cellulose was added to the filtrate, stirred ½ hour and filtered. The filter cake was washed with water while still on the filter and the washings were collected with the filtrate. The filter cake contained mostly non-isomerase material. The purified filtrate contained 9.2 IGIU/ml.

30 g of DEAE cellulose was suspended in 1,500 ml of water, stirred vigorously and the suspension allowed to set for 30 to 45 minutes. The supernatant suspension was decanted to remove the DEAE cellulose fines. This was repeated 4 times and then the suspension was filtered under vacuum. 20 g (dry basis) of the DEAE cellulose filter cake were added to the purified filtrate containing 9.2 IGIU/ml and stirred for ½ hour at room temperature. The suspension was filtered and a moist filter cake containing the DEAE cellulose with the absorbed isomerase was recovered.

The moist filter cake contained 215 IGIU/g. 6.25 g of the moist filter cake was slurried in 50 ml of an aqueous solution containing 0.001 mol of $CoCl_2$ per liter, 0.005 mol of $MgSO_3$ per liter and 0.1 mol of NaCl per liter. This slurry was poured into a column having a diameter of 1 cm and allowed to settle. The bottom portion of the column contained a shallow layer of glass wool to retain the DEAE cellulose. The bottom outlet of the column was opened and when the bed height of the DEAE cellulose was 25 cm, 350 ml of an aqueous solution containing 0.001 mol of $CoCl_2$ per liter, 0.005 mol of $MgSO_3$ per liter and 0.1 mol of NaCl per liter passed through the bed to remove colored materials and other impurities from the DEAE cellulose.

A glucose solution at a pH of 6.5 containing 3 mols of glucose per liter, 0.001 mol of $CoCl_2$ per liter and 0.005 mol of $MgSO_3$ per liter was passed through the column at a rate of 0.2 ml/min. The temperature of the column was maintained at 60°C. The conversion of glucose to fructose after 6 hours was 49.6% and after 186 hours was 45.0%.

Alumina-Bound Glucose Isomerase

Another application of immobilized enzymes is given by *R.A. Messing; U.S. Patent 3,868,304; February 25, 1975; assigned to Corning Glass Works* for use with isomerases to convert glucose-containing syrups to fructose. The enzymes are absorbed on a porous alumina carrier and placed in a flow-through column for isomerizing conditions.

It is important that the average pore size of the alumina carrier does not exceed about 1000 A for several reasons. As the average pore size of the inorganic carrier exceeds about 1000 A, there occurs a considerable decrease in surface area available for adsorbing and, hence, loading the enzyme. Further, as the average pore size increases beyond about 1000 A, the rigid internal structure of the pores affords less protection for the enzyme from environmental stresses which tend to detach the enzyme. For example, if the immobilized enzyme composite were used in a flow-through column, the pressure drop through the column represents one such turbulent environment which tends to detach enzymes adsorbed within the pores of carriers having an average pore size larger than about 1000 A. Because of the requirement of a relatively high surface area for loading the enzyme (e.g., greater than about 5 m^2/g) and because of the need to protect the adsorbed enzyme from external stresses which tend to detach the enzyme, the average pore diameter of the porous alumina carrier should be no larger than about 1000 A. The preferred average pore size for the porous alumina carrier is between about 140 and 220 A.

After the porous alumina bodies have been ground and sieved to the desired particle size, they are hydrated or preconditioned with a suitable buffer or salt system; hydration at a pH of 7.0 is preferred for the alumina carrier. The buffer system external to the pores is then removed but the porous alumina bodies are kept wet prior to the adsorption step.

The glucose isomerase (in solution) is then added to the wet carrier at a concentration per unit weight of porous alumina carrier which will permit optimum adsorption (loading) of the glucose isomerase. The adsorption process is facilitated by stirring, circulation, inversion, use of a fluidized bed reactor containing the carrier, or other known means suitable for forming an agitated reaction environment which encourages mass diffusion of the enzyme through the pores. The time required for optimum adsorption depends, of course, on the carrier particle size, the size of the enzyme vis-a-vis the average pore size, and other factors such as temperature and pH, but the adsorption steps should be for at least about one-half hour.

Example: A crude glucose isomerase preparation consisting of about 444 IGIU/g derived from a *Streptomyces* organism was used for the adsorption step. About 5 g of the glucose isomerase preparation was added to 28.7 ml of 0.1 M magnesium acetate solution in a 50 ml beaker. The slurry was stirred for 25 minutes at room temperature and then filtered through filter paper. The residue on the paper was washed with 14.3 ml of 0.1 M magnesium acetate followed by washes with 14 ml of 0.5 M $NaHCO_3$. The washes were collected directly into the original enzyme filter. The total volume of the enzyme-wash solution (filtrate) was 50 ml.

500 mg of porous alumina bodies were placed in a 50 ml round bottom flask. 10 ml of the above glucose isomerase solution was then added to the flask. The flask was attached to a rotary evaporator. Vacuum was applied to the apparatus and the flask was rotated in a bath maintained between 30° and 45°C over a 25-minute interval. An additional 10 ml of glucose isomerase solution was then added to the flask, and evaporation was continued over the next 35-minute interval under the same conditions. This addition and evaporation were carried out over a 4-hour period under the same conditions.

The procedure was repeated 2 more times with separate 10 ml glucose isomerase solutions. A final 7 ml aliquot of glucose isomerase solution was then added to the flask and evaporation was continued for an additional 1 hour and 10 minutes at 45°C. The flask and contents were removed from the apparatus and placed in a cold room over the weekend. A total of 47 ml of glucose isomerase solution had been added to the alumina. 50 ml of buffer (0.01 M in sodium maleate,

0.001 M in cobalt chloride, 0.005 M in magnesium sulfate, pH 6.8 to 6.9) was added to the composite and the sample was extracted over the next hour at room temperature. The extract, volume 50 ml, was saved for assay. The composite was then washed with 200 ml of water, followed by 10 ml of 0.5 M sodium chloride. The final wash was performed over a fritted glass funnel with 50 ml of water. The composite was then transferred to a 50 ml Erlenmeyer flask and stored in buffer at room temperature with periodic assays of the total sample over a 106-day interval. The results were as follows:

Enzyme extract assay (50 ml)–39.8 IGIU/ml (enzyme activity recovery in extract–90%);
Composite average assay value–130 IGIU/500 ml sample (enzyme activity recovery on composite–6%); and
Average loading value–260 IGIU/g.

The storage stability determinations over the 106-day interval involved 21 periodic assays in which 16 assays on the same sample indicated an activity between 100 and 150 IGIU/500 mg or 200 to 300 IGIU/g.

Xylose Isomerase

Detailed descriptions for production of xylose isomerase are given by *F.C. Armbruster, R.E. Heady, and R.P. Cory; U.S. Patent Reissue 29,152; March 15, 1977; assigned to CPC International Inc.* A mutant strain is used for production of the enzyme which does not require the use of xylose to induce production of the enzyme.

Production of Mutant Strains: *S. olivochromogenes* ATCC 21114 was selected as a parent strain for mutation because it was known to be a good source of xylose isomerase. This microorganism was exposed in the spore state to a dose of ultraviolet light sufficient to kill 97% of the exposed microorganisms. The irradiated spores were plated out, and each colony was tested for enzyme activity. Since the medium on which the colonies were grown contained no xylose to induce the formation of isomerase enzyme, only that colony that had the desired characteristics showed a positive enzyme reaction. This colony was isolated and was extensively tested to demonstrate that the selected mutant strain did indeed produce appreciable quantities of xylose isomerase without the need for xylose in the nutrient medium as in inducing agent.

The steps involved in this procedure were as follows. *S. olivochromogenes* was grown on Difco starch agar, made to 2.0% agar with added agar, until the culture had sporulated. These spores were then harvested and suspended in 20 ml of a 0.1% solution of the surfactant, Tween 80, and 0.1% of a dispersing agent, Marasperse C was added. The suspension was sonicated for two bursts of 15 seconds each to break up chains and clumps of the spores.

This spore suspension was exposed in a shallow dish, with stirring, to ultraviolet light. This exposure produced a 97% kill of the spores. The suspension was then diluted and, at the proper dilution to give about 100 colonies per plate, was streaked out in Petri dishes containing the following medium: 15 DE corn syrup solids, 1.0%; Difco yeast extract, 0.05%; Difco BactoPeptone, 0.05%; Difco BactoTryptone, 0.05%; agar, 1.5%; and pH to 7.5 with NaOH. About 40 to 80 colonies per plate did grow and were subsequently isolated. These isolates were screened for the ability to isomerize xylose after being grown on a medium that did not contain xylose or any material supplying xylose.

Characterization of the Enzyme: A determination of the Michaelis constants (Km) for reaction on xylose and on dextrose was made. The purpose was to reveal the relative affinities of the enzyme preparation for these substrates. All isomerases examined that convert dextrose directly to levulose are xylose isomerases. The determination was made utilizing sonic extract of the culture. It was found that a lower Km was obtained when the enzyme preparation acted on xylose than when it acted on dextrose. This established xylose as the natural substrate of the isomerase, and the enzyme as a true xylose isomerase.

The capacity of the xylose isomerases of the mutants to accept dextrose as a substrate is believed to be due to close structural similarities between xylose and dextrose. Since the mutants produce the isomerase enzyme either in the presence of xylose or in its absence, the enzyme is constitutive rather than necessarily induced only by the presence of xylose in the culture medium. When the culture medium does contain xylose, the enzyme may be both constitutive and induced.

Example: Enzymatic Isomerization – To demonstrate the effectiveness of the isomerase produced by the mutant strains for converting dextrose to levulose, several conversions were conducted. The enzyme preparation used consisted of frozen whole cells of either *S. olivochromogenes* mutant CPC 3 or mutant CPC 4.

A series of conversions of 95 DE corn starch hydrolyzate was conducted, using several different enzyme dosages. The conversions were conducted at 70°C, at a pH of 6.25. Magnesium sulfate was added to the hydrolyzate at the level of 0.01 M. The dry substance level of the hydrolyzate was at about 600 mg/ml. During isomerization, the hydrolyzate was maintained under an atmosphere of nitrogen, and the pH was maintained by titration as necessary. The hydrolyzate was agitated throughout by stirring with bars in the conversion reactor, driven by a magnetic stirring motor. The results are reported below:

Dosage Series Without Cobalt

	Age of Conversion in Hours			
	16-18	40-44	64-66	88-90
Dosage (u/g)	Conversion of 95 DE Hydrolyzate to Ketose (%)			
0.8	17.7	30.2	35.2	38.1
1.0	21.9	35.6	40.4	43.0
1.2	24.7	37.5	41.3	42.0
1.4	26.0	40.1	43.4	–
1.6	28.1	41.0	43.2	–
2.0	33.4	43.2	45.2	–

As these data demonstrate, an enzyme dosage of 1.0 unit per gram is adequate to produce enough levulose to provide a 40% ketose level after a 66-hour conversion without cobalt; and an enzyme dosage of 1.4 units per gram is adequate for the production of 40% ketose after 42 hours of conversion time. Several additional conversions were conducted successfully at 65°C. In addition, some of the conversions were conducted successfully at somewhat higher pH values. The results observed are summarized below:

Conversion at 65°C Without Cobalt

		Age of Conversion in Hours			
		22	43	69	88
	Dosage	Conversion of 95 DE Hydrolyzate to Ketose			
pH	(u/g)	(%)			
6.25	0.7	16.5	24.1	30.0	36.9
6.25	1.0	20.3	30.0	36.6	42.1
6.50	0.7	16.9	25.9	31.9	36.3
6.50	1.0	23.4	33.3	38.8	42.6
6.75	0.7	18.9	28.2	33.6	37.2
6.75	1.0	22.1	30.1	36.0	39.5

The results observed appear to demonstrate that the use of the higher conversion temperature of 70°C is somewhat preferable. The higher pH levels did not appear to produce any advantage. To summarize, it would appear that conversions conducted at 70°C, at a pH of about 6.25, and at a dry substance level in the range from about 600 mg/ml to about 800 mg/ml, with an enzyme dosage of about 1.2 to 1.4 units per gram, for from 40 to 48 hours approximately, and in the presence of about 0.01 M magnesium, in the absence of cobalt and with nitrogen sparging, will produce highly satisfactory sweet syrup products. Compositions of the following kind can be expected:

Ingredient	Weight Percent (ds basis)
Levulose	40-44
Dextrose	45-50
Higher saccharides	7-8

These data demonstrate that a 40% levulose syrup can be made economically from a 95 DE hydrolyzate utilizing enzyme preparations derived from the mutants without the use of added cobalt.

Solid Fructose Sweetener

An extension of the basic process of converting dextrose (glucose) to fructose is the manufacturing procedure to prepare a solid sweetener as described by *R.G.P. Walon; U.S. Patent 4,049,466; September 20, 1977; assigned to CPC International Inc.* The levulose (fructose) containing sweetener is prepared by spray drying an aqueous mixture of a starch hydrolysate carrier and a levulose syrup containing dextrose.

One source of levulose and dextrose is invert sugar containing about equal proportions of dextrose and levulose and prepared by the action of invertase on sucrose. Another source of starting material is a product obtained upon partial or total isomerization of dextrose, often referred to as invert dextrose. A further starting material source can be the isomerization product of a starch hydrolysate having a high dextrose equivalent, such as about 60 to 90. A variety of additional starting materials, such as those obtained by synthetic blending of individual sugars or compositions as described, are also possible. The starch hydrolysate carriers are starch hydrolysates having a dextrose equivalent between about 1 and 25 and a polysaccharide content, having a degree of polymerization greater than 5, of at least about 60 weight percent. While any starch hydrolysate having properties falling within this range can be successfully used, those starch hydrolysates having lower dextrose equivalents and a greater proportion of polysaccharides, having a degree of polymerization above 5, are preferred. The starch hydrolysate carriers can be prepared by conventional means such as acid or base hydrolysis, pyrolytic hydrolysis, enzyme conversion or a combination from various starches such as cornstarch, waxy maize starch, tapioca starch, potato starch, wheat starch, sago starch, arrowroot starch, rice starch and barley starch.

To prepare a starch hydrolysate having the required properties for preparing the sweetening compositions, an aqueous slurry of starch of a convenient concentration, usually ranging from 2 to 4 parts by weight of starch per 10 parts of water, containing an enzyme such as α-amylase, is heated to a temperature ranging from about 70° to 90°C for a period sufficient to result in the desired degree of conversion. The conversion is then stopped by inactivating the enzyme by increasing the temperature of the conversion mixture under superatmospheric pressure to yield the desired starch hydrolysate.

Example 1: Starch Hydrolysate Preparation Using a Single Step Enzyme Conversion – An aqueous starch slurry is prepared containing 30% by weight of corn starch. The slurry is heated to a temperature of between about 85° and 95°C, and a bacterial α-amylase preparation is added to an amount of about 0.025% by weight of the starch over a period of about 30 minutes. The mixture is then stirred for a period of about 30 minutes while maintaining the temperature between about 80° and 90°C. After this time, the temperature is reduced to 80°C, and the enzyme conversion is carried out until the desired dextrose equivalent is reached. The temperature of the mixture is then quickly raised to about 120°C to inactivate the enzyme and terminate the conversion.

Typical starch hydrolysates, which can be prepared by the method described in this example, are shown in the table below where DE designates dextrose equivalent, DP_1 designates a total quantity in percent by weight of monosaccharides present in the starch hydrolysate, DP_2 designates the total quantity of disaccharides present in the starch hydrolysate and so forth.

	Starch Hydrolysate No.				
	1	2	3	4	5
DE	5.0	10.0	15.0	20.0	25.0
DP_1	0.1	0.3	0.7	1.4	2.4
DP_2	1.3	3.4	5.3	7.6	9.7
DP_3	1.8	4.3	6.9	9.4	12.0
DP_4	1.8	3.5	5.2	6.9	8.6
DP_5	1.8	3.6	5.5	7.4	9.3
DP_6	3.3	7.0	10.6	14.3	18.0
DP_7*	89.9	77.9	65.6	53.0	40.0

*And higher

Example 2: Preparation of a Starch Hydrolysate Utilizing a Two Step Enzyme Conversion – Unmodified cornstarch is slurried in water to provide an aqueous suspension containing from about 28 to 32% by weight of the starch. Bacterial α-amylase is added to the aqueous suspension in an amount of about

0.05% by weight based on starch. The starch suspension is then heated with agitation at a temperature of about 90°C for a period of about 10 minutes to inactivate the enzyme and terminate the conversion. After this time, the conversion product is cooled and filtered and a second quantity of α-amylase in an amount of about 0.02% by weight based on starch solids is added to the liquefied starch. The resulting mixture is then heated at a temperature from about 80° to 85°C to result in the desired starch hydrolysate. A typical starch hydrolysate, prepared by the procedure of this example, has a dextrose equivalent of about 20 and a degree of polymerization of 7 and greater of about 50% by weight.

Example 3: Preparation of a Solid Sweetening Composition – An aqueous syrup containing 18% by weight levulose, 63% by weight dextrose and having a dextrose equivalent of about 89 and a solids content of 80% was mixed with a starch hydrolysate having a dextrose equivalent of 20.5 and a solids content of 70% and a degree of polymerization of 5 of at least about 60% by weight to yield an aqueous mixture having a 75.5% by weight solids content, a levulose content on a dry basis of about 10% and a dextrose equivalent of about 58.4.

This aqueous mixture was preheated to a temperature of about 80°C and was spray dried in a heated moving-air chamber having an air inlet temperature of about 130°C and an air and product exit temperature of about 60°C, and equipped with a turbine spray head revolving at about 22,000 rpm to yield a white powder having the following properties: moisture content, 2.5%; dry substance, 97.5%; dextrose equivalent, 58.6; dextrose content, 40.4%; and levulose content, 9.8%.

SPECIAL IMMOBILIZED AMYLASE

A simple and inexpensive means has been found for the reversible immobilization of proteins, peptides, enzymes, antibodies, antibiotics, and whole cells (nitrogen-containing organic substances) which involves the formation of water-insoluble metal chelates. This process can be used for the recovery of the nitrogen-containing organic substance as well as its reversible immobilization. Chelates of the nitrogen-containing organic substances are those formed with the hydrous oxide of zirconium and to a lesser extent the hydrous oxides or hydroxides of tin, iron, vanadium and titanium.

The chelates of the nitrogen-containing organic substances possess the biological properties initially present in the organic substances and can, therefore, be used for the same purposes, and once the biological activity of the chelate complex has been used to a desired extent in a process involving, for example, a catalytic, antibiotic or antibody purpose, the nitrogen-containing organic substance can easily be separated from the chelate complex by treatment of a suspension of the chelate in an aqueous medium at an alkaline pH with an aqueous bicarbonate or carbonate solution (e.g., aqueous sodium or potassium bicarbonate). Details of the process are given by *S.A. Barker, I.M. Kay, and J.F. Kennedy; U.S. Patent 3,912,593; October 14, 1975; assigned to Gist-Brocades NV, Netherlands.*

To an aqueous solution of a zirconium salt (e.g., chloride or sulfate), there is added a hydroxide, ammonium or sodium hydroxide, until a pH of from 6 to 8 is attained and the hydrous oxide of the precipitated hydrous oxide(s) or hydroxide(s) is then mixed with the nitrogen-containing organic substance in an aqueous medium to yield a solid metal chelate of the nitrogen-containing organic substance.

The process is carried out at ambient temperature with the pH of the reaction medium in which the chelate is formed at pH 6 to 8. Generally, 0.5 to 3 hours is sufficient for maximum retention of biological activity of the organic compound. The higher the initial concentration of the nitrogen-containing organic compound in relation to the hydrous metal oxide or hydroxide in the mixture the better the biological activity of the resultant chelate.

For maximal biological activity of the chelate, mixing of the nitrogen-containing organic substance and the hydrous metal oxide or hydroxide should be done in the absence of carbon dioxide, carbonates, bicarbonates and certain other ions (e.g., citrate) which can affect chelation of the nitrogen-containing organic substance as they modify the nature of the hydrous metal oxide or hydroxide and prevent effective chelation. It is possible to form metal chelates of enzymes such as glucose oxidase, glucose isomerase, lactase, catalase, invertase, α- and β-amylase, pullulanase, penicillin acylase, bacterial protease trypsin, chymotrypsin, glucoamylase, dextranase, and glucosidase.

Particular merits of the enzyme-chelate complexes are (a) the water-insoluble immobilized enzymes retain to a high degree the specific activity of the free enzyme, (b) the enzyme-chelate complexes usually possess greater heat stability than the water-soluble enzyme giving a longer shelf-life and an extended temperature range of operation, and (c) the support matrix for the enzyme may be regenerated when the activity of the enzyme is exhausted or decreased to such an extent that it is unsatisfactory.

Example 1: Conversion of Starch by Glucoamylase Bound to Hydrous Zirconium Oxide – 18.36 g of zirconium tetrachloride were dissolved in 120 ml of hydrochloric acid (1 N). To this solution was added at once, with stirring, 200 ml of sodium hydroxide (2 N) resulting in a dense, flocculant precipitate of hydrous zirconium oxide. After 18 hours incubation at 4°C with a solution of 3 g of glucoamylase in 150 ml of distilled water, the suspension was centrifuged and washed several times with distilled water. The enzymatic activity was determined with pNPG (p-nitrophenyl-α-D-glucopyranoside). The activity was 510 pNPG units calculated on 1 g of dry ZrO_2. The percentage adsorption was about 84%. The conversion of starch by the immobilized glucoamylase suspension as such is shown below:

Time	% Conversion	
First 24 hours	94	100 ml starch solution with 100 g wet suspension
Second 24 hours*	94	50 ml starch solution with 80 g wet suspension

*After the first conversion, the suspension was centrifuged and washed. New substrate was added.

The conversion of starch by the immobilized glucoamylase suspension after lyophilization (2 g of lyophilized material per 50 ml of substrate) is shown in the table on the following page.

Time	% Conversion
First 24 hours	66
Second 24 hours	63

Conversion of starch by immobilized glucoamylase, after drying the suspension at 20°C (2.5 g of dry material per 50 ml of substrate), is shown below. On subsequent immersion in water the dry material broke down to smaller granular particles.

Time	% Conversion
3.5 days	23
6 days*	42

*Between the two runs, neither new substrate was added nor the solids were washed.

The substrate was prepared by adding bacterial α-amylase (1 g of 4,400 BAU/g per 500 g of starch) to a 30% starch solution. The suspension was heated to 90°C in about 1 hour. After a 15-minute boiling period, the suspension was filtered. The percentage of conversion was 18.6.

Example 2: Elution of Glucoamylase from Hydrous Zirconium Oxide — Samples of hydrous zirconium oxide were prepared. A glucoamylase solution was added and the samples were stirred for 18 hours at 4°C (in the manner described in Example 1). The samples were centrifuged and washed several times and to the residues there were added several buffer or other solutions in a ratio of 1:1, as detailed below. The mixtures were stirred for 1 hour at about 30° to 35°C. After centrifugation, the supernatants were removed and the residues were assayed for glucoamylase activity. The activity of the original solution was 482 pNPG/g, calculated on dry ZrO_2.

Sample	Buffer Solution	Activity Residue in pNPG Units per Gram ZrO_2	Remaining Activity Residue as Percent of Original Suspension
1	water	482	100
2	phosphate buffer (0.1 M, pH 5.0)	410	84.5
3	phosphate buffer (0.1 M, pH 6.9)	418	85
4	acetate buffer (0.005 M, pH 5.0)	534	100
5	di-Na H phosphate (0.5 M)	244	51

POLYACROLEIN BOUND INVERTASE

Binding of enzymes to a sulfited aldehyde or ketone polymer was developed by *P.S. Forgione; U.S. Patent 3,753,861; August 21, 1973; assigned to American Cyanamid Company*. Insolubilization is effected by formation of a polymeric network via reaction with a sulfited polymer adduct through available sites such as vinyl groups, OH groups, etc. For example, polyacrolein, a water-insoluble polymer which contains some groups with which most enzymes are reactive, is first contacted with a sulfite such as sodium bisulfite in order to render it water-soluble and more susceptible to enzyme reaction. In such a condition, however, the polymer cannot be reacted with an enzyme because recovery of any product is relatively impossible. Crosslinking of the sulfite-polymer product, however, renders it gel-like in consistency and effectively hydrophilic so as to allow reaction with the enzyme. As a result, the sulfite-polyacrolein product is crosslinked with a diamine such as hexamethylene diamine.

Based on theory, the result of these two reactions is that the sulfite breaks some of the heterocyclic rings of the polyacrolein creating more enzyme-reactive aldehyde groups in addition to a series of sulfite groups. The diamine reacts with some of these aldehyde groups with the formation of –CH=N– linkages between two polymer molecules, thereby crosslinking the polyacrolein. Reaction of the enzyme, e.g., invertase, forms an adduct or covalent bond between the enzyme and the other available aldehyde groups, and also may result in reaction through the sulfite groups. The resultant adduct has a series of crosslinked groups and bound enzyme groups. One proposed structure of such a product is the following, sodium bisulfite and hexamethylene diamine having been used:

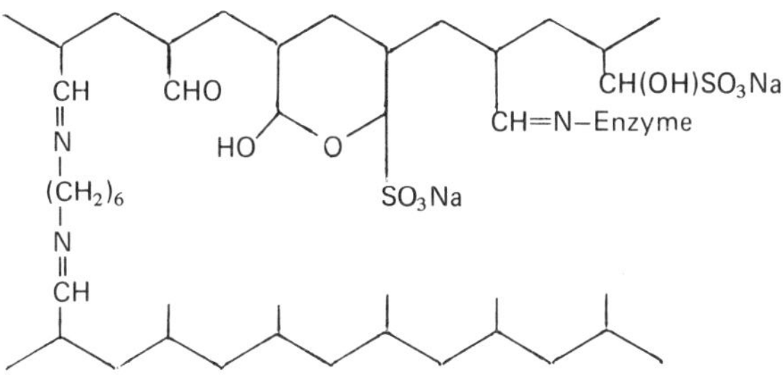

Examples of carbonyl polymers which may be utilized include those produced from such aldehyde monomers as acrolein; α-alkyl acroleins, e.g., methacrolein, α-propylacrolein; crotonaldehyde; 2-methyl-2-butenal; 2,3-dimethyl-2-butenal; 2-ethyl-2-hexenal; 2-decenal; 2-dodecenal; 2-methyl-2-pentenal; and 2-tetradecenal.

The products are used in the conversion of enzymatically convertible substrates to their conversion products. More specifically, enzymes are known to function catalytically in nature to convert various materials known as substrates to other materials. For example, since lipase converts fat to glycerol and fatty acid, invertase converts sucrose to invert sugar, lactase acts upon lactose, the process enzyme compositions can be used to treat fat, sucrose, lactose, etc. Further examples of substrates which may be treated include milk, soy isolate, cheese whey, fruit juices such as apple juice, grape juice, etc., wines, blood serum, beer, tea, mixtures of fructose and glucose.

The polymer bound enzyme products, being substantially water-insoluble, enable the substrate to be converted while in aqueous solution, while still retaining the catalytic activity of the enzyme. Thus, for example, an aqueous solution of sucrose can be continually contacted with polymer bound invertase to convert the sucrose to invert sugar. The catalytic activity of the invertase remains high for a long period of time. Similarly, milk can be contacted with bound rennin continuously over long periods, e.g., over 6 months, and the rennin is still active.

Although the enzyme may not remain as highly active as when initially bound, it does not lose all its activity after extended periods of use and, in fact, systems with bound invertase have been used for over 5 months while losing only one-tenth of their activity.

Example 1: (A) Crosslinked Polyacrolein-Sodium Bisulfite Addition Complex – 0.88 part of 1,6-hexanediamine containing 3 parts of water is slowly added, with stirring, to 44 parts of a 10% solution of polyacrolein-sodium bisulfite adduct (MW 80,000). The mixture is then heated to 65°C for 10 minutes and the yellow hydrophilic product which forms is then washed until neutral with distilled water and filtered.

The wet hydrophilic adduct thus obtained is suspended in 50 parts of water and reacted with 0.104 part of invertase (twice recrystallized) which had been first dissolved in 4 parts of water. The enzyme reaction mixture (pH 6.8) is gently stirred for 18 hours at 10°C and the resulting hydrophilic, covalently bound enzyme polymer adduct is then washed free of unreacted enzyme. Assay of the hydrophilic enzyme-polymer adduct shows high activity and about 80% of the charged enzyme has been bound.

(B) Conversion of Sucrose to Invert Sugar – 5% by weight of the isolated wet cake (72% H_2O) converts 100 ml of a 10% sucrose solution at pH 4.8 to 81.0% invert sugar in 15 minutes at 42°C. Similarly, 5% by weight of the freeze dried polymer enzyme adduct prepared from the wet cake results in a 75.8% conversion to invert sugar using the same experimental conditions. Five months later, the activity of the bound enzyme is still about 85% of its initial value.

Example 2: (A) 150 parts of wet (20% solids) crosslinked polyacrolein-sodium bisulfite addition complex (as prepared in Example 1) are suspended in 200 parts of technical grade invertase solution (k = 0.6) at pH 6.3. The reaction mixture is stirred for 48 hours at 6°C and, at the end of this time, the resulting covalently bound invertase is washed until free of unbound enzyme and filtered (180 parts).

(B) Assay of 0.05% by weight of the total wet (80% H_2O) product with standard 10% sucrose solution [same experimental conditions as Example 1(B)] results in a 8.5% conversion. This represents 150 mg invertase bound per gram of dry support. After 2 months, the bound enzyme retains 93% of its original activity.

Example 3: 0.5 ml of rennin is slowly dissolved in a suitable reaction vessel with minimum agitation into a phosphate buffer (0.01 M, pH 5.8). This solution is then slowly added to 1 part of wet (80% H_2O) polyacrolein, modified as in Example 1, slurried in phosphate buffer (0.01 M, pH 5.8). The pH of the mixture is adjusted to 5.8 and the reaction stirred at 4°C for 18 hours. The adduct is washed with large amounts of phosphate buffer (pH 5.8, 0.01 M) and then with deionized water until there is no detectable activity in the washings. The adduct is filtered and a sample assayed for activity. The bound enzyme loses only 25% of its activity upon treatment of milk over a period of 3 weeks.

FRIED FOOD AMYLASE TREATMENT

In the unique process of *C.F. Roan; U.S. Patent 4,058,631; November 15, 1977; assigned to GB Fermentation Industries, Inc.*, α-amylase treatment is used prior to deep fat frying to reduce fat absorption by starch foods. α-Amylase derived from *Bacillus subtilis* is a food grade high temperature starch dextrinizing and liquefying enzyme. The enzyme is capable of converting native or modified starches to dextrins or syrups having a dextrose equivalent (DE) up to 40. A 40 DE syrup prepared from corn starch will contain 5 to 7% dextrose, 15 to 17% maltose and higher polysaccharides.

Enzymes of this type also attach Lintner starch and other soluble starches, thereby converting them to dextrins and sugars. The optimum pH range for the dextrinization of soluble starch at 37°C is 5.2 to 6.0, although the enzyme system is useful at this temperature in the range of 4.9 to 6.8. The optimum pH for starch liquefaction at 80°C is 7.0, although the enzyme is useful in the pH range of 6.5 to 7.8 at this temperature. Plant and animal amylases such as, for example, malt amylase and pancreatic amylase, also are suitable sources of α-amylases.

The food products which can be treated with the enzyme solution include foods such as French fried potatoes, potato chips, donuts, and cereal-containing snack food items which are customarily fried in fats and oils prior to retail packaging or by the consumer prior to use. In general, the raw food product is treated with an aqueous solution of the α-amylase for a period of time sufficient to penetrate and coat the surface of the food. Soaking of raw potato pieces for about 15 minutes to 1 hour generally is sufficient in an aqueous solution containing from about 0.1 to 1% of α-amylase enzyme having an activity of from about 10,000 to 20,000 Bacterial Amylase Units per gram or equivalent α-amylase. A Bacterial Amylase Unit (BAU) is defined as that quantity of the enzyme that will dextrinize 1 mg of starch per minute under standard experimental conditions.

In the case of products prepared from processed cereals and starches, an aqueous solution of the α-amylase enzyme can be applied to the surface of the food pieces either by immersion for a short period of time or by means of spray equipment. The enzyme coating of the food product is facilitated by holding at ambient room temperature or at elevated temperatures not exceeding 100°F prior to frying. The aqueous solution of the α-amylase enzyme used for treating the raw starchy food product contains a small amount of sulfur dioxide, for example, up to about 100 ppm, as an antioxidant.

Example 1: French fried potatoes were prepared by giving one sample of potatoes a preliminary treatment with enzyme prior to frying while a control sample was fried without the enzyme treatment as follows. Materials: (1) U.S. No. 1 Idaho Potatoes; and (2) *Bacillus subtilis* α-amylase having about 15,000 BAU per gram.

Procedure: A 0.2% aqueous solution of the bacteria α-amylase enzyme was prepared by dissolving about 2.5 grams of the enzyme in 1 quart of water. 1 lb of the U.S. No. 1 Idaho Potatoes were peeled and cut into a regular French fry form about 6" long by ½" by ½". The French fries were allowed to absorb the bacterial α-amylase by soaking in tap water. Soaking was conducted for 1 hour and 5 minutes. A control sample of 1 lb of U.S. No. 1 potatoes of similar size and form were placed in tap water without any enzyme and allowed to soak for a similar period of 1 hour and 5 minutes. The enzyme treated potatoes and the control potatoes were then both heated at 160°F for 7½ minutes followed by heating at 190°F for 7½ minutes. The treated potatoes and the control potatoes were then placed in a fryer at 375°F containing a commercial frying fat. The treated potatoes and the control potatoes were fried until they both appeared to be done by observation of similar browning color. The treated potatoes were thus fried for 10 minutes whereas the control potatoes were fried for 8 minutes. The two samples were then tasted. In a panel of 4 individuals who tasted the potatoes, it was the unanimous conclusion that the enzyme-treated French fry was preferred since it had less grease in the potato and tasted better than the control sample.

Example 2: French fried potatoes were again prepared and fried in a deep fat fryer, one sample being enzyme-treated and the other sample being untreated prior to frying to serve as a control sample. Each sample was then subdivided into several parts for replicate determinations of fat content and moisture according to the following procedure: weighed potato samples (25 to 30 grams) were macerated and placed in a vacuum desiccator (with desiccant) for several days until a constant weight was obtained. The deposit of grease on the implements used for each sample maceration was quantitatively removed by thorough wiping with filter paper which was included with the potato sample for the moisture and fat determination.

The weight difference between the initial sample and the desiccated sample was used to calculate the percentage of moisture removed. Fat was extracted from the entire amount of desiccated sample in a conventional Soxhlet apparatus using petroleum ether as a solvent (20 hours extraction time). The weight of oil collected was compared with the desiccated weight to calculate the percentage of fat removed.

Duplicate determinations were made on all samples. In this analysis, the duplicate enzyme-treated samples were calculated to contain 13.9 and 15.1% fat content on a dry basis or an average of 14.5% fat whereas the duplicate control samples were calculated to contain 16.8 and 17.6% fat content on a dry basis or an average of 17.2% fat.

ENZYME TREATED EXTRUDER PASTE

A. Buffa and A. Holliger; U.S. Patents 3,861,293; January 21, 1975; and 3,950,543; April 13, 1976; both assigned to UNICEF and Gebruder Buehler AG, Switzerland give a process which utilizes heat-resistant α-amylase to convert starch in raw materials such as cereal grains and oil seeds to dextrins and simple sugars. The mixture is in the form of a paste that is passed through an extruder to complete the starch breakdown. The equipment for the process is described as shown in Figures 3.2a and 3.2b.

The steps of the process include mixing water and at least one high temperature-resistant starch hydrolyzing enzyme with the raw materials to form an extrudable paste, in which the moisture content of the initial mix is from about 30 to 50% of the total weight; obtaining a partial gelatinization and initiating fermentation by extruding the paste in an extruding apparatus heated to bring this paste to a temperature of from about 65°C (149°F) to 115°C (239°F); converting the starch into dextrins and sugars by treating the extruded product in a fermentation apparatus for about 40 to 90 minutes at a temperature around 55° to 90°C in a high humidity atmosphere, as near as possible to saturation; and drying the product after fermentation. Under these conditions, as much water as possible is used as makes it still possible to obtain an extrudable paste. This relative high water content favors the hydration and conversion of starch. Heating the paste in the extruding apparatus ensures that the fermentation will start as the product is malaxed and the starch structure is pregelatinized, facilitating thereby the penetration and action of the enzymes, in such a manner that the following fermentation phase brings about the desired conversion of starch into dextrins and sugars.

Figure 3.2: Extrusion Apparatus

a.

b.

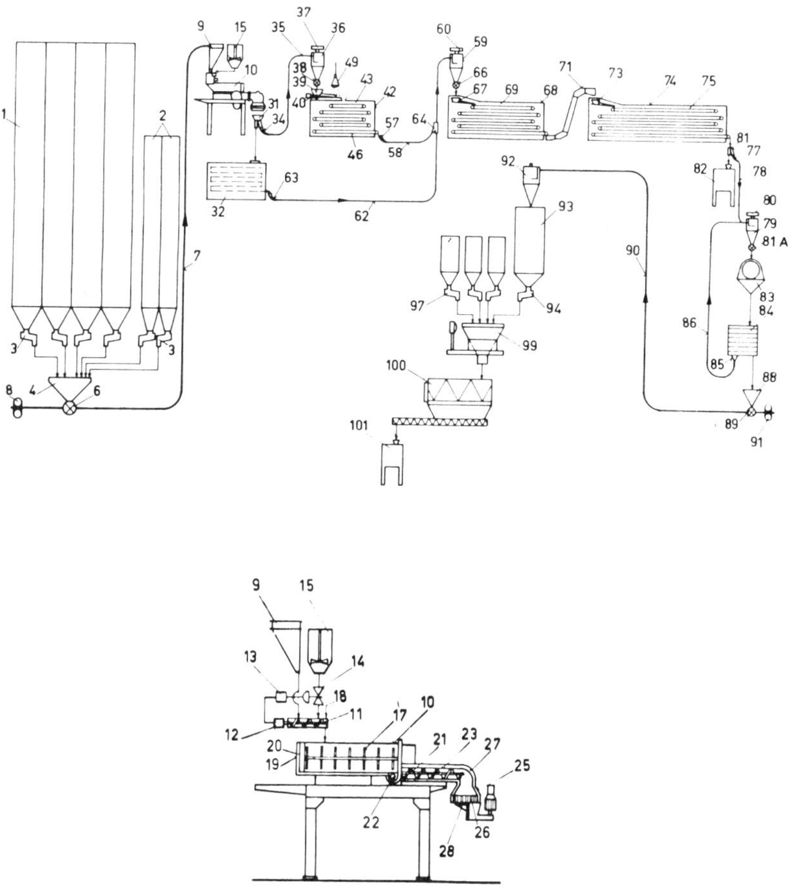

(a) Diagrammatic illustration of the apparatus
(b) Extruding apparatus of Figure 3.2a in longitudinal section

Source: U.S. Patent 3,861,293

The process, if used to manufacture dietetic products, can be pursued until a predetermined content of dextrin and reducing sugars is obtained. In the case of baby foods and weaning flours, the fermentation can be continued until a dextrin content of at least 25% (preferably 50%) and a content of reducing sugars of at least 15% of the initial starch content is obtained, after which the product is dried, comminuted to a flour and mixed with additional ingredients to render a product of low viscosity and high solubility.

Acceptability and tolerance and nutrition tests for infants' diets were successful: no vomiting, flatulence, abdominal pain or lack of desire to take the food mixture. Also, the acceptance of the product was more universal in the sense that children accepted products made of raw materials for which they were not atavistically prepared. As an example, a product based on a mixture of North African raw materials, such as hard wheat (durum) flour, chick pea flour, lentil flour, was accepted without difficulty by Central American children.

To increase the effective amount of reducing sugars, hydrolysis of the starch may be effected with a two enzyme process, at least one of which is a high temperature-resistant α-amylase (enzyme-enzyme-conversion) until a predetermined content of dextrin and reducing sugar, especially mono- and disaccharides, is obtained. Although the apparatus is designed mainly for the manufacture of weaning flours, it may be used to make other products based on these raw materials. While it may be technologically advisable to use very high temperatures, limiting factors, such as stoppage of enzymatic activity and degradation of amino acids have to be kept in mind.

Example: The measuring extracting devices **3** (Figures 3.2a and 3.2b) are set to produce a mixture consisting of 30% hard wheat flour, 40% chick pea flour, 15% lentil flour, 8% skim milk, 6% sucrose and 1% additives. These additives consist of a premix of vitamins, aromas, salts of calcium, phosphorus, iron and of amino acids. The pneumatic conveyor **7** carries this mixture to the pneumatic separator **9** which discharges it onto the measuring worm-conveyor **11**. The geared motor **12** sets the valve **14** so as to add water for obtaining a moisture content of 32% of the total weight of the initial mixture. In the blending vessel **15**, a quantity of α-amylase has been added to the water so as to obtain an α-amylase content of 0.04% of the total weight.

The initial mix is continuously discharged by the worm conveyor **11** into the mixer **17**, where the pH is regulated to 6.5 by adding small quantities of additives, such as citric acid or tartaric acid. The continuous mixer **17** is set so that the average dwelling time is 15 minutes. The paste produced in the mixer **17** is then conveyed by worm-conveyor **22** in a vacuum atmosphere which helps in breaking up the starch structure, then into the extruder **21** where it reaches a temperature of 70°C (158°F) in the course of a through-put time of 3 minutes. As a result of vigorous working-up and of the high temperature, the starch structure is partly broken down, and a good effect from the α-amylase is obtained, resulting in a priming of fermentation. A partial gelatinization of the starch is also obtained.

The product extruded through the die plate **26** is cut **28** and conveyed continuously by pneumatic conveyor **35** to the highest belt conveyors **46** in the closed vessel **42** of the converter **43**. The speed of the belt conveyors **46** is set to obtain a through-put time of 40 minutes in an atmosphere having a relative humidity of 97% at a temperature of from 55°C (131°F) to 65°C (149°F). In this manner, 25% of the initial starch contents are converted into dextrins and 15% into reducing sugars (such as maltose).

The product is continuously conveyed from the lowest of the belt conveyors **46** into the first stage drier **68** where the through-put time is 20 minutes in an atmosphere of 80% relative humidity and a temperature of 95°C (203°F), so the enzyme is inactivated. The product is then conveyed from the first stage drier **68** into the second stage drier **74** where the through-put time is 180 minutes in an atmosphere of 30% relative humidity at a temperature of 95°C (203°F), thereby sterilizing the product and drying it to a total moisture content of approximately 5 to 8% of its total weight.

From the second stage drier **74**, the product is continuously conveyed to impact mill **83** and the flour issuing from the classifier **84** is conveyed continuously to storage element **93** where it will remain until needed. It is then discharged by extractor **94** into mixing scales where it will be blended with supplemental products, such as sugar, protein-concentrate, dry powdered milk and mixes of vitamins, amino acids and mineral salts from the storage elements **97**. The blended products are mixed in the batch mixer **100** and then packaged in the package filling machine.

The analysis of the product in storage element **93** is as follows: approximately 5 to 8% water; 20% proteins; 55 to 58% carbohydrates (starch or starch derivatives) including 15% of the whole product of dextrins and 10% reducing sugars, such as maltose; 7% sucrose (that is a little more than initially); 2.8% ash; 1% crude fiber; 3.8% crude fat. The pH value is 6.5.

When the plant is used for manufacturing conventional short-cut macaroni products, the switches **31** and **64** are changed over to lead the products from the press **10** to the predrier **32** and from the predrier **32** into pneumatic conveyor **62** to the first stage drier **68**. Likewise, the switch **81** is changed over to lead the dried products from the second stage drier **74** to the packaging machine **82**. The through-put, the temperatures, air humidities and through-put times of the plant are changed to values suitable for this type of production.

LOW DEXTROSE EQUIVALENT CONFECTIONS

Starch Aerated Method

Basic whipping compositions can be formulated to contain a protein whipping agent and a low DE (dextrose equivalent) starch hydrolysate. The low DE starch hydrolysate replaces about 70% of the conventional whipping agent. The preparation of the low DE starch hydrolysate by an enzyme process is given by *H.E. Horn, E.R. Jensen, and B.A. Kimball; U.S. Patent 3,586,513; June 22, 1971; assigned to CPC International Inc.*

These starch hydrolysates are made by subjecting a source of starch to enzyme or acid treatment or a combination of both. It is important that the starch hydrolysate have a relatively low DE in the range from 5 to 25. Starch hydrolysates of this type have been found to be excellent agents useful in reducing moisture pickup of normally hygroscopic edible foods whereas use of other hydrolysates having a DE outside this range results in inferior products which have a tendency to become sticky.

In one method, referred to as Method A in the example, an aqueous slurry of a waxy starch, having a solids content less than 50%, is subjected to the hydrolytic action of bacterial α-amylase under suitable conditions of fermentation to produce a starch hydrolysate. The hydrolysate may be further characterized as having the sum of the percentages (dry basis) of saccharides with a degree of polymerization of 1 to 6 divided by the DE to provide a ratio of at least about 2.0. This ratio is referred to as the characteristic or descriptive ratio. Those low DE products having a descriptive ratio less than about 2 are somewhat undesirable in that they exhibit less water solubility and also tend to form haze in solution as compared to those products with a ratio of at least 2.

The same product as described above may also be made via a number of other routes. For example, a mixture of starch and water having a solids content less than 50% may be first subjected to the hydrolytic action of a bacterial α-amylase followed by a high temperature heating step to solubilize any unsolubilized starch. Since this temperature tends to inactivate the enzyme, it is then necessary to cool the solubilized partial hydrolysate and subject it to a second hydrolysis by treatment with additional bacterial α-amylase to obtain the final starch hydrolysate. This method is referred to as Method B in the example.

A third method of making a class of low DE starch hydrolysates, referred to as Method C in the example, consists of hydrolyzing a mixture of starch and water by the action of acid to reach a DE between about 5 and 15. The partial hydrolysate is subsequently subjected to the action of bacterial α-amylase to obtain a starch hydrolysate having a DE of from about 10 to 25.

The term whipping agent composition is used to define the basic and unique combination of a conventional whipping agent and a low DE starch hydrolysate where the starch hydrolysate replaces a portion of the conventional whipping agent. It is, therefore, used to define a combination of ingredients which allow the volume of a mass to increase upon aeration, despite interfacial surface tension.

It is generally accepted that conventional whipping agents such as colloidal proteins allow aeration to take place because they are surface active molecules (or surfactants) and thus allow the interface between air and mass to increase despite interfacial tension. The exact function of the low DE starch hydrolysate is not precisely known. One belief forwarded is that it too serves as a whipping agent in that it also may act as a surfactant in the mass. Such a belief seems to be borne out by the fact that such a high amount of hydrolysate is capable of being used in place of the known whipping agent, i.e., up to about 70% by weight.

However, the low DE starch hydrolysate, when used alone or in amounts above about 70% in whipped confections, results in a dramatically inferior product. It would, therefore, appear that the whipping agent compositions of this process are synergistic in nature and the results achieved by their use are not explainable as merely the combination of two whipping agents.

Example: Method A, One Step Enzyme Technique — An aqueous starch slurry was prepared containing 30% solids by weight of waxy milo starch. The temperature of the slurry was raised and held between 85° and 92°C. A bacterial α-amylase preparation was added in an amount just above 0.025% by weight of the starch over a period of slightly more than 30 minutes. The mixture was then held at the same temperature for an additional period of 30 minutes. The temperature was then reduced to below 80°C and the conversion was allowed to continue until the desired DE was reached. The temperature of the mixture was then suddenly raised to about 120°C in order to inactivate the enzyme and terminate the conversion.

Method B, Two Step Enzyme-Enzyme Technique — Unmodified corn starch was slurried in water to provide an aqueous suspension containing 28 to 32% by weight of the unmodified corn starch. The pH was at 7.5 to 8.0. To this mixture was added HT-1000 bacterial α-amylase in an amount of 0.05% based on starch solids. This starch suspension was added over a 30-minute period to an agitated tank maintained at a temperature of 90° to 92°C. After completion of starch addition, liquefaction was continued for 60 minutes, at which time the hydrolysate was within the DE range of 2 to 5. The liquefied starch was then heated to 150°C and held at this temperature for 8 minutes. The heat treatment destroyed residual enzyme activity and resulted in improved filtration rates and in decreased yield losses upon filtration.

Further saccharification to the final DE was accomplished by the addition of more HT-1000 bacterial α-amylase after cooling the liquefied starch hydrolysate to a suitable temperature for conversion. The liquefied starch was cooled to 80° to 85°C and HT-1000 added in an amount of 0.02% by weight starch solids. After 14 to 20 hours of conversion, the desired terminal DE of 20 was obtained.

Method C, Two Step Acid-Enzyme Technique — Several samples of corn starch (A, B and C) were slurried in water providing slurries ranging from 14° to 22°Bé. These slurries were partially acid hydrolyzed to a maximum of 15 DE. The DE achieved by acid hydrolysis in each of the samples is set forth in the first table below.

After acid hydrolysis, the slurry was neutralized to a pH between 6 and 7. The neutralized liquor was cooled to between 80° and 85°C, and dosed with bacterial α-amylase in the quantity set forth below. A final DE of 19 to 21 was obtained in each of the samples in a period of time between 1 and 3 hours. The final conversion liquors are low in color. These liquors are easily refined and evaporated to about 42°Bé to provide syrups. Dry products may also be obtained. Set forth in the tables below are the reaction conditions for conversion and the product analyses respectively. The hydrolysates from methods A, B or C can then be used in whipping agent compositions.

Enzyme Conversion Conditions

Sample	DE of Acid Hydrolysate	Dry Substance, %	Temp., °C	pH	Enzyme Dose, %	Time, hr	Final DE
A	15.2	38	80	6.5	0.01	1	19.7
B	12.9	37.5	85	6.5	0.05	2	20.2
C	10.3	38.1	85	6.5	0.1	2	21.8

Product Analyses

Sample	Final DE	Dry Substance, %	DP_1	DP_2	DP_3	DP_4	DP_5	DP_6	DP_7+	Descriptive Ratio
A	19.7	72	3.9	5.8	8.2	7.2	7.3	10.2	57.4	2.1
B	20.2	72	2.3	4.9	8.4	6.4	6.6	12.6	57.7	2.1
C	21.8	75	2.3	8.3	10.9	8.1	9.2	16.9	44.3	2.5

Gum Confections

Gum confections are usually formed by first cooking starch with water and a sweetening constituent until the starch is gelatinized, thereby forming a liquid. The hot syrup so formed by the cooking step is next shaped and dried to set up a gel as a final product. Even under the most rigorously and efficiently controlled commercial techniques used to effect the above described process for making gum confections, drying times are excessively long. Such inordinately prolonged drying procedures not only reduce production but necessitate the use of excessive space and drying molds which could otherwise be recycled into the production line.

For example, in one typical commercial technique for making gum confections, the starch is cooked at atmospheric pressure in a sugar solution containing excess water, and the mixture is then subjected to a rolling boil until sufficient water is lost by evaporation to yield a syrup of the proper consistency. The end points are judged by a skilled candy maker by dipping a paddle into the hot syrup and observing the clarity and the consistency of the syrup as it drains from the paddle in a flat, sheet-like form.

In more modern methods for making gum confections, the requisite amounts of starch, sucrose, corn syrup and water are cooked under pressure to super temperatures by steam injection or by use of a heat exchanger. This technique permits cooking on a continuous basis rather than by batches and additionally only requires the amount of water necessary to gelatinize the starch so as to form a syrup. Thus, the cooking time may be reduced.

After cooking, the subsequent steps in the process have remained virtually unchanged and unimproved for many years. The cooked syrup is continuously deposited in small measured portions into depressions in a bed of dry powdered molding starch which is contained in shallow trays. The surface of the bed of molding starch is first smoothed over and then imprinted with depressions of the desired shape, i.e., the crescent shaped orange slices or round hemispherical form of the conventional gum drop. The trays of molding starch generally are carried on a continuous belt and the hot syrup resulting from the cooking operation is automatically injected, as by depositing hopper, into the preformed depression. The freshly deposited gum confection is relatively soft and semifluid, and even when cooled, a considerable aging period is necessary for the development of desired gel structure. Desired gel structure is developed by formation first of an interior core and then by a light skin on the exterior.

Upon sufficient skinning and core development, the confection may be removed from the mold but generally will still require further setting time to retain a permanent shape. Therefore, the gum confection frequently is allowed to remain in the mold until final shape is attained whereby the gum confection is sanded (sugared) and packaged. The necessary total drying time frequently ranges from one to several days. As stated above, such prolonged drying times are expensive and space consuming. During drying, several desired changes occur. For instance: (1) the deposited starch-sugar syrup gradually sets up to a firm gel, (2) the starch-sugar syrup gradually loses water to the bed of molding starch, and (3) the surface of the gum confection (i.e., the skin) gradually develops the desired properties of gloss and smoothness as well as a firmness sufficient for subsequent panning or sugaring operations in final packaging.

To facilitate water exchange between the gel and the starch, the molding starch may be predried to approximately 5 to 8% moisture content, under which circumstances it gradually removes water from the confection, thus facilitating the development of a gel structure therein. If the temperature of the conditioning room is exceptionally high, the predrying of the molding starch is not necessary since moisture in the confection will migrate through the starch into the hot air. After conditioning for the necessary period of time to effect the above desired characteristics (usually in the range from 24 to 72 hours at temperatures from ambient to about 70°C), the gum confections are screened from the molding starch, panned or sanded, i.e., given a coating of crystalline sugar, and packaged. The molding starch must then be redried to the requisite low moisture content for reuse. In order for the final products so formed to be commercially acceptable, they must exhibit the necessary levels of tenderness, firmness, resiliency, and strength.

Three basic methods for making low DE starch hydrolysates (described in the previous process, U.S. Patent 3,586,513) are:

> Method A—one step enzyme technique, starch slurry treated with bacterial α-amylase;
>
> Method B—two step enzyme-enzyme technique, starch slurry treated with bacterial α-amylase followed by further saccharification to the final DE; and
>
> Method C—two step acid-enzyme technique, acid hydrolysis of starch slurry to 15 DE, and, after neutralization, treatment with bacterial α-amylase.

According to *M.M. Godzicki and B.A. Kimball; U.S. Patent 3,589,909; June 29, 1971; assigned to Corn Products Company*, improved compositions may be formed by substitution of a combination of dextrose and a low DE starch hydrolysate for the sucrose, such that the total weight of the sweetening constituent, including the hydrolysate, equals the amount of the sucrose and corn syrup sweetening constituent.

Combinations of starch hydrolysate and dextrose may be formulated using any ratio of these two ingredients as desired. However, it has been found that gum confections having materially reduced drying times and acceptable levels of resiliency, strength, firmness, tenderness and skin formation may be formed using mixtures of about 20 to 60% by weight hydrolysate and about 80 to 40% by weight dextrose. Most preferably, the mixtures contain about 60% by weight hydrolysate and about 40% by weight dextrose. In those instances where less than 35% dextrose is used, it has been found desirable to add a small amount of a conventional artificial sweetener to the mixture in order to insure that the final products have the necessary degree of sweetness.

The starch hydrolysates used in the confections of this process are a relatively new class of starch materials. These starch hydrolysates are made by subjecting a source of starch to enzyme or acid treatment or a combination of both. It is important that the starch hydrolysate has a relatively low DE within the range of 5 to 15. Starch hydrolysates of this type have been found to be excellent agents useful in reducing moisture pickup of normally hygroscopic edible foods whereas use of other hydrolysates having a DE substantially outside this range results in inferior products which show a tendency to become sticky.

Example: The first gum confection was formulated from a standard sucrose-based confection recipe using by weight composition, 47.4% corn syrup (82% solids), 31.2% sucrose, 10.4% (dry basis) cornstarch (67 fluidity), and 11.0% water. This confection is referred to as the control. The second type of gum confection was formulated by using different mixtures of dextrose and a starch hydrolysate produced according to Method A from waxy milo starch and having a DE of about 10. Different combinations were used as a total replacement for the sucrose of the control composition and all other ingredients remained the same.

All confections were formulated by the same prescribed technique of initially forming a solution of the corn syrup in the water. Next, the sucrose and mixture of hydrolysate and dextrose were dispersed in the solution and thoroughly mixed. The starch was then added and the resulting slurry was mixed with constant agitation for 45 minutes at a temperature ranging from 185° to 195°F to insure complete mixing and wetting of the starch. The slurry was then cooked in a continuous cooker at 295°F and at a slurry feed rate of 0.2 gpm. The cooked compositions at approximately 79% solids were then placed in predried starch molds and were dried at ambient room temperature until the starch gum confections were capable of being removed and sanded in a conventional manner.

In all instances, the drying times of the gum confections containing low DE starch hydrolysates were less than half the drying times of the sucrose-based control which used the exact same confection formulation except for its sweetening constituent. Such improved drying characteristics result in substantial savings in time, money, production space, and tray equipment. All of the confections with starch hydrolysates exhibit superiority in one or more of their resiliency, strength, firmness, and tenderness characteristics. This, coupled with such substantially improved drying times, makes them more desirable than the control.

In this respect, it is noted that the confection, which used 80% hydrolysate and 20% dextrose, exhibited a tough skin formation. This confection is particularly suitable for the chewy-type gum confection. The gum confection, which used as its sweetening constituent corn syrup along with a mixture of 40% dextrose and 60% hydrolysate, was superior in every way to the control.

SUGAR ANTICAKING ENZYME

Powdered sugar is highly apt to show a unique solidification phenomenon, so called caking, which leads to not only difficult problems in the storage but also many serious problems in the production, marketing and application. Various attempts have been previously made in order to efficiently prevent such caking, for instance, by incorporating starches, dextrin, glucose, fructose, dextran, sucrose esters, casein, albumin, powdered milk, calcium lactate, magnesium stearate, tricalcium phosphate, calcium and silicate into powdered sugar, but favorable results could not be attained. *T. Tanaka and H. Hayashi; U.S. Patent 3,720,521; March 13, 1973; assigned to Sankyo Company Limited, Japan* report that incorporation of invertase into powdered sugar can effectively prevent the caking of sugar.

It is necessary to select and use a purified form of invertase having a potency as high as possible in order to keep the original color, taste and flavor of the sugar, but any enzymatic preparation containing a suitable amount of invertase may also be used as long as the preparation does not impair the natural states of the sugar. Representative of the enzymatic preparation, which may be used, is dry yeast containing invertase in a powdery or granular form. Usually, the enzyme may be added in an amount to provide 10,000 to 30,000 units per kg of the powdered sugar.

Example: With 1.3 kg of powdered sugar was homogeneously mixed 0.1 g of powdery invertase (Sumner's unit: 270,000 units per gram) and the resulting mixture was packed into a polyethylene bag having a thickness of 0.05 mm and heat sealed. As a control, an equal amount of powdered sugar was similarly packed into a bag without incorporation of invertase. These bags were subjected to an aging test by holding them in a vessel maintained at a constant temperature of 40°C and humidity of 80% for 48 hours.

The results from the aging test clearly show that the powdered sugar having invertase does not exhibit any degree of caking, whereas the control powdered sugar exhibits a remarkable degree of caking and produces a great number of sugar blocks, each having a diameter of 10 to 20 mm. Further, the bags were allowed to stand at room temperature for 1 week. Caking was observed in the control, whereas no degree of caking was observed in the test bags.

THICKENED SOUP ENZYME PROCESS

A unique process was developed by *P.J. Spek; U.S. Patent 3,615,589; Oct. 26, 1971; assigned to Lever Brothers Company* to decompose starch thickeners in food products so that these products have immediate ready-to-eat consistency without dilution as for regular concentrated soups and gravies. The process uses amylase which can be used also to decompose starch thickening agents for frozen food products.

The amylase may be added, for example, in capsulated form, so that it will exert its disintegrating action by raising the temperature. It is also possible to make a frozen dessert or custard (with, for example, cherries or other fruits as the solid material) with an adequate thickening agent and spraying amylase on the frozen product which will also freeze to a solid film or droplets on the frozen product. As soon as the product is melted and especially warmed up, it will disintegrate the thickening agent so as to obtain a thin dessert or custard with fruit.

In order to prevent the enzyme from decomposing any solid starch-containing material which is part of the composite bulk material, the enzyme is inactivated after its action of decomposing the thickening agent. This inactivation can be achieved by heating of the portions above the inactivation temperature of the enzyme, which temperature may be above room temperature, for example, during a cooking or sterilizing operation of the portions, or the inactivation can be achieved by an inactivation agent which is added in such a way that it will exert its action at the right moment. For inactivating an enzyme such as α-amylase, the pH may be lowered below 4.5.

Enzymes which are suited to decompose a thickening agent are, for example, amylase, mannohydrolase, cellulase, alginate glucanohydrolase, proteinase and pectinase. Combinations of thickening agent and enzyme which have proved to be operative are starch and α-amylase, guar gum and β-1,4-mannohydrolase, which can be obtained, for example, from sprouted guar beans, locust bean flour and β-1,4-mannohydrolase, carboxymethylcellulose and β-1,4-glucan-4 glucanohydrolase, alginates and alginate glucanohydrolase, gelatin and pancreas proteinase and low methoxy pectin and pectinase.

Sometimes, as in the case of soups which contain noodles or other particulate starchy material and in which α-amylase is used, care has to be taken that the effect of the enzyme activity is not too great (overdosing has to be avoided), otherwise the noodles, etc. may disintegrate and cause an unwanted turbidity or may stick together. For this reason, the quantity of α-amylase has to be judiciously chose. Therefore, in each case in which the composite bulk contains particulate solid material which is desired in the end product to be unaffected by the enzyme, care has to be taken that the enzyme cannot decompose the solid material; to this end, an enzyme used in a judiciously chosen quantity has to be selected which is inactivated, for example, by sterilization of the filled containers.

In the case of soups, the bulk material is first cooked in kettles. Then, generally the temperature is allowed to drop before the filling operation, in most cases to about 70°C. After the filling operation, the α-amylase will be active on the gelatinized starch in the containers till the inactivation temperature is reached during the warming-up period of the sterilization process. For malt α-amylase, this inactivation temperature is about 80° to 85°C at a neutral pH. Consequently, when the filled containers are sterilized, for example, at 110°C, the α-amylase will be inactivated. The inactivating temperature depends on the type (origin) of the α-amylase, pH, the salt concentration, etc. Also, when the inactivating temperature of the α-amylase is high, usually less activity will be required, since during the warming-up operation to the sterilization temperature the enzyme can exert its action for a longer period.

Generally, the quantity of starch used in soups will be between 10 and 40 g/l. The quantity of the enzyme to be used depends on the kind of enzyme and the process as such. Usually the process is conducted in such a way that the viscosity of the liquid phase in ready-for-consumption condition of the end product will be decreased to between 1 and 5 cp measured at 60°C with an Epprecht viscometer.

A very important advantage of the process is obtained when, as in the case of soups, for example, after the filling and closing of the containers the sterilization time can be reduced by lowering the viscosity of the contents. A reduced sterilization time does not only increase the economy of the process, but also decreases organoleptic impairment by the action of heat. When starch is used as a thickening agent, another advantage presents itself in the case of clear soups. Normally, such soups become turbid possibly due to retrogradation of the starch after sterilization, and the turbidity tends to increase during the shelf life of the soup. When the starch has been enzymatically broken down, this retrogradation does not take place and the soup will remain clear.

Example: A concentrated chicken soup, containing broth, spices, salt and meat extract, was prepared in a normal way with vermicelli, 60 g cubes of chicken meat and 28 g potato starch per liter. After preparation and before filling into the half-liter cans, the temperature of the soup was 70°C.

Just before filling each can of 500 ml capacity, 10 ml solution containing 3 mg of bacterial α-amylase, activity 2,500 SKB units/g amylase was injected. After filling with the soup, the cans, each containing the same amount of ingredients, were closed and in the normal way conveyed to the loading station. During this transport over several conveyors, a good mixing was obtained. At 20 minutes after the first can and 10 minutes after the last can had been closed, autoclaving was started at a temperature of 118°C. The required sterilization time was 15 minutes. (A control without addition of α-amylase required 60 minutes sterilization.) After sterilization, the soup was water-thin and clear, whereas the control was turbid and rather thick.

BABY CEREAL ENZYME APPLICATIONS

Starch Gelatinization and Enzyme Treatment

Dehydrated cereal products can be produced by preparing a paste or thick suspension, i.e., a slurry or puree, obtained by heating the raw materials to form a pulp which may be strained or sieved to ensure uniform particle size. The slurry or puree is then applied to the surface of conventional drying equipment such as a drum dryer, where substantially all (90 to 98%) of the water is removed. The dried product is conveniently flaked and packaged for use.

One conventional process for precooked, dehydrated grain cereal production involves the initial controlled heating of a slurry of raw (or steamed) cereal grain to a temperature suitable for the simultaneous gelatinization of the starch molecules present and where enzyme activity can accomplish the conversion of the gelatinized starch to dextrins and sugars.

This enzyme process results in an increased cereal production rate and a product that is reconstitutable with an amount of liquid that is readily ascertainable and, most importantly, reproducible. However, such a conventional process suffers from the disadvantage of requiring close control of processing variables such as enzyme concentration, concentration of the solids in the cereal slurry, reaction temperature, reaction time and ultimate dryer speed when dehydrating the slurry into a continuous cereal sheet. Furthermore, in such conventional processing, it has been shown that the salt in the formula does more than merely enhance the flavor; it also aids in the drying operation, particularly for rice cereal.

A process to prepare a dehydrated instant cereal for infant feeding with reduced salt content is described by *V.J. Kelly, L.L. Cloud, and W.J. Smalligan; U.S. Patent 3,930,027; December 30, 1975; assigned to Gerber Products Company*. The process requires the use of malt which contains diastase enzymes for starch conversion to maltose and higher sugars. This is accomplished within certain limits for maltose and dextrins to avoid problems of caking in the ultimate packaged product.

It appears that enzyme hydrolysis is limited to that portion of the cereal slurry starch that has previously been gelatinized. Furthermore, by heating the starch slurry to a temperature significantly above the gelatinization temperature of the starch, i.e., in about the range of 160° to 220°F, not only will all the starch present be gelatinized, but any naturally occurring enzymes will be in part permanently inactivated or destroyed. The process is particularly useful where the grain requires high temperature gelatinization, higher than the inactivation temperature of most diastatic enzyme systems. Thus, grains such as long grain rice can now be utilized more readily in cereal preparation.

As a result, subsequent quantity control of the amount of gelatinized starch that is allowed to be subjected to enzyme conversion will enable the hydrolysis to be conducted without stringent process conditions, i.e., hydrolysis of the gelatinized starch can go to completion. The ultimate effect of such uniformly complete hydrolysis appears to be that the dextrins are more completely hydrolyzed. Therefore, dehydration is not dependent on the presence of salt as an aid.

Two alternative procedures have been found to be satisfactory. The first involves dividing a slurry of ungelatinized starch into a first and second stream. The second stream is independently subjected to high temperature gelatinization. The hot gelatinized starch slurry is cooled by incorporating the first stream of unheated ungelatinized starch. The combined starch slurry is then contacted with an enzyme and the gelatinized portion converted to the hydrolyzed product. Finally the converted starch-containing slurry is subjected to conventional dehydration such as on a drum dryer.

The alternative procedure which has been found to be effective involves the initial high temperature gelatinization in the total cereal slurry. Thereafter, the gelatinized starch is divided into a first and second stream, the second stream cooled, for example, in conventional heat exchange equipment such as a line heat exchanger to a temperature of approximately 90°F. The cooled stream is then subjected to enzyme conversion. Upon completion of enzyme hydrolysis, the streams are recombined and subjected to conventional dehydration such as on a drum dryer.

An enzyme that hydrolyzes gelatinized grain starch can be used in amounts up to about 1.0% by total weight of slurry formulation. The enzyme can be introduced into the cereal slurry either prior to gelatinization (with the result that a portion of the enzyme is inactivated by the high temperature) or subsequent to gelatinization. Holding times for enzyme conversion have been varied from 10 to 25 minutes with satisfactory results.

The rate and completeness of hydrolysis of the gelatinized slurry is dependent upon temperature conditions as well as time of conversion (digestion). Although it is preferred to accomplish the conversion at about 120°F, the temperature can be varied, depending upon the amount of starch that has been gelatinized. For example, when only 30% of the starch is gelatinized, it is desirable to maintain a temperature of about 120°F. Conversely, when the gelatinized portion approaches or exceeds 50%, a lower temperature down to 90°F can be used.

Example 1: A formulation was prepared in the following proportions:

Ingredient	Quantity Percent by Weight
Rice flour	70.0
Rice polish	25.0
Dicalcium phosphate	2.6
Sodium iron pyrophosphate	0.8
Rice oil	0.7
Vitamin mix	0.2
Malt*	0.7

*Source of amylase

Sufficient water was added to reduce the overall solids content to about 22.6%. The slurry at about room temperature was pumped through a line strainer having a screen size of 0.060" to an agitating heater at a temperature of about 210°F and held for about 15 minutes. The slurry was then removed and approximately two-thirds, by weight thereof, transferred to a holding tank and the temperature adjusted to about 90°F.

After about 17 minutes at this temperature, the slurry containing hydrolyzed starch was recombined with the portion containing gelatinized starch and fed to a conventional double drum dryer, each drum operating at an internal pressure of 80 psig. The drums were rotated at 5 rpm and the dried sheet easily removed with doctor blades. The resulting sheet was nonplastic, continuous and had a film thickness of about 5 mils. Similar tests were conducted with corn, barley, wheat and oat and mixtures thereof.

Example 2: Using the slurry formulation of Example 1, a quantity of slurry was divided into 2 portions in about a 2:1 weight ratio. The larger portion was subjected to gelatinization as in Example 1 at 170°F. Thereafter, the remaining ungelatinized slurry, still at room temperature, was added thereto, lowering the temperature to about 120°F. Hydrolysis was allowed to occur for about 20 minutes. The resulting slurry was again drum dried and the dehydrated sheet easily removed from the drum surfaces. In each of Examples 1 and 2, the resulting cereal was found to have excellent reconstitution characteristics and a highly appetizing taste.

High Malt (Diastase) Treatment

In another process, *L.L. Cloud, V.J. Kelly, and W.J. Smalligan; U.S. Patent 3,956,506; May 11, 1976; assigned to Gerber Products Company* developed a coarse textured pregelatinized infant cereal. An infant cereal having more adult mouthfeel, yet still being particularly suitable for infant feeding, can be provided by modifications in the concentrations of conventional ingredients and the utilization of unique processing conditions. It has been found that by providing for reducing sugars in an amount of from about 12 to 30 weight percent of the grain cereal slurry and subjecting the slurry to sterilization conditions of greater temperature and line pressure, the resulting drum dried product possesses the desired coarse texture, yet remains fully assimilable by an infant between 6 months and 2 year of age. The product is made from a slurry having a significantly higher malt concentration than that usually used in this industry. A malt concentration in excess of 3% by weight, such as 3 to 7% by weight is preferred. Starch hydrolysis is accomplished through the use of malt enzyme (diastase) digestion at 140° to 160°F for 10 to 30 minutes in order to convert at least part of the starch to reducing sugars, preferably 18 to 24%. Other diastatic enzymes that appear to be useful for starch reduction include the mixture of α-amylase and β-amylase from fungal, bacterial or animal sources.

An infant cereal of the instant type is provided that, upon reconstitution with water (or milk or formula), retains a desired coarse texture and desirable flavor for the toddler age infant. While various grains can be used, oat and wheat are the most popular grains for instant cereals. However, any cereal grain flour normally used in this art, for example, rice, corn, barley, may be employed.

The pregelatinized instant cereal flake will normally be made from appropriate cereal flour and various mineral supplements. Salts, such as calcium sulfate, calcium phosphate or calcium carbonate, and other minor additives including vitamins, electrolytic iron, phosphatides, e.g., lecithin, may be included. Optionally, a minor amount of a fruit puree may be added, such as bananas, prunes, strawberries, apples, or imitation flavors may be incorporated. Still further, dried fruit can be physically blended with the dried mixture.

Conveniently, a mixture containing 80 to 95 weight percent of the appropriate grain cereal flour or flours, from about 2 to 4 weight percent of a mineral supplement such as calcium sulfate, and from about 3.0 to 7.0 weight percent of malt are mixed. In addition, minor amounts of food supplements may be added, normally being in the range of from about 0.1 to 2 weight percent in total amount, individual amounts varying from 0.01 to 0.5 weight percent.

The dry mixture may then be slurried with from about 2.5 to 10.0 times its weight of water to provide a slurry which is then heated in the range of about 140° to 160°F in conventional equipment such as atmospheric, vacuum, or pressure cooking tanks. The time used is necessarily sufficient to insure the hydrolysis of the cereal slurry. A temperature of about 160°F is required to inactivate starch enzymes.

The slurry is subjected to additional heating, for example, in a tangential heater, under a pressure of more than 30 psig. The secondary high temperature, pressurized heating not only accomplishes sterilization and deactivation of the enzymes, but apparently creates the slurry conditions which ultimately result in a coarse texture when the dehydrated product is ultimately rehydrated.

The object of dehydrating is to remove as much of the water present in the slurry as possible without detrimentally affecting its quality. Therefore, the final products usually contain at least 90% solids and preferably from 95 to 98% solids. Drying is accomplished by any of the readily available conventional equipment, such as single or double drum dryers. The dehydrated base material can be removed from the drying surface as a continuous sheet, usually having a thickness of about 4 to 7 mils. The particles may then be flaked in accordance with the size desired.

By increasing the normal malt percentage in products of this type beyond the conventional 1 to 2% by weight, the resulting reconstituted product has been found to exhibit the desired texture without adverse effect upon other characteristics. In particular, it has been found that malt concentration in the range of 3 to 7 weight percent is essential, in conjunction with sterilization line pressure, to accomplish the result desired. Data demonstrating the effect of malt concentration (in percent) on texture is provided below:

Enzyme Concentration

Malt (%)	Slurry Solids (%)	Line Temp. (°F)	Line Back-Pressure (lb)	Texture
1-2	16	230	50	poor
3	16	230	50	good
4	16	230	50	good
5	16	230	50	very good
6	16	230	50	very good
7	16	230	50	very good
8	16	230	50	poor
10	16	230	50	poor

Example: Oatmeal Cereal — In a dry ingredient blender were mixed 100 lb of oat flour, 5 lb of malted barley flour, 1.6 lb of calcium sulfate, 1.5 lb of vanilla sugar, 0.25 lb of vitamin mix (containing niacinamide, riboflavin, pyridoxine hydrochloride, and thiamine mononitrate) and 0.06 lb of electrolytic iron. The blended ingredients were then passed through a finisher where 60 gallons of water were slurried.

The slurry was heated for about 20 minutes at 140°F in a cook tank and transferred through a tangential heater maintained at 240°F where the slurry was subjected to a sterilization line pressure of 50 psig. The pressure was accomplished by providing a girdler valve within the sterilization line and adjusting to 50 psig. The resulting gelatinized sterilized slurry was fed to a conventional double drum dryer, with the drums operating at about 50 psig internal pressure. The drums rotated at 4 rpm, and the dried sheets were removed with doctor blades. The sheets were then flaked to the desired size, screened and packaged. The size of the flakes was in the range of 0.25" to 1.0".

The cereal was prepared for eating by placing 1 ounce of the cereal in a bowl and adding ½ to ⅔ cup hot or cold water, milk or formula. Upon addition of water, the instant cereal rapidly reconstituted to a porridge having the desired flavor and coarse texture.

CARBONATED DRINK MALT PROCESS

The enzyme amyloglucosidase is used in the process for a kola carbonated soft drink. The enzyme is used to convert maltose into glucose after malt action on the starch. The steps of the process are described by *J. Clark; U.S. Patent 4,001,436; January 4, 1977; assigned to Banks Barbados Breweries Ltd., Barbados.*

Sugar Solution: 2,100 lb of cane sugar were dissolved in spring water and pumped to a stirred and heated cooker vessel. 40 lb of sodium chloride and 5 lb of potassium chloride were then added. The solution was brought to 2,500 U.S. gallons at a gravity of 1040.8° (water = 1000°); the solution was brought to the boil. 3 liters of 1.75 specific gravity phosphoric acid were then added and the solution boiled gently for 1 hour to convert sucrose to glucose and fructose. 2,400 U.S. gallons were then transferred to another vessel where the solution was allowed to stand for 15 minutes. The solution was then cooled without aeration to yield 2,250 U.S. gallons at a gravity of 1045°.

Malt Extract: 1,200 lb of 6-Row malted barley and 2,800 lb of 2-Row malted barley were mashed in spring water at 120°F. 18 lb of Burton Sales were added to adjust the water hardness and the mash stirred for 1 minute, followed by standing for 20 minutes. The temperature was then raised to 142°F, followed by standing at this temperature for 10 minutes. The temperature was then further raised to 156°F with stirring every 15 minutes until starch conversion to sugars was completed.

The mash was then transferred to a lauter vessel for clarification from solids, the lauter vessel having been precooled by the introduction of cold water. The mash was allowed to stand in the lauter vessel for 10 minutes and then circulated until clear and substantially free of solids. The clear wort was then transferred to another heated vessel and 110 lb of Ambazyme PC 25 was added with stirring at an early stage in the filling of the heated vessel. Ambazyme PC 25 is a solid amyloglucosidase enzyme having an activity of 25 AG units per gram.

The temperature was carefully controlled at 131°F for 4 hours by adding sparging water with stirring every 15 minutes. 3,500 U.S. gallons of malt extract was collected at a gravity of 1043.2°, allowed to stand for 4 hours and then boiled gently for 1 hour. The final yield was 3,300 U.S. gallons at a gravity of 1046.0°. The malt extract was then allowed to stand for 15 minutes followed by cooling without aeration to 38° to 42°F.

4,500 U.S. gallons of sugar solution and 1,500 U.S. gallons of malt extract were mixed followed by stirring for ½ hour and cold storage for at least 48 hours to allow cooling and cold break precipitation of protein. The mixture was then roughly filtered through a diatomaceous earth filter. The mixture was carbonated with CO_2 to a level of 2.3 volumes (NTP) CO_2 per volume of liquid. The mixture was dosed throughout filtration with a proteolytic enzyme for shelf-life purposes, and 1 lb of potassium metabisulfite as an antioxidant.

An artificial flavoring essence is added having the following composition by weight: 1 part kola nut extract (unflavored), 2 parts Honey AM, and 3 parts Redcurrant No. 1. Rough filtration was followed by fine filtration again through a diatomaceous earth filter. The mixture was carbonated to 2.9 volumes (NTP) CO_2 per volume of liquid. The mixture was dosed throughout filtration with 15 lb of ascorbic acid (Vitamin C).

Bottling: The product was bottled with pasteurization at 160°F. The product was found to be a tangy, refreshing drink having excellent properties as to thirst quenching, restoration of energy and stimulation of physical activity.

BREAD AND BAKED GOODS

REDUCED FERMENTATION TIME

Straight Dough Process

The two basic and conventional methods for the manufacture of yeast raised bread products are known as the "straight dough" method and the "sponge dough" method. In both methods, long periods are devoted to the fermentation processes which are essential to achieve workability or processability of the dough and to develop the desired qualities of grain, texture, loaf volume and flavor.

In the straight dough method all the ingredients of the formula, e.g., flour, sugar, shortening, milk powder, salt, yeast and water, are mixed into a pliable dough until it is fully developed or clears the sides of the mixer. The dough is then allowed to ferment for 1½ hours to 3 hours under controlled temperatures of 80° to 86°F. The dough is then divided, rounded, and given intermediate proof. It is then molded into designated shapes.

The molded doughs are then panned and then proofed for 50 to 70 minutes at temperatures of 100° to 105°F and 90 to 95% humidity. The proofed dough pieces are then baked at varying temperatures from 375° to 450°F for times varying from 18 minutes to 30 minutes based on their size and quality level. Although the chemical activity associated with fermentation continues from the time of original mixing until baking of the loaf, it is nonetheless necessary to devote an elapsed period of between 1½ and 3 hours solely to the development of the proper dough properties through the fermentation process.

In the sponge dough method, there is initially formed a sponge batch utilizing only part of the ingredients. The sponge is fermented over a long but wide ranging time, and is subsequently mixed with the remaining ingredients prior to baking. For example, approximately 70% of the flour in the batch is mixed for about 6 minutes in a horizontal mix or along with the yeast and equivalent amount of water to produce a stiff dough with very little development. This is the sponge. The sponge temperature is set at approximately 78°F and allowed to ferment in troughs for 2½ to 5 hours. The sponge is then remixed with the other 30% of the flour and the rest of the ingredients, e.g., shortening, sugar, milk powder, salt and water.

After the dough is again formed in slow speed in about 1 minute, it is then mixed in high speed for approximately 10 minutes to full development. Then the dough is allowed to rest or relax in troughs for 15 to 30 minutes, depending upon the product to be produced, before automatic dividing. This rest period is called floor time. The dough is then processed as in the straight dough method for white bread through dividing, rounding, intermediate proofing, molding, panning, proofing and baking.

L.J. Smerak and J.A. Miller; U.S. Patents 3,494,770; February 10, 1970 and 3,578,462; May 11, 1971; both assigned to Caravan Products Co., Inc. present a process using enzymes which significantly reduces fermentation time and improves bread characteristics (volume, grain, texture, flavor and shelf life). The reduced fermentation time is sufficient to eliminate the advantages of the sponge dough process. The new process can be called a "fast straight dough method".

Unlike ordinary prolonged straight dough procedures, the fast process is characterized by doughs that maintain good extensibility (as evidenced by their ability to sheet out readily without tears or weak spots) and excellent easy processing properties on machinery. The loaf itself may absorb 1 to 3% more moisture than ordinary bread and have a volume that is as much as 5 to 10% larger than comparable breads made by conventional methods and recipes. Moreover, the breads are found to possess texture, grain and keeping qualities that exceed that of typical good loaves prepared by conventional means.

These results are associated with the combined effect of the three components: ammonium salt; oxidizing agent; and selected enzymes. The ammonium salt apparently serves as a nutrient in the growth and reproduction of the yeast cells used in fermentation of bread dough and accelerates the production of carbon dioxide. A particularly effective combination of oxidizing agents is a 1:4 ratio of potassium iodate to potassium bromate, with 0.001% potassium iodate and 0.004% potassium bromate based on the weight of total flour.

It is believed that the reason that this combination is particularly effective is because the potassium iodate tends to show its effectiveness very early whereas that of the potassium bromate, which is sensitive to pH is delayed somewhat by the slow development of fermentation acids. As a result, the iodate is effective predominantly right after the mixing stage whereas the bromate extends its effectiveness through the dough stage, the proofing stage and the first portion of the baking stage.

It would appear that the oxidizing agent functions to oxidize the natural reducing agents found in yeast, milk, and most importantly, flour. Oxidizing doughs are aided in their development to tolerate machine handling through improved extensibility and dryness. Such doughs produce bread of fine grain, good texture, increased volume, improved tenderness and longer shelf life.

The third necessary component in a dough mix is an enzyme mixture. In particular, the enzymes that have been found to be effective are combinations of fungal amylase and fungal protease. These two enzymes, when incorporated into a bread dough formula together with the oxidizing agent and the ammonium salt, appear to act as biochemical catalysts that accelerate the rate of the chemical reactions, particularly the reactions which produce carbon dioxide gas, modify the flour starch, condition the gluten and retain gas and leaven the bread for baking.

Flour naturally contains cereal enzymes known as alpha-and beta-amylase. Together they are known as diastase. The beta form is usually present in effective amounts, but the alpha form is not. This deficiency has, in the past, been remedied by the incorporation of ground malted wheat. Such a procedure has often resulted in sticky doughs and excessive softening of the bread crumb structure through excessive dextrinization. This is brought about by the fact that during the baking process, cereal alpha-amylase reaches its peak activity at the same temperature (160°F) that starch gelatinization is accelerating.

However, supplemental enzymes eliminate this drawback and at the same time promote the proper chemical reactions at the time they are needed. One of these is the enzyme fungal alpha-amylase, which is derived from the mold *Aspergillus oryzae*. It functions to modify the amylase and amylopectin fractions of the flour starch and effects a reduction in dough viscosity.

The initial action occurs prior to baking, but during fermentation, the fungal alpha-amylase hydrolyzes the small percentage of flour starch (2 to 5%) damaged in the milling process to dextrins, and the beta-amylase converts the dextrins to sugar. Since the quantity of damaged flour starch is so low, no over-dextrinization occurs at this stage.

Subsequently, during the baking, the undamaged flour is gelatinized (beginning at 150°F) and only then does it become susceptible to enzymatic action. However, at the temperatures encountered during this process, the activity of the fungal alpha-amylase is decreasing. Indeed, the fungal alpha-amylase is so much more heat sensitive than cereal amylase that it is only 50% active at 150°F and is only about 3% active at 176°F. By reason of the lower inactivation temperatures of the fungal alpha-amylase, there is avoided the risk of over-dextrinization and the consequent undesirable gummy bread crumb.

The fungal alpha-amylase, in conjunction with the naturally occurring beta-amylase, carry out the function of converting starch to sugars so as to increase the availability of maltose and dextrose to the yeast cells, which in turn convert them to carbon dioxide, alcohol and flavor components. The resulting bread is resistant to firming on aging and has improved volume, grain, texture and crust color.

Fungal protease enzyme is not normally found in flour and wheat protease and has practically no effect on modifying the flour gluten, and even when malted wheat flour is added to compensate for lack of alpha-amylase has far less protease activity than is needed. However, effective amounts of a fungal protease are used in combination with the fungal alpha-amylase, ammonium salt and the oxidizing agent. It is believed that, in part, the fungal protease catalyzes the hydrolysis of flour gluten by splitting the peptide linkages between the amino acids comprising the complex and massive molecular structure of the protein. The fungal protease does not destroy the hydrogen and disulfide bonds of the amino acid chains, and it therefore complements and balances the effect of mixing and of oxidizing agents on the gluten. Although a protease level of 250,000 HU (hemoglobin units) per 100 pounds flour is high, the activity is a function of time, and high levels are suited to the shortened processing schedule permitted by the process, particularly in view of the inhibiting effect of salt on proteolytic action.

The inhibiting effect of salt may be as high as 60% or more and for this reason, variations in the salt levels may alter the optimum requirements of protease enzyme. A typical and satisfactory recipe is given in the table below.

Ingredients	Amount Added to Dough Batch*
Potassium iodate, lb	0.001
Potassium bromate, lb	0.004
Diammonium phosphate, lb	0.160
Fungal alpha-amylase, SKB units	54,400
Fungal protease, HU	250,000

*Containing 100 lb of flour

U.S. Patents 3,494,770 and 3,578,462 describe similar processes except for use of ammonium and calcium salts respectively, in combination with fungal protease, fungal alpha-amylase and an oxidizing agent.

Example: Two bread doughs were prepared, both with 700 lb of flour. One of these was a standard dough formulation containing additives (cysteine, whey and potassium bromate) intended to reduce fermentation time and improve product quality and which had incorporated therein a "brew" (water, sugar, yeast, Brewloid yeast food) which had been given a preferment of 1 hour.

The other was a standard dough containing the process composition described in the above table. The ingredients for both doughs were initially mixed at low speed for 1 minute and at high speed for 11 minutes. No independent fermentation or sponge time was allotted to either batch (except that the control bread contained a brew as mentioned above). Both doughs were processed on the schedule below.

Floor time, min	30*
Dough temp, °F	84
Make-up time, min	30
Proof time, min	55
Bake time, min	23
Bake temp, °F	420

*40 for control

The dough prepared by the process demonstrated superior dough sheeting (extensibility) and pan flow as compared with the standard quick dough. After baking, both breads were found to be acceptable. However, the loaf made by the process had 10% larger volume. Moreover, the bread was found superior in grain, texture and crust color and exhibited superior shelf life. In general, the relaxation or floor time of a yeast leavened bread dough need be no greater than 40 minutes. It has been found that the freshness of bread can be extended for considerable periods of time by the addition of small quantities of certain surface active agents or emulsifiers.

Of particular interest are calcium stearyl-2-lactylate and sodium stearyl fumarate. Both of these emulsifiers are known as being useful in extending the shelf life of bread. However, even normal levels of such emulsifiers exhibit unusual activity in the presence of the oxidizer-ammonium salt-enzyme additives. The effect is apparently synergistic, and the result has been to extend the useful shelf life of the bread by as much as 3 days as contrasted with breads having no emulsifiers and by as much as 2 days for breads having equivalent amounts of emulsifying agents, but not containing the three-component enzyme system.

The effectiveness of the emulsifiers is greater for larger quantities; however, significant effects are achieved predominantly with concentrations in the range of 0.25 to 0.50 part per 100 parts of flour. A convenient way of measuring bread freshness and hence the shelf life of the bread is by means of a compressimeter. This is an instrument that measures the deformation of the bread slice by a predetermined applied force. Readings are high for more tender slices of bread and consequently are considered to be of higher quality.

As the readings are recorded over a period of days, the loaves of high quality have a slower rate of increased firmness and the compressimeter readings would not decrease as rapidly as those for loaves of lower quality. It has been established that compressimeters are a measure of the consumers "squeeze" test and eating quality of bread loaves since the majority of consumers perfer bread of a soft silky texture.

A bread utilizing the combination of an oxidizing agent, an ammonium salt, amylase-protease enzymes, and an emulsifier selected from the group consisting of calcium stearyl-2-lactylate and sodium stearyl fumarate, was found to have a freshness as measured by a compressimeter that was markedly improved over that of a standard bread dough containing the same emulsifiers.

Monoglyceride Enzyme Activation

The Federal Standards of Identity for bread permit the fungal enzyme material derived from *Aspergillus oryzae* to be included as an optional ingredient. The proteolytic activity of this enzyme material has a desirable effect on dough characteristics by decreasing optimum mixing time and improving dough machineability. This is particularly important in the case of unusually strong flours and often constitutes a valuable aid in adapting varying flours to the production schedules of mechanized bakeries.

Further, doughs treated with this enzyme material have been widely observed to produce baked products with improved grain, texture and crumb structure, compared with the products of untreated doughs. The use of this enzyme material in dough produces partial hydrolysis of the gluten proteins and in this way reduces mixing requirements, however, the other improvements are believed to be related, at least in part, to additional reactions which are not well understood.

Nonfungal enzyme materials are known to attack the gluten protein system rapidly and, therefore, reduce mixing requirements considerably, but they do not exhibit the tolerance of enzyme material of fungal origin and, thus, tend to produce excessive proteolysis, leading to extensive impairment of product quality. A process incorporating the enzyme material derived from *Aspergillus oryzae* is described by *B.W. Landfried and J.R. Moneymaker; U.S. Patent 3,527,644; September 8, 1970; assigned to Top-Scor Products, Inc.*. The essential step in the process is the addition of a saturated monoglyceride which allows wide tolerances for mechanized bakeries in their use of flours with day-to-day or shipment variations.

A significant increase in desirable enzyme material activity is achieved by the proportional incorporation of 0.05 g of the monoglyceride with a quantity of the enzyme equivalent to 1,000 HU (hemoglobin units) protease. Increased levels of the monoglyceride produce increased enzyme material function in yeast raised bakery products up to 0.30 g/1,000 HU protease. Greater proportions of the monoglyceride do not appear to further improve the function of the enzyme material, but produce the known desirable effects of monoglyceride addition in dough. The treated enzyme material has been observed to more efficiently improve dough characteristics, decrease dough formation time and increase average quality with or without shortened fermentation time.

Example 1: A 90% α-saturated monoglyceride in a high surface dry dispersion was prepared by hot-blending with flour. The flour and monoglyceride were heated, with mixing to avoid destructive overheating, to 160°F, then cooled with agitation to 115°F. The blend or plated flour was sifted, resulting in a particle size analysis as follows: 31% over 149 μ; 23% over 105 μ but under 149 μ; 20% over 74 μ, but under 105 μ; and 26% less than 74 μ. The blend was mixed with a standardized, diluted, powdered enzyme material of *Aspergillus oryzae* in different proportions so that the level of α-monoglyceride per 1,000 standard HU of protease activity varied from 0.03 g to 0.32 g. The enzyme material used contained 37,000 HU/g. The various proportions of enzyme material and monoglyceride follow.

Preparation	Alpha Monoglyceride/ 1,000 HU, g
1	0.03
2	0.06
3	0.13
4	0.20
5	0.26
6	0.32

These preparations, along with three controls respectively containing no additive, enzyme material only, and the highest level monoglyceride without enzyme material, were included in a standard commercial white bread formula at levels adjusted to introduce 1,400 HU of protease/lb of formula flour. This quantity of enzyme material is within the range which may be used commercially with normal mixing; however, to illustrate the process, all the doughs were intentionally mixed less than the known optimum for the flour used.

A conventional sponge dough preparation method was used and the preparations were introduced with the sponge ingredients in the normal manner. The condition of the dough following mixing and rounding was carefully examined by a trained observer and judged. Each test was performed in duplicate. Following baking in 1 lb units, the loaves were allowed to cool under controlled conditions and stored in polyethylene bags for 18 hours. Loaf volume was then measured by rape seed displacement and the test bread was cut and scored subjectively using a system which separately takes into account the major internal and external characteristics of the loaves. The bread formula is as follows.

Ingredient	Parts by Weight
Patent flour (14% moisture basis)	700.00
Bakers yeast	17.50
Sucrose	28.00
Cerelose	28.00
High heat nonfat milk solids	28.00
Lard	17.50
Salt	14.00
Arkady yeast food	3.50
Water	469.00

Evaluated characteristics of these loaves include loaf volume, crust color, crust character, break and shred, grain, texture, crumb color, aroma, taste and eating quality. In this scoring system each factor is considered separately and penalized according to the degree of deviation from a hypothetical perfect loaf which would score 100. An acceptable commercial loaf will receive a score of 90 or higher, with a score of 94 or higher considered excellent. Variation in quality scores within a single test series of 1 point or greater is considered significant. The results are as follows.

Dough Characteristics

Variables	At Mixer	At Rounder	Average 1-lb Loaf Volume, cc	Average Quality Score
No additive	under mixed	stiff	2,625	92.0
Enzyme material only	under mixed	stiff	2,687	92.0
Preparation 1	under mixed	stiff	2,675	92.0
Preparation 2	well developed	normal	2,675	93.5
Preparation 3	well developed	normal	2,725	95.0
Preparation 4	well developed	normal	2,663	94.0
Preparation 5	over developed	soft	2,650	93.0
Preparation 6	over developed	soft	2,700	93.0
High monoglyceride only	under mixed	stiff	2,625	93.0

Improved dough development was recorded with the addition of 0.06 g of α-monoglyceride/1,000 HU of activity. Further enhanced activity was achieved by greater proportions of the ester, up to 0.26 g/1,000 HU. The higher level of 0.32 g/1,000 HU produced no discernible further increased enzyme material activity. It is significant that the enzyme material without the ester was insufficiently functional, with the less than optimum mixing, to significantly alter dough characteristics. Further, the monoglyceride at the highest level tested, which amounts to about 0.1% based on the formula flour, did not alter dough characteristics. Higher than control score in this case was due to improved softness and texture.

Example 2: A representative sweet dough formula, detailed on the following page, was produced by a conventional sponge dough procedure.

	Parts by Weight
Sponge ingredients:	
Patent flour	420.0
Arkady yeast food	3.5
Bakers yeast	28.0
Water	270.0
Dough ingredients:	
Patent flour	175.0
Cake flour	105.0
Granulated sucrose	126.0
Nonfat dry milk solids	28.0
Vegetable shortening	70.0
Monoglyceride softener	14.0
Salt	14.0
Egg yolk solids	14.0
Bakers yeast	28.0
Water	166.0

The enzyme material used contained 38,600 HU/g and 6,700 SKB α-amylose units per gram. A saturated, distilled monoglyceride, 90.6% alpha, derived from tallow was dispersed as described in Example 1. A control preparation containing only enzyme material and no activator was prepared along with a preparation containing enzyme material activated with 0.30 g of the monoglyceride/1,000 HU of protease. Both preparations were added with the sponge ingredients to a level of 1,760 HU/lb of formula flour. The results are as follows.

Dough Characteristics

Variables	At Mixer	At Molder	Average 1-lb Loaf Volume, cc	Average Quality Score
Enzyme material only	normal	fair	2,612	91
Activated enzyme material	normal	excellent	2,700	94

In this test the increase in dough-improving effects by the enzyme material when activated as described was less obvious at the mixer due to the nature of the high sugar dough. Effects were first noted at molding where the treated enzyme material in the dough was observed to exhibit improved development leading to superior makeup and subsequent quality.

Oxidizing Agent and Acid Formulation

In the process of *N.C. Duquette; U.S. Patent 3,650,764; March 21, 1972; assigned to H.C. Brill Company, Inc.* fermentation time is significantly shortened by the combined effect of an enzyme, aging agent, and carboxylic acid on the yeast present in the dough. Various other ingredients are used in order to supply desirable characteristics to the baked products. The enzymes used are essentially proteolytic in character. That is, they are capable of breaking down proteins into peptides and amino acids. They manifest their proteolytic activity in the temperature range of 70° to 125°F and the pH range of 4 to 6. Such enzymes should have little or no diastatic, i.e., amylolytic, activity within these ranges. Amylolytic activity in the enzyme produces a dough which is undesirably "young" and which produces a tough reddish crust on the finished baked goods.

The proteolytic enzymes, or proteinases, particularly suitable are those derived from plant sources such as bromelain, papain and ficin. The dough aging agents are those generally regarded as oxidizing materials and are thought to have an effect on the gluten in the flour. Typical among these are alkali metal and alkaline earth metal bromates and iodates such as potassium bromate, potassium iodate, calcium bromate and calcium iodate.

The edible acids used are aliphatic carboxylic acids, unsaturated polycarboxylic acids, di- or tricarboxylic acids such as fumaric acid, maleic acid, or hydroxy-substituted aliphatic carboxylic acids such as lactic acid and citric acid. Various other ingredients can also be included, such as ammonium and calcium salts and fatty acid esters. Such salts have been discovered to exert a beneficial effect on the flour and protein degradation product. For example, ammonium chloride appears to exert a beneficial influence on the elasticity of gluten in wheat flour, and ammonium sulfate aids in yeast activity. The calcium salts are useful for their effect on flour, especially where white products are being prepared.

The fatty acid esters serve at least two purposes. Firstly, they aid in producing sturdy sidewalls in the finished product, and secondly, they can be used to reduce the oxidizing activity of the maturing agents on the enzyme, especially when the agent or agents have been blended with the enzyme in a liquid medium. Suitable fatty acid esters are mono- and/or diglycerides of fatty acids such as glyceryl monostearate and propylene glycol monostearate.

It has been found that the presence of glucono-delta-lactone in amounts up to 1.5% can serve to impart a character to the dough which manifests itself during the baking process. For example, a good degree of symmetrical development and minimal wildness of shred is obtained when the material is present as indicated.

Example: The following ingredients are selected.

	Ingredient	Weight, lb	Percent
(1)	Bromelain	1.75	0.087
(2)	Potassium bromate	8	0.399
(3)	Potassium iodate	2	0.0996
(4)	Ammonium sulfate	2	0.0996
(5)	Ammonium chloride	17.25	0.859
(6)	Calcium sulfate	31	1.54
(7)	Monocalcium phosphate	94	4.67
(8)	Glyceryl monostearate	125	6.21
(9)	Acetic acid (99% purity)	60	2.99
(10)	Lactic acid	125	6.21
(11)	Salt	500	24.9
(12)	Dextrose	375	18.6
(13)	Soy flour	660	32.8
(14)	Silicone emulsifier	10	0.497

Ingredients (1) through (8) are combined and blended in a stardard ribbon blender mixer. The acetic acid and lactic acid are mixed with each other and blended with the combined ingredients (1) through (8) for 2 minutes. A reaction is observed to commence and the mix begins to turn yellow. The salt and dextrose are added immediately and the mix blended for 1 minute; there is no evidence of reaction after these materials are added. Then the soy flour and silicone are blended and mixed for about 5 minutes. The resulting enzymatic blend is a fine granular dry material. 14 g of this material are then added into a white bread dough mix of the following components, thus providing an enzyme level of 0.39 g per 100 lb of flour.

Ingredient	Weight, oz
White flour (wheat)	50
Water	33
Yeast	1.5
Sugar	3
Salt	1
Milk	1
Liquid vegetable shortening	2
Total	91.5

The mixture is blended in a laboratory mixer at 83° to 85°F for 1 minute at 50 rpm, and 11 minutes at 112 rpm. The blended dough is allowed to rest for 40 minutes (floor time), is then scaled into bread pieces and the pieces allowed to rest for 10 minutes. Loaves are then formed and allowed to rise in a steam box having approximately 95% humidity and at a temperature of 112° to 120°F. The rise time (also called proof time) is 65 minutes.

The raised loaves are then baked at 400°F for 25 minutes or until done. There results a sturdy white bread product having excellent external strength and internal grain characteristics, good taste, flavor and odor. The fermentation time of 40 minutes is a substantial improvement over the fermentation times of 4 to 5 hours normally encountered. In an alternative procedure, the above description is followed for the preparation of the enzymatic blend except that the potassium bromate and iodate are added after the salt and dextrose.

DOUGH TREATED WITH PROTEOLYTIC ENZYME

A proteolytic enzyme effective in manufacture of baked goods is prepared in the process of *A. Belloc; J. Florent; D. Mancy and J. Verrier; U.S. Patent 3,875,006; April 1, 1975; assigned to Rhone-Poulenc S.A., France* from *Streptomyces hygroscopicus* culture. The enzyme, designated in the process description as enzyme 25,462 RP, shows activity on a large number of protein substrates.

The results obtained for the hydrolysis of casein by the enzyme 25,462 RP at different pHs and different temperatures are given below. For each experiment, 22 μg of enzyme, of strength 2,250 KU/g, that is, 0.05 KU were used. KU are Kunitz units as determined by the procedure described in Kunitz, *J. Gen. Physiol,* 30, 291 (1947).

Hydrolysis of Casein at 30°C by 25,462 RP–Effect of pH

pH of Reaction Medium	Nature of Buffer	Peptides Set Free in 20 Minutes (mg Tyrosine Equivalents)
5.5	0.1 M phosphate	0.230
6.0	0.1 M phosphate	0.288

(continued)

pH of Reaction Medium	Nature of Buffer	Peptides Set Free in 20 Minutes (mg Tyrosine Equivalents)
6.5	0.1 M phosphate	0.435
6.8	0.1 M phosphate	0.455
7.0	0.1 M phosphate	0.365
7.5	0.1M phosphate	0.148
8.0	0.2 M borate	0.100
8.5	0.2 M borate	0.085
9.0	0.2 M borate	0.060
9.5	0.2 M borate	0.060

Hydrolysis of Casein at pH = 6.8 by 25,462 RP–Effect of Temperature

Temperature (°C)	Peptides Set Free in 20 Minutes (mg Tyrosine Equivalents)
25	0.460
30	0.455
35	0.400
40	0.135
45	0.060
50	0.055
55	0.050
60	0.005

The enzyme 25,462 RP is thus active in a rather narrow range of pH. The optimum pH is 6.8 and the enzyme retains an activity which is greater than 50% of its maximum activity in the range of 5.5 to 7.0. At pH = 6.8, the optimum temperature is between 25° and 30°C, at 40°C the enzyme shows only about 30% of its maximum activity. The kinetics of formation of peptides from casein by the enzyme 25,462 RP and by trypsin are given in the table below. It is found that the enzyme 25,462 RP does not hydrolyze more bonds than trypsin in this substrate.

Hydrolysis of Casein by Enzyme 25,462 RP and by Trypsin*

Reaction Time (Minutes)	Peptides Set Free . . . (mg Tyrosine Equivalents) by. . . 25,462 RP**	Trypsin***
1	0.078	0.050
3	0.191	0.130
5	0.270	0.224
10	0.430	0.312
15	0.485	0.400
20	0.560	0.500
40	0.705	0.635
60	0.780	0.770

*at pH 7.0 and 37°C

**47 μg of enzyme 25,462 RP of strength 2,250 KU/g (0.103 KU)

***23 μg of crystalline trypsin, of strength 4,500 KU/g (0.103 KU)

The process enzyme can be used in bread-making when added to flour in amounts of 5 to 10 mg/kg. The enzyme makes it easier to work the dough, decreases fermentation time and improves the quality of the bread produced (by ensuring, in particular, a more uniform distribution of air cells in the soft part of the bread). The organism which produces 25,462 RP belongs to the *Streptomyces* genus; it is called *Streptomyces hydroscopicus* NRRL 3,999. The following example gives details on the use of the enzyme in a bread dough.

Example: A bread dough is prepared with the following mixture: wheat flour, 1,000 g; yeast, 20 g; salt, 22 g; and water, 640 cc; and, in parallel, with an identical mixture to which there has been added the enzyme 25,462 RP (5 mg), of strength 1,150 KU/g, in the form of a concentrate of 0.1% by weight in the wheat flour.

After intensive kneading and a first fermentation of 60 minutes, shaping is carried out and it is found that, in the presence of the enzyme, the working of the dough is greatly facilitated; in effect, the dough can be stretched much more easily and is much less elastic than the control dough. The preparation time (second fermentation) in the presence of the enzyme is identical to that of the control (135 minutes) and the dough obtained has a content of carbon dioxide which is identical to that of the control dough.

However, the tolerance of the dough is less. If this second fermentation were extended too much, a loss of carbon dioxide would be observed. After cooking, in the experiment with the enzyme, a loaf is obtained, the volume of which is identical to that of the control loaf, but which has a much more uniform distribution of air cells in the soft part of the bread.

REDUCED DOUGH VISCOSITY

In another process *A. Belloc, J. Florent, D. Mancy and J. Verrier; U.S. Patent 3,906,113; September 16, 1975; assigned to Rhone-Poulenc S.A., France* prepared a new amylase useful in the manufacture of baked goods. The enzyme, referred to as 27,591 RP is obtained from the culture of *Penicillium expansum* (NRRL 5437).

The activity of the enzyme 27,591 RP is exerted on a certain number of polysaccharides, such as starch, glycogen, and limit dextrins (residues obtained by treating starch with β-amylases). Its action is thus at least partially of the α-amylase type.

The results obtained for the hydrolysis of starch by the enzyme 27,591 RP at different pHs are given in the table below. For each experiment, 150 μg of enzyme, of strength 2,660 I.U./g were used. The enzyme 27,591 RP is especially active in a slightly acid medium; the optimum pH is 4.0. The region of activity is very asymmetric relative to this value: the activity decreases rapidly below 4.0 (40% of the maximum activity at pH = 3.5; 7% at 3.0) or above 7.0 (50% at 7.0, 0 at 7.5). At pH = 4.0, the optimum temperature is about 55°C; at 65°C, the enzyme still exerts about 50% of its maximum activity. The enzyme can be used advantageously as an adjuvent in bread-making.

Hydrolysis of Starch by 27,591 RP
as a Function of the pH

pH of Reaction Medium	Nature of Buffer	Reducing Sugars Liberated per Minute (μeq of maltose)
3.0	0.2 M acetate*	0.029
3.5	0.2 M acetate*	0.115
4.0	0.2 M acetate*	0.400
4.5	0.2 M acetate*	0.379
5.0	0.2 M acetate*	0.350
5.5	0.02 M phosphate**	0.264
6.0	0.02 M phosphate**	0.260
6.5	0.02 M phosphate**	0.250
7.0	0.02 M phosphate**	0.208
7.5	0.02 M phosphate**	0.000
8.0	0.02 M phosphate**	–
8.5	0.05 M borate***	–
9.0	0.05 M borate***	–

*Mixture of 0.2 M acetic acid and 0.2 M sodium acetate.
**Mixture of 0.02 M KH_2PO_4 and 0.02 M Na_2HPO_4.
***Mixture of 0.2 M H_3BO_3 and 0.05 M $Na_2B_4O_7 \cdot 10H_2O$.

Example 1: The following are introduced into a 170 liter fermentor.

Ingredient	Gram
Corn steep (50% of dry material)	4,800
Glucose monohydrate	2,400
Calcium carbonate	600
Sodium chloride	600
Magnesium sulfate heptahydrate	120
Tap water (to make up to 110 liters)	

The pH is adjusted to 7.6 by adding 10 N sodium hydroxide solution (580 cc). The medium is sterilized by bubbling steam at 122°C through it for 40 min. After cooling, the volume of the broth is 120 liters, the pH is 6.75. It is inoculated with a culture (200 cc) of *Penicillium expansum*, strain F 1046, produced in a shaken Erlenmeyer flask. The culture is developed at 27°C for 30 hours, while shaking and aerating with sterile air. It is then suitable for inoculating the production culture. The production culture is prepared in a 350 liter fermentor into which the following substances are introduced.

Ingredient	Gram
Corn steep (50% of dry material)	8,000
Glucose monohydrate	4,000
Calcium carbonate	1,000
Sodium chloride	1,000
Magnesium sulfate heptahydrate	200
Tap water (to make up to 170 liters)	

The pH is adjusted to 7.6 by adding 10 N sodium hydroxide solution (950 cc) and then the medium is sterilized at 122°C for 40 minutes. After cooling, the volume of the broth is 200 liters; the pH is 6.80. It is inoculated with the inoculum culture (20 liters) from the 170 liter fermentor described above. The culture is developed at 27°C for 20 hours, while stirring with a turbine revolving at 205 rpm and while aerating with 10 m^3/hr of sterile air. From the 20th hour and over the next 20 hours, a 40% solution of glucose hydrate (10 liters) is added under sterile conditions at the rate of 500 cc/hr, while maintaining the same aerating and stirring conditions. After the addition of all the glucose, the pH of the culture is 5.30. The total duration of the culture since inoculation is 7 days. At the end of the culture, the pH is 8.30 and the amylase activity (at pH 4) is 2 I.U./cc.

A must (8,000 cc) prepared under the above conditions is adjusted to pH 7 by adding 5 N hydrochloric acid, Supercel (diatomaceous earth, 800 g) is added and filtered through a fluted filter. A filtrate (8,000 cc) of strength 2.5 I.U./cc is obtained. The filtrate obtained is concentrated under reduced pressure (2 mm Hg) at 25°C until it is ⅛ of its original volume. A concentrate (1,050 cc) of strength 17.7 I.U./cc is obtained.

The concentrate is dialysed for 17 hours against distilled water (10 volumes) at +4°C and then it is clarified by filtration (centrifuging is also suitable). A clear concentrate (1,440 cc) of strength 12.6 I.U./cc is obtained. The clarified concentrate is cooled to +4°C; methanol (4,500 cc) which has previously been cooled to -20°C is then added with stirring and stirring is then continued for a further 2 minutes. The precipitate is collected by centrifuging at +4°C and 2,000 *g* for 5 minutes.

The precipitate is stirred for 15 minutes in distilled water (1,000 cc) which has been cooled to +4°C. The solution is clarified by centrifuging at 2,000 *g* and +4°C for 5 minutes. A solution (1,050 cc) of strength 7.7 I.U./cc is obtained. Crystalline ammonium sulfate (496 g) is added to the clarified solution which is still kept at +4°C and the whole is stirred for 20 minutes to dissolve the salt. The solution is left to stand for 3 hours and is then centrifuged at 2,000 *g* for 10 minutes at +4°C.

The precipitate is dissolved in distilled water (250 cc) at +4°C and the solution is dialysed overnight against distilled water (2,500 cc). A dialysate (300 cc) of strength 37 I.U./cc is obtained. Isopropanol (450 cc) which has previously been cooled to -20°C is poured into the dialysate which is kept at +4°C and stirred, and then stirring is continued for a further 5 minutes. The enzyme is collected by centrifuging at 0°C for 10 minutes at 12,000 *g*. After drying the product obtained, under reduced pressure and at a temperature of about 20°C, the enzyme 27,591 RP (1.7 g) of strength 2,900 I.U./g is obtained.

The enzyme 27,591 RP can be used at various concentrations, as a mixture with other compatible products which are either inert and thus act as diluents, or which possess a definite use, e.g., dispersing agents, flavoring substances, dyestuffs or denaturing agents. A highly concentrated material for industrial use in the textile industry can, e.g., contain: 99% by weight of enzyme 27,591 RP (approximately 2,900 I.U./g); 0.9% by weight of pyrogenic silica; and 0.1% by weight of denaturing dyestuff. A less concentrated composition for bread-making, can, e.g., contain: 1% by weight (approximately 30 I.U./g) of enzyme 27,591 RP of strength 2,900 I.U./g; and 99% by weight of wheat flour or starch. These compositions can be diluted at the time of use, at the choice of the user depending on the desired results. The compositions usually contain an amount of enzyme representing 2,900 to 2 x 10^{-4} I.U./g.

Example 2: This example illustrates the action of the amylase 27,591 RP on aqueous suspensions of flour. Suspensions of flour are prepared according to the following formulation.

Wheat flour (14% moisture content)	80 g
Enzyme 27,591 RP of strength 2,600 I.U./g	x mg*
Water	450 cc

*x having one of the following values: 0, 8, 80 or 800.

The gelling of the suspensions produced is investigated, as a function of the time and the temperature, using a Brabender amylograph, a torsion viscometer which can record variations in viscosity of a paste. The temperature is initially 25°C and it increases uniformly by 1.5°C/min. The amylogram makes it possible to determine the temperature at which gelling starts, corresponding to an increase in viscosity of 20 BU (arbitrary units defined by the designer of the apparatus), the maximum gelling temperature and the maximum viscosity. The table below gives the values obtained for these 2 variables as a function of the concentration of the enzyme.

Enzyme 27,591 RP Conc. Relative to Flour		Gelling Temperature, °C		Maximum Viscosity
(g/kg)	(IU/g)	Start	Maximum	(BU)
0	–	62	86	590
0.1	0.26	62	87	490
1	2.6	62	88	280
10	26	–	–	80

It can be seen that the amylase added decreases the viscosity of the suspensions. The amylase can thus be used in bread-making or in biscuit-making when hypo-amylase flours are used.

STABILIZED DIASTASE CONCENTRATE

Baking and related industries increase the naturally occurring diastase in wheat and wheat flour by the addition of a diastase during processing. A minimum level of diastase is necessary to promote adequate gas production and proper starch modification during fermentation of the dough. α-Amylase may be added to baking flours in the form of dry powders or in tablets. In either form, it is very important that the enzyme preparation has a highly stabilized diastatic activity.

By diastatic enzyme or diastase is meant the glycogenic principle which hydrolyzes starch to dextrins and ultimately to glucose and which contains principally α-amylases, but there may also be present other enzymic complexes such as esterases, carbohydrases, maltases, peptidases, cellulases, pectinases, and proteases. A classic source of α-amylase was barley malt. However, the amount of diastase in barley malt was uncontrolled and excess proteases were present which hydrolyzed the gluten and caused poor rising of the dough due to an excess loss of gases. A uniform source of α-amylase was therefore needed. *B.B. Drake; U.S. Patent 4,010,073; March 1, 1977; assigned to Rohm and Haas Company* developed a free-flowing enzyme preparation from a diastase concentrate suitable for use in the bakery industry. The enzyme has 5,000 SKB units per gram and is stable for at least 3 months. Under use conditions, the diastatic activity for these preparations is standardized at 5,000 SKB units per gram of material. This unit expresses the α-amylase activity as measured by the method of Sandstedt, Kneen and Blish.

The criterion for storage stability was that it should preferably not lose more than 5% of its diastatic activity per month when stored at 30°C over an extended period. It was found that this could be achieved only if the water content was less than 8.5%. The general procedure which has been found to give a satisfactory product is to blend an α-amylase concentrate having a known water content and SKB units with a sufficient amount of an edible starch having a maximum of 3% water in a proportion such that there is a minimum of 1,000 SKB units per gram and a maximum of 8.4% water. Such a product will be free-flowing, will have acceptable storage stability and can be used by the baking industry in an amount which is reasonable for augmenting the α-amylase content of the flour.

Example 1: One part of Rhozyme 87-L concentrate, a fungal diastase, was mixed with 2 parts of redried cornstarch (specification, 2% water) and spread out in a thin layer and dried at atmospheric pressure at 45° to 50°C for 22 hours. The dried product was sieved through a 40 mesh screen and the uniform dried powder was analyzed for diastatic SKB activity. The starting liquid concentrate contained 26,700 SKB units per gram. The prepared dried powder had an activity of 11,100 SKB units per gram which was a 94% recovery of the input SKB activity of the Rhozyme 87-L concentrate used.

Example 2: To 26.5 g of low moisture constarch (3% water by analysis) was added 3.0 g of Rhozyme 87-L concentrate, a liquid fungal diastase, containing 43,500 SKB units per gram. The fungal diastase was dispersed into the starch with a spatula, care being taken to break up the large moist lumps. To this was added 0.25 g of silicon dioxide in the form of an expanded silica and 0.25 g of tricalcium phosphate.

After blending, this was a fairly free-flowing powder containing a few small lumps. The blend was ground in a mortar with a pestle after which it was sifted through a 30 and then a 40 mesh seive. The sifted blend was placed in a container and tumbled for 15 minutes to insure uniform enzyme dispersion. A free-flowing, lump-free, uniform powder resulted. By analysis, this was found to contain 8.1% moisture. The blend was stored at 30°C for 110 days with the following results.

Time (Days)	SKB Units per Gram	% Retention
0	4,390	(110)
30	3,820	87.1
60	3,860	87.9
90	3,846	87.6
110	3,820	87.1

Example 3: A pilot plant preparation was made by placing 2,560 g of low moisture corn starch (3% moisture), 50 g of silicon dioxide and 90 g of tricalcium phosphate in a liquid-solids twin shell blender and mixed for 5 minutes.

After this, 300 g of Rhozyme 87-L concentrate containing 43,500 SKB units per gram was added to the tumbling, dry ingredient blend in 1½ minutes. The blender was rotated for an additional 8½ minutes for a total blending time of 15 minutes. The blend obtained was of uniform particle size, free-flowing and lumpless. By analysis it was found to contain 8.0% moisture. The free-flowing powder was stored at 30°C for 110 days. The following stability data were obtained.

Time (Days)	SKB Units per Gram	% Retention
0	3,700	(100)
30	3,180	86.1
60	3,200	86.5
90	3,050	82.5
110	3,180	85.9

The blends described were quite satisfactory for use in the baking industry. Similar preparations made from higher water content starch and containing 8.5% or more of water were unstable on storage and lost so much of their SKB units that they could not be used as an auxiliary source of α-amylase for baking purposes. The conditioning agents, such as silicon dioxide and tricalcium phosphate, used in the examples are not critical with respect to making an α-amylase preparation useful for the baking industry. It is often convenient, however, to add such conditioning agents as an aid in preventing caking and in contributing to the free-flowing properties.

PENTOSANASE ANTISTALING PROCESS

A. Cooke and H. Johannson; U.S. Patent 3,512,992; May 19, 1970; assigned to Delmar Chemicals Limited, Canada uses pentosanase or pentosan-degrading enzymes in the bread-making process to improve resistance of the final loaf to hardness or staling. The maximum amount of pentosanase preparation to be added is inversely proportional to the fermentation time allowed. Thus, an amount of enzyme preparation equivalent to 103 units of pentosanase activity (PU) per 400 g of flour, is considered excessive in a sponge dough formula if the fermentation is allowed to take place overnight.

This same amount, however, is effective in a sponge and dough formula if fermentation time is reduced to 4½ hours. Similarly, if the fermentation time is reduced to 2 hours using the straight dough method, as much as 216 units of pentosanase enzyme preparation can be used with good results. Use of 18 units in the same 2 hour fermentation recipe give a detectible improvement, but it is not recommended to use less than this amount.

In a continuous process, the amounts are approximately those of the 2 hour straight dough formula. The enzyme process can be successfully applied to any of the several commercial procedures including the conventional batch procedure for preparing bread. The enzyme can be added to the flour, or it can be added and mixed in with the water. In all cases the enzyme preparation should be added at the beginning or as close to the beginning of the bread-making process as possible for best results. In a continuous process, the ingredients are first mixed in a mixing zone. The mixed products are then moved to a developing zone where the dough is subjected in a continuous manner to mechanical work so as to develop the gluten. After passing through the developing zone in continuous manner, the dough is forced through an orifice into a series of baking pans which are moved into a proofer, where the dough is allowed to rise. When it has risen sufficiently, it goes to the oven for baking. For commercial applications, where multiples of 100 lb of flour are used in formulas, the required amounts of enzyme preparation are as follows.

Overnight sponge and dough formula:
preferred amounts: 2,000 to 4,000 PU/100 lb
4½ hour sponge and dough formula:
preferred amounts: 4,000 to 8,000 PU/100 lb
2 hour straight dough formula:
preferred amounts: 8,000 to 16,000 PU/100 lb
Continuous formula:
preferred amounts: 8,000 to 12,000 PU/100 lb

A large excess of pentosan-degrading enzyme would be operative in the sense that it would increase the softness of the bread. However, such a large amount would also cause collapsing of the bread under normal handling. The pentosanase preparation is mixed with dried flour having not more than 8% moisture in a proportion of 2.25 g of pentosanase preparation having 1,800 units of pentosanase activity per gram to dried flour to make up 4 oz, care being taken to insure uniform blending.

To this effect, it has been found expedient to add a free-flowing agent such as, e.g., tricalcium phosphate in proper amounts, e.g., 3% by weight of flour previous to mixing with pentosanase preparation. The resultant pentosanase-flour concentrate is then used at the rate of from 2 oz to as high as 20 oz per 100 lb of flour in standard bread-making formulas depending on the fermentation time. The pentosanase additive can obviously be added directly in equivalent controlled amounts to the flour to be used, or it can be added to an edible carrier such as, e.g., starch, flour, NaCl, $CaSO_4$ or sugar, and the mixture then added to flour and the other ingredients.

Example: The following example illustrates the results of using different amounts of pentosanase in a standard bread-making formula. Five loaves were made using the sponge dough procedure, loaf 1 being the control loaf and loaves 2, 3, 4, and 5 containing respectively 0.0025%, 0.005%, 0.01% and 0.015% by weight of Rhozyme HP-150 having 1,800 units of pentosanase activity per gram.

For each formula, 400 g of baker's grade bread flour was used as can be seen from the following table of ingredients to which was added and thoroughly mixed 0 g, 1 g, 2 g, 4 g and 6 g respectively of a previously prepared mixture obtained by mixing together 990 g flour and 10 g pentosanase preparation having 1,800 PU/g. The formula for the standard ingredients was as follows.

Ingredient	Sponge, g	Dough, g
Flour	267	133
Water	174	87
Yeast	8	–
Sugar	–	24
Salt	–	8
Shortening	–	24

For all loaves, the procedure was as follows. The sponge ingredients were mixed in a Hobart laboratory scale mixer at a controlled temperature for about 2 min. The sponge was discharged into a trough and placed in a fermentation room maintained at about 80°F and 75% relative humidity, where it was allowed to remain overnight. The sponge was then returned to the mixer and the dough ingredients added and mixed in. After mixing, the dough obtained was divided, rounded, proofed, molded, panned and pan-proofed, in accordance with conventional practice.

The loaves were then baked at 450°F for approximately 25 minutes. The baked bread was allowed to cool, wrapped and stored in a moisture proof container overnight before judging and scoring. The loaf having the lowest concentration of enzyme (loaf 2) was significantly softer than the control and with increasing concentration (loaves 3 and 4), the crumb became progressively softer. Loaf 5 was so soft that it could not withstand any significant pressure applied to it. Grain was not materially affected, but the texture (feel of the crumb) was softer with increasing amounts of pentosanase preparation.

ENZYME-OXIDIZED-FAT BREAD FLAVOR

Flavor and body of baked goods, particularly bread, is improved by the addition of lipoxidase and a fat capable of being peroxidized in situ in at least part of the water for the dough during mixing in the presence of oxygen. The composition and process for bread with improved flavor is described by *A.W. Kleinschmidt and S.T. Viren; U.S. Patent 3,506,448; April 14, 1970; assigned to J.R. Short Milling Company*. A powdery material such as dairy whey solids is employed as a carrier for the fat.

Such dry materials are very finely particulate, with an average particle size of not more than about 100 μ, and therefore, present a very large surface area for supporting the enzyme-peroxidizable fat. Of equal importance, however, is the high solubility and dispersibility of the constituents of such materials in aqueous media. Dairy whey solids are superior because of their very high rate of solution and dispersion, and their use results in a very prompt and stable emulsion of the enzyme-peroxidizable fat in the aqueous medium involved.

It is thought that the speed of emulsification of the fat, and the unusually stable nature of the emulsion results from the highly soluble and dispersible nature of the whey solids. Further, possibly because of the nature of the emulsion obtained, use of the dairy whey solids results in a quicker, and apparently more extensive, enzyme peroxidation of the fat with the result that, for a given time of agitation and a given proportion of the additive composition, a more marked improvement in body and flavor of the baked goods is attained.

The enzyme material can be lipoxidase in any edible form, suitable for introduction into an aqueous system. For practical purposes, particulate legume materials are the best enzymatic materials for use. Thus, enzymatically active soybean flour, soybean meal and soybean flakes are all excellent lipoxidase-containing materials. Equivalent materials derived from peas, peanuts, lima beans, navy beans and lentils are suitable. When a naturally oil-rich material such as soybean flour is used, it can be either full fat, low fat or fat free, so long as it exhibits adequate lipoxidase activity.

Though the dairy whey solids or other predominantly sugar and protein containing powdery material can be used as the sole carrier for the fat, or as the sole carrier other than the enzymatic material, it has been found that best results are obtained when the dairy whey solids constitute only a portion of the total carrier material. The balance of carrier can be made up by any finely particulate edible material which is essentially insoluble in the fat and is inert with respect to both the fat and the enzymatically active material. Partially gelatinized or partially dextrinized corn endosperm flour is suitable, as is also wheat farina. Preferred compositions can comprise 20 to 30 wt % enzymatic material, 10 to 20 wt % fat, 5 to 15 wt % dairy whey solids and the balance made up of other carrier material.

The compositions can be prepared by any conventional blending procedures, with the fat being distributed primarily on the carrier material or materials or both on the carrier and the enzymatically active materials. The compositions can be packed, shipped and stored in conventional bags and remain stable, exhibiting no undue loss in enzyme activity, over storage periods as long as 6 months at room temperature. Compositions can be used in both continuous mix baking procedures and batch procedures.

In the continuous mix methods, the composition is best added to all or part of the water for the brew, and the water then agitated until the composition is well dispersed. Agitation is required both to accomplish adequate dispersions of the additive composition and to provide good contact between the lipoxidase, fat and oxygen to accomplish peroxidation of the fat, the oxygen being present because of the air introduced during agitation.

The time required depends upon the effectiveness of the agitation and, if only mild agitation is used, the time period can be extended, the upper limit being about 6 hours. In batch dough-making procedures, the additive composition is simply dispersed in all or a portion of the water to be used in making the dough. To assure proper in situ peroxidation of the fat, the temperature of the aqueous medium in which the composition is dispersed is maintained in the range of 40° to 110°F, and the pH is kept in the range of 4.0 to 8.5.

Example 1: Thirty parts by weight of full fat, enzymatically active soybean flour, 45 parts by weight of partially dextrinized corn endosperm flour, and 10 parts by weight of "extra grade" commercially available dairy whey solids are charged to a conventional conical blender equipped with an intensifier bar for the introduction of liquids, the blender being in the nature of a tumbling mixer.

The blender is then operated for 5 minutes to assure uniform mixing of the 3 dry ingredients. 10 parts by weight of soybean oil and 5 parts by weight of safflower oil are then introduced, via ducts and ports in the intensifier bar, over a period of 10 minutes with continual mixing, and the composition is then mixed for an additional 10 minutes. The finished composition, identified as Composition A, is dry, lump-free and free-flowing, despite its high fat content.

With the composition prepared in this fashion, the oil is distributed over the surfaces of the particles of all three of the particulate constituents. The same procedure can be used in preparing the additional compositions listed in the table on the following page. In all of the compositions described in this example, the corn endosperm flour is a partially dextrinized product having an average particle size of 50 to 175 μ, and the enzymatically active soy flour has an average particle size in the range of 50 to 100 μ and a native fat content of approximately 19% by weight.

Composition	Full Fat Enzyme Active Soy Flour	Dairy Whey Solids	Soy Oil	Corn Endosperm Flour	Safflower Oil
	....	(parts by weight)	....		
B	30	20	10	35	5
C	30	30	10	25	5
D	42.5	42.5	10	–	5
E	40	40	13.33	–	6.67
F	30	48	10	7	5
G	5	5	3	87	–
H	30	–	10	55	5

Composition H can be considered as a control composition, containing no dairy whey solids. All of Compositions A through G are distinctly less oily and more free-flowing than Composition H. Compositions A, B and C are progressively drier, due to the increasing whey solids content, with Composition C being completely powdery. All of Compositions A through G disperse in water more promptly, with less agitation required, than in the case of Composition H.

Example 2: The bread improving effects of Compositions A and H were compared under continuous mix conditions, using a pilot plant continuous mix unit.

Ingredient	Percent of Flour Weight
Brew formula:	
Water	62.0
Sugar	8.0
Milk solids, nonfat	2.0
Yeast	2.5
Salt	2.3
Nonphosphate yeast food	0.5
Calcium acid phosphate	0.2
Calcium propionate	0.1
Dough formula:	
Bleached white bread wheat flour	100.0
Additional water	4.0
Shortening	3.25

An aqueous solution of potassium bromate and potassium iodate was used to provide 50 ppm of potassium bromate, based on the total flour, and 12.5 ppm of potassium iodate. The brew was fermented at 86°F for 2.25 hours. A baking time of 18 minutes at 435°F was used.

One dough was prepared with the addition of 1%, based on the total flour weight, of Composition H. Another dough was prepared with 1% of Composition A. In each case, 1 gallon of the water for the brew was placed in an open top container and stirred with a propeller agitator operated at 1,720 rpm, both of the additive compositions (Composition H for one dough, Composition A for the other) being added during such agitation, and agitation then being continued at

1,720 rpm for 10 minutes to accomplish dispersion and enzyme peroxidation of the fat content of the additive composition. The entire aqueous composition was then employed in preparing the brew. The bread obtained was scored comparatively 1 day after baking, and the bread from the dough prepared with Composition A proved to be observably better, as to both body and flavor, than the bread from the dough prepared with Composition H.

SOURDOUGH FLAVOR–RENNET ADDITION

Sour dough flavor is developed in a stepwise manner, starting with skim milk in the process of *A.J. Luksas; U.S. Patent 3,615,695; October 26, 1971; assigned to Beatrice Foods Co.* The skim milk or other casein source is acidified directly with lactic acid to a pH of 5.2. There can also be used sodium lactate or potassium lactate. When skim milk or other source of lactose is used microorganisms known to produce lactic acid from lactose can be used.

Thus, there can be used *Streptococcus lactis, Lactobacillus lactis, Lactobacillus bulgaricus, Lactobacillus delbrueckii, Lactobacillus thermophilus, Lactobacillus casieii, Lactobacillus leichmanii, Streptococcus citrovorus, Streptococcus cremoris, and Lactobacillus helveticus*, etc. To the acidified skim milk or other casein source there is added rennet and the mixture incubated for 0.5 hour or until the system sets, i.e., a curd is produced. The curd is separated either by centrifuging, filtering or cooking the whey out.

The curd is then resuspended in either water or a saline solution, e.g., 5% sodium chloride or 0.85% sodium chloride in an amount of 0.5 to 10 times the original volume. The curd is reconstituted to the original volume. Then it is inoculated with a nontoxic organism of either the genera *Citrobacter* or *Micrococcus*, aerated, agitated as a fluid system for 3 days. Suitable organisms include *Citrobacter freundii, Citrobacter intermedic, Micrococcus caseolytics,* (e.g., ATCC 13548), *Micrococcus conglomeratus*, (e.g., ATCC 401), or *Micrococcus freudenreichii* (either ATCC 407 or ATCC 8459).

All fermentations are carried out at 32°C. The finished product has the odor and flavor of sour dough. It can be left in the liquid form or dried, for example, by spray-drying or roller-drying. The sour dough flavor concentrate (either liquid or dry) can be used to impart sour dough flavor to bread and nonbread foods, e.g., rye bread, white bread, whole wheat bread, French bread, pumpernickel, biscuits, rolls, bagels, rice stuffing for turkeys and other fowls, dips made from cheese or sour cream salad dressing.

Example 1: 100 cc skim milk was acidified with lactic acid to a pH of 5.2. Rennet was added and the system incubated until a curd was produced (about 30 minutes). The mixture was centrifuged and the curd removed. The curd was resuspended in 100 cc of water and inoculated with *Citrobacter freundii* and incubated for 3 days at 32°C with aeration and vigorous agitation to give a finished product having the odor and flavor of sour dough. The product was spray dried to give a powder which retained the flavor and odor of sour dough.

Example 2: The procedure of Example 1 was repeated using *Micrococcus caseolytics* (ATCC 13548) in place of the *Citrobacter*. The final fluid product, rather than being spray-dried, was heated to 160°F to kill the *Micrococcus* and give a stable fluid product having a strong odor and flavor of sour dough.

Example 3: The procedure of Example 1 was repeated, but instead of lactic acid, there was employed *Lactobacillus lactis* which was incubated at 32°C until the pH of the skim milk fell to 5.2. Then the cultured milk was pasteurized at 161°F, and the rennet added. Further processing was as in Example 1 to produce a fluid product having the characteristic odor and flavor of sour dough. The fluid was roller-dried. The dry product retained its sour dough odor and flavor.

SOURDOUGH–*Lactobacillus* STRAINS

Procedures for growing *Lactobacillus sanfrancisco* bacteria in high cell count in cultures, and for storage of harvested cells for long periods of time are discussed by *E.L. Sing; U.S. Patent 4,021,581; May 3, 1977.* Described below is a standard type formulation and procedure for preparing a culture for sour dough bacteria. An aqueous solution of the following ingredients was prepared:

> 1.5% (solids content) of fresh yeast extractives prepared by merely using a heat treated suspension of yeast cells in unpurified form with the insoluble solid content allowed to settle and be removed;
>
> 5% of enzyme-digested nonfat dry milk solids (enzyme digested with a protease enzyme produced by *B. subtilis*, Pabst Protease L-423);
>
> 5% malt syrup extract containing 80% solids at a pH of from 6.8 to 7;
>
> 0.03% Tween 80 (sorbitan polyoxyethylene monooleate);
>
> 0.05% sodium thioglycolate;
>
> 20 parts per million of manganese in salt form;
>
> 20 parts per million of magnesium in salt form; and
>
> strain 114-3 of *L. sanfrancisco*.

The pH of the above mixture was maintained at 4.7 by the addition of a solution of potassium hydroxide and the temperature was maintained at 30°C. The mixture was incubated for 18 to 19 hours. When the percent transmission of light measured at 650 mμ reached a minimum, the fermentation was terminated and the temperature was reduced to 13°C. Minimum percent transmission over a 1 cm path would range from 30 to 35%. The minimum was reached for only a short time and fermentation needed to be terminated when or slightly before the minimum was reached.

Eight parts of the culture were then added to six parts of a diluent containing the following ingredients at pH 5.2: 0.05% Tween 80; 2% monosodium glutamate; 9 to 10% glycerol; and the balance being reconstituted nonfat dry milk having 10.5% solids. The resultant mixture was then cooled and stored at about -18°C. Activity at the end of one month was approximately 50% of the activity of the cells immediately after storage.

In this process a medium containing a culture of *Lactobacillus sanfrancisco* is incubated at 30°C while maintaining the pH at about 4.7 by the continuous addition of alkali. The medium contains an enzyme-digested nonfat dry milk, malt syrup, and unsaturated fatty acid derivative and fresh yeast extractives. After incubation for a period of time, the culture is then mixed with a diluent which includes glycerol, nonfat dry milk, a saccharide (fructose, glucose or sucrose), and unsaturated fatty acid derivative and monosodium glutamate.

The mixture is then cooled to a temperature of about 0°C and maintained at that temperature for at least about 20 minutes and then cooled to below -30°C. Ideally the bacteria are then maintained at a temperature between about -30°C and about -40°C until just prior to the product's use. It has been found that by the use of a combination of two different strains of *Lactobacillus sanfrancisco* produces higher activity and viability after storage. In particular, strain 114-3 and strain F-3 have been found to function extremely well in combination. These strains have been deposited in the Stock Culture Collection of the U.S. Department of Agriculture from which samples may be obtained. Strain 114-3 has been assigned the accession number of NRRL B-8116. Strain F-3 has been assigned the accession number NRRL B-8117.

Example: An aqueous solution of the following ingredients was prepared:

1.75 to 2.0% (solids content) of fresh yeast extractives prepared by merely using a heat treated suspension of yeast cells in unpurified form, with the insoluble solid content allowed to settle and be removed;

5% of enzyme-digested nonfat dry milk solids (enzyme-digested with a protease enzyme produced by *B. Subtilis*, Pabst Protease L-423);

5.5% malt syrup containing 80% solids at a pH of from 6.8 to 7;

0.03% Tween 80 (sorbitan polyoxyethylene monooleate);

0.05% sodium thioglycolate;

20 parts per million of manganese in salt form;

20 parts per million of magnesium in salt form; and

a 50/50 blend of strain 114-3 and strain F-3 of *Lactobacillus sanfrancisco*.

The pH of the mixture was maintained at 4.7 by the addition of a solution of potassium hydroxide and the temperature was maintained at 30°C. The mixture was incubated for 17 to 18 hours. When the percent transmission of light measured at 650 mμ reached a minimum, the fermentation was terminated and the temperature was reduced to 5°C. Minimum percent transmission over a 1 cm path would range from 26 to 30%. The minimum was reached for only a short time and fermentation needed to be terminated when or slightly before the minimum was reached.

Nine parts of the culture were then added to five parts of the diluent containing the following ingredients at pH 6.5: 0.05% Tween 80; 2% monosodium glutamate; 9 to 10% glycerol; 10.5 to 15% Iso Sweet fructose syrup (enzymatically converted corn syrup); and the balance being reconstituted nonfat dry milk having 10.5% solids.

The mixture was adjusted to pH 6.5 and was then cooled to 0°C and maintained at that temperature for 30 minutes. At the end of the 30 minute period, the resultant mixture was cooled to between -30° and -40°C and maintained at that temperature until use.

Activity at the end of 6 months was 90% of the activity of the cells immediately after storage. The above procedure produced a cell count in the final product of from 3.8×10^9 to 4.1×10^9 cells per milliliter. In place of the fructose syrup, other saccharides may be substituted to achieve improved storage, however, it has been found that fructose is more effective than other saccharides in prolonging shelf life for *L. sanfrancisco* products stored at low temperatures.

The cooling of the mixture to about 0°C prior to final cooling to the storage temperature apparently allows the cells to become acclimated to a low temperature environment and minimizes the detrimental effect which freezing otherwise would have on the cells.

PIE CRUSTS–COATED PAPAIN

Pie crust doughs are ordinarily prepared in a two-step process. In the first step, shortening, flour and optional dry ingredients such as salt are blended until a substantially uniform mixture is achieved. This mixture is denoted a dry mix even though it can contain up to about 10% by weight moisture which is present principally in the flour ingredient. In the second step, water is added to the dry mix, and this combination of ingredients is mixed until the water is absorbed by the flour.

During this second step, proteins in the flour in the dry mix are hydrated. The mixing action in the second step causes hydrated protein strands to become ordered. The hydrated protein arranged in such fashion forms a tough, rubbery, elastic structure which is the essential framework of dough on which other constituents, for example, shortening, are supported. The hydrated protein is denoted gluten, and the protein framework is denoted a gluten network.

The mixing action is known as gluten development. When mixing action causes the dough to be characterized by maximum strength and elasticity, the gluten network is said to be fully developed. Dough with a strongly developed gluten network has substantial strength and elasticity.

During baking, water is evaporated from the dough, and the gluten coagulates. This coagulation is an irreversible denaturization characterized by decreased solubility and extensibility of the protein strands. After coagulation is completed, the baked dough is referred to as crust. The gluten begins to coagulate at a temperature on the order of 165° to 200°F. A relatively strong or highly developed gluten network is preferred in unbaked dough so that it has sufficient strength to resist tearing during rolling out, placing in a pie pan and shaping. Dough with a weak gluten network has a tendency to tear during such handling.

On the other hand, the stronger the gluten network is during baking, the more the crust shrinkage is. This is because a strong gluten structure induces contraction in the gluten framework during baking when water in the dough is evaporating and the gluten is coagulating, resulting in high crust shrinkage. A weaker gluten network allows less contraction during baking and provides a lesser amount of crust shrinkage. Pie crust shrinkage is a problem, especially with cream pies where the pie shell is cooked before filling addition, since the pie shell often shrinks below the level of the top of the pie pan, whereby not as much filling is able to be used, as compared to where there is reduced shrinkage. Moreover, due to crust shrinkage, the filling can overtop the crust, leaking out when individual slices of pie are cut and served and detracting from the general appearance of the served portion.

Thus, in dough processing, the extent of gluten development is ordinarily a compromise, providing a strong enough gluten network to resist tearing during handling while attempting not to result in an excessive amount of shrinkage in the baked crust.

J.E. Luebering and E.R. Hair; U.S. Patent 3,561,975; February 9, 1971; assigned to The Procter & Gamble Company gives an enzyme method for weakening the gluten network in unbaked dough to reduce crust shrinkage, and at the same time maintaining good dough handling properties before baking. An essential step of the process is the coating of the enzyme with shortening so that the gluten network can develop slowly during the mixing of the dry ingredients with water.

With papain as the proteolytic enzyme, a coating shortening with a complete melting point ranging from about 115° to about 140°F has been found to be suitable, with a melting point of about 125°F considered to be optimum. With these melting points, the enrobing shortening melts to release the papain readily for maximum exposure to the gluten network at a temperature well below the enzyme's deactivation temperature.

Moreover, the coating shortening in addition to possessing the above required melting characteristics, preferably has a Solids Content Index (SCI) value at 70°F ranging from about 30 to about 100, an SCI value at 105°F ranging from about 15 to about 60, and an Iodine Value (I.V.) ranging from about 20 to about 85.

The particles of crust shrinkage-reducing agent can be produced by various conventional methods for coating particulate solids with plastic or solid coating material. Suitable coating techniques include, for example, fluidized bed coating, where the particulate enzyme to be coated is suspended in a fluidized bed and contacted with melted shortening, and pan-coating so that melted shortening is sprayed onto the particulate enzyme while the enzyme is agitated in a spherical coating pan.

The preferred method for producing the spherical particles is by spray-cooling, i.e., by prilling a dispersion of enzyme in melted shortening. In this prilling process, the melt dispersion is forced through a nozzle, formed into drops, and then these drops are allowed to solidify while being cooled, for example, by a cooling gas such as cold air.

Example: A 6 lb batch of shrinkage-reducing agent is made. The shrinkage-reducing agent is in the form of spherical particles having diameters ranging from about 200 to about 500 μ. The particles contain shortening and enzyme. The batch contains 89.3% by weight shortening and 10.7% by weight enzyme. Thus, the weight ratio of shortening to enzyme in the batch is 8.3:1. The composition of each spherical particle is similar to the composition of the batch. The shortening is cottonseed hardstock, i.e., cottonseed oil hydrogenated to an I.V. of 8. It has a complete melting point of 135°F. The enzyme is papain. It is present in the form of particles each having its longest dimension ranging from 10 to 50 μ. Its particle size is further characterized in that it can be sifted through a 350 mesh screen. The batch of shrinkage-reducing agent is prepared with the processing apparatus depicted in Figure 4.1.

Figure 4.1: Spray Cooling Process

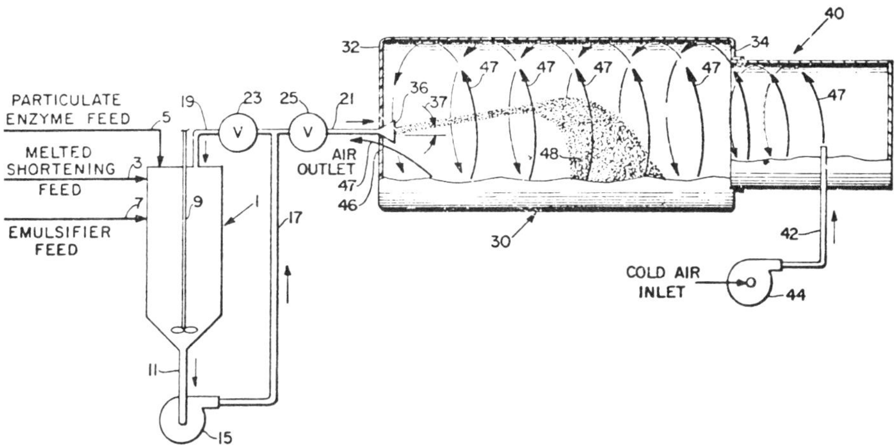

Source: U.S. Patent 3,561,975

With reference to Figure 4.1, the chamber **30** is constituted of polyethylene and is 4.5' in diameter and 8.5' long. The chamber **40** is a 55-gallon carbon steel drum. The air inlet **42** enters chamber **40** tangentially at a distance of 11' from the feed end **32** of the chamber **30**. In this preparation, 5.35 lb of the shortening is melted by heating in a jacketed stainless steel tank at 160°F. The melted shortening is then pumped into dispersion tank **1** and cooled to 138°F by the addition of dry ice. Then 0.65 lb of particulate papain is added to tank **1**. The shortening and papain are then moderately agitated in tank **1** for 45 seconds to form a dispersion.

The dispersion is pumped from the dispersion tank into the spray nozzle at the rate of 100 lb/hr. No dispersion is recycled back into tank **1**. The dispersion is sprayed through the nozzle at a pressure of 40 psig with a V-spray angle of 25°. The spray pattern is directed at an angle of 10° with and above the horizontal. The spray pattern thickness is approximately 0.06". The spraying action of the nozzle forms the dispersion into spherical droplets having diameters ranging from 200 to 500 μ.

The droplets are solidified in the form of spherical particles by cooling air which enters the apparatus through pipe **42**. The cooling air enters the apparatus at a temperature of 0°F and at a rate of 3 cfm. The ratio of air flow rate to dispersion spray rate is 0.03 cfm/lb/hr. The spherical particles contain on the average 89.3% shortening and 10.7% enzyme.

A pie crust dry mix is then prepared as follows: 140 g of soft winter wheat flour containing 11% protein, and 3.6 g of salt are premixed with 2.0 g of the above prepared shrinkage-reducing agent by mixing for 3 minutes in a paddle mixer at 150 rpm. This mixture is then combined with 92.4 g of enzyme-free plastic shortening by mixing for 3 minutes in a paddle mixer at 150 rpm.

The resulting dry mix contains by weight 58.82% flour, 38.82 % enzyme-free shortening, 1.52% salt, and 0.84% shrinkage-reducing agent. The enzyme-free shortening is present as discrete lumps having diameters not exceeding ½" and also as a constituent in greased flour. The papain is 0.09% by weight of the mix and is present in the mix in an amount providing 0.7 Northrop unit of activity per gram of flour.

At this point, the 238 g of the dry mix at 75°F, are weighed into a standard dough mixing bowl having a capacity of 0.5 quart. To the dry mix is then added 42 g of water so that the wetted mix contains 15% by weight water. The wetted mix is then mixed for 0.5 minute with a paddle mixer with a mixer head speed set at 150 rpm to form dough. The dough is then subjected to a one hour conditioning period, i.e., a one hour rest period at room temperature. From this batch of conditioned dough, 200 g are separated to be formed into crust. The rest of the dough is saved for sheet strength testing.

The 200 g batch of dough is rolled to ⅛" thickness, placed in a pie pan, trimmed, and shaped to fit the pan. No tearing occurs during this handling. The dough has a sheet strength of 1,550 g. The dough is then baked for 11 minutes in a 450°F oven. In another case, crust is baked from control dough prepared and conditioned and handled as above except that the control dough does not contain a shrinkage-reducing agent, but contains instead an equal weight of additional flour. Shrinkage in crust made from the agent-containing dough is reduced approximately 25% compared to the crust made from the control dough.

In still another case, dough is prepared and conditioned as above except that uncoated enzyme is used as a shrinkage-reducing agent and flour replaces the weight of the coating. This dough is estimated to have sheet strength of 800 g, and tears during handling.

UPGRADED SEMOLINA FLOUR

The conventional method of making pasta consists of hydrating the semolina with about 30% of water, preparing the paste and then extruding the latter to produce pasta of the desired kind which is then dried. This method is carried out continuously in a trough in which a bladed shaft rotates for the production of a pasty mixture. This mixture is then extracted from the trough by a screw terminating in a cylinder in which rotates another screw which serves to knead and compress the paste, which is then extruded through a die. The paste shreds are then laid on rods which pass through a drying tunnel in which the air moisture becomes gradually lower.

Dried pasta has, among cereal derived foodstuffs, a very distinct microscopic structure. It has a continuous protein mixture phase (the gluten or the protein network) where the starch granules are dispersed. While cooking in hot water, the starch granules gelatinize, i.e., absorb water, swell and turn into starch paste.

The gluten (the protein network) is denatured through cooking and if it is not sufficiently resistant, the starch granules, when swelling, can tear the meshes of its continuous phase, giving rise at the periphery of the pasta, to a viscous layer of starch paste. The state of the protein network after cooking can also affect the elasticity of the pasta. The main problem which has to be solved to obtain elastic and nonsticky pasta consists of increasing the resistance of the protein network to cooking. The addition of eggs to the semolina strengthens this network and it is known that pasta prepared with eggs withstands cooking better.

R. Menzi; U.S. Patent 3,520,702; July 14, 1970 adds any in a group of enzyme materials, singly or in mixtures, to semolinas to improve the elastic and nonsticky characteristics of pasta even when poor quality hard wheat semolina is used. The method for improved pasta consists of adding to either of two starting substances, semolina and water, at least one enzymatic substance capable of setting up intermolecular bonds in the protein network, followed by preparation of the paste, kneading, extrusion and drying.

The enzymatic substance used comprises at least one of the following industrial enzymatic preparations: pancreatin; glucose-oxidase; catalase of bacterial origin; Takamine proteinase; SP 160 Takamine amylase; and wheat germ lipase. The quantity to be added, before the preparation of the paste, to the semolina or to the water, is about 2.5 g/100 kg.

In view of the fact that these enzymatic preparations, which preparations are all of natural origin, are added to the semolina in such extremely small quantities, they will cause no change in the pasta except an increase in the resistance to cooking of their protein network.

Example 1: 2.5 g of pancreatin are added to 100 kg of poor quality hard wheat semolina and the pasta is prepared in the usual manner. The dried pasta, once cooked, after being kept stored for a week, exhibits an excellent elasticity and hardly sticks. Pasta made from the same semolina, but by the conventional method, i.e., without adding pancreatin, has poor elasticity and is very sticky.

Example 2: The same semolina is used as in Example 1, but the pancreatin is replaced with the same quantity of glucose-oxidase. The resulting pasta, once cooked, exhibits a good elasticity and does not stick.

Example 3: The same semolina is used as in Example 1 but the pancreatin is replaced with 5 g of catalase of bacterial origin. The resulting pasta, once cooked, exhibits a good elasticity and does not stick.

Example 4: The procedure is the same as in Example 1 except that the pancreatin is replaced with a mixture comprising 1.25 g of glucose-oxidase and 1.25 g of catalase of bacterial origin. The resulting pasta, once cooked, exhibits a good elasticity and does not stick.

BAKERY WASTES UTILIZATION

Recovery of sugar and reprocessing of starch-protein mixtures from bakery wastes are facilitated with the aid of enzyme processes. The process, as illustrated in Figure 4.2, is described by *H. Müller and L. Grob; U.S. Patent 4,046,789; September 6, 1977; assigned to Chemap AG, Switzerland.*

Figure 4.2: Bakery Wastes Utilization

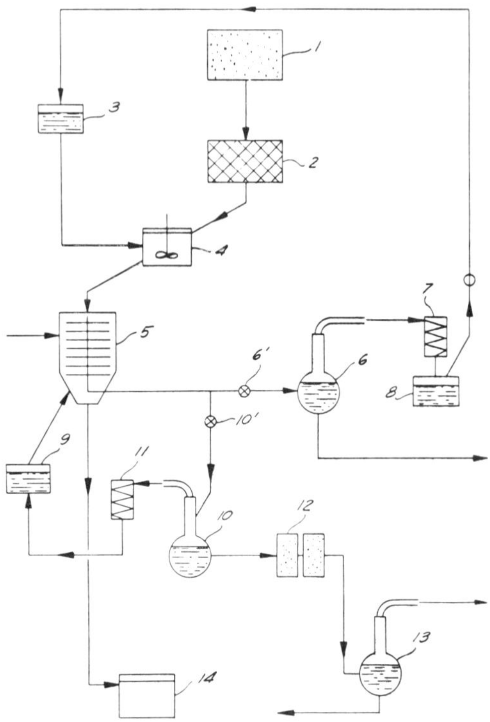

Source: U.S. Patent 4,046,789

Waste products, for example, from biscuits are indicated by the reference numeral **1**. They are first subjected to a comminution in the vessel **2**. The comminuted material is then mixed in a mixing vessel **4** with a solvent which, for instance, may be hexane obtained from the storage vessel **3**. The mixture of waste product and solvent is then subjected to a solvent extraction and subsequent filtration in one and the same vessel **5**. Between successive extraction steps, the solvent is blown out with steam (see the arrow at left of the vessel **5**) and is passed through a valve **6'** to a regeneration vessel **6**.

The main product obtained in the extraction and filtration step, the solvent-oil mixture, is likewise subjected to the distillation indicated at **6**. The recovered solvent is then passed from the top of the distillation device to a condenser **7** and from there to a temporary storage vessel **8** from which it is passed back into the solvent storage vessel **3** and further is recycled in the process as indicated. The oil-fat mixture itself is drawn from the bottom of the distillation and subjected to a refinement step. In a second extraction step the sugar remaining in the filtration extraction device **5** is extracted, for instance, with an alcoholic solution at **9** of a 70% concentration.

After this extraction step is complete, the filter cake is subjected to a steam extraction from the alcohol (see the arrow left of the frame **5**). The sugar-alcohol solution then is passed by the valve **10′** into the separation device **10** where the alcohol is distilled off and recycled, after passing a condenser **11**, into the alcohol storage vessel **9**. The aqueous solution, which is now free of alcohol, then is passed, for instance, to an ion exchanger **12** in order to separate the salt present which mainly is sodium chloride salt. The salt-free sugar solution is then further processed in vessel **13** to obtain sugar, particularly saccharose.

The filter residue in this case will still contain mainly starch. It is subjected to a regeneration and recovery to a fermentation indicated at **14**. For instance, the starch-containing residue may first be subjected to a hydrolysis and subsequently to an aerobic fermentation. The starch-containing material can also be subjected to a fermentation without prior hydrolysis, the fermentation being carried out, for instance, in a symbiotic process where two different microorganisms are used simultaneously. The first one, which may be *Endomycopsis fibuliger*, opens up the starch while a yeast of the *Saccharomyces* or of the *Candida* type converts the mono- or disaccharides into yeast.

The protein-containing starches and cellulose material are first subjected to a hydrolysis of the starch and subsequently to an ultrafiltration and aerobic fermentation. The hydrolysis of the starch may be carried out with a dilute acid, for instance, hydrochloric acid, sulfuric acid or phosphoric acid. The hydrolysis may also be carried out with just one enzyme. The hydrolysis may, for instance, be effected with an enzyme-generating microorganism which then is separated out prior to the ultrafiltration. On the other hand, it is also possible to separate the microorganism itself from the product of the hydrolysis by ultrafiltration.

The filtrate coming out of the ultrafiltration step may then be subjected to the aerobic fermentation. However, the ultrafiltration may also be effected subsequent to the hydrolysis of aerobic fermentation. The sugar-starch mixture which remains after extraction of the fat may be subjected also directly to an enzymatic degradation with amylases followed by an isomerization step. The starch residue may, for instance, be hydrolyzed in a first step where it may then be subjected to enzymatic degradation with the sugar solution until glucose is obtained, which later may then be subjected to an isomerization.

Following the fat extraction, the protein and other undesired material may be separated. The isomerose obtained of equal parts of glucose and fructose may be further purified by ultrafiltration. It is, however, also possible to effect the entire isomerization subsequent to the ultrafiltration. The isomerization may be effected after separating by precipitation the protein and the salts that may be present. However, the isomerization can also be effected prior to such separation. The removal of the salts, particularly sodium chloride, preferably is effected in an ion exchanger.

Example 1: In this example a bakery product which had been obtained in the production of biscuits was processed. The fat found in the waste from this product was extracted in conventional manner with solvents such as petroleum ether or hexane. The mixture with the solvent was mixed in the tank **4**. The extraction was carried out at a temperature of 25°C during 2 hours in the filter extractor **5**. The temperature could also have been anywhere between 20° and 50°C and the extraction may have been carried out at a time anywhere between 1 and 3 hours.

The amount of hexane used was 1.5 liters, but the general range would be between 2 and 3 liters per kilogram of bakery waste. The fat-containing solution was separated from the solid residue by filtration following the extraction in the filter extractor **5**. The solvent was subsequently separated from the fat by a fractionation indicated at **6**. The yield in fat relative to the original bakery waste was 12 to 15%. The recovered solvent at **8** was then recycled and used for a further extraction in the mixing tank **4** and extractor **5**.

The fat-free residue was mixed with 3 liters of water per 1 kilogram of bakery waste. The general range would be 3 to 5 liters water in this step. Thus, a suspension was formed from which, by means of passage through an ultrafilter, sugar and salt were separated. The ultrafiltrate now contained the sugar and salt. The total sugar contents of the filtrate was 12 to 15% relative to the original bakery waste. The concentration of sugar in the solution of the filtrate was 4 to 5% sugar and 0.4 to 0.5% salt.

The filtrate was then subjected to a fermentation. The fermentation conditions were 37°C at a pH of 3.5. The microorganisms were of the genus *Saccharomyces,* preferably *S. cerevisiae, S. carlsbergensis* and *S. ludwigii*. As nutritive salts were added, per liter of substrate, 3.5 g KH_2PO_4, 0.7 g $(NH_4)_2SO_4$ and 0.35 gram $MgSO_4 \cdot 7H_2O$. The aeration was effected at a rate of 1.2 volumes per volume per minute (vvm), (the range could have been between 0.5 and 1.5 vvm). The fermentation resulted in a yeast containing 45 to 48% crude protein. The yield of yeast was 6 to 7.5% relative to the original bakery waste.

The concentrate remaining after ultrafiltration contained between 65 and 70% solids of which 6 to 7% were protein and 50 to 60% were starch. All of these percentage figures relate to the original bakery waste. The entire concentrate was then subjected to drying in a roller or spray-dryer.

Example 2: The separation of the fats in this case was effected as described in Example 1. The residue which remained after the removal of the fats was then subjected to an extraction in the same filter extractor **5** with different alcohols. The filter extracter was provided for this purpose with a centrifugal purification filter.

The alcohol was 70% ethanol; isopropanol or methanol could also be used. For each kilogram of fat-free residue, 2 liters of ethanol were added and the extraction was carried out during 1.5 hours (a range between 1 and 2 hours). The residue was then removed and again mixed with 1 liter of 70% ethanol and the final residue was separated from the solution by repeated filtration and eventually was washed with one half liter of 70% ethanol.

The residue remaining after the alcohol extraction was subjected to an air current to remove any alcohol residue. They were subsequently subjected to drying in conventional manner and a high value protein-starch product was obtained. The alcoholic phase of this step which contained the entire amount of sugar was passed through a valve **10'** into the separator **10** where the alcohol was distilled off in conventional manner. The remaining aqueous sugar solution was then mixed with an aliquot of water and passed into a cation-anion column exchanger. The cation exchanger compound was of the type LeWatit S100, while the anion exchanger was LeWatit MP62. The salt-free sugar solution was then concentrated by evaporation to a syrup.

SOY AND VEGETABLE PROTEINS

PROTEIN HYDROLYZATES

Beverage-Soluble Peptides

An acidic beverage prepared from defatted soybean can be prepared by enzyme hydrolysis of soybean protein to form soluble peptides or amino acids. The process provides for control of the amount of free amino acids formed which cause a disagreeable taste in the beverage. The process is described in detail by *T. Yokotsuka, Y. Aoyama, T. Kikuchi, S. Ishii and M. Matsuura; U.S. Patent 3,897,570; July 29, 1975; assigned to Kikkoman Shoyu Co. Ltd., Japan.*

The denatured defatted soybean, that is, defatted soybean containing denatured protein, is prepared by heating or steam cooking the defatted soybeans in the presence of water or lower alcohols, and the thus obtained, denatured, defatted soybean can readily undergo the enzymatic action. Further, it is an essential condition that the decomposition product of the denatured, defatted soybean be tasteless and odorless.

As a result of various studies of enzyme preparations and methods for decomposition, it was found that to produce an acidic beverage in a high yield it is necessary to react denatured, defatted soybeans with an acid protease produced by microorganisms or an enzyme preparation containing acid protease as an essential enzyme under an acidic condition and to stop the reaction before a ratio of formol-state nitrogen to total nitrogen in the filtrate of reaction mixture reaches 20%. In that case, too short a reaction period lowers yields of peptides, whereas too long a reaction period contributes to over-decomposition. Thus, it is necessary to properly adjust the period of reaction.

In the case that a protease other than an acid protease such as, for example, alkaline protease or neutral protease is used, there is obtained a product having a FN/TN (formol-state nitrogen to total nitrogen) ratio more than 20%.

The denaturing is carried out by heating the defatted soybean protein in the presence of water or a water-soluble organic solvent, or soaking the defatted soybean protein in a water-soluble organic solvent.

The end point of denaturing of the protein can be determined by conventional methods, such as an enzymatic method or an acidity titration method. According to the enzymatic method, a definite amount of denatured soybeans is placed in a test tube, and 10 ml of 0.5 M buffer solution, 10 ml of an enzyme solution from yellow *Aspergilli* and 1 mol of toluene are added to the test tube. Then, the test tube is tightly stoppered and shaken at 30°C for 20 hours.

After the shaking, the reaction mixture is subjected to centrifugal separation, and the supernatant liquid is collected and trichloroacetic acid is added. The degree of light extinction at 660 mμ of the solution is measured. When there remains nondenatured protein in the solution, precipitates are formed with trichloroacetic acid, but when the denaturing is complete, the degree of light extinction at 660 mμ is zero. The denatured defatted soybean protein having zero degree of light extinction is used in this process. The flavor of the denatured, defatted soybean is quite different from that of raw material soybeans, with the so-called "green note" completely eliminated.

Typical of acid proteases having a high thermostability which can be used is that produced by *Penicillium duponti* ATCC 20186. This enzyme shows an enzymatic activity at a pH of 1.5 to 6.5 and its optimum pH is 2.0 to 3.0. This enzyme can work in a wide temperature range, e.g., from room temperature (about 20°C) to 90°C. The optimum temperature is 60°C at a pH of 2.5, 75°C at a pH of 3.5 and 75° to 80°C at a pH of 4.5. The optimum temperature is higher than that of any known acid proteases.

The decomposition period, when acid protease is used, depends on the decomposition temperature but a period of the decomposition from 3 to 10 hours at a temperature of 40° to 70°C with a means of agitating the mixture is usually suitable for a mixture of 1 part by volume of raw material denatured, defatted soybean, 5 to 7 parts by volume of water and an organic or inorganic acid for controlling the pH of the mixture.

Example 1: To 300 kg of defatted soybean meal was added 2 kl of water, and denaturing of the defatted soybean meal was conducted by steam cooling at 100°C for 20 minutes. The solution extracted from cooked defatted soybean, with an enzyme from yellow *Aspergilli,* underwent no precipitation when trichloroacetic acid solution was added. It was confirmed that complete denaturing was attained. Then, 30 kg of citric acid was added to the cooked solution to adjust pH to 3.8 and then 3 kg of an enzyme preparation containing acid protease obtained by subjecting koji of wheat bran, which is obtained by culturing *Aspergillus inuii* ATCC No. 14333, to water extraction, subjecting the extract to ethanol precipitation and isolating the precipitate according to the conventional method.

Enzyme reaction was conducted at 50°C for 6 hours with stirring, and then reaction mixture was subjected to pressure filtration. To 1.9 kl of the acidic peptides solution having a pH of 4.0 was added 5 kg of active carbon for decoloring and then the decolored solution was subjected to spray drying whereby 200 kg of powders was obtained. The powders were odorless and tasteless, but had acidity and 55% peptides content in terms of protein. The formol-state nitrogen was 16% of the basis of total nitrogen. 5 kg of the powders were thoroughly mixed with 5 kg of glucose, 0.2 kg of tartaric acid, 0.5 kg of sodium bicarbonate and small amounts of powdery cola essence, caramel and artificial sweeteners to prepare a sparkling cola-type powdery beverage product. When 11 g of the prepared product was dissolved in 100 ml of water, a cola-type beverage containing 2.8% peptides in terms of protein was obtained.

Example 2: 30 kg of defatted soybean meal was treated with 90% ethanol at 80°C for 20 minutes to denature the meal. After removal of ethanol, 200 liters of water and 3 kg of citric acid were added to the denatured meal to adjust the pH to 3.8, and then 300 g of commercial acid protease was added. The decomposition was carried out at 50°C for 7 hours with stirring.

After the completion of the reaction, the reaction mixture was subjected to pressure filtration and 185 liters of an acidic peptides solution was obtained. The peptide solution was spray-dried, whereby 21 kg of powders were obtained. The powders were tasteless, though acidic and had 58% peptides content in terms of protein. The formol-state nitrogen was 17% on the basis of total nitrogen. 2 kg of the powders were mixed with 4 kg of glucose, 0.1 kg of tartaric acid and small amounts of cola essence, and coloring agent, and a nonsparkling, cola-type powdery beverage product was obtained. When 1 part by volume of the product was dissolved in about 10 parts by volume of water, a cola-type beverage containing about 2% peptides in terms of protein was obtained.

Double Enzyme Method

Two enzyme systems are used in the process of *F.F. Noe and W.T. Faith, Jr.; U.S. Patent 3,761,353; September 25, 1973; assigned to Rohm and Haas Company* to (1) solubilize isolated protein and (2) hydrolyze such proteins to make them more useful in food applications. These steps are accomplished with an enzyme for solubilization obtained from a bacterial organism. Hydrolyzing is realized by an enzyme from a fungal source.

While it is possible in many instances to solubilize isolated protein by alkaline treatment, this procedure is generally undesirable especially when the resulting protein is to be utilized as a food or food supplement. Alkaline treatment often destroys certain essential amino acids which are present in the protein. Thus, this tends to detract from the nutritional characteristics of such proteins. Furthermore, such alkaline treatment is often uneconomical and the protein resulting is lacking in many of the properties which would render it suitable for satisfactory utilization as a food material.

Consequently, there has been a continual search for a means whereby isolated protein could be readily and economically solubilized. Similarly, substantial efforts have been expended in seeking a means whereby isolated protein could also be hydrolyzed.

The enzyme preparation used in solubilizing isolated protein may be referred to as "Solutase," which is derived from a bacterial organism and is further characterized by being able to render an isolated protein dispersible in water to the extent that a colloidal system will result. Furthermore, the protein so dispersed will be essentially recoverable by precipitation.

This ability of a Solutase enzyme preparation to solubilize isolated protein shall be referred to as "Solutase Activity." The degree is expressed in "Solutase Units" (SU). One SU is that amount of enzyme preparation which will disperse 1 g of isolated protein in 100 ml of water in 15 minutes at a pH of 7.8 when the mixture is agitated at 275 rpm at 40°C. By disperse is meant, of course, the forming of a colloidal system with the solubilized protein being precipitable. Thus, the Solutase Activity of an enzyme preparation may be determined by observing the amount required to disperse a known amount of isolated protein recoverable from solution by precipitation.

Conversely, the Solutase Activity of an enzyme preparation may be determined by observing the amount of protein which will be dispersed by a known amount of enzyme preparation and recoverable by precipitation. On the other hand, hydrolysis of the isolated protein and particularly isolated protein which has been solubilized is achieved by contacting such with a fungal enzyme preparation. This fungal enzyme preparation is referred to as "Hydrolase" and is characterized by being capable of altering the protein to the extent that such will not precipitate from an aqueous solution thereof.

This ability of a Hydrolase enzyme preparation to hydrolze isolated protein and solubilized isolated protein may be termed "Hydrolase Activity." The degree is expressed in "Hydrolase Units" (HU). One HU is that amount of enzyme preparation which will render 1.0 g of solubilized protein unprecipitable when such is contacted for 16 hours at 40°C while being agitated at 275 rpm and maintained at a pH of 6.2. Thus, the Hydrolase Activity of an enzyme preparation can be readily determined.

Both the Solutase- and Hydrolase-producing organisms may be grown in either a liquid-submerged or surface culture on a medium containing a utilizable source of energy, assimilable carbon and nitrogen. Of course, any of the regularly used growth factors and mineral salts or combinations may also be incorporated in such culture.

The process is especially adapted to solubilizing and/or hydrolyzing isolated protein derived from soybeans. Consequently, it is now possible to utilize soybeans in the preparation of various foods, food additives and food supplements. More specifically, it is now possible to solubilize and hydrolyze isolated protein derived from soybeans and to formulate such into drinks both carbonated and noncarbonated, baby foods, whipping materials, frostings and toppings. In addition, such protein may now be rendered useful for the preparation of adhesives, binders and numerous other industrial products.

A particularly advantageous aspect of the process involves the preparation of nutritious drinks containing soybean protein. These drinks were previously lacking in uniformity and aesthetic appeal since the protein tended to settle out of the suspension. Now, however, by treating the isolated protein derived from soybeans with an enzyme, such protein is readily solubilized and/or hydrolyzed and then the necessary additives may be incorporated.

Example 1: 100 ml aliquots of an aqueous medium composed of 2% corn distiller's dry solubles, 6% cornstarch, 0.6% monoammonium phosphate, 0.2% calcium chloride, 2% potassium chloride and 1% sodium citrate are added to 1-liter wide-necked Erlenmeyer flasks which are capped with 3 milk filter discs.

Following sterilization of the medium at 15 psig for 30 minutes, the flasks are cooled and inoculated to a final 2% concentration with an aqueous suspension prepared from a *Bacillus subtilis* culture grown on potato dextrose agar for 48 to 72 hours. The flasks are then incubated on a shaker at 35° to 37°C at 50% relative humidity. The initial pH is 7.0.

After 48 hours, the cultures are removed from the shaker ready for harvesting. A concentrate of the enzyme preparation is obtained by adding with constant mixing three volumes of ethanol at 4°C to the crude filtrate of the above described whole culture. The enzyme preparation is precipitated, then collected by filtration and dried in vacuo at 30° to 35°C. A 5 g portion of the enzyme Solutase is combined with 20 g of lactose. The resulting composition has an activity of 300 SU/g.

Example 2: Approximately 200 parts by weight of wheat bran and 100 parts of wheat middlings are mixed with 400 parts of water. This mixture is then sterilized by heating, and subsequently cooled. A heavily sporulated culture of *Aspergillus oryzae* (ATCC No. 14,605) in an amount equal to 0.1% by weight of the above mixture is thoroughly dispersed. The temperature is raised to 35°C and maintained in that manner for 16 hours to promote rapid growth of the fungus and held for 48 hours at 28°C.

The resulting culture is dried in a current of warm air and then extracted with water. Four volumes of ethyl alcohol are added per volume of extract. The resulting precipitate is centrifugally settled, rinsed with alcohol and dried in a current of warm air at 60°C. This fungal enzyme preparation is then combined with sufficient lactose so as to result in a composition having an activity of 10 HU/g.

Example 3: In an Erlenmeyer flask containing 200 ml of water there is suspended 6 g of isolated soy protein. To this suspension there is added 0.1 g of the Solutase composition (300 SU/g) of Example 1 and 0.1 g of the Hydrolase composition (10 HU/g) of Example 2. This suspension is then maintained at 40°C for 16 hours at a pH of 6.2 and rotated at 275 rpm. To 50 g of the filtrate there is added 10 g of sucrose and 0.1 g of food coloring. After mixing, this yields a highly nutritious and tasty drink. Another 50 g sample of the filtrate prepared above is also combined in the same manner with sucrose and coloring. However, in addition this mixture is then treated with carbon dioxide so as to result in a tasty and nutritional carbonated drink.

Solubilized and hydrolyzed protein such as used above in the preparation of nutritious soft drinks are similarly utilized to fortify the following: acidic citrus juices, breakfast cereals, baby formulas, soups, beers, dietary foods, and bread, macaroni and noodles.

Alkaline Protease Method

An alkaline proteolytic enzyme complex, namely pancreatin, is used by *F. Perini; U.S. Patent 3,928,630; December 23, 1975; assigned to Syntex (USA), Inc.* to prepare hydrolysates.

The pH of the reaction mixture is controlled during the treatment by the addition of calcium hydroxide to maintain a pH in the range of 7.2 to 8.0. Also as the initial substrate is typically slightly acid or neutral, it is desirable to increase the pH of the initial substrate liquid medium mixture to 7.5 to 9.5 by the addition of calcium hydroxide. Although the enzyme functions most effectively below pH 8 (e.g., about pH 7.5) the excess alkalinity is quickly consumed by the acid-generating character of the hydrolysis and, in fact, the excess alkalinity provides a slight grace period in controlling the pH within the prescribed range during the hydrolysis.

It is also best to conduct the treatment using a biostatic agent such as, for example, chloroform or toluene or mixtures of chloroform or toluene. Any proteolytic enzyme which causes the hydrolysis of proteins into peptides and/or amino acids, under alkaline conditions can be used. Typically, a mixture of enzymes obtained from the pancreas of cattle or hogs, commonly referred to as pancreatin, is used as it yields good results and is readily commercially available and relatively inexpensive. Also isolated proteolytic enzymes such as trypsin, chymotrypsin, carboxypeptidases A and B, collagenase and compatible mixtures can also be used.

Suitable protein sources which can be used include insoluble nonfunctional fish protein concentrate, fish meals, slaughterhouse animal wastes, and animal blood. The processes are particularly useful with respect to fish protein concentrate substrates and animal blood substrates.

After the desired hydrolysis is completed, the pH of the reaction is brought down to a pH of about 7 by the addition of phosphoric acid. This results in the formation of a precipitate of calcium phosphate, typically a mixture of mono- and dibasic calcium phosphate. The reaction mixture is next filtered using the precipitated calcium phosphate as the sole filter aid, thus avoiding the need for an external filter aid. Because the calcium phosphate precipitate is a nutritionally acceptable material, the filter cake can be used directly as an animal feed supplement or recycled, whereas the filter cakes obtained using external filter aids must be subjected to additional treatments to remove the filter aid in order to be acceptable for animal consumption.

The process also facilitates the removal of undesirable metallic impurities such as mercury in the case of fish proteins and iron in the case of animal proteins, and also fluorides. The iron impurities and fluoride ions are respectively precipitated as iron phosphate and calcium fluoride and removed with the calcium phosphate precipitate. The mercury contaminants are also removed with the calcium phosphate precipitate either as insoluble salts or by absorption or inclusion within the calcium phosphate precipitate. Where unacceptable quantities of toxic components are recovered with the calcium phosphate precipitate, the resulting filter cake cannot, of course, be directly used as an animal feed supplement.

If desired, the resulting product filtrates, which are rich in protein hydrolysate such as peptide and amino acids can be filtered through or mixed with particulate activated charcoal, then filtered to remove any remaining color impurities. The protein hydrolysate can then be recovered as an essentially white powder from the filtrate by conventional procedures such as, for example, evaporation, freeze drying, spray drying, etc. The resulting product is readily soluble in water forming a flavorless and colorless solution as compared with other products which typically contain undesirable flavors and/or colors characteristic of the particular protein substrates used, e.g., fish flavor, or blood flavor and/or color.

Example: This example illustrates the process on a scaled-up commercial size operation and further illustrates the improvement in filtration rate obtained as compared to a corresponding process using a conventionally added filter aid. In this example 462 lb of fish protein concentrate is mixed with 4,600 lb of deionized water at a temperature of about 113°F. The pH of the mixture is adjusted to about pH 8.5±0.2 by the controlled addition of calcium hydroxide; about 5 lb is required. Four pints of toluene and four pints of chloroform are then added to the mixture as biostatic agents and the resulting mixture is heated to about 110°F; then 13.9 lb of pancreatin 4XNF is added.

Upon digestion of the fish protein concentrate, the pH of the reaction mixture decreases and is controlled at about 7.7±0.3 by the addition of calcium hydroxide at one-half hour intervals for approximately four to five hours. About 15 to 20 lb of calcium hydroxide is required. The digestion is allowed to continue for an additional 14 to 15 hours, at the end of which time the pH has dropped to about 7.4±0.2. The pH is then adjusted to 7.0±0.1 by the addition of 85% by weight aqueous phosphoric acid. The reaction mixture is then heated to 143°F for 30 minutes to deactivate the pancreatin enzyme complex and is then pumped through a filter press at the rate of 25 to 40 gal/min using only the basic calcium phosphate precipitate formed after the addition of the phosphoric acid as the filter aid.

The filtrate is recovered and mixed with 60 lb of powdered activated charcoal and mixed for 30 minutes and then filtered to remove the charcoal. The charcoal filter cake is washed with 100 gal of hot water (about 150°F) and combined with the filtrate. The combined filtrate and washing is heated and condensed by evaporation to solids content of approximately 10%. The concentrate is cooled to 40°F and then spray dried affording approximately 260 lb of protein hydrolysate as substantially white powder. A sample of the product is then dissolved in water forming an essentially colorless clear solution having no discernible taste.

The above procedure is then repeated but using ammonium hydroxide in place of calcium hydroxide and using about 260 lb of diatomaceous earth as a filter aid during the initial filtration process. In this case, the maximum filtration rate which can be obtained is 5 gal/min and a sample of the resulting product forms an aqueous solution having a discernible yellow tinge and a brackish or off-taste.

Nonbitter Low Protein Acid Food Products

J.L. Adler-Nissen; U.S. Patent 4,100,024; July 11, 1978; assigned to Novo Industri AS, Denmark uses an alkaline protease labelled "Alcalase" to prepare soy protein hydrolysate suitable for low protein acid food products, for example, carbonated soft drinks.

An important step of the process involves halting hydrolysis before it proceeds to a level where the bitter taste of low molecular weight polypeptides appears in the hydrolysate. Since the enzymatic action solubilizes otherwise insoluble protein matter as well as hydrolyzes the solubilized proteins it is noteworthy that complete solubilization of the soy protein does not take place. As much as 40% of the initial protein content remains undissolved, and is removed from the (supernatant) polypeptide solution, e.g., by centrifugation, filtration or decantation.

The DH (degree of hydrolysis) plays an important part in control of the process. The hydrolysis is monitored by means of the DH. When DH reaches a critical value, hydrolysis is interrupted. The DH is both the monitoring parameter of the hydrolysis, and a measure of product suitability, i.e., DH of the product is preferably 9.5 to 10.5%.

Studies made with different proteins, notably casein, corn gluten, soy protein isolate, and cottonseed protein, show that the kinetics of hydrolysis are strongly dependent upon the origin of the protein. During this study, the enzyme (Alcalase) itself, the enzyme concentration, the substrate concentration, the pH and the temperature, i.e., all hydrolysis parameters other than the kind of protein were held constant. Casein and corn gluten were hydrolyzed to a greater degree than soy protein. Cottonseed protein was hydrolyzed to a lesser degree than soy protein.

Example 1: 4,000 ml of a suspension of soy protein isolate containing about 8% protein (calculated by multiplying %N by 6.25) was hydrolyzed with 0.2% of Alcalase S6.0 [calculated with respect to the weight of the protein as corresponding to a proteolytic activity of 12 Anson units per kilogram (AU/kg)] at pH 8.0 and a temperature of 50°C. Alcalase S6.0 has a proteolytic activity of 6.0 AU/g. During hydrolysis, which was followed by means of the pH-Stat (Radiometer), the pH was kept constant by addition of 4 N NaOH.

After 2 hours hydrolysis time, a degree of hydrolysis of 10% corresponding to 0.8 meq/g was obtained. (Complete hydrolysis, DH = 100, of soy protein corresponds to 7.9 meq of base consumption per gram of soy protein.) 4 M citric acid was then added until the pH reached 3.5. The hydrolysis mixture was then allowed to stand for 30 minutes before the supernatant was decanted through a paper filter using diatomaceous earth as filter aid. The supernatant was treated twice with 0.01% w/v activated carbon powder for 30 minutes at a temperature of 30°C. This produced a protein hydrolyzate in 68% yield and with a nonbitter, pleasant taste.

A beverage with 3% of protein hydrolyzate, 10% of sucrose and 0.005% of lemon oil was produced and found to be organoleptically acceptable. The beverage was pasteurized in closed vessels and stored in a refrigerator for several weeks. No precipitation or growth of microorganisms occurred, and only a slight discoloration because of Maillard reactions was observed.

Example 2: A hydrolysis was performed as in Example 1, but with soy concentrate instead of soy isolate. The resultant mixture was a clear liquid which remained stable after addition of 0.1% w/w sorbic acid as a preservative. The liquid could easily be diluted with water, and a standard drink, 2.62% protein (N x 6.25) and 9% sucrose, was prepared. The drink was evaluated for bitterness and was reported to be nonbitter.

Example 3: Soy isolate was mixed with water in a blender to constitute a suspension of 8% protein (N x 6.25). 120 liters of the suspension was pumped to a 150-liter stainless steel vessel which was fitted with a thermostated heat-jacket, a stirrer and a pH electrode. The suspension was hydrolyzed with 0.2% Alcalase

S6.0 (calculated with respect to the weight of the protein as corresponding to a proteolytic activity of 12 AU/kg protein) at pH 8.0 and a temperature of 50°C. During hydrolysis pH was maintained at pH 8.0 by addition of 50 to 100 ml quantities of 5 N NaOH at regular intervals. The hydrolysis was terminated at a DH of 10% by addition of dry citric acid until pH had dropped to 4.2.

After 30 minutes, the reaction mixture was centrifuged in a continuous centrifuge. The supernatant was adjusted to pH 5.0 and mixed with 0.1% w/v activated carbon powder. After 30 minutes at 50°C, the reaction mixture was filtered in a filter-press and concentrated in a vacuum evaporator until a polypeptide concentration of 23.3% (N x 6.25) was reached. After evaporation, a small precipitate was removed by filtering. The resultant mixture was a clear brown liquid which remained stable after addition of 0.1% w/w sorbic acid as a preservative. The liquid could easily be diluted with water, and a standard drink 2.62% protein (N x 6.25) and 9% sucrose was prepared. The drink was evaluated for bitterness and was reported to be nonbitter.

Acid Protease Method

It is known that mold proteases are generally used for protein hydrolysis. *K. Mitsugi, T. Nakase and Y. Hasegawa; U.S. Patent 3,677,898; July 18, 1972; assigned to Ajinomoto Co., Inc., Japan* report that an acid protease can also be produced from yeast of the genus *Rhodotorula.* The acid protease is particularly useful in the hydrolysis of soybean proteins and is not inactivated by sodium lauryl sulfate.

Rhodotorula glutinis AJ-5193, for example, had the following characteristics:

Growth in yeast-malt broth at 25°C for 3 days: Cells are round, short oval to oval, sometimes long oval, 2.5 - 5 x 3 - 7.5 μ and occur singly or in pairs. An incomplete ring and sediments are formed.

Growth on yeast-malt agar at 25°C for 10 days: A streak culture is light reddish orange to pink, smooth, soft, shining, and has an entire margin.

Sporulation: Absent.

Formation of pseudo- and true mycelia: Absent.

Fermentation: Absent.

The culture medium should be kept acid at pH 1.5 to 4, and significant amounts of protease are not produced in a neutral medium although the growth of the microorganisms is not impaired by the higher pH of the medium, as is evident below, relating to *Rhodotorula glutinis* AJ-5193.

Culture Period (days)	Initial pH 3.0			Initial pH 7.0		
	Final pH	Growth	Activity (u/ml)	Final pH	Growth	Activity (u/ml)
2	2.05	0.860	41.2	6.2	0.730	3.2
3	1.9	0.820	62.8	6.3	0.850	0.8
4	1.9	0.750	73.2	6.7	0.766	2.0
5	1.9	0.750	74.2	–	–	–

In the tests for which results are listed above, the culture media contained 5 g/dl glucose, 1 g/dl yeast extract, 0.5 g/dl KH_2PO_4, 0.04 g/dl $MgSO_4 \cdot 7H_2O$, and 2 ppm Fe and Mn ions. They differed in pH only. The growth of *Rhodotorula glutinis* AJ-5193 was determined from the light absorbancy of the broth at 560 mμ when diluted with 25 volumes of water. The protease activity was determined as follows:

3 ml 1.5% milk casein solution in ⅒ M lactic acid-sodium lactate buffer of pH 3.0 was mixed with 1 ml diluted protease solution. The protease solution was kept at 37°C for six minutes prior to mixing, and the mixture was incubated at the same temperature for ten minutes. Then, 4 ml 0.4 M trichloroacetic acid solution were added, and the mixture was filtered after 25 minutes. 1 ml of the filtrate was mixed with 5 ml 0.4 M sodium carbonate solution and 1 ml Folin reagent.

The amount of arylamino acid released into trichloroacetic acid-soluble fraction at 37°C in ten minutes was determined from the light absorbancy of the mixture at 660 mμ. A protease unit is the activity of a protease which releases 1 μg of tyrosine into trichloroacetic acid-soluble fraction per minute.

The protease may be recovered from the culture medium in a conventional manner as by salting out, precipitation by water-miscible organic nonsolvents, or chromatographic separation. The acid protease recovered from culture broth of *Rhodotorula glutinis* AJ-5193 and *Rhodotorula aurantiaca* AJ-4860 has characteristic properties not present in other acid proteases.

The optimum pH for proteolysis of a milk casein substrate is 2.0 to 2.5, and the optimum temperature is 60° to 65°C at which the optimum pH is 2.5. The enzyme is not inactivated to a measurable extent by heating in aqueous solution to 50°C for two hours at pH 1.5 to 6.0. It is not inactivated in aqueous solution by sodium lauryl sulfate, and thereby distinguished from mold-derived acid protease. The specific protease activity of the enzyme in units per gram is four times that of commercial pepsine and 2.5 times that of "Molsine," the trade name of an acid protease derived from a mold.

Example: 50 ml batches of a culture medium containing 5 g/dl glucose, 0.3 g/dl KH_2PO_4, 0.04 g/dl $MgSO_4 \cdot 7H_2O$, 2 ppm each Fe and Mn ions, and 1 g/dl yeast extract, and adjusted to pH 3.0 were placed in 500 ml shaking flasks, sterilized at 115°C for ten minutes and inoculated with *Rhodotorula glutinis* AJ-5193 which had previously been cultured on a slant of the same composition. Each flask was kept at 31°C with shaking for four days. The culture broth then contained 120 units/ml acid protease.

1.5 liters of the culture broth were centrifuged to remove the yeast cells, and the liquid so obtained was cooled and mixed with 55% (by volume) acetone to precipitate impurities which were removed by centrifuging. The supernatant liquor was then mixed with enough acetone to make its acetone content 75%, whereby the acid protease was precipitated. It was separated from the liquid by centrifuging and dried in a vacuum. The crude enzyme powder weighed 1.45 g.

1 g enzyme powder was dissolved in 200 ml citric acid-HCl buffer solution of pH 4.0, and the solution was passed over a chromatographic column packed with

crosslinked dextran (Sephadex G-100) which provided a hydrophilic, insoluble molecular sieve, the diameter of the column being approximately 8 cm. The chromatographic medium had been pretreated with the buffer solution. The effluent was collected in fractions, and the fractions having acid protease activity were combined, mixed with acetone, and kept overnight in an ice box whereby a colorless, crystalline precipitate was formed. After repeated recrystallization from acetone, 0.62 g crystalline acid protease was obtained. It showed maximum absorption bands at 280 mμ and minimum absorption bands at 230 mμ. The specific acid protease activity of the crystalline material was 5.24 times that of crystalline pepsine obtained from Sigma Chemical Co.

High Salt Tolerance System

Although hydrolysis of proteinaceous materials is generally conducted by either an acid process or an enzymatic process utilizing a proteolytic enzyme preparation or an aspergillus mold culture, the latter process is superior in product quality. However, it is also known that the enzymatic process utilizing a proteolytic enzyme preparation will give a hydrolyzate in which the ratio of free amino acids to total solubilized nitrogen in the hydrolyzate of protein materials (this ratio is referred to as free amino acid ratio) is low.

It is conceivable, therefore, that in order to attain sufficient splitting of proteinaceous materials by the method of enzymatic hydrolysis, it is necessary to use a proteolytic enzyme having a wide range of substrate specificities or a great variety of enzymes having different substrate specificities among them. However, since ordinary enzyme preparations contain neither such an enzyme of broad-range substrate specificity nor a great variety of enzymes in amounts sufficient for practical use, the low free amino acid ratio has been considered as a defect of the enzymatic hydrolysis.

It is further known that as compared with the above process, utilizing a proteolytic enzyme, a relatively high free amino acid ratio is attained by a so-called koji process, a variety of the enzymatic process, whereby a proteinaceous material is hydrolyzed by means of "koji" obtained by inoculation and cultivation of an aspergillus mold capable of producing proteolytic enzymes on the protein material as substrate, as in the brewing of soy. The free amino acid ratio of the product obtained by such a method is also not quite satisfactory.

T. Yokotsuka, T. Iwaasa, T. Okami, M. Noda and M. Fugii; U.S. Patent 3,932,671; January 13, 1976; assigned to Kikkoman Shoyu Co., Ltd., Japan found that salt-resistant lactic acid bacteria isolated from soy mash do not exhibit good hydrolyzing effects in the presence of high concentrations of sodium chloride as in soy brewing when ordinary proteolytic enzymes are used. The new process produces an increase in free amino acid content in the presence or absence of almost 12% sodium chloride.

In order to obtain desired results by using salt-resistant lactic acid bacteria, either the bacteria are inoculated into the protein hydrolyzing system and cultured or a large amount of cells of the bacteria are added to the system. The inoculation and cultivation of the salt-resistant lactic acid bacteria or the addition of cells can be conducted either before or after substantial termination of hydrolysis of the proteinaceous material with proteolytic enzyme, the choice being made by taking account of the conditions under which the proteinaceous material is hydrolyzed, such as temperature, pH, etc., though usually conducted after substantial termination of the hydrolysis of proteinaceous materials caused by proteolytic enzymes.

In using the salt-resistant lactic acid bacteria, it is necessary that the concentration of sodium chloride in the protein hydrolyzing system be maintained within the range of 0 to 12% by weight based on an amount of an aqueous medium. Concrete examples are shown below concerning the experiments conducted to examine relationships between the sodium chloride concentration and free amino acid content in a protein hydrolyzate obtained by use of a proteolytic enzyme when the salt-resistant lactic acid bacteria are inoculated and cultured or the bacteria are added directly.

By use of a proteolytic enzyme preparation obtained from *Aspergillus oryzae* grown on wheat bran in the form of koji (a dried powder of precipitates obtained from an aqueous extract of the wheat bran koji on addition of ethyl alcohol three times in volume), defatted soybean which had been subjected to denaturation treatment was hydrolyzed to obtain a protein hydrolyzate of a total dissolved nitrogen content (TN) of 2.0% (w/v), containing 0.86% (w/v) of free amino acids in terms of amino nitrogen (AN), the free amino acid ratio (AN/TN) being 0.43.

The above-obtained protein hydrolyzate was then subjected to heat treatment at 90°C for 5 minutes to deactivate the proteolytic enzyme and filtered. Sodium chloride was added to the clear filtrate to prepare protein hydrolyzate solutions (pH 6.0) containing sodium chloride in various concentrations. *Pediococcus soyae* ATCC 13621 was inoculated into each solution and allowed to undergo stationary culture at 20° to 30°C for 10 days, when the percentage increase in free amino acid content (AN) was determined to obtain the results as shown below.

The percentage increase in free amino acid content was found to be substantially the same after the protein hydrolyzate solution (in which the salt-resistant lactic acid bacterium had been inoculated and cultured) had been left standing for a longer period of time.

NaCl Concentration %, (w/v)	Increase in Free Amino Acid Content, %
0	22.5
3.5	16.0
7.0	18.0
10.5	20.0
12.0	16.0
13.0	4.0
14.0	4.0
15.0	2.0

Example 1: To 4 kg of thermally denatured defatted soybean, were added 1 kg of wheat bran koji prepared with *Aspergillus oryzae* in a customary way and 25 liters of tap water. While the mixture was stirred at 55° to 60°C for 15 hours, hydrolysis was allowed to proceed. Then the hydrolyzed mixture was filtered to obtain a clear solution (TN = 1.01%, w/v). The ratio of amino nitrogen to total nitrogen, i.e., AN/TN, of the solution was 0.42.

To 1 liter of the hydrolyzate solution, was added sodium chloride to give a concentration of 7.5% (w/v). After having been adjusted to pH 6.0 and sterilized at 120°C for 10 minutes, the hydrolyzate solution was inoculated with 10 ml of a culture liquor prepared by culturing *Pediococcus halophilus* ATCC 21786, a lactic acid bacterium isolated from soy, in a soy fermentation medium (a diluted solution containing 5.0% (w/v) of NaCl and 0.4% (w/v) of TN). The inoculated hydrolyzed solution was subjected to a stationary culture at 30°C. On the twelfth day of culture, the hydrolyzate solution showed that AN/TN = 0.52, and percentage increase in free amino acid content was 23.8%. The protein hydrolyzate obtained had a flavor far superior to that of conventional products.

The protein hydrolyzate was treated and left standing at 30°C for 12 days in the same manner as mentioned above, except that the above-noted salt-resistant lactic acid bacterium was not inoculated. The free amino acid ratio (AN/TN) of the resulting protein hydrolyzate was not changed from that of the clear hydrolyzate solution immediately after hydrolysis.

Example 2: In a 2% (w/v) solution (pH 7.0) of milk casein, was dissolved a proteolytic enzyme to provide a concentration of 2,000 units/ml. Hydrolysis was allowed to proceed at 45°C for 10 hours to obtain a hydrolyzate solution (TN = 0.28%, w/v) having a free amino acid ratio (AN/TN) of 0.36.

After having been adjusted to pH 6.5 and sterilized, 100 ml of the hydrolyzate solution was inoculated with two platinum loopsful of cells of *Pediococcus soyae* ATCC 13623 and incubated at 30°C for 10 days. At the end of incubation, the hydrolyzate solution showed a free amino acid ratio (AN/TN) of 0.45 and had an excellent flavor.

The hydrolyzate solution was treated and left standing at 30°C for 10 days in the same manner as mentioned above, except that the above-noted salt-resistant lactic acid bacterium was not inoculated. The free amino acid ratio (AN/TN) of the resulting hydrolyzate solution was not changed from that of the hydrolyzate solution immediately after hydrolysis.

CARBOHYDRASE FLATULENCE REDUCTION

Yeast Carbohydrase

Certain carbohydrates are not readily hydrolyzed in the intestine; hence, they tend to ferment when subjected to the normal bacterial flora present. Such fermentation results in the production of gas, that is flatulence. Needless to say, the presence of any of these carbohydrates in any foodstuff, food additive or food supplement significantly detracts from its value as such. In point of fact, many foodstuffs containing such carbohydrates are rendered essentially useless as foods. This is due to the fact that the discomfort caused by flatulence is highly undesirable, not only in human beings but likewise in animals.

One foodstuff of particular value which is especially illustrative of this problem is the soybean. Since soybeans are grown in great quantity and also constitute significant sources of protein, their potential nutritive value is immense. Nevertheless, their food value is somewhat limited because of their flatulence-producing saccharide content.

The flatulence-producing saccharides in soybeans and soybean derivatives are primarily stachyose and raffinose. Stachyose which constitutes the major portion is a tetrasaccharide which is best designated as galactose-α-1,6-galactose-α-1,6-glucose-β-1,2-fructose.

Stachyose and raffinose like other flatulence-producing saccharides tend to remain unhydrolyzed in the upper intestine of human beings and animals. Thus, when the saccharide is fermented by bacteria in the lower bowel, it produces gas which results in substantial discomfort. Before soybeans or soybean derivatives are to be utilized as foods, or food additives, it is desirable to render innocuous the stachyose, raffinose and any other similar materials.

S.E. Sherba; U.S. Patent 3,632,346; January 4, 1972; assigned to Rohm and Haas Company describes the use of a carbohydrase enzyme which is effective in reducing flatulence-producing saccharides.

The enzyme preparations are derived from cultures of *Aspergillus* and *Rhizopus* including various strains and mutants. Some typical members of these genera include, for example, *A. niger, A. oryzae, A. parasiticus, A. awamorii, R. oryzae* and *R. oligosporous.* A special strain of *A. niger*, which is typical of those that might be used, has been deposited in the American Type Culture Collection and designated as ATCC No. 20107.

Many yeasts are also excellent sources of carbohydrase. Of particular significance are cultures of *Saccharomyces* and *Candida.* Within these genera, *S. cerevisiae, S. italicus, S. carlsbergensis* and *C. pseudotropiclas* are effective. Likewise, numerous *Bacillus* cultures constitute suitable sources of carbohydrase. *B. subtilis* is a typical organism which may be used.

The carbohydrase activity of any particular enzyme preparation may be determined by observing the number of mg of enzyme preparation required to produce 61.2 mg of reducing sugars from an aqueous solution containing 100 mg of essentially pure stachyose after 16 hours. The stachyose solution should be maintained at 50°C and at a pH of 5.0 for 16 hours. A formula in determining the number of carbohydrase units per gram in any enzyme preparation is as follows:

$$\text{Carbohydrase units/g} = \frac{1{,}000}{\begin{array}{l}\text{Mg of enzyme preparation}\\ \text{required to obtain 61.2 mg}\\ \text{of reducing sugars.}\end{array}}$$

Numerous means are available for determining when the flatulence-producing saccharides in any particular foodstuff have been rendered innocuous. Generally, this may be done by quantitatively and qualitatively evaluating the sugars with gas liquid chromatography or by chemically measuring the amount of reducing sugars obtained.

The process is especially adapted for rendering innocuous the flatulence-producing saccharides in soybeans, and particularly, soybean derivatives such as soy milk. Full fat soy milk is obtained as a filtrate from water-extracted soybeans which have been soaked, dehulled and flaked. Alternatively, the dehulled soybeans may be ground and defatted before water extraction so as to result in defatted soy milk. Soy milk obtained by either process contains a substantial amount of flatulence-producing saccharides primarily in the form of stachyose. By incorporating sufficient carbohydrase in this milk, the flatulence-producing saccharides are rendered innocuous. Similarly, the solid residue resulting from soy milk produced by either of the above methods may also be treated by the process.

Example 1: A surface medium containing 20 g of whole wheat bran, 13 g of full fat soybean flakes, 2 g of cornsteep water and 20 g of distilled water is added to 1-liter wide-mouth jars and capped with cotton stoppers. Following steam sterilization of the medium at 15 psi for 30 minutes, the flasks are cooled and inoculated with 30 g of an aqueous suspension of a culture of *Aspergillus niger* ATCC No. 20107 which was grown on Nutrient Agar for 72 hours.

The flasks are then incubated in a water bath while being maintained at a temperature in the range of 25° to 30°C. The initial pH of the culture is 4.8. After 96 hours, the jars are removed from the water bath and the culture is harvested and extracted with water to solubilize the enzymes. The enzymes in the extract are precipitated with ethanol and dried.

This enzyme preparation is then tested for carbohydrase activity. Accordingly, three 10 ml aliquots of a 1% stachyose solution are maintained at 50°C at pH 5.0 having incorporated 10, 20 and 30 mg, respectively, of the enzyme preparation prepared above. After 16 hours all three solutions are analyzed for reducing sugars. The results of such analysis are as follows:

Amount of Enzyme Added (mg)	Reducing Sugar Produced (mg)
10	44.0
20	60.5
30	68.6

Thus, by interpolation of this data it follows that 20.8 mg of enzyme preparation produced 61.2 mg of reducing sugars. Accordingly, the enzyme preparation has a carbohydrase activity of 48 carbohydrase units/g.

Example 2: Two aliquots, 1,000 ml each of soy milk derived from whole ground soybeans and containing 3% by weight of total solids are adjusted to a pH of 5.0 with HCl and equilibrated at 50°C. The enzyme preparation of Example 1 is incorporated in one aliquot of the soy milk in an amount equal to 0.15% by weight of the solids in the soy milk. After 16 hours at 50°C both samples are analyzed by gas liquid chromatography. The control sample of soy milk, that is, the solution not having the enzyme preparation incorporated, evidenced upon analysis the presence of a mixture of stachyose, raffinose, and melibiose constituting 0.05% by weight of the total soy milk. These three saccharides, namely, stachyose, raffinose and melibiose, are flatulence-producing saccharides. Analysis of the soy milk aliquot having the enzyme preparation incorporated does not show the presence of any significant amount of these flatulence-producing sugars.

Enzyme/Yeast Fermentation

It has been established that apart from proteasic inhibitors, the presence in soy of certain carbohydrates, particularly oligosaccharides (stachyose, raffinose, etc.), is responsible for digestive disorders (flatulance, diarrhoea, etc.) and for the imperfect assimilation of the soy proteins in young children and young animals, even when the soy has undergone heat treatment. These carbohydrates are mostly insoluble in water (less than 1% are water-soluble).

Oil-free purified soy concentrates have been produced on an industrial scale for several years and their carbohydrate content has been greatly reduced by the solvent action of an alcohol in a heated state.

It is possible to transform a considerable fraction of the carbohydrate in soy and other plant materials to obtain products low in carbohydrate and high in protein. A process for this conversion by enzymatic methods is described by *M.M. Gay; U.S. Patent 3,958,015; May 18, 1976.* Enzymes are used to hydrolyze or modify carbohydrates so that the resulting products can be metabolized by yeasts to reduce further the carbohydrate content. The essential steps of the process are:

(1) The partially or totally deoiled soybean and/or other plant material like legume beans (field beans, broad beans, white beans, peas, etc.), cereals (wheat, corn, rice, etc.), roots and tubers (potatoes, manioc), either with or without being thermally treated, dehulled, peeled ground or crushed, or flattened, is subjected to the simultaneous action of enzymes. The action of the enzymes like amylases, pectinase, cellulases, invertases, β-gluconases is specific to all the carbohydrates. The enzymatic action takes place in an aqueous medium at suitable pH, temperatures, agitation for an appropriate period of time.

(2) The aqueous suspension of soy and/or other plant material is heated for a short period of time to deactivate the enzymes.

(3) The aqueous suspension of soy and/or other plant material is subjected to fermentation by an alimentary yeast at suitable pH, temperature, air, nitrogen supply for an appropriate period of time.

(4) The yeast-fermented suspension is then pasteurized and

(5) The pasteurized suspension is dried by an appropriate means.

Example: (a) 100 g of fine soy flour from which the oil has been removed and which is toasted and placed in suspension in 400 g of pure water is heated to a temperature of 45°C. The pH of this suspension is adjusted to 5.2 by the addition of sulfuric acid.

(b) 150 mg of a commercial pectinase are added to this suspension measuring: polygalacturonase having a PG activity of 800 units/gram, pectinmethylesterase having a PG activity of 20 units/gram, a considerable uncalculated activity of pectintransaminase and a hemicellulasic activity of 11,400 units/gram of CMC activity (Cx). 50 mg of an invertase having an inversion power of 560 PI are also added.

(c) The soy suspension is stirred continuously at a constant temperature of 45°C for 5 hours.

(d) The soy suspension is brought to a temperature of 70°C, with constant stirring for 30 minutes in order to deactivate the enzymes.

(e) 20 g of a 20% suspension of bakers' yeast, *Saccharomyces cerevisiae* are added to the soy suspension heated to 30°C.

(f) For 6 hours the fermenting liquor is stirred continuously and nitrogen is added in the form of 200 mg of ammonium phosphate.

(g) The suspension of soy and yeast is pasteurized.

(h) The suspension of soy and yeast is concentrated by evaporation in vacuo and dried by atomization in a spray tower.

At the beginning of the operation, the soy flour has 50% dry weight raw protein and 0.80% reducing sugars. After the stage (c) of treatment with enzymes, the proportion of reducing sugars reaches 15%. After the final drying stage (b) the proportion of raw proteins in the dry concentrate reaches 60%. Reducing sugars are less than 1%.

Penicillium α-Galactosidase

It is known that α-galactosidase is capable of hydrolyzing oligosaccharides containing galactose, i.e., sugars with at least one α-D-galactopyranosyl bond, in other words "flatulent" sugars such as stachyose and raffinose.

Among the leguminosae, soy in particular is used in large quantities for feeding cattle and, to an increasing extent, is replacing all or some of the staple foods, such as meat and milk-based products, intended for human consumption. Most of the assimilation problems attributable to the presence of nonresorbable oligosaccharides responsible for flatulence may be resolved by treating the basic materials such as defatted soy flour with enzymatic preparations of microbial origin. It is known that, in order to make up for the absence of α-galactosidase in human beings and cattle, it is possible to use microorganisms capable of synthesizing this specific enzyme.

Microorganisms known for their ability to produce α-galactosidase, such as *Mortierella vinacea,* variety *Raffinoseutilizer,* have their effect in producing α-galactosidase induced by the presence of a galacto-oligosaccharide, for example, raffinose or lactose, in the fermentation medium. It is for this reason that various culture media have been proposed, ranging from a conventional basic medium enriched with raffinose to a medium consisting of soy powder, rice bran and water.

N. Arnaud and D.A. Bush; U.S. Patent 3,966,555; June 29, 1976; assigned to Societe d'Assistance Technique pour Produits Nestle SA, Switzerland found that it is possible to culture the mold *Penicillium duponti* for the production of α-galactosidase.

The enzyme is synthesized in the cells of the microorganism. If the mycelium is separated from the culture medium by filtration, for example, no trace of extracellular α-galactosidase can be found in the filtrate. It is for this reason that either the mycelium itself or an extract may be collected as the enzymatic preparation.

In the first case, the mycelium is pulverized in a mortar, separated from the medium and washed. In the second case, the mycelium may be homogenized in a buffer at pH 5.5 to 7.5, centrifuged and the supernatant phase collected. The α-galactosidasic activity of the supernatant phase is almost as great as that of the mycelium itself. Slight differences may be attributable to certain losses incurred during homogenization. The mycelium as the enzyme may be maintained in lyophilized form and loses only a very small part of its properties as a result of this treatment.

It is possible by the process to prepare a biomass with a dry weight per liter of culture broth of from 2 to 3 g. If a unit (U) of α-galactosidasic activity is defined as the quantity of enzyme which releases 1 μg of reducing groups from a 0.05% stachyose solution in 30 minutes, the activity of the nonhomogenized mycelium obtained by the process may reach 45,000 U/g of dry mycelium, and the activity of the mycelium after homogenization may reach 42,000 U/g of dry mycelium. This activity has numerous commercial uses. Conclusive results have been obtained in practical tests in which, on the one hand, the lyophilized enzyme was used for treating a soy milk, and, on the other hand, the lyophilized mycelium was used for degrading the raffinose present in the juice of sugar beet.

Example 1: A water from the blanching of haricot beans (*Phaseolus vulgaris*) has the following composition in g/l:

Stachyose	1.0
Raffinose	0.2
Other sugars	1.6
Total nitrogen	0.4
Lipids	1.7
Ash	1.1

The pH value of this water is adjusted to pH 6.5, after which the water is introduced into 2 liter flasks in quantities of 500 ml per flask. An inoculum of a strain of *Penicillium duponti (Talaromyces thermophilus)* ATCC 10518 is introduced into the flasks. The inoculated flasks are agitated at 45°C in agitators turning at 120 rpm over a period of 24 hours.

The culture broth is filtered by drawing it through ordinary sintered glass and the mycelium is collected, representing 2.7 g of dry material per liter of broth. A test on stachyose shows that the α-galactosidasic activity of mycelium thus obtained amounts to 45,000 U/g of dry mycelium, one unit of activity being defined as the quantity of enzyme which liberates 1 μg of reducing groups from a 0.05% stachyose solution in 30 minutes.

Example 2: 100 mg of the enzyme or supernatant phase prepared in the same way as in Example 1 and then lyophilized, are added to 500 ml of a soy milk containing 3.7% of stachyose. The enzymatic treatment is continued for 30 minutes at 55°C. After the treatment, the stachyose content of the soy milk is reduced by 40% with hardly any change in the taste of the milk.

Example 3: 1 g of mycelium produced in Example 1 is added to 200 ml of a sugar beet juice containing 22% of saccharose and 2% of raffinose. The enzymatic treatment is continued for 24 hours during which the degradation of the raffinose is observed. A percentage degradation of 40% is observed after approximately 1 hour, rising to 60% after about 2.5 hours, to around 80% after 4 hours, ultimately stabilizing at around 85% after approximately 8 hours.

HEAT-DENATURED FUNCTIONAL PROTEIN

R.C. Jolly; U.S. Patent 4,107,334; August 15, 1978; assigned to Pfizer Inc. describes a process for preparation of a functional protein prepared from whey, microbial or vegetable protein which can be used as a replacement for nonfat dry milk, sodium caseinate, gelatin and egg albumin. The impure protein in aqueous medium is adjusted to the desired pH, at about the isoelectric pH of the protein, and is then subjected to heat treatment for precipitation of the protein.

The precipitated heat-denatured protein is separated from the aqeuous medium by any convenient method such as filtration or centrifugation. The separated precipitate is washed with water to minimize the water-soluble impurities in the precipitate. These impurities include, for example, lactose and salt in the case of whey protein and ribonucleic acid in the case of microbial protein. The separated precipitate may also be washed with a water-soluble food grade alcohol such as ethanol prior to the water wash, if desired.

The heat-denatured protein is slurried in aqueous medium and converted to functional protein with proteolytic enzyme. Any acid, neutral or alkaline protease of microbial, plant or animal origin is suitable. Preferred are microbial proteases of bacterial origin such as those derived from *B. subtilis* and *B. licheniformis* and those of fungal origin such as those derived from *A. oryzae, A. flavus, A. niger* and *S. griseus;* plant proteases such as papain, ficin and bromelain; and animal proteases such as pepsin, trypsin, chymotrypsin and rennin. The sole requirement is that the protease be present in sufficient amount to insure conversion of the denatured protein to functional protein within 30 minutes without creating objectionable bitterness in the final protein product.

At the completion of the reaction, the resulting hazy to clear functional protein solution is treated for enzyme deactivation. The method of treatment will depend upon the nature of the enzyme, but the deactivation is usually accomplished by heating the reaction solution to from 80° to 100°C. Depending upon the enzyme employed, such treatment may be accompanied by pH adjustment.

The functional protein solution following enzyme deactivation is normally cooled to about room temperature, adjusted to pH 6 to 8 and then either used directly in food compositions or dried by conventional means such as spray or freeze drying prior to such use.

The functional protein, because of its acid solubility, is readily suitable for use in acid food compositions such as carbonated and acidic noncarbonated beverages where traditional proteins cannot be used. The functional protein can replace up to twice its weight of nonfat dry milk in many food compositions including baked goods such as yeast-leavened bread, confections such as frostings and milk nougat, desserts such as puddings and ice cream, instant breakfast foods, processed meats such as meatloaf, dressings such as salad dressing and mayonnaise, and fermented milk products such as yogurt. Sodium caseinate is totally replaced in such food compositions as coffee whiteners, nondairy products such as cream cheese, nondairy processed cheese and nondairy cheese spreads, whipped toppings and sour cream as well as in both milk and citrus-type instant breakfasts and in processed meat products. Gelatin may be wholly replaced in such food compositions as instant frozen pudding and bakers' marshmallow, while egg albumin is partially to totally replaced in such products as meringues, confections such as soft-centered candies, angel food cake, textured vegetable protein, pasta and noodles.

Suitable vegetable protein sources include, for example, soybeans, wheat gluten, cottonseeds, okra, corn gluten, peanuts, potatoes, alfalfa, oats, rice, rapeseeds, sesame seeds and sunflower seeds. Preferred sources are soybeans, wheat gluten and cottonseeds. Especially preferred to soybeans in such forms are soy grits, solvent-extracted soybean flakes, soy flour, alcohol-treated soy flakes, soy concentrate and soy whey. The protein source is usually slurried in water, any undissolved material is separated, and the liquid phase is introduced into the process. Vegetable protein whey, like cheese whey, may be processed directly.

Example 1: A sample of fresh sweet cheese whey (94.6 liters) was adjusted to pH 5.0, heated to 90°C and held at that temperature for 5 minutes. The resulting slurry was cooled to room temperature and centrifuged. The centrifuge cake was washed with water to yield 1,415 g of wet heat-denatured whey protein solids (436 g dry).

A portion (32.7 g) of the wet heat-denatured whey protein solids (10.1 g dry) was dispersed in 100 ml water. The slurry was adjusted to pH 8.5 and 50°C, 20 mg of Alcalase S-6, which is microbial alkaline protease derived from *B. licheniformis,* Novo Enzyme Corp., (2,970 Delf units/g protein) was added, and the mixture was stirred at 50°C and pH 8.5 for 30 minutes, the pH being maintained by the addition of 1 N NaOH. The resulting solution was heated to 80°C for 2 minutes to deactivate the added enzyme. The solution was then cooled to room temperature, adjusted to pH 7.0 with 1 N HCl and freeze dried. The dry product passed the protein functionality test and had acceptable taste with no objectionable bitterness or saltiness.

The fresh whey may also be adjusted to pH 4.0 and 40°C, or to pH 7.0 and 150°C, pH 0.5 and 100°C or pH 9.0 and 120°C, and held for an appropriate time to substantially precipitate the whey protein solids. Proteolysis of the heat-denatured whey protein solids may also be run at 20°C or at 65°C, using 2,300 Delf units or more of Alcalase S-6 per gram of dry protein to insure solution of the protein within 30 minutes.

Example 2: S. cerevisiae yeast cells (800 g wet, 172 g dry) were dispersed in 1,000 ml water and the pH of the slurry was adjusted to 9.5 by the addition of 1 N NaOH. The cells were homogenized at 9,000 psi (633 kg/cm^2) and 30°C for 80 minutes. The resulting mixture was centrifuged and the supernatant liquor decanted. The residual solids were repulped in 1,000 ml water and recentrifuged. The supernatant liquors were combined, adjusted to pH 6.0, heated to 70°C and held at that temperature for 1 minute. The resulting slurry was centrifuged and the centrifuge cake was washed with water to yield 392 g wet heat-denatured yeast protein solids (72 g dry).

A portion (50 g) of the wet heat-denatured yeast protein solids (9.2 g dry) was dispersed in 200 ml water. The slurry was adjusted to pH 8.5 and 50°C, 100 mg subtilisin (16,300 Delf units/g protein) was added and the mixture was stirred at 50°C and pH 8.5 for 30 minutes, the pH being maintained by the addition of 1 N NaOH. The resulting solution was heated to 80°C for 5 minutes to deactivate the added enzyme, and the solution was then cooled to room temperature, adjusted to pH 7.0 with 1 N HCl and freeze dried. The product passed the protein functionality test and had acceptable taste with no objectional bitterness or saltiness.

The hydrolysis may be repeated using fungal protease at a level of 240 HUT/g (Hemoglobin Units on the Tyrosine basis) protein at 50°C and pH 5.0 for 30 minutes to yield a functional protein with acceptable taste. Substitution of *S. fragilis* or *C. utilis* yield cells, *P. methylotropha, L. bulgaricus* or *S. lactis* bacterial cells, or *T. viride, F. solani* or *A. oryzae* fungal cells for *S. cerevisiae* cells in this preparation will also result in a functional protein with acceptable taste.

CATALASE SOLUBILITY IMPROVEMENT

Peroxidase or catalase enzymes are added to a soybean mixture in the process of *P.U. de Paolis; U.S. Patent 3,682,646; August 8, 1972; assigned to Hunt-Wesson Foods, Inc.* to decompose peroxides which may have denaturing and insolubilizing effects on protein constituents. The resulting product has high solubility as well as acceptable flavor characteristics.

The starting defatted and dehulled soybean material for making soy protein products is immersed and mixed in boiling or near boiling water at a temperature of between 175° and 212°F for 15 to 35 minutes as shown in step **10** of Figure 5.1.

A pH of at least 7 is maintained by addition of sodium hydroxide. By maintaining a pH of about 7.0 and as high as 9 at this stage of processing and throughout the process, a substantial increase in the yield results since the solubility of soy protein in aqueous solution is more than 50% with a pH of 7.0 in contrast to the solubility of about 38% when the water solution is held at a pH of 6.6.

By mixing the soybean material in water at a high temperature, the trypsin inhibitor in the soybean material is destroyed while very little denaturization of the soy protein occurs. Further, other off-flavor potentiating materials are, to a great extent, destroyed.

The aqueous soybean mixture formed in step **10** is cooled to a temperature between 100° to 140°F. As optional step **12**, a peroxidase enzyme, for example, catalase, is introduced into and mixed with the aqueous soybean mixture. About 0.1 ml to 2 ml of a solution of the enzyme catalase (each milliliter of solution contains 1,000 Baker units of the enzyme catalase) are added per 100 grams of starting defatted, dehulled soybean material. This amounts to 100 to 2,000 Baker units of the enzyme catalase. A Baker unit decomposes 264 mg of hydrogen peroxide under assay conditions.

Figure 5.1: Process for Production of Soy Protein Products

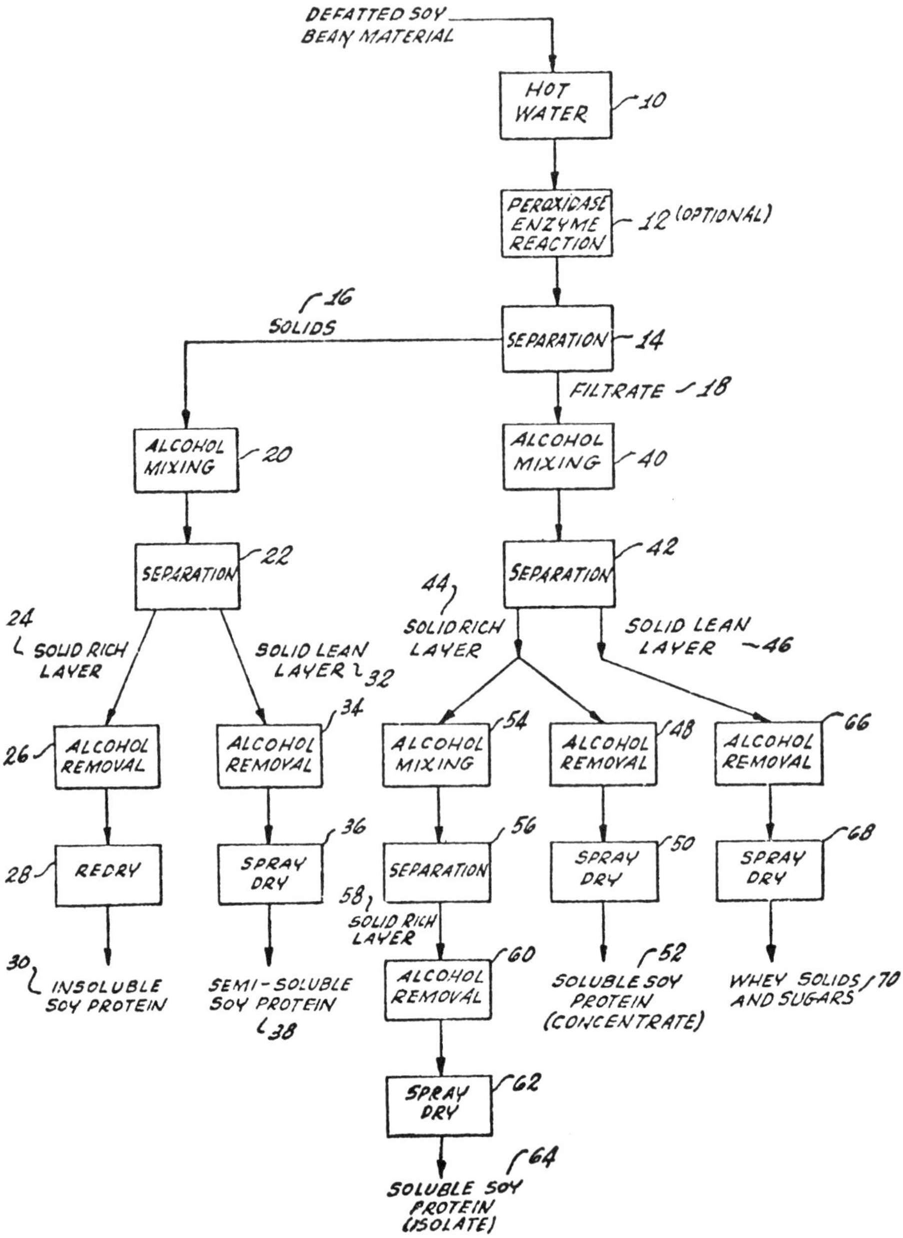

Source: U.S. Patent 3,682,646

Examples of other enzymes that can be used are: catechol oxidase, p-diphenol oxidase, ascorbate oxidase, NAD peroxidase, NADP peroxidase, fatty acid peroxidase, cytochrome peroxidase, glutathione peroxidase, and iodinase.

It is believed that some hydrogen peroxide is produced during the boiling phase of the process. In order to minimize the denaturing and insolubilizing effect of the peroxide on the protein constituents, catalase is introduced at a temperature below 150°F. The optimum range of the mixture temperature for the introduction of catalase lies between 130° and 135°F. This optional step **12** is allowed to proceed for a period of twenty minutes. The remaining steps of the process outlined in Figure 5.1 are illustrated by means of the example which follows.

Example: 1,000 lb of defatted, dehulled soybean flakes was mixed with 9,000 lb of water at 185°F for twenty minutes. The pH was raised to 7.0 by the addition of sodium hydroxide. The mixture was then allowed to cool to 130°F and then sufficient Fermocomolase was added to provide 1,000 Baker units per 100 g of flakes and reaction was allowed to proceed for twenty minutes. Separation of the insoluble material (solids line **16**) from filtrate **18** takes place by means of rotary vacuum filter **14**.

A 70% methanol solution is then added to the insoluble material (step **20**), the insolubles line containing 667 lb solids and 2,612 lb water. About 7,600 lb of 70% methanol is added. After about twenty minutes, the supernatant (solid-lean) layer is then stripped of alcohol (step **34**) and dried. The resultant product weighs 112 lb, contains about 6% moisture and 80.50% protein of about 52.1% NSI (Nitrogen Solubility Index) solubility.

Alcohol is also removed from the sludge (solid-rich) layer and dried. The resultant product weighs 538 lb and contains about 3.4% moisture and 63.10% protein, of 11.46% NSI solubility.

The filtrate **18** contains about 333 lb solids and 6,476 lb water making a total of 6,809 lb entering a methanol reactor or alcohol mixing step **40**. 11,720 lb of 70% methanol is added to the filtrate. Mixing continues for about twenty minutes in step **40** and then separation (step **42**) as by centrifuging results in two phases, the solid lean layer **46** and the solid rich layer **44**.

The solid rich layer **44** has alcohol removed by a flask desolventizer (step **48**), and is spray dried (step **50**). The resultant product weighs 185.6 lb, containing about 5.2% moisture, and 72.50% protein (dry basis) having an NSI solubility of about 98.8%.

In order to obtain the isolate, the solid rich layer **44** rather than being processed as just set forth, is mixed with 230% of 70% methanol by weight and separated by centrifugation, into a further solid rich layer **58** and a solid lean stream. The layer **58** is then stripped of alcohol and dried. The resultant product weighs 236.4 lb and contains 5.9% moisture and 90% protein (dry basis) of 93.1% NSI solubility.

Finally, the solid lean layer **46** separated from step **40**, containing about 157 lb solids, 6,500 lb water and 12,500 lb methanol is passed through a methanol stripping column and the product is spray dried (step **68**).

The resultant product **70** weighs 166.4 lb and contains 5.4% moisture, 82.6% polysaccharides and about 12% proteins.

GELABLE EGG WHITE SUBSTITUTE

Soy protein can be treated with enzymes to produce a functional protein which is used as a substitute for egg white for gelling and whipping properties. The process was developed by *A. Pour-El and T.C. Swenson; U.S. Patent 3,932,672; January 13, 1976.*

"Gels" are commonly regarded as two-phase systems with a high degree of interface between a continuous, or at least intermeshed, system of solid material holding an aqueous (or other solvent) phase which may also be continuous or finely dispersed. The solid material is usually thought of as fibrillar in form and may consist either of strongly solvated molecules, such as proteins, or of thread-like crystals, as in gels of silicic acid or crystalline cellulose. A certain degree of structural stability is contributed by the extended skeleton of the continuous phase so that the system as a whole possesses many of the properties of a solid, e.g., rigidity and elastic response to applied distorting forces.

Gels commonly found in food systems include starch gels, protein gels, and polyuronic acid gels. The principal protein gels are derived from collagen or connective tissues of animals. The distinguishing feature of gelatin is its ability to form rigid gels from cooled aqueous sols at concentrations of 1% or below. Gelatin occurs as the result of the binding together of protein molecular chains, principally by hydrogen bonding. The major uses of gelatin are in dessert and salad bases, canned meats, and several kinds of confectionary.

Regarding the whipping characteristics, it is merely necessary to have a foamable composition. While many whippable materials are known, few are both whippable and gelable. In particular, the food industry has used egg white or egg albumen to form a gelling and foaming material. Egg white contains about 12% solids, principally protein. Thermal denaturation of egg albumen results in formation of a firm gel. The use of egg white in the food industry is based on its ability to form foams rigid enough to support large amounts of other ingredients. Egg white coagulates rapidly at 130° to 135°F. High concentrations of sucrose, some inorganic salts and high protein concentrations tend to elevate the coagulation temperature. Coagulation or precipitation of egg albumen requires the presence of salts to precipitate the protein.

Modification of soy protein to approximate the egg albumen can be achieved by modifying the size of the protein molecule (soy protein is a larger molecule), modifying the number and availability of sulfhydryl groups and modifying the intermolecular bonding forces or "tightness" of the protein molecule (soy protein is generally "tighter").

In terms of these characteristics, egg albumen has a molecular weight of about 46,000, about five free sulfhydryl groups per molecular weight unit, and a relatively loose molecular structure. The function of sulfhydryl groups in gelling proteins such as egg albumen is to form protein-protein bonds through disulfide linkages, thereby forming a protein matrix which can be solvated by water to form a gel structure. The formed gel structure, by disulfide bonding, is generally thermally irreversible (in contrast to gelatin gels, where proteins are bound together primarily by hydrogen bonding and which are consequently thermally reversible).

Soy protein is treated enzymatically to reduce the average molecular weight of the soy protein (naturally about 105,000 to 300,000), and at the same time forms a more loose arrangement (which is required in order that the molecules have sulfhydryl groups within reach of one another) capable of bonding to form a matrix during gelation. This change in molecular structure is accomplished by enzymatic degradation of a previously separated protein (isolate) at alkaline pH. The degradation must be controlled in order to achieve the correct structure approximating that of egg albumen. Subsequently, sulfhydryl groups are introduced by the addition of, for example, cysteine, or other sulfhydryl-containing compounds.

More specifically, the two essential steps of this process will be further considered. The first is an enzymatic treatment which loosens and reduces the size of the molecular unit. A protein, for example, Promine D, a commercially available isolated soy protein, or Nutrisoy 7-B, an extracted soy flake can be treated with proteolytic enzymes, for example, ficin or bromelin and the protein fractions reduced are separated on the basis of molecular size and solubility in acid solution.

This separation produces an active gelling fraction which is insoluble in acid and is called acid-insoluble protein fractions (AIPF). This material forms a solid gel when heated at 60° to 90°C for 5 to 60 minutes at a pH of about 6.9 in the presence of ½ to 1 M (0.5%) sodium chloride and 2.46 mg cysteine per gram of protein. A 5 to 10% concentration of AIPF forms the best gels. By comparison, 5% egg white forms a slightly cloudy precipitate at 55°C. The addition of sodium chloride greatly increases the rate of gellation by precipitating the protein, but the AIPF still gels in the absence of sodium chloride, whereas egg white requires such salt to gel.

The steps of the process to be used generally involve a pretreatment of the protein material which can either be direct digestion, where the whole extract of the base material is used after separation from insoluble material, or isolated protein digestion, in which the protein was at first isolated by isoelectric precipitation at pH 4.75 followed by two water washings.

The pH for digestion in the process is alkaline; for ficin it is in the range of 9 to 9.6. If other enzymes are used, their optimum pH of activity should be used, always, however, using an alkaline-active enzyme. The optimum temperature during digestion is in the range of 90° to 110°C, but slightly higher temperatures can be used. It is generally found that the gel strength is reduced if too high a digestion time and/or digestion temperature is used.

After digestion, an isolation step is required for whole extract materials where isoelectric precipitation techniques are used, generally at acid pH, especially pH 4.75, with centrifugation to separate the protein. Precipitation should be at a fairly high temperature, and precipitation is to be carried out right after digestion when the digestion medium is at approximately 100°F. It has been discovered that a nongelling protein results if this precipitation is not accomplished.

The second essential step is treatment of the gelable protein after enzymatic digestion by exposure of the isolated and digested protein to a relatively high pH (higher than that used in digestion). Exposure to high pH results in a better gelling protein after only a short time (approximately 5 minutes); in the case where ficin is digested at a pH of 9 to 9.6, the optimum pH is 10.5 for this exposure, but slightly higher pH values are acceptable (11 to 11.5). It is necessary also to conduct this exposure to high pH after digestion since if it is exposed to this pH before digestion, an insoluble protein results.

Following exposure to this pH for a brief interval, the pH is reduced to about neutral (6.8 pH for ficin) before the protein is dried. The final pH adjustment is necessary, for if the pH of the material prior to drying is above 7, it is found gelation is retarded. Similar characteristics are noted when dealing with egg white proteins.

The dried material is now treated by the addition of salt (0.1 M solution) to a strength of about 0.5% of the protein concentration. Salt is required to precipitate the protein and effect formation of the matrix required in gelation. The ultimate step in forming a gel is to heat to 140° to 190°F for a period of about 10 minutes.

It is desirable to introduce sulfhydryl groups into the protein material so produced before it is formed into the gel. A desirable material for this purpose is cysteine, at concentrations of about 18 mg per gram of protein, which can be added to the protein just after or along with addition of salt; also blends with about 20% egg white solids could be made which solids are added before or with the salt, etc. Upon addition of salt, these blends will gel in a solution of 3 to 10% protein when heated to 140° to 190°F for 10 minutes.

Example: Production of Gelling Protein By Ficin Digestion — (1) 100 g of commercially available Nutrisoy 7-B soy flakes (undenatured soybean flakes that are hexane extracted and flash desolventized) are extracted with 2,000 g of water at pH 9 at a temperature of 103°F for one hour. Following this extraction, the mixture is centrifuged and decanted. The extracted flakes are subjected to a repeated extraction under the above conditions, using 1,000 g water at pH 9 for one-half hour.

(2) The combined supernatant liquors from the extractions are centrifuged at high speed to remove insoluble fines and suspended exhausted flakes.

(3) The clear supernatant is filtered through glass wool and brought to a pH of 4.5 to 4.75 at 45°F and centrifuged at low speed (it may also be filtered), and the supernate (the whey) is discarded.

(4) The precipitated protein material (curd) is washed three times with 1,000 g of water, collecting the precipitate in each instance by filtering through five mesh orlon screen using vacuum.

(5) The washed precipitate is redissolved in four times its weight of water at pH 9, whereupon, the total volume is approximately 4,000 cm^3 and is stored overnight in a refrigerator. It is also possible to use freeze drying prior to redissolution.

(6) The dissolved protein is digested with approximately 1% (0.01 g/g protein) ficin at 120°F and a pH of 9.6 for one hour. Cysteine can be added during or after ficin digestion.

(7) The digested protein solution is brought to a pH of 4.75 to precipitate protein and the medium is then centrifuged and the supernate [whey (2)] is discarded.

(8) The precipitated digested protein is washed twice with five times its weight of distilled water, separating each precipitate as before.

(9) The washed protein is suspended in water having a total volume of about 600 cm^3 and adjusted to a pH of 7 until dissolved. Then the pH is raised to 10.5 for 5 to 15 minutes. Cysteine may be added at this point if desired.

(10) The mixture is neutralized by addition of sodium hydroxide to approximately pH 6.85.

(11) The protein material is then dried by conventional techniques. The molecules have been broken into a heterogeneous mixture of or about the size of analbumin molecules.

(12) The material is gelled by redissolving in distilled water at 4 to 5% protein, and, NaCl and cysteine are added followed by heating to either 140°F for about 20 minutes or 190°F for about 10 minutes. Gels formed by this method are found to produce a medium gel. When a protein isolate is used as starting material a strong gel results.

PEPSIN MODIFIED WHIPPING PROTEIN

In the food industry, proteinaceous materials, such as egg whites, hydrolyzed milk proteins and soy albumen, have been used for many years as aerating agents. By an aerating agent is meant a material capable of entrapping air in an aqueous mix, such as the basic sugar-syrup mix used in the confectionery trade or the flour-sugar mixes used in the bakery field. A basic requirement of a good aerating material is, that it possesses the ability to entrap the maximum quantity of air in the minimum length of time. In other words, such a product should whip rapidly to a high volume. The economies realized are rather obvious when such a product is used since the desired degree of aeration can be achieved with a minimum amount of whipping agent.

Egg albumen, hydrolyzed milk proteins and soy albumen, although effective in whipping air into certain type mixes, are lacking in total whipping efficacy and consequently must be utilized at relatively high concentrations to produce the desired degree of aeration.

An enzyme process for the preparation of a whipping agent prepared from hydrolyzed protein material is given by *R.C. Gunther; U.S. Patent 3,814,816; June 4, 1974; assigned to A.E. Staley Manufacturing Company*.

Oil-free soybean meal or flakes is extracted with an aqeuous alkaline solution (e.g., sodium carbonate, sodium hydroxide), at a pH in the range of 7.5 to 10.0

and a temperature of 30° to 50°C. The insolubles may be removed from the extract solution either by filtration or centrifugation and the clarified extract liquor is given a relatively mild alkaline hydrolysis by treatment with an alkali (e.g., sodium hydroxide or calcium hydroxide) at a temperature of 50° to 85°C for one to two hours (the amount of alkali used will ordinarily be from 5 to 10% by weight of the protein content of the extract liquor). The amount of alkali used, the temperature and length of time of the treatment will be determined by the degree of hydrolysis desired.

It is most important that the hydrolysis be terminated before the protein is hydrolyzed to the point that it would be soluble at its usual insoluble isoelectric pH. Following hydrolysis, the pH is adjusted to the isoelectric pH of the protein, pH 4.1 to 4.5, and the precipitated curd is washed free of soluble carbohydrates, minerals, etc., by decantation.

The enzymatic modification is carried out by first treating a purified alkaline modified protein curd with a suitable acidifying agent such as sulfuric or hydrochloric acid to produce a pH ordinarily in the range of 1.2 to 2.5 and at a temperature of 25° to 40°C. The enzyme, pepsin (1:10,000 strength), is then added in an amount usually of 0.05 to 0.5% by weight of the protein content of the solution.

Enzyme modification is allowed to proceed for 18 to 24 hours, after which the solution may be heated to 60° to 70° C to stop enzyme action. The pH is then adjusted to 5.0 to 5.5 with caustic, and the solubles separated from the insolubles by filtration or centrifugation. The clear aqueous solution containing the solubles is then spray-dried to produce a very light colored, free-flowing product (referred to for convenience as E-500).

The applications and uses of a whipping agent with the whipping efficiency and relatively bland flavor of the E-500-type protein are many and varied. One of the outstanding properties of the E-500-type whipping protein is its ability to aerate whole egg mixes. Egg albumen, soy albumen and hydrolyzed milk protein will not produce this effect. For example, 260 g fresh whole eggs and 5 g E-500-type protein can be whipped to a specific gravity of 0.19 in two minutes in a Mixmaster beater. This is a stiff, rigid whip which can be folded into batters, etc.

This same amount of fresh whole eggs with as much as 10 to 15 g of egg albumen (or soy albumen or hydrolyzed milk protein) showed no aeration even after 10 to 15 minutes whipping. Consequently, one of the principal specific applications of this protein product is in a prepared sponge cake mix in which whole eggs are added to the dry, prepared mix and whipped to a light aerated batter. Also, an E-500-type protein, because of its low flavor level and whipping efficiency, finds usage in other prepared (or packaged) cake mix products. In packaged angel food cake mixes, this whipping protein greatly improves the egg white foam whipping rate and volume.

Example: To 125 lb of water at 30°C in a 30-gal tank was added 10 lb of solvent-extracted soybean flakes. Caustic soda (15% solution) was added to obtain a pH of 8.5. The flakes slurry was then agitated for 1 hour during which time the pH was maintained at 8.5 by the addition of caustic soda solution. Agitation was stopped and the slurry allowed to settle for two hours. The clear supernatant liquid (comprising approximately one-half the original volume of the slurry) was removed and filtered through a 60 mesh screen to separate small amounts of very finely dispersed flake material.

To the remaining slurry in the tank was added an equal volume of water at 30°C and the slurry agitated for 30 minutes at a pH of 8.5. Again the suspension was allowed to settle for two hours and the clear supernatant removed as before. The two supernatant solutions were combined and then heated to 80°C for the alkaline hydrolysis. Caustic soda (40% solution) was added to obtain a pH of 11.8 and the solution gently agitated for 25 minutes. Hydrochloric acid (20%) was then added to obtain a pH of 4.4. The precipitated protein was allowed to settle for four hours and the clear supernatant removed by decantation. The protein curd was washed once with cold tap water in an amount equal to the original volume of the solution.

A sample of the washed curd was removed and further concentrated to 20% solids. After dissolving at pH 9.5 and heating to 48°C for one hour, a viscosity determination was made. The viscosity was 135 cp. The solids of the washed curd in the tank were adjusted to 6 to 7% by the addition of water, and 20% hydrochloric acid added to obtain a pH of 1.5. The slurry was then heated to 30°C and 0.4% of 1:10,000 pepsin (based on weight of dry solids) added. The mix was stirred for five hours (pH adjusted twice during this period to 1.5) and then allowed to stand at 30°C for seventeen hours, or a total enzyme modification time of twenty-two hours.

At the end of this period, the solution was heated to 70°C, and caustic soda solution added to raise the pH of the solution to 5.1. Filter aid was added and the solubles separated from the insolubles by filtration. The clear liquor from the filter press was spray dried in a stream of heated air to yield a light colored, free-flowing powder.

Using the chocolate chiffon whip test, the minimum density of the whipped product was determined to be 0.305 g/ml. The time required to obtain a density of 0.44 g/ml was one minute and forty-five seconds. A pepsin modified soy protein made by a similar process without hydrolyzing the protein, when tested in the same manner, had a minimum density of 0.409 g/ml and a whip rate of three minutes and fifty seconds to achieve a density of 0.44 g/ml.

The chocolate chiffon whip test previously referred to in the example is carried out as follows: dry blend 154 g of chocolate chiffon base (a mixture of sucrose, nonfat milk powder, cocoa, starch and gelatin) and 9.0 g of whipping agent in the small bowl of a Mixmaster beater. Add 240 ml (1 cup) of boiling water and beat on highest speed (No. 12) for four minutes. The density of the aerated mix is determined by weighing a sample in a calibrated beaker.

PARTIALLY HYDROLYZED FOAMING PROTEIN

Partial hydrolysis of soy protein by an enzyme procedure is used by *K. Sawada, M. Kajikawa and K. Kotani; U.S. Patent 4,015,019; March 29, 1977; assigned to Takeda Chemical Industries, Ltd, Japan* to prepare a protein that can be whipped or used to prepare a foam for use in food products such as frappes, fudge, cream candies, etc.

Such soybean products can be prepared through series of steps (1) subjecting soybean materials to enzymatic partial hydrolysis, (2) heating the hydrolyzed product for a short time to inactivate the enzyme, (3) cooling the heated product at neutral or weakly acid pH, (4) removing the precipitates formed, and (5) concentrating the resultant solution followed by pulverization.

The first step of the process is enzymatic partial hydrolysis of soybean material which is carried out in an aqueous dispersion, suspension or solution. The pH and temperature of the mixture are adjusted to the optimum levels according to the properties possessed by the proteolytic enzyme to be added. The acid or alkali in pH adjustment may be hydrochloric acid, sulfuric acid, phosphoric acid, acetic acid, sodium hydroxide, sodium sulfite, sodium carbonate and so on.

Neutral or alkaline protease may be used, and it may be of the vegetable, animal or microbial origin. Examples are pepsin, papain, or acid protease obtained by the cultivation of the microorganism belonging to the family *Polyporacease,* for example, *Trametes sanguinea, Trametes cinnabarino* and *Poria vaporaria.*

The partially hydrolyzed product is heated to a temperature of about 90°C, maintained at this temperature for about 1 to 10 minutes, preferably for 3 to 5 minutes, then cooled to a temperature of not more than 60°C, preferably between room temperature and 60°C. The pH adjustment is an operation to be conducted if necessary to bring the pH of the product to 5-7. This pH adjustment may be carried out either before or after the heat treatment, or either before or after the cooling step.

As for the procedure for removing the insolubles, one may employ one of the routine procedures, for example, a centrifugal separator of the decanter type or a filter press with the addition of a suitable filter aid such as diatomaceous earth. The filtrate is then concentrated and/or dried in a conventional manner to obtain the desired soybean product.

Example: Production of Foaming Soybean Product Samples – To 2.5 kg of defatted soybean flour was added 10 kg of lukewarm water (about 40°C) and, after stirring to homogeniety, the mixture was adjusted to pH 2 with concentrated hydrochloric acid and an internal temperature of about 40°C.

Then, a solution of 5 g of pepsin (protease activity 100,000 units/g) in a small amount of water was added to the above mixture. The enzymatic hydrolysis was carried out for five hours, with gentle stirring and while the above internal temperature was maintained. The procedure yielded a partially hydrolyzed product with an analytical value of 0.54.

The hydrolyzed product was divided into five aliquot portions. Four aliquots were heated to 90°C for 1, 3, 10 and 30 minutes, respectively, and then rapidly cooled to 30°C. One aliquot was untreated. Then, each processed soybean product was adjusted to pH 5 by gradual addition of an aqueous solution of sodium hydroxide and centrifuged to remove the insolubles.

The filtrate was concentrated at an internal temperature of 60°C with a vacuum concentrator to obtain a concentrate with a solids content of about 20%. The concentrate was spray dried to obtain a foaming soybean product. A test for foaming property and a test for flavor of the foam formed were performed on samples prepared in the above manner.

Test Results: It was clearly demonstrated that a soybean product with an excellent flavor and excellent foaming property can be obtained by subjecting the enzymatically hydrolyzed product to a heating at an internal temperature of 90°C for 1 to 10 minutes and a subsequent cooling to 30°C.

GLUTAMIC ACID PROCESSES

Specific Glutamyl-Peptides

Glutaminase reacts only with free glutamine and is unable to deamidate the γ-amide group of glutamine when bonded to peptides. In the production of glutamic acid using glutaminase, it is necessary to first decompose the protein with a protease and then liberate glutamine to react with glutaminase for deamidation.

In the production of beverages and foods such as soy sauce, miso (bean paste), and cheese using protein-containing substances as raw material, it has been the practice to use proteolytic enzymes, i.e., protease or peptidase, to react with the raw material, thereby to obtain peptides or amino acids.

However, when protein is first hydrolyzed by aid of protease or peptidase, a free glutamine or a glutamine being bonded to N-terminal of peptide is produced. Such glutamines are liable to form tasteless pyroglutamic acid because γ- and α-positions of glutamine tend to bond cyclically to each other.

When glutamine is liberated once, the rate of forming pyroglutamic acid accelerates, and hence, it is preferable to effect, if possible, deamidation of glutaminyl-peptides to form glutamyl-peptides with no free glutamine. *M. Kikuchi, K. Sakaguchi and E. Nakano; U.S. Patent 3,796,633; March 12, 1974; assigned to Kikkoman Shoyu Co., Ltd. and Noda Institute for Scientific Research, Japan* describe an enzyme which is capable of deamidating γ-amide group of glutamine in peptides irrespective of the adjacent amino acids, producing a large amount of glutamic acid from protein-containing raw materials without forming pyroglutamic acid.

A strain No. 466 which has been isolated from soil and is considered to belong to the genus *Bacillus circulans* is found to be capable of producing the enzyme peptidoglutaminase. This enzyme reacts directly with glutamine being bonded to peptide and deamidates γ-amide group of glutamine without splitting the peptide bond. This enzyme peptidoglutaminase is hereafter referred to as "PGase." A culture of the strain is on deposit under accession No. ATCC 21590 with American Type Culture Collection.

The amount of enzyme to be added to foods, time required for effecting reaction of the enzyme and other conditions may vary depending on the kind of proteinous final products. In preparing seasonings such as soy sauce containing as a principal ingredient, amino acids, particularly glutamic acid, where a protein-containing raw material is treated in a known method and the protein is decomposed, the enzyme is added with stirring to the raw material system just before or during the fermentation and ripening operation so that the amount of enzyme either in the form of cell containing enzyme, its extracted crude enzyme liquid or purified enzyme preparation may be 0.01 to 100 units per 1 ml of the fermentation broth.

In the cases of solid foods such as miso and cheese, the enzymes may be added with thorough mixing in an amount of 0.01 to 100 units per 1 g of the solid raw material just before or during the common ripening process step.

The enzymes react with glutaminyl-peptides which had been produced with protease from protein, forming glutamyl-peptides, and then this compound is hydrolyzed in the presence of peptidase to free glutamic acid. In producing beverages and foods using a protein-containing raw material having a high molecular weight, the presence of protease is indispensable to form such glutamyl-peptides or free glutamic acid by the action of the enzyme. Usually, however, in the case of producing beverages and foods, the enzyme can sufficiently and effectively exhibit its ability if such amount of proteinase or peptidase as may usually be used in the production of such beverages and foods, is present.

Example 1: 5 kg of defatted soybean, which had been subjected to water spray at a rate of 125%, was cooked under steam pressure of 1 kg/cm^2 for 40 minutes. The cooked soybean was mixed with 5 kg of roasted and powdered wheat, and soy sauce seed koji was added by spraying. The culture was continued for four days to obtain 12 kg of koji of excellent quality. The obtained koji was charged in 12 liters of a 24% saline water and the saline koji broth was subjected to fermentation and ripening at about 30°C for three months with aeration from time to time.

Separately, strain No. 466 of *Bacillus circulans* was cultured in a liquid medium (pH 7.2) prepared by adding 0.3% of yeast extracts, 1% of peptone, 0.5% of lactose, 0.025% of magnesium sulfate, 0.001% of ferrous sulfate and 0.17% of dipotassium hydrogen phosphate to 400 liters of water, at 30°C for sixteen hours with aeration-agitation. The bacterial cells, in which peptidoglutaminase-I and peptidoglutaminase-II had been accumulated, were ground by means of a sonic wave disintegrator and centrifuged to collect a supernatant solution, i.e., a crude enzyme solution. To the crude enzyme solution was added gradually a solution containing streptomycin and a sediment of nucleic acid was removed.

The solution was further saturated with ammonium sulfate to a degree of 38 to 54% and a sediment fraction was collected. The sediment was lyophilized to obtain 4 g of powdery crude enzyme (a mixture of PGase-I and PGase-II). This crude enzyme contained 4,300 units of PGase-I and 3,650 units of PGase-II. The powdery peptidoglutaminase was added in an amount consisting of 2,000 units of PGase-I and 1,800 units of PGase-II to the above-prepared saline koji broth just before the broth was subjected to fermentation and ripening.

After the 3-month ripening at 30°C, the ripened mash was pressed to obtain 23 liters of a soy sauce containing the total nitrogen content of 1.7%, of which the amount of glutamic acid was 18.9 mg/ml. This glutamic acid content represents 10.56% of the product soy sauce taken as the proportion of the resultant glutamic acid nitrogen to the total nitrogen content. In view of the fact that a soy sauce, which had been prepared in the same manner as above but without addition of the peptidoglutaminase, contained glutamic acid nitrogen at 7.69% of the total nitrogen content, it is made clear by this example that glutamic acid nitrogen was markedly increased.

Example 2: 100 g of wheat gluten was charged with tap water to total up to 380 ml, stirred and cooked under a pressure of 1 kg/cm^2 for 30 minutes. Then the cooked gluten was charged with 5 g of a proteolytic enzyme (an enzyme commercially available, which is produced from *Aspergillus oryzae* and is partially refined) and 2.4 g of a peptidoglutaminase preparation (the ammonium sulfate-precipitated powder obtained in Example 1. The mixture was subjected to hydrolysis at 30°C for 3 days in the presence of toluene, where the pH was 5.0.

After the hydrolysis was over, the reaction mixture was pressed to give 360 ml of a gluten hydrolyzate which was excellent in palatability. The liquid seasoning thus obtained contained 30 mg/ml of glutamic acid, whereas the same seasoning but without addition of the PGase preparation, contained only 8 mg/ml of the same. This gluten hydrolyzate may be usable for the preparation of various seasonings and soups, either as such or after being concentrated or, if necessary, dried and powdered in per se known manner.

Glutamine-Glutamic Acid Conversion

Glutamic acid derived by hydrolyzing soybean or wheat gluten is widely used as a flavoring material for prepared foods. *T. Yokotsuka, T. Iwaasa and M. Fugii; U.S. Patent 3,717,470; February 20, 1973; assigned to Kikkoman Shoyu Co., Ltd., Japan* uses a glutaminase for effective conversion of glutamine to glutamic acid for higher yields of glutamic acid.

Glutamine can be transformed also into pyroglutamic acid devoid of flavoring effect during the course of conventional processing treatments such as heating, and during the storage of processed materials.

The content of free glutamine and free glutamic acid in the raw materials for preparation of foods and beverages and the amount of free glutamic acid in the processed commercial products prepared are shown below.

Content in Various Food and Beverage Materials

	Free Glutamine	Glutamic Acid
	..........(mg/ml)...........	
Cucumber	0.40	0.18
Eggplant	0.78	0.16
Chinese cabbage (Hakusai)	1.45	0.24
Japanese radish	1.58	0.07
Green asparagus	1.65	0.79
Tomato	1.35	0.90
Grape	0.42	0.14

(continued)

	Free Glutamine	Glutamic Acid
	..........(mg/ml)............	
Mushroom	1.27	1.67
Enokitake *(Flammulina veltipes)*	0.48	0.71
Pork	0.37	0.31
Beef	0.12	0.28
Whale meat	0.11	0.11

Content in Various Commercial Foods and Beverages

	Free Glutamic Acid (mg/ml)
Canned asparagus	0.20
Pickled radish (Japanese-style)	2.09
Tomato juice (a)	1.62
Tomato juice (b)	2.11
Tomato catchup (a)	2.08
Tomato catchup (b)	3.16
Canned mushroom	1.30
Canned beef	2.90

To take the case of radish as an example, the Japanese-style pickle products (so-called Takuan) are sometimes required to contain up to 2.09 mg/ml of free glutamic acid, and a large amount of L-glutamic acid is usually added to meet the requirement. The content of free glutamic acid in the radish for use as the material is as low as 0.07 mg/ml, whereas the content of free glutamine is as high as 1.58. However, the glutamine has never been utilized as the flavoring component, because of its conversion into pyroglutamic acid.

In this process, free glutamine in the raw materials is converted into free glutamic acid by hydrolysis of the amide linkage. The hydrolysis of the amide linkage can be effected either enzymically or chemically by use of acids, etc. In the preparation of foods, the enzymic means is used because chemical decompositions accompanying the formation of by-products are undesirable in most of the cases.

Glutaminase for hydrolyzing the amide linkage of glutamine may be of any origin. There is glutaminase produced by microorganisms such as yeasts, molds, bacteria and actinomycetes.

Example 1: According to the ordinary procedure, tomato was washed, crushed, heated, and squeezed to obtain tomato juice. To 1 kg of the juice, was added 1 g of a glutaminase preparation [acetone-dried fungus mass of cultured *Cryptococcus albidus* (ATCC 20293)] and mixed thoroughly. Then the mixture was vacuum-concentrated in the ordinary way to obtain 0.52 liter of tomato puree. The content of free glutamic acid in 1 ml of the supernatant liquor obtained by centrifuging the puree was 3.26 mg, whereas the content was 1.82 mg in the case of puree prepared in the same manner except that no glutaminase was used.

Example 2: 4 kg of Chinese cabbage cut in halves or quarters and 200 g of common salt were placed in a 10-liter polyethylene vessel and a weight was applied. When sufficient juice was squeezed out of the Chinese cabbage to steep the pieces, 2 g of a glutaminase preparation (an extracted enzyme obtained from the microorganism mass of cultured *Escherichia coli)* was added to the juice. Then the steeping was further continued under a somewhat lighter weight. A control steeping was run in the same way as that mentioned above, except that no glutaminase was used.

After having been steeped for 2 months at the room temperature, the Chinese cabbage was crushed in a juicer to obtain juice. The content of free glutamic acid in 1 ml of the juice was determined to be 1.02 mg, whereas the content was 0.21 mg in the case of control steeping.

Example 3: 1 kg of radish cut into pieces was sprayed with 30 g of common salt in a vessel, and a weight was applied. When sufficient volume of juice was squeezed out, 1 g of the same glutaminase preparation as that in Example 2 was added into the vessel, and the steeping was further continued. After having been steeped for several days at room temperature the radish was crushed in a juicer to obtain juice. The content of free glutamic acid in 1 ml of the juice was determined to be 1.06 mg, whereas the content was 0.07 mg in the case where no glutaminase was added.

BEANY AND ASTRINGENT SOY FLAVOR

Soy Cheese Flavor Improvement

There are many difficulties in obtaining a curd with soy cheese that has the desirable texture, flavor and body characteristics. The curd must be developed as slow as possible and additives are necessary for proper coagulation.

H. Tsumura, Y. Hashimoto and H. Kubota; U.S. Patent 3,857,970; December 31, 1974; assigned to Fuji Oil Co. Ltd, Japan developed an improved process for preparation of soy cheese which contains soy milk, animal fat and uses a protease enzyme in the milk before and after cooking of the curd for improved flavor characteristics. The primary function of the protease is to eliminate the beany odor without generating a bitter taste in the cheese-like product.

Soy milk generally can be prepared from any soy protein-containing material which is derived from soybeans and is suitable for making a cheese-like food. Commercially available soy protein concentrates which are extracted from the soybeans by conventional techniques can be used. For example, the soy milk can be formed by dissolving or dispersing a soy flour, a defatted soy flour, or isolated soy protein in water, the weight of the water used being 7 to 10 times that of the protein-containing material.

When soy flour or defatted soy flour is used as the soy protein-containing material, the water is heated to 50°C. Extraction temperatures higher than about 60°C should be avoided to prevent denaturation of soy protein in the soy milk. In order to minimize undesirable taste characteristics in the final product, the bean refuse compounds are removed from the resultant soy milk before it is used. Generally, the concentration of soy protein in the soy milk is within the range of 3.5 to 4 wt %.

The soy milk is mixed with a fat such as milk fat or other fat compositions having physio-chemical properties resembling those of milk fat. Generally, any synthetic fat derived from animal fats (other than milk fat) and/or vegetable fats and which has physio-chemical properties approximating the well recognized physio-chemical properties of milk fat could be used.

Representative examples of natural fats which can be used, includes animal fats, such as beef tallow, lard, hardened fish oil, and hardened whale oil, and vegetable fats, such as palm oil, coconut oil, palm kernel oil, soybean oil, rapeseed oil, and cottonseed oil. The natural fats used can be hydrogenated or fractionated. Generally, a natural cheese-like flavor is not imparted to the resultant soy cheese when the total amount of butyric acid and/or caproic acid in the fat composition is below 1 wt %. On the other hand, a bitter flavor is imparted to the resultant soy cheese when the total amount of these two short chain fatty acids in the fat composition is above 20 wt %.

An appropriate amount of the fat is blended with the soy milk to form an oil-in-water emulsion. The proportions of fat and soy milk used to form the emulsion can vary widely. Generally, the amount of fat used in the emulsion can be about 3 to 4 wt %, based on the total weight of the emulsion. Although a satisfactory emulsion can be prepared without an emulsifier because of the excellent emulsifying of soy protein, one or more suitable edible emulsifiers, such as lecithin, mono- and diglycerides of higher fatty acids, polyethylene sorbitan and higher mono-fatty acid esters can be added to aid in the emulsification if desired.

The mixture of fat and soy milk is maintained at a temperature of 50° to 60°C during emulsification in order to obtain good emulsifying. Other additives, such as sugar, edible dyes, vitamins, preservatives, flavoring additives, etc. can be used if desired.

The emulsion is heated to a temperature in the range of 98° to 100°C for about 5 minutes; cooled (e.g., about 50°C); and then homogenized to achieve a maximum dispersion of fat globules into a stable emulsion. This treatment improves the elastic mouth-feel of the resultant soy cheese product which is a common defect of a fat-deficient cheese product. Either single- or two-stage homogenization can be used.

The emulsion is inoculated with a suitable lactic acid-forming cheese starter culture and incubated. One or more heat tolerant lactic acid-forming bacteria, such as *Streptococcus thermophilus, Lactobacillus bulgaricus* or similar lactic acid-forming starter cultures conditioned to the fermentation temperature used, can be utilized as the starter. The remaining steps in the process are illustrated by the following examples.

Example 1: 1,000 g of soy milk extracted from a defatted soy flour with warm water and having crude protein content of 3.6% were mixed with 20 g of glucose and 30 g of a synthetic fat composition having physio-chemical characteristics simulating milk fat to form an emulsion. The fat composition had a softening point of 32.4°C and the following short chain fatty acid composition: 3.04 wt % C_4, 1.8 wt % C_6, 1.3 wt % C_8, 2.6 wt % C_{10}, 2.9 wt % C_{12}, all based on the total weight of fatty acid in the composition, calculated as the acid.

After being heated to 90°C for 10 minutes and cooled, the emulsion was innoculated with 20 ml of a cheese starter culture derived from *Streptococcus thermophilus* (live bacteria count about 10^7/ml to 10^8/ml) and maintained at a temperature of 50°C until the acidity of the emulsion increased to 0.20 (i.e., about 1 hour). 2 g of D-glucono-δ-lactone, which eliminates the beany odor without imparting a bitter taste, was added to the resultant broth which was then held at 50°C until a curd was formed. The curd was then cooked at 55°C.

After draining and light pressing of the curd to remove the whey, the curd was salted with 2 to 3 wt % salt, molded, pressed, coated with paraffin, and aged for 2 to 3 months at 15°C to produce a soy cheese sample. A separate sample was prepared in substantially the same manner, except 3 g of calcium sulfate was used as the coagulant instead of D-glucono-δ-lactone.

After two months aging, the physical properties of the two soy cheese samples were compared. The soy cheese sample exhibited uniform texture and body, while the texture of the control soy cheese sample was very fragile and uneven.

Example 2: A sample of soy cheese and a control sample of soy cheese were prepared in the same manner as described in Example 1 except about 0.04 wt %, based on the total weight of the cooked curd, of papain was added to the cooked curd at molding.

The texture and body of the enzyme-treated control sample was substantially weaker than that of the untreated control sample of Example 1. On the other hand, the texture and body of the enzyme-treated sample of soy cheese made in accordance with the process was substantially the same as the untreated sample of Example 1 and exhibited a superior natural cheese-like flavor profile.

Incubation Method

Incubation of soybean curd or defatted soybean flour with certain highly specific proteolytic enzymes liberates extraction-resistant beany and astringent flavor constituents and reversion-sensitive lipid materials from their apparent extremely close association with the proteinaceous constituents. The removal of the enzymatically liberated objectionable taste and oxidation susceptible constituents by washing with aqueous ethanol then provides stabilized soybean materials having little if any astringent and beany flavor and a distinctly diminished tendency to flavor reversion.

The details of this process are given by *M. Fujimaki, H. Kato, S. Arai and M. Yamashita; U.S. Patent 3,585,047; June 15, 1971; assigned to the U.S. Secretary of Agriculture.*

Hexane-defatted soy flour or soy curd is subjected to 2 to 6 hours of selective digestion by a microbial protease selected from the group consisting of the acidic aspergillopeptidase-A produced by *Aspergillus saitoi,* the neutral protease produced by *Aspergillus oryzae,* and the alkaline protease produced by *Bacillus subtilis,* or the carboxypeptidase-A isolated from bovine pancreas. The liberated or degraded astringent and beany principles are removed and the also liberated hexanol, hexanal and saponin. The reversion-susceptible bound lipids are then extracted with highly aqueous ethanol.

Free amino acids and particularly the peptides liberated from soy protein by extensive hydrolysis with pepsin contain significant amounts of leucine, isoleucine, phenylalanine, and valine, which specific amino acids per se are known to have a bitter flavor, which bitterness was found to be even more intense in the concurrently formed or liberated diffusible peptides. Bitter amino acids and the more intensely bitter peptides are listed in the table below. These are listed as examples of bitter peptides formed by the action of proteolytic enzymes on soybeans.

Peptide	Amino Acid Sequence
A-1	H-Gly-Leu-OH
A-2	H-Leu-Phe-OH
B-1	H-Leu-Lys-OH
C-1	H-Ser-Lys-Gly-Leu-OH
D-1	H-Phe-(Ile, Leu_2)-Gln-Gly-Val-OH
D-2	H-Arg-Leu-Leu-OH
D-3	H-Arg-Leu-OH

It was found that the desired flavor improvements of defatted soybean flour or curd are obtained by subjecting the defatted soybean protein material in an aqueous medium to from 2 to 6 hours of selective hydrolysis by a protease member of the group consisting of aspergillopeptidase-A produced by *Aspergillus saitoi,* a mixture of the aspergillopeptidase-A and crude takadiastase-SS formed by *Aspergillus oryzae,* or carboxypeptidase-A from bovine pancreas, and then removing any released adverse flavor principles by washing with 50% aqueous ethanol.

The soybean curd substrate used per se in the example and, after peptic digestion, as the substrate in previous work on bitter peptides by the same authors, was prepared by soaking 100 g portions of soybean in 300 ml water for 12 hours at room temperature, crushing the wet beans to a paste, adding 200 ml water, boiling for 20 minutes, filtering, and adding to the filtrate 3 g $MgCl_2$ whereby was precipitated 30 g of wet soy curd. Compositional analysis of the curd was 88.7% water; 1.04% N; 0.32% crude fat; 3.70% carbohydrate; and 0.61% ash.

Likewise, defatted soybean flour substrate was prepared by pulverizing hexane-extracted soy meal to a flour. The flour analyzed 9.0% water, 8.0% nitrogen, 0.7% crude fat, 31.6% carbohydrate, 6.4% ash, and 2.3% crude fiber.

Example: 10 g (dry basis) of the above soybean curd suspended in 1,000 ml of dilute HCl (pH 2.8) was digested for 2 hours at 50°C with 0.1 g of commercially obtained aspergillopeptidase-A, produced by *Aspergillus saitoi.* The enzymatic digestion was then stopped by neutralizing with NaOH, and the precipitate obtained by filtration or centrifugation was treated with a 10-fold weight of 50% ethanol to remove the liberated beany and astringent factors; the wash liquid was removed by filtration, and the treated curd was lyophilized.

Compared with identically treated soybean curd, excepting for the addition of the enzyme, the product was found to be free of odor and almost completely free of beany and astringent flavor (completely free when the enzymatic digestion was extended to 4 hours). Furthermore, the presence in an ether extract of the supernatant from the above neutralization step of greatly increased percentages compared with an untreated control extract, of phospholipids, phosphatides, n-hexanol, genistein, ninhydrin positive substances, and carbonyl compounds, e.g., n-hexanal, confirms the specific enzymatic liberation of apparently implicated adverse flavor and autooxidation instability components present in the untreated soybean curd.

Plastein Synthesis

Another enzyme process to eliminate the offensive taste and odor of hydrolyzed proteins obtained from soy, fish and petro-yeasts is outlined by *M. Fujimaki, H. Kato, S. Arai and M. Yamashita; U.S. Patent 3,803,327; April 9, 1974; assigned to Idemitsu Petrochemical Co., Ltd, Japan.* The process has as an end product an oligopeptide, defined as a plastein.

In prior processes, a protein is hydrolyzed to a peptide having an extremely low degree of bonding; this peptide is used as a starting material and is subjected to plastein reaction. Accordingly, the yield of the resulting plastein is extremely low. Moreover, a long period of time is required for the hydrolysis of protein. For the above reasons, the process has not been utilized for commercial scale production of plasteins.

In this process a protein obtained from a protein source is hydrolyzed to a hydrolysis degree of 60 to 90% to prepare an oligopeptide having an average molecular weight of 1,200 and 2,000, and the oligopeptide is subjected to plastein reaction, whereby a plastein, which is free from offensive taste and odor and is suitable for use as a foodstuff, is obtained in a high yield and, moreover, the time required for the production of plastein can be greatly shortened.

The hydrolysis degree of protein shows the proportion of a 10% aqueous trichloroacetic acid-soluble peptide fraction in the hydrolyzate of protein, and is a value calculated according to the following equation:

$$\text{Hydrolysis degree} = \frac{\text{N in 10\% aqueous trichloroacetic acid-soluble fraction}}{\text{Total N}} \times 100$$

The 10% aqueous trichloroacetic acid-soluble fraction is an oligopeptide having an average molecular weight of 1,200 to 2,000, while the 10% aqueous trichloroacetic acid-insoluble fraction is an oligopeptide having an average molecular weight of more than 2,000. These oligopeptides polymerize according to plastein reaction to form a plastein. The yield of plastein is particularly high in the case of a hydrolyzate having a hydrolysis degree in the range of 60 to 90% which contains a large amount of the oligopeptide having an average molecular weight of 1,200 to 2,000.

In case the hydrolysis degree is less than 60%, a large amount of an oligopeptide having an average molecular weight of more than 2,000 is formed, while in case the hydrolysis degree exceeds 90%, the amount of an oligopeptide having an average molecular weight of less than 1,200 becomes larger. In both cases, the yield of plastein is lowered.

Protein sources include vegetable proteins such as soybean protein and wheat gluten, animal proteins such as codfish protein and pig hemoglobin, and microbial proteins such as petro-yeast proteins.

Examples of the enzyme include α-chymotrypsin, pepsin, neutral proteinase obtained from a strain of *Rhizopus* sp., subtilisin obtained from a strain of *Bacillus subtilis,* and Asperogillopeptidase-A obtained from a strain of *Aspergillus* sp. A protein is first extracted from a protein source by use of an aqueous alkali hydroxide solution. The extract is adjusted to an acidic pH, and is then hydrolyzed with pepsin to a hydrolysis degree of about 80%. The hydrolyzate is washed with an organic solvent such as ether and then concentrated under reduced pressure, so that the concentration of oligopeptide having an average molecular weight of 1,200 to 2,000 becomes 20 to 40%. To the solution is added a plastein synthesis enzyme. The solution is adjusted to proper pH for each enzyme as shown below and is then incubated at about 37°C for at least 8 hours.

Enzyme	Optimum pH Value
α-Chymotrypsin	5-7
Pepsin	4-5
Neutral proteinase	3-6
Subtilisin	5-7
Aspergillopeptidase-A	4-6

After completion of the plastein reaction, an alcohol or acid is added to the reaction product to gel the plastein, which is then separated and recovered by use of a centrifugal separator.

The plastein has a molecular weight within the range from 10,000 to 20,000, and the general composition, nutritional value, essential amino acid pattern and digestibility are identical with those of the starting protein. However, the plastein has such properties of being colorless, tasteless and odorless. For use as a foodstuff, the plastein is properly seasoned or is added for the purpose of enrichment to other foodstuff. Furthermore, the plastein does not contain harmful substances contained in the starting material and scarcely contain fats, so that it shows a high storage stability without suffering from rancid flavor or browning due to oxidation of fats.

Example: To 300 g of defatted soybean was added 3 liters of water to extract a soybean protein. The extract was adjusted to pH 4.5 by addition of hydrochloric acid and the soybean protein was precipitated. The precipitated protein was separated by use of a centrifugal separator to obtain 100 g of soybean protein. The soybean protein was dissolved in 10 liters of dilute hydrochloric acid at pH 2, and the resulting solution was adjusted to pH 1.6 by addition of a small amount of concentrated hydrochloric acid.

The solution was equally divided into 5 portions to form 5 solutions. Each of the solutions was incorporated with 1 g of pepsin, and then successively hydrolyzed at 37°C over 8, 10, 12, 15 and 48 hours to form 5 aqueous solutions of soybean protein hydrolyzates having hydrolysis degrees of 60, 70, 80, 90 and 98%, respectively. Each of the 5 aqueous solutions was washed 3 times with 10 liters of ether, and then concentrated under reduced pressure at 30° to 50°C to form 6 aqueous solutions having oligopeptide concentrations of 10, 20, 30, 40, 50 and 60% by weight, respectively.

In the above manner, a total of 30 kinds of aqueous soybean hydrolyzate solutions were obtained. Each of these solutions was adjusted to pH 7.0 by addition of sodium hydroxide, incorporated with 1 g of α-chymotrypsin as a plastein synthesis enzyme, and then incubated at 37°C for 24 hours to form a plastein. Each reaction product containing plastein was gradually charged with 99% purity ethanol so as to attain a final ethanol concentration of 95%. The resulting plastein gel was separated by means of a centrifugal separator, and was then dried to obtain a powdery plastein.

The yields of plasteins obtained from the 30 kinds of aqueous soybean protein hydrolyzate solutions were as set forth in the table below.

	Hydrolysis Degree (%)				
	60	70	80	90	98
Peptide	Average Molecular Weight				
Concentration	2,000	1,800	1,500	1,200	1,000
(%)	Plastein Yield (%)				
10	47	38	29	34	5
20	70	77	80	69	31
30	66	80	89	72	40
40	55	70	91	63	39
50	49	50	55	45	30
60	38	38	28	32	24

It can be seen that the maximum plastein yield is attained when the hydrolysis degree of protein is 80%, and that particularly favorable results are obtained when the oligopeptide concentration is in the range from 20 to 40%. The yield of plastein obtained from the control aqueous protein solution having a hydrolysis degree of 98% is very low.

CASSAVA PROTEIN SUPPLEMENT

H.L. Scott; U.S. Patent 3,979,522; September 7, 1976 gives a process for production of a high protein food supplement from cassava. Cassava is deficient in protein, a fact that is generally true for tubers, aroids and dioscorea. Tubers from various sources analyzed 1.5 to 3.0% by dry weight of protein. Plants that store large amounts of starch and moisture, as do these plants, invariably are deficient in protein. These plants, however, can be grown on a large scale on land not presently used for crops.

The starch-containing plant is treated to hydrolyze the starch to sugars preferably by means of a hydrolyzing enzyme, then a food yeast is added to the mixture and the yeast is grown aerobically on the sugars present in the mixture while introducing ammonia into the composition during growth of the yeast for a time sufficient to produce a substantial amount of protein in addition to the protein originally present in the plant. Air is introduced in addition to the ammonia during the yeast growth to insure aerobic fermentation and to avoid anaerobic fermentation which produces alcohol.

A very important part of the process is the removal of the prussic acid and any CN^- radical from the composition prior to yeast growth. Maximum production of yeast is not accomplished unless the CN^- radical is completely eliminated. A preferred way of doing this is to release the steam pressure during heating of the slurry. The release of the steam carries with it the prussic acid in the plants.

Any food yeast may be used in the process. Highly satisfactory results have been attained with *Candida utilis,* var. major, including mutants thereof, *Saccharomyces carlsbergensis* and *Saccharomyces cerevisiae*.

Candida utilis var. major, is the preferred strain of yeast because it is a large cell yeast which proliferates very rapidly under aerobic conditions and utilizes ammonia efficiently to synthesize all of the essential amino acids. It utilizes more different monosaccharides and disaccharides than any other species or variety of yeasts, which means higher yields of starch per pound of yeast. It is also the highest producer of the water-soluble vitamins of the three yeasts, but the other two produced adequate amounts.

The true yam as well as cassava may be used as a raw material, but yields probably could not be made to approach cassava's, and the yam is much more exacting in its soil characteristics and rainfall demands. It is also subject to rotting under damp conditions. The giant yam, *Dioscorea escuelenta,* has possibilities. Under conditions of good rainfall and ideal fertilization, high yields have been reported. Another dioscorea, *Pachyrrhizus erosus,* a native of Mexico and Central America, has been naturalized in various tropical soils. It prefers sandy loams like cassava, and yields up to 95 tons per hectare have been reported. It reproduces from seeds and attains maturity in about 8 months.

Aroids such as taro are grown and eaten by people, but would only be of interest if it were necessary to use water-logged or swampy soil. Yields are not high. Tubers, aroids and yams are eaten throughout the world for their starch, but in general, they range in protein from 2 to 3%. All of the tubers are high in moisture when first harvested.

In processing for protein synthesis, the entire plant is put through a mill and ground to a coarse meal, coarse enough to prevent dusting, yet fine enough to make the starch readily available. A Wiley mill has been used in the laboratory with great success in grinding the wet meal to pass a 20 or 30 mesh screen.

The damp meal goes continuously through a two-roll press to squeeze out the water. This squeezing removes part of the water and about half of the prussic acid. The juice tested about 71 ppm prussic acid plus about 6% protein, based on a solids content of 10%. Cassava holds moisture tenaciously. Some cassava under 12,000 lb pressure was squeezed until it formed a solid cake. It still contained 30 to 35% moisture. The still damp meal goes from the press into a tunnel drier on a continuous conveyor where it is dried to 10 to 12% moisture. The length of the tunnel is determined by the amount of moisture left after squeezing, the thickness of the bed of meal, and the final degree of moisture desired. The dried meal may then be shunted to storage bins or sent directly to the plant for further processing.

There are three ways of processing cassava for the biochemical synthesis of protein. They are the discontinuous feeding batch process, the continuous process, and the true batch process. Discontinuous feeding is simply adding media containing sugar to the fermenter as the yeast uses up the sugar. The reason for this method is that above 10% concentrations of sugar inhibit the multiplication of yeast. This method requires no extra equipment if the plant is designed for batch processing.

The batch process will be described in detail because the continuous process simply applies modern technology to the fundamental method. A slurry is prepared by mixing 1 part by weight of cassava with 3 to 5 parts by weight of water in a large pressure cooker equipped with rake-type agitation. It is cooked under pressure for 20 to 30 minutes, depending on the amount of prussic acid (HCN) in the meal and also upon the amount of starch present. The amount of water may also vary with the starch content, because very high amounts of starch tend to gel, and that means dismantling the entire system and cleaning it.

This pressure cooking is an important step, because it not only makes the starch available by breaking down the walls of the starch cells, but it completely eliminates the prussic acid. Repeated tests have failed to show any HCN residues. HCN is bound to the glucosides in the cassava and the heat and sudden release of pressure causes the HCN to volatilize. This is usually accomplished by fermenting, boiling or cooking after maceration, but it was found that these methods always yield measurable residues of HCN.

The elimination of prussic acid is doubly important because it has been found impossible to get a yeast count above 6×10^6/ml with 71 ppm HCN present. The CN radical blocks the Krebs cycle and interferes with the respiration of the yeast. Maximum production of yeast is not accomplished unless CN is completely eliminated.

After the slurry is pressure cooked, it is pumped to one of two converters where it is cooled to 148° to 150°F. The converters are equipped with rake-type agitators. Two converters are used to permit uninterrupted operation. While the temperature of the converter is maintained between 148° to 150°F, a solution containing equal parts of pure A and B amylase is added in amounts depending on the amount of starch present and upon the dilution of the amylases.

As an alternate to the use of the concentrated enzymes, there may be used up to 10% of barley malt, which was ground to pass a 30 to 60 mesh screen. Barley malt converts the starch into sugars within one-half hour, with efficiencies up to 100%. Monosaccharides and disaccharides are produced, with glucose predominating by a wide margin. This is why *Candida utilis,* var. major, is preferred because it can utilize a greater number of disaccharides, thus squeezing a little more efficiency out of the fermentation process.

After converting the starch to sugars, the slurry is pumped to giant fermenters through heat interchangers. The temperature is brought down to 60° to 70°C, at which time the pH of the slurry is carefully adjusted to 4.0. Lactic acid is preferred but larger amounts of other organic acids may be used if not too expensive. The yeast is prepared and grown separately. Preparation may be either batch or continuous, and may be grown either in a closed system or in open tubs.

When cell counts reach maximum, the fermenter is "dropped" to a well to permit washing out and sterilizing the fermenter with steam to prevent contamination. Maximum cell counts are usually 6×10^{10}/ml. Continuous operation is accomplished with this method.

In one method, the contents of the well are pumped into either a five-step or seven-step series of evaporators which are heated by steam. A vacuum is pulled, either with a steam jet or high pressure water jet. The evaporation is by thin film which comes in contact with countercurrent steam. It takes roughly a pound of steam to remove a pound of water. The slurry is evaporated down to a thick syrup which will not solidify as long as it is kept hot. The syrup is 15 to 20% water.

The syrup is pumped over dehydrators (drum-driers), which are steam heated drums rotating inward on each other that pick up a thin film of the syrup. A knife edge on the front end of the drum cuts the film off and it drops into a screw conveyor which chops it into flakes of assorted sizes. It comes off of the rolls just like a sheet of paper. The final product is 8 to 12% moisture.

The high protein supplement may be used immediately in this country as cattle feed, hog feed, chicken feed and cat and dog food. This would take the pressure off of corn and soy meal for cattle feeders, thus permitting greater amounts for export.

This product may also be formulated for human food, especially in the developing countries. The supplement may be mixed with its own starch, and prepared as a balanced flour, or into any of the prepared food products of today. The product may be processed into cereals in the customary ways.

GELATIN AND WHEY PEPTONES

Two protein sources, namely gelatin and milk whey, are used as raw materials to produce hydrolyzed proteins which are rich in essential low molecular weight peptones. Gelatin solutions with a protein content of up to 40% can be hydrolyzed over a period of six hours to such an extent that on cooling to 2°C gelling is no longer detectable. The gelatin molecule, which can be assumed to have a mean molecular weight of 60,000 to 80,000, is split into peptide fractions of 30 to 40 amino acid units.

The enzyme reaction does not cause the formation of compounds which impair taste. The only drawback to its use as a foodstuff is the low content of essential amino acids in the gelatin. In order to improve on the biological value of the gelatin, the attempt was made to find new protein source. Whey is found to be such a suitable source. The whey protein essentially comprises lactoglobulin with a molecular weight of around 17,000. When such low molecular weight is involved, the protein is digested and resorbed with sufficient rapidity.

The wettability with water of the dry powder recovered from nonpretreated whey is extremely poor. Lumps form on stirring the dry powder into water. This disadvantage can be overcome by enzymatic pretreatment of the whey protein. Native protein molecules are split into low molecular fractions by highly active proteolytic enzymes in the enzymatic reaction. These low molecular fractions can no longer be flocculated by thermal treatment. However, since the proteolytic reaction results in compounds which are intensely bitter tasting, it is impossible to use whey treated in this manner in the food industry. Other natural and synthetic protein substances present similar difficulties and problems.

According to *D. Schwille, H. Seiz, E. Sorg and U. Sommer; U.S. Patent 3,974,294; August 10, 1976; assigned to AGFA-Gevaert, AG, Germany* an enzyme mixure prepared from fresh pancreas is used for the improved process. The proteolytic enzyme mixture is sold under the name of pancreatin. The individual enzyme groups in the mixture have the following activity values: 3,500 units/g protease, 18,000 units/g lipase, and 30,000 units/g amylase.

To qualify for use in this process, a proteolytic enzyme has to fullfill the requirements of the following test. To a 20% aqueous suspension of the protein substance to be treated, is added the enzyme to be tested in a weight ratio of 1:750, based on the dry weight of the protein. This mixture is heated for 3 to 4 hours to a temperature between 50° and 60°C. The optimum pH of the enzyme under study is maintained. In the case of pancreatic enzymes of low activity or the preferred enzymes of the trypsin group, this pH value is in the range of between 8 and 8.5.

To qualify as suitable, the proteolytic enzymes must meet two conditions: the enzymatically treated protein should be easily suspendable or soluble in water, that is flocculation should not occur upon heating to higher temperatures; and the enzymatically treated protein should not be bitter-tasting after treatment.

The following natural protein sources may be used in the process: whey, milk (casein), soy protein concentrates or protein isolates, maize protein, potato protein, wheat protein, oil-seed protein substances, nut kernel protein substances, sunflower kernel protein, or egg proteins.

It has been found that if the enzymatic treatment of the proteins with proteolytic enzymes of low activity is carried out in the presence of gelatin or decomposition products of gelatin, the process can be performed without difficulty to produce protein substances which meet the above conditions and are suitable as food additives. The gelatin or gelatin decomposition products would appear to influence the proteolytic action of the enzymes so that the resulting protein products are split into lower molecular protein products which have the required properties as regards flocculation temperature and bitter substance content.

The protein products have excellent food-technological properties. They show excellent wettability and very good solubility in water. Almost clear, aqueous solutions without residues are obtained which are miscible with water or other water-containing preparations in practically unlimited ratios. Even if the solution is left to stand for hours, or heated to temperatures over 80°C, still no signs of flocculation are detectable. The products have an almost neutral taste and are hence eminently suitable for use as additives in all types of foods, e.g., in the preparation of fruit juice concentrates.

It is of particular importance that even concentrated aqueous solutions, which contain the protein substances in concentrations of up to about 50%, can be heated for hours without flocculation occurring to temperatures suitable for effecting sterilization.

Example: A 20% aqueous solution of whey protein is heated to 45°C and adjusted with sodium hydroxide solution to a pH value of 8.5. At this temperature there is added a mixture of the enzymes proteinase, lipase and amylase, for example, the product sold under the name of pancreatin in a weight ratio of 1:1,000 (based on the whey protein contained in the solution). Stirring is carried out for four hours at a pH value of 8.5 and a temperature of 45°C and a further amount of pancreatin (weight ratio 1:2,000) added. The temperature is then slowly increased to 60°C. The solution becomes viscous, but the whey protein does not flocculate. After a further two to three hours, the protein is hydrolyzed to such an extent that the protein solution can be pasteurized in the sterilizer without flocculation occurring.

Example 2: A 20% gelatin is mixed at 45°C with a 20% solution of whey protein in a ratio of 1:1. By means of sodium hydroxide solution a pH value of 8.5 is obtained. Pancreatin is added at 45°C in a weight ratio of 1:1,000 (based on the dry protein substance contained in the solution). Stirring is continued at a pH value of 8.5 and a temperature of 45°C until a distinct decrease in viscosity is no longer detectable. Additional pancreatin (1:2,000) is then added and the temperature slowly increased to 60°C. After a further two to three hours, the protein is so far hydrolyzed that the solution can be pasteurized in the sterilizer at 70°C without flocculation occurring.

The protein solution of Example 2 is dried in the usual way by spray-drying. Analysis revealed the products thereafter obtained to have the following composition:

Product Composition	Gelatin	Whey	Gelatin + Whey
	........(percent by weight)........		
Weight loss on drying	5.9	5.3	4.5
Carbohydrates	–	4.4	3.9
Ash	5.7	7.4	5.4
N content	14.1	13.2	13.7
Protein content (natural protein N x 6.25)	88.1	82.5	85.6

The decomposition products of the protein of the gelatin, whey and gelatin + whey (1:1) in percent by weight, are as follows:

	Gelatin	Whey	Gelatin + Whey
	........(percent by weight)........		
Lysine	4.55	8.82	5.92
Histidine	1.78	2.89	1.66
Arginine	5.21	3.38	6.88
Tryptophane	2.42	–	1.33
Aspartic acid	6.82	7.88	7.17
Threonine	1.96	4.66	2.77
Serine	3.77	5.23	4.23
Glutamic acid	8.43	11.32	14.54
Proline	15.20	10.13	12.51
Glycine	6.03	1.85	10.46
Alanine	5.12	3.58	7.17
Cysteine	1.35	1.61	0.82
Valine	2.53	6.09	4.04
Methionine	0.81	3.12	1.57
Isoleucine	1.46	5.38	2.83
Leucine	3.14	10.13	5.47
Tyrosine	–	4.97	1.89
Phenylalanine	1.93	5.22	3.25
Glycocoll	24.97	3.84	4.44

ENRICHED GELATIN DESSERT

It is possible to add enzyme converted nongelatin protein to a gelatin-type dessert or dry beverage mix to nutritionally improve the product. According to *J.R. Feldman, G.J. Haas and J.C. Lugay; U.S. Patent 3,970,520; July 20, 1976; assigned to General Foods Corporation* the steps involved entail extracting and washing the proteinaceous material, proteolytically hydrolyzing the isolate proteinaceous material, deactivating the enzyme, and separating and drying the hydrolyzed proteinaceous material prior to adding it in certain quantities to the particular foodstuff.

Enzymatic hydrolysis of proteinaceous isolates is less drastic and not destructive compared with acid hydrolysis, and leaves nutritive low molecular weight polypeptides intact. Whereas, during acid hydrolysis of proteinaceous isolates of the nutritive polypeptides, essential amino acids such as serine, threonine, tryptophan and methionine are destroyed, thereby lowering the overall nutrition value of the resulting hydrolyzate.

Generally, in preparing a protein isolate, the proteinaceous source material is first extracted with water to prepare an aqueous protein solution. For proteins such as soy, the pH of the extractment should be adjusted to significantly above the isoelectric point. The resulting protein solution is then treated by acidifying, heating or both to obtain a precipitate which is separated from the supernatant. This recovered precipitate is the protein isolate. It is necessary to wash the isolate with effective volumes of water to substantially remove off flavors and odors. With proteins such as soy which have a degree of solubility at near neutral pH, it is necessary to make the wash water dilutely acid to reduce protein loss due to solubilization. The isolate is then hydrolyzed as described in the examples.

Two enzyme systems have been identified as being particularly effective in producing bland, soluble protein hydrolyzates. One of these systems is ficin, papain and neutral protease from *B. subtilis.* The total enzyme combination should contain from about 25,000 to 500,000 protease units/100 g of protein. A protease unit, in this instance, is defined as that amount which acts on milk casein for one minute at 30°C and produces a quantity of folin color-producing substances, not precipitated by trichloroacetic acid, that is equivalent to 1 mg of tyrosine.

Individually, the enzyme should be at levels of from 10,000 to 180,000 protease units for each 100 g of protein. While it is generally preferred to add all the enzymes at the beginning of the reaction and carry out the reaction at a temperature selected for optimum overall effectiveness, e.g., 50°C, they may be added sequentially using the optimum temperature for each enzyme for a period of time following its addition. Where the enzymes are added sequentially, it is best that the last enzyme in the sequence is added no later than after 50 to 75% of the total reaction time.

The other of the two enzyme systems is ficin plus neutral protease from *B. subtilis.* The total enzyme activity will be the same as for the other. The individual enzymes will be present at levels of from 10,000 to 250,000 units for each 100 g of protein.

Example 1: Preparation of Soy Isolate — A 10% by weight aqueous slurry of soy grits is prepared and homogenized in a Waring Blendor for three minutes.

The pH of the slurry is slowly adjusted to about 8.0 with sodium hydroxide and the slurry is stirred at ambient temperature for about 30 minutes. Next the mixture is centrifuged or filtered and the residue is discarded. The pH of the supernatant is adjusted to about 4.5 with hydrochloric acid and the acid mixture is brought to boiling and held there for approximately six minutes to destroy protease inhibitors, then the mixture is cooled and centrifuged. The precipitate is then suspended in water and the pH is again adjusted to about 4.5. The acid mixture is centrifuged and the supernatant is discarded. The acid washing is repeated once more and the residue is then freeze dried for storage or resuspended to make a slurry for hydrolysis.

Hydrolysis of Soy Isolate — A 10% by weight slurry of soy isolate is prepared and the pH is adjusted to 6.0 to 6.5. A proteolytic enzyme combination of ficin, papain and neutral protease from *B. subtilis* (in respective amounts of 1:1:2) is then added at 0.5% by weight (total enzyme activity of about 125,000 protease units/100 g of protein) of the isolate and the slurry is incubated for about two hours at 50°C. Alternatively, a mixture of ficin and neutral protease from *B. subtilis* (1:1 ratio) may be used. At the end of the incubation period hydrolysis is terminated by raising the temperature to 90°C to inactivate the enzymes and precipitate the heat-insoluble fractions. The mixture is centrifuged and the residue is used in the next batch for further hydrolysis.

The supernatant is adjusted to pH 3.9 with phosphoric acid (2.6 N), if it is to be used in acidic foodstuff formulations such as fruit drinks, gelatin containing desserts, etc. (For nonacidic formulations the pH is left as is.) The supernatant is then winterized at 0°C overnight to remove the unstable cold labile fraction. The solution is then centrifuged or filtered and freeze dried. The molecular weight of this hydrolyzed protein ranges from 200 to 2,900.

In the instance where the specific foodstuff is gelatin dessert and the freeze dried hydrolyzed protein is derived from soy isolate, the nutritionally fortified dessert is prepared by adding the hydrolyzed isolate to conventional orange-flavored gelatin dessert mix (gelatin, sucrose, citric acid, and buffer salts) in a ratio which provides from 2.0 to 4.3 g of the hydrolyzed protein per serving (which is about 21 to 25 g of the conventional dessert solids). The dessert is then dissolved in hot water (70°C) while stirring, and permitted to cool (at about 10°C) and gel in the same manner as nonnutritionally modified gelatin desserts.

The setting time, gel strength, and the organoleptic taste qualities are also substantially the same as nonnutritionally modified gelatin desserts. The clarity is better than conventional orange-flavored gelatin dessert, as evidenced by optical density readings. For example, a Beckman DBG spectrophotometer is used at a wavelength of 490 mμ to obtain optical density readings on conventional gelatin dessert and the nutritionally modified dessert. The control or conventional gelatin had an optical density at 490 mμ, which is 0.07 unit greater than nutritionally modified gelatin containing 2.0 g of hydrolyzed protein per serving.

It is to be noted that protein from egg albumen and milk whey when washed and hydrolyzed according to this process, may be used either alone, with each other or in combination with soy protein, with comparable results.

Example 2: Preparation of Whey (Milk) Isolate — Fresh milk whey or dried whey powder is used in preparing a 5 to 10% slurry of whey protein powder.

The slurry is then adjusted to a pH of about 5.3 to 5.5 and then heated to boiling to precipitate the proteins. The mixture is cooled and centrifuged, whereupon the supernatant is discarded. The precipitate is resuspended in water having a pH of 5.3 to 5.5, after which it is again centrifuged and the supernatant discarded. This washing step is repeated once more and at this stage the lactose content, salts and milky taste are substantially reduced. The precipitate is then freeze dried for storage or resuspended to make a slurry for hydrolysis.

Hydrolysis of Whey Isolate — A 5 to 10% by weight slurry of whey isolate is prepared and the pH is adjusted to 6.0 to 6.5. A proteolytic enzyme combination of 0.5% ficin, 0.5% papain and 1.0% neutral protease from *B. subtilis* (total enzyme activity of about 500,000 protease units/100 g of protein) is then added (based upon the weight of whey isolate). Ficin and papain are first added and the hydrolysis is allowed to proceed for about one hour at 40°C, then the temperature is raised to 50°C and neutral protease from *B. subtilis* is added and the hydrolysis continued for another hour.

At the end of the incubation period, hydrolysis is terminated by raising the temperature to about 90°C to inactivate the enzymes and precipitate any possible heat-insoluble fractions, which may be used in the next batch for further hydrolysis. The supernatant is winterized at 0°C overnight if it is to be used in a foodstuff formulation requiring clarity, otherwise it is centrifuged and freeze dried. If it is to be incorporated in acidic formulations, the pH must be adjusted to about 4.0 prior to freeze drying. This hydrolyzed protein will have a molecular weight range between 200 and 2,900.

MODIFIED CORN GLUTEN

Corn gluten appears as a by-product from the wet milling process for production of maize starch. Corn gluten is predominately used as animal feed; but it can also be used as a cork binding agent, as an additive for printing dyes in the pharmaceutical field, and for other purposes. It would also be desirable to use corn gluten for human food, but the properties of unmodified corn gluten are not ideal for this purpose.

Thus, due to its unbalanced amino acid pattern, corn gluten is not very suitable as a protein source in human nutrition; however, mixing corn gluten with soy protein (which is deficient in methionine) produces a product, superior in nutritional qualities both to soy protein and maize protein. The practical use of corn gluten in human food has been limited by the fact that the proteins in corn gluten are largely insoluble in water. In fact, the gritty mouthfeel experienced upon ingestion of corn gluten is ascribed to its pronounced insolubility.

In liquid or semisolid feed preparations, the corn gluten will settle out like fine sand. Raw corn gluten has an unsatisfactory extrusion capability, consistent with the fact that it also has a low water absorption capability. The gluten forms a two-phase system if suspended in an aqueous medium and then left alone. A treatment process is thus needed to modify raw corn gluten to improve its water absorption capability and to enable it to form a one-phase system when suspended in an aqueous medium.

In the process of *J.L. Adler-Nissen; U.S. Patent 4,100,151; July 11, 1978; assigned to Novo Industri AS, Denmark* microbial proteases are used to improve the solubility and functionality of corn gluten for use in foods. One of the advantages of the modified corn gluten is the improved water absorption capability and the concomitant improved extrusion capability and increased tendency to form a single-phase system. In the examples which follow, three different microbial proteases are illustrated, viz, Alcalase, which is a protease produced by means of *Bacillus licheniformis;* Neutrase, which is a protease containing a neutral and an alkaline constituent and originating from *Bacillus subtilis;* and SP 99, which is a preparation containing the neutral constituent of Neutrase. All three enzyme preparations exhibited a proteolytic activity of 1.5 Anson units/g.

Example 1: Approximately 150 g maize protein concentrate containing approximately 68% protein (% N x 6.25) was treated with Neutrase under the following conditions: substrate, 8% w/w protein in aqueous suspension; enzyme concentration, 0.2% of substrate protein (corresponding to 0.24 Anson units/l); pH (approximately), 8; and temperature, 50°C. After two hours treatment, a uniform, thick paste was obtained which was pseudoplastic in its rheological behavior. The paste did not separate into two phases under the influence of gravity. The taste was pleasant, bland and nonbitter.

Example 2: Several hydrolyses of corn gluten were carried out as shown in the table below. All experiments were run at 50°C. The viscosity is measured at 15°C on a rotational viscosimeter with an internal, rotating cylinder. The viscosity is measured at a constant shear rate of 441 sec^{-1}. The viscosity was measured on an aqueous suspension of the protein. Suspensions containing the % of dry protein indicated in the table (calculated as % N x 6.25) were obtained by diluting the more concentrated hydrolysates with the appropriate amount of distilled water.

In order to avoid further action of the proteolytic enzyme after the treatment of the corn gluten, which could cause a bitter taste in the modified corn gluten, the proteolytic enzyme is inactivated. This inactivation can be performed in any known manner, e.g., by heat treatment or pH adjustment. This inactivation is most easily carried out as a heat inactivation during the extrusion together with the soy protein when such a mixed product is made.

No.	Protein (%)	..... Enzyme ..... Type	% of Protein	pH	Final Viscosity (cp)
1	6	SP 99	0.1	7.00	21
2	8	SP 99	0.1	8.00	164
3	8	SP 99	0.1	7.00	148
4	8	SP 99	0.1	6.45	78
5	8	Neutrase	0.1	8.00	138
6	8	Neutrase	0.1	7.00	148
7	8	Neutrase	0.1	6.45	131
8	8	SP 99	0.2	7.00	107
9	8	SP 99	0.05	7.00	92
10	8	SP 99	0.1	7.50	139
11	8	Alcalase	0.1	8.00	140
12	8	Alcalase	0.05	9.00	185
13	8	Alcalase	0.4	9.00	98
14	8	Alcalase	0.1	9.00	125

MEAT AND FISH PRODUCTS

ENZYME TENDERIZATION PROCESSES

Enzyme Activator Method

The tenderness of meat, and particularly beef, is a prime consideration to the consumer and thus the desire for increased tenderness has influenced livestock feeding procedures. Although improved feeding procedures can influence the tenderness of meat, the feeding procedures have not provided the desired increase in tenderness, with the result that auxiliary tenderizing processes have been utilized in an attempt to further increase the tenderness of meat.

In the past, it has been proposed to increase the tenderness of meat by maintaining the carcass under refrigeration for several weeks, during which period the tenderness is enhanced by autolysis of the meat proteins. In another common method of tenderization, the meat is held at an elevated temperature for several days and subjected to ultraviolet radiation while at the elevated temperature to prevent microorganism growth. While these processes have resulted in an improved tenderness in certain individual carcasses, these processes have not resulted in uniform tenderization of all cuts of meat and have the further disadvantage of incurring weight loss due to shrinkage and surface deterioration.

The ability of proteolytic enzymes to break down the fibrous material in meat and provide tenderization is well established. With this in mind, attempts have been made to tenderize meat by injection of proteolytic enzymes post-mortem into the animal carcass. While the addition of proteolytic enzymes to the carcass does aid in increasing the tenderization of meat, the primary problem in such a process is effecting a uniform distribution of the proteolytic enzyme throughout the carcass. Processes of this type normally utilize a pumping apparatus for distributing the enzyme through the vascular system of the carcass after bleeding. Even though a pumping system is used, the primary cause of nonuniform tenderization lies in the failure to penetrate the finer capillaries and arterioles with the enzyme solution. An earlier process was developed for antemortem injection of proteolytic enzyme solution into the vascular system of the animal. This ante-mortem injection has the advantage that the proteolytic enzyme will be uniformly distributed throughout the vascular system of the animal before slaughtering and, therefore, aids in producing more uniform tenderness of the meat. While the ante-mortem injection of proteolytic enzymes into the animal does aid in increasing the tenderness of the meat, the process is relatively expensive due to the high cost of the enzymes involved.

M.M. Weber; U.S. Patent 3,561,976; February 9, 1971; assigned to Midwest Biochemical Corporation takes a different approach by injecting ante mortem into the vascular system of the animal certain substances capable of activating the natural occurring proteolytic enzymes. The activators function to increase the activity of the natural proteolytic enzymes so that the enzymes break down the fibrous material at a faster rate and thereby tenderize the meat. The activator solution is injected into the animal about 15 to 30 minutes before slaughter, and it has been discovered that the activator, in this relatively short period of time before slaughter, is capable of activating the proteolytic enzymes to a degree necessary to provide a marked increase in tenderness of the meat.

In larger animals such as cattle, sheep and hogs, injection is made into the jugular vein, although in some cases the injection can be made directly into the heart or arteries. On smaller animals such as chickens, injection is made by means of a needle or hypodermic syringe into one of the exposed veins such as the humeral or directly into the internal metatarsal vein.

The activator to be injected into the animal can be any material capable of activating proteolytic enzymes. Proteolytic enzymes are those which hydrolyze proteins and include pancreatin, ficin, papain, bromelin, fungal and bacterial proteases and cathepsin. The activators are generally classified as reducing compounds and may take the form of sodium hydrosulfide, sodium hydrosulfite, sodium sulfide, sodium sulfite, sodium bisulfite, sodium thiosulfate, glutathione, cysteine, cysteine hydrochloride, methionine, thioacetic acid, thiolactic acid, thiobenzoic acid, thiomalic acid, sodium thioglycolate, thioglycerol, benzyl mercaptan, N-butyl mercaptan, 2-diethylaminoethanethiol, ethylthioglycolate, 3-mercaptopropionic acid, methoxyethyl thioglycolate, o-mercaptobenzoic acid, dithioerythritol, glycol dimercaptoacetate, pentaerythritol tetra(3-mercaptopropionate), pentaerythritol tetrathioglycolate and glycol dimercaptopropionate.

Ammonium sulfate which activates pancreatin can also be utilized as the activator. It is believed that the activators function to open up SH groups in the enzyme molecule. The activator solution is injected into the animal ante mortem up to 120 minutes before slaughtering and generally within 15 to 30 minutes before slaughtering. During this period after injection and before slaughter the activator is uniformly distributed throughout the vascular system of the animal and is capable of activating the naturally occurring proteolytic enzymes.

After the animal is slaughtered, the blood is drained by conventional procedures, and the draining of the blood prevents an excessive concentration of the activator in the larger blood vessels. The smaller blood vessels consisting of the venules, arterioles and capillaries are drained only slightly in the bleeding step and thus contain blood which includes the enzyme activators. It has been found that the most significant improvement in tenderization of the meat is achieved if the meat is refrigerated for a period of 4 to 7 days after slaughtering and before freezing. This aging period aids in improving the tenderizing effect by providing additional time for the activator to activate the proteolytic enzymes which in turn act to break down the fibrous materials and increase the tenderness of the meat when cooked.

Example 1: Ante-mortem injections of an aqueous solution of proteolytic enzyme activators were made with a series of cattle. In each case the injection was made into the jugular vein of the animal with a 100 ml syringe containing the activators listed in the following table. The table also lists the weight, age and sex of the cattle, the amount of the activator employed, and the time period after injection and before slaughter.

Animal	Weight (lb)	Age (yr)	Sex	Activator	Amount of Activator (g)	Minutes Held Before Slaughter
1	590	1	steer	cysteine HCl	20	5
2	850	2	female	cysteine HCl	20	5
3	740	3-4	female	cysteine HCl*	20	10
4	715	5	female	cysteine HCl*	20	10
5	895	2	female	cysteine, free base	15	10
6	685	1	male	cysteine HCl	20	10
7	810	1	male	cysteine HCl	45	5

*Neutralized with $Na_3(PO_4)_2$.

Meats from the above-treated cattle were subsequently roasted for tenderness panel tests and the results of the tenderness tests were as follows. Animals 1 through 6 were rated good and animal 7 was rated very good. The results of the above tests indicate that the injection of proteolytic enzyme activators ante mortem into the vascular system of the animal substantially improves the tenderness characteristics of the meat.

Purified Enzyme Preparation

Aqueous solutions of crude enzymes or food grade enzymes are heat treated to inactivate proteolytic enzymes without affecting total enzyme activity. The process of *J.F. Beuk, W.D. Warner and C.K. Kang; U.S. Patent 4,024,285; May 17, 1977; assigned to Swift and Company Limited* involves use of enzymes to tenderize meat by ante-mortem injection into live animals.

Briefly, this technique involves the injection of a proteolytic enzyme solution into the vascular system of a living animal, then holding the animal for a time sufficient to obtain uniform distribution of the enzyme throughout the animal's system, and thereafter slaughtering the animal. Meat cuts derived from animals injected in this manner are very tender and exhibit good textural properties. While this technique is of substantial value in obtaining meat which, when cooked, is uniformly tender, there are undesirable animal side reactions which occur if some commercially available enzymes are utilized. Specifically, injection of the enzyme into a live animal often causes internal hemorrhaging and edema of the internal organs.

It has formerly been thought that the adverse physiological reactions observed in live animals were primarily due to the presence of certain undesirable components (impurities) present in the commercially available enzyme. Much work has been done to remove or render harmless these undesirable components from commercial enzyme preparations.

Commercial food grade and crude enzyme powders contain a mixture of active enzyme molecules, reversibly inactive enzyme molecules, and irreversibly inactive enzyme molecules. The ratios of these types of enzyme molecules vary from one batch of enzyme to another. Active enzyme molecules produce meat tenderization through proteolysis, but also may cause severe physiological reactions when injected into live animals. For example, intravascular injection of live animals with solutions of commercially available crude enzymes, such as papain or ficin, produces such symptoms as labored breathing, nasal congestion, depression, frothing at the mouth, and in severe cases the animal will become cyanotic and die. The autopsy findings of such animals usually include hemorrhaging in the kidneys, heart, liver, intestines, gall bladder and larynx.

As a result of these symptoms, such animals are condemned by governmental inspectors. Irreversibly inactive enzyme molecules are those which, due to oxidation, hydrolysis, etc. of the active enzyme molecule, have permanently lost their activity, at least under the processing conditions of ante-mortem enzyme solutions, and therefore produce no tenderization effect or animal reaction.

Reversibly inactive enzyme molecules produce no adverse physiological reactions upon intravascular injection into live animals. However, these enzyme molecules are subsequently reactivated within the animal's vascular system by naturally occurring reducing compounds and by heat, thereby resulting in subsequent tenderization of the meat. Therefore, the primary desideratum in preparing enzyme solutions for ante-mortem animal injection is to obtain an enzyme solution containing a maximum amount of reversibly inactive enzyme molecules, and a minimum amount of active and irreversibly inactive enzyme molecules.

Essentially, the amount of active enzyme molecules in a given solution is determined by using a milk clot assay of available enzyme activity. The units for measuring active enzyme molecules are generally called nonreduced milk clot units, and are referred to as the available milk clot activity or available milk clot units (AMCU). Where a measure of the reversibly inactive enzyme molecules is desired, it is necessary to first determine the total enzyme activity, and then subtract the amount of active enzyme (AMCU). To determine total enzyme activity, the enzyme solution is treated with an activator or reducing agent such as cysteine, cyanide or bisulfite, whereby all the reversibly inactivated enzyme is restored to the active state.

The milk clot assay using cysteine or other activators measures the sum of active and reversibly inactive enzyme molecules. This measurement is referred to as either reduced milk clot units or total milk clot units (TMCU). Any reduction in this value (TMCU) during processing indicates a loss in the overall enzyme activity of the solution being assayed. However, reduction in the AMCU, without a corresponding reduction in the TMCU value, indicates the amount of active enzyme molecules which have been converted to reversibly inactivated molecules.

Pure papain tends to exhibit stability at alkaline pH values of 8.0 to 12.0, at least for a few minutes, and denatures at pH values above 12.0. However, in both crude and food grade commercial papains, there are present several proteolytic enzymes other than pure papain. For example, chymopapain proteases may constitute up to 90% of the total proteases in food grade and crude papain commercial powders. Moreover, with respect to these food grade and crude enzyme powders, it has long been felt that such commercial enzyme powders could not be held over substantial periods of time at high pH values; however, it was found possible to process crude and food grade papain and ficin proteases over extended periods of time at pH values of 11.0 to 12.5 without enzyme denaturation and without causing any substantial loss in total enzymatic activity. Generally, the method involves holding aqueous solutions of papain or ficin at a pH of from 11.0 to 12.5 over an extended period of time to reversibly inactivate the enzymes. The reversibly inactivated enzyme may then be injected ante mortem into animal stock without resulting in adverse physiological reactions in the animal.

The process is specifically directed to the treatment of the plant sulfhydryl proteases papain and ficin. The initial enzyme solution is prepared by first forming a paste of the food grade or crude papain or ficin enzyme by mixing with an organic water-soluble liquid such as glycerine, glycols, or other water-soluble or water-dispersible wetting agents. The paste formed is then dissolved or suspended in water having a temperature of about 10° to 15°C by merely adding the water to the paste with constant stirring.

The pH of the solution is then adjusted to within from 11.0 to 12.5, by the addition of a water-soluble base, preferably sodium hydroxide. The aqueous enzyme solution is then held at a low temperature, usually 0° to 10°C for about 12 to 72 hours. The holding time required is that time sufficient to reversibly inactivate active enzyme content (AMCU) of the solution to a point whereby intravascular ante-mortem injection into animals does not result in adverse physiological reactions. As a general rule of thumb, reduction of enzyme activity to below 2 AMCU, will result in a final enzyme solution which is suitable for ante-mortem intravascular injection into animal stock.

Another important advantage of the process, not obtainable by prior procedures, is that bacterial contamination (aerobic count) is drastically reduced at such high pH values. It is believed that the effectiveness of the process is due to a unique configuration of the enzyme molecule at very high pH values. It is known that, depending on the particular plant protease, it will denature at a pH in excess of about 12.0 to 12.5. It is believed that when the enzyme is exposed to a pH of about 12.0, the enzyme chain is in an incipient breaking configuration, and that therefore, the enzyme molecule is more receptive to reversible inactivation.

By definition, as used in the following examples, a milk clot unit (MCU) is that amount of activity in one ml that will cause 5 ml of milk substrate to clot in 60 seconds at 40°C. Thus, the formula for calculating the milk clot unit of a properly diluted enzyme solution is as follows where t equals clotting time in seconds: MCU/ml = 60/t.

Example 1: A 10% by weight solution of crude papain was prepared. pH was adjusted to 12.0 and held for 24 hours. The pH was then lowered to 8.0 and 1 g thiamine disulfide was added per 25 g of enzyme powder in solution. The solution was held for an additional 24 hours, and then analyzed for enzyme activity. Total milk clot units (TMCU) measured 51 and available milk clot units (AMCU) measured 2.0. Solutions prepared with the high pH treatments of this procedure were injected into old ewes with no animal reaction, and cuts of meat obtained exhibited increased tenderness.

Selective Disulfide Inactivators

It has generally been thought that the adverse physiological reactions in live animals were primarily due to the presence of certain undesirable components (impurities) present in the commercially available enzyme. Much work has been done to remove or render harmless these undesirable components from commercial enzyme preparations. Unfortunately, these purification procedures result in irreversible inactivation of a substantial portion of the active enzyme.

C.K. Kang, W.D. Warner and E.E. Rice; U.S. Patent 3,818,106; June 18, 1974; assigned to Swift & Company worked on methods of reversibly inactivating the proteolytic enzymes without reducing total enzymatic activity. One method formerly used was treatment of the enzyme fraction with sodium tetrathionate (disulfide reagent) at pH 6.5 to 8.0. A refined method yields a reversibly inactivated enzyme which retains 100% of its original enzymatic activity and the disulfide-modified enzymes possess sufficiently slow reactivation rates such that adverse physiological reactions in the injected animals are avoided. Generally, the method involves treating a sulfhydryl protease, held at a low temperature and at a pH of 6 to 12, with one or more of certain disulfide reagents that reversibly inactivate the active enzyme fraction. The reversibly inactivated enzyme may then be injected ante mortem into animal stock.

Commercial enzyme powders of plant-derived enzymes contain a mixture of sulfhydryl proteases. For example, commercial papain contains pure papain enzyme, along with two or more chymopapain enzymes. Commercial food grade enzyme powders contain mixtures of active enzyme molecules, reversibly inactive enzyme molecules, and irreversibly inactive enzyme molecules. The ratios of these three types of molecules vary from one batch of enzyme to another.

Active enzyme molecules produce the meat tenderization through proteolysis, but also may cause severe physiological reactions when injected into live animals. For example, intravascular injection of live animals with crude enzyme solutions of either papain, bromelain, ficin, or combinations thereof, containing a substantial portion of active enzyme molecules, produce such symptoms as labored breathing, nasal congestion, depression, frothing at the mouth, and in severe cases the animal will become cyanotic and die. The autopsy findings of such reactors usually include hemorrhaging in the kidneys, heart, liver, intestines, gall bladder, and larynx. As a result of these symptoms, the reactors are condemned by governmental inspectors. Fully active enzymes may cause very rapid expiration of the animal.

Irreversibly inactive enzyme molecules are those which, due to oxidation, hydrolysis, etc., of the active enzyme molecule, have permanently lost their activity, at least under the processing conditions of ante-mortem enzyme solutions, and therefore, produce no tenderization effect or animal reaction. Reversibly inactive enzyme molecules appear to produce no tenderization or adverse animal reactions upon intravascular injection. However, these enzymes may be reactivated within the animal's vascular system, thereby producing subsequent tenderization of the meat.

Therefore, it is most desirable in methods of ante mortem meat tenderization by intravascular injection to inject an enzyme solution containing as high a concentration of reversibly inactive enzyme molecules as is possible. That is to say, the problem of eliminating animal reaction is to reversibly inactivate the active enzyme molecules in such a manner that the enzyme will be reversibly inactivated at the time of injection, and will not be reactivated between injection and slaughter, or at least will be reactivated at a slow rate that can be tolerated by the animal. An accepted method was developed for a quantitative determination as to the enzyme activity of active molecules and as to the total enzyme activity, which is made up of both the active enzyme molecules and the reversibly inactive enzyme molecules.

Essentially, the amount of active enzyme molecules in a given solution is determined by employing a milk clot assay of the available enzyme activity. The units for measuring active enzyme molecules are generally called nonreduced milk clot units (NRMCU), and are sometimes referred to as the available milk clot activity (AMCU). Where a measure of the reversibly inactive enzyme molecules is desired, a second portion of the enzyme solution is treated with an activator or reducing agent such as cysteine, cyanide or bisulfite, whereby all the reversibly inactivated enzyme is restored to the active state.

The milk clot assay using added cysteine or other activators measures both the active and the reversibly inactive enzyme molecules. This measurement is referred to as either Reduced Milk Clot Units (RMCU) or Total Milk Clot Activity Units (TMCU). Any reduction in this value during processing indicates a loss in the overall enzyme activity of the solution being assayed. However, reduction in the NRMCU without a corresponding reduction in the RMCU value, indicates the amount of active enzyme molecules which have been converted to reversibly inactivated molecules.

As has been previously mentioned, it was formerly believed that adverse animal reactions upon injection of proteolytic enzymes was due in part to impurities caused by enzyme autolysis and to the presence of impurities in the enzyme formulation, and specifically nonenzyme proteins and low molecular weight compounds. It was found, however, that with proteolytic enzymes, the animal reaction is apparently caused solely by the active enzyme per se. That is, when the enzyme is purified and then activated, it causes severe reaction when injected into animals. Even though most of the nonenzyme protein and low molecular weight compounds have been removed, the purified active enzyme causes as severe a reaction as an equivalent amount of crude enzyme.

Thus, no amount of purification of an active enzyme will eliminate animal reaction. In view of these findings, it became apparent that to obtain truly efficient enzyme formulations for ante-mortem injection, an inactivating reagent must meet all of the following requirements: (a) it must reversibly inactivate proteolytic enzyme, such that the enzyme will be inactive at the time of ante-mortem injection; (b) it must provide a modified enzyme that will reactivate in situ at such a rate that active enzyme will not be released between injection and slaughter so as to cause adverse physiological reactions, but active enzymes for tenderization will be released between slaughter and consumption of the meat; (c) it must not be toxic to humans at the levels that it will be present in the tenderized meat.

The reaction of sulfhydryl proteases with the symmetrical disulfide compounds proceeds in accordance with the following equation:

$$ESH + RSSR \underset{\text{activation}}{\overset{\text{inactivation}}{\rightleftharpoons}} ESSR + RSH$$

where ESH is the active sulfhydryl enzyme, such as papain, bromelain, or ficin; RSSR is the symmetrical disulfide inactivator; ESSR is the reversibly inactivated enzyme (a mixed disulfide of the sulfhydryl enzyme); and RSH is a low molecular weight thiol compound formed during the reaction. The reaction proceeds best at neutral or alkaline pH. Apparently, the proteolytic enzyme (ESH) must be ionized (ES-) before the reaction proceeds. The reaction can be driven to completion by oxidation of the thiol (RSH) to the disulfide form by air or with other oxidizing agents. This type of reaction generally proceeds at low temperatures (ice bath temperature) and occurs with virtually no side reactions. Thus, the yields of reversibly inactive enzyme (ESSR) approach 100%.

The animal should be slaughtered within about 1 hour from injection. If the animal is slaughtered too quickly after injection, the enzyme will not have had a chance to be uniformly circulated throughout the animal's vascular system. On the other hand, if the animal is not slaughtered fairly soon after injection, the tenderization effect of the enzyme will not be achieved. Usually it is preferred to dispatch the animal about 10 to 30 minutes after injection. The process depends on two primary factors for success. First, the enzyme must be in a reversibly inactive state at the time of injection. Secondly, the rate of reactivation by reducing agents (mainly glutathione) in the blood must be suitably slow. Of course, the reactivation rate is in part dependent on the level of glutathione in the blood; the higher the level, the faster the reactivation.

Glutathione level is known to vary from one animal to another, and from one species to the next. Thus, the rate of reactivation would vary accordingly. Several techniques exist for nullifying this blood glutathione factor. As previously mentioned, adding an excess of disulfide inactivator to the enzyme will act to tie up blood glutathione. In addition, injecting oxidizing or disulfide reagents prior to injection of the enzyme makes it possible to lower the activating capacity of the blood, and thus lower animal reaction. Potassium iodate or tetrathionate function as satisfactory reagents for lowering the activating capacity of the blood.

The uniqueness and many advantages of the process are readily apparent from the following considerations. Firstly, the reaction time required to reversibly inactivate the proteolytic enzyme is greatly reduced. Other methods which reversibly inactivate the enzyme by holding the solution at an alkaline pH require from about 12 to 24 hours to achieve inactivation of the enzyme; and many preparations can never be fully inactivated by such treatment. Also, inactivation techniques using hydrogen peroxide, while relatively fast, result in irreversible inactivation of the originally active enzyme fraction.

Secondly, the process gives results in yields of reversibly inactive enzymes approaching 100%. It is practically impossible to obtain these high yields of reversibly inactive enzyme when hydrogen peroxide technique is used, as addition of hydrogen peroxide causes an irreversible loss of activity.

Thirdly, the process results in a more stable enzyme solution. The mixed disulfides (ESSR) produced are resistant to further oxidation and self-digestion (autolysis). Thus, normal enzyme activity losses during enzyme processing (filtering, freezing and storage) are minimized. Storage tests indicate that the enzyme activity of the disulfide-inactivated enzymes is higher than that of companion solutions identical except for absence of disulfide inactivator. Fourth, and probably most important, the modified enzymes obtained are slowly reactivated in the live animal. Animals will tolerate a certain level of active enzyme. If the reactivation rate of the enzyme in the blood is such that this level of active enzyme is not exceeded, an adverse physiological reaction will not materialize. By testing various disulfide-inactivated enzymes with various levels of blood glutathione, one can accurately predict which disulfide inactivator will result in the proper reactivation rate for the particular needs.

This control of reactivation rate of the reversibly inactivated enzyme is unique. Reversible inactivation by other techniques, such as holding at high pH and/or hydrogen peroxide treatment, do not afford any control over the reactivation rate of the enzyme once injected into the animal bloodstream.

Example 1: By definition, a milk clot unit (MCU) is that amount of activity in 1 ml that will cause 5 ml of milk substrate to clot in 60 seconds at 40°C. Thus the formula for calculating the milk clot unit of a properly diluted enzyme solution is as follows where t = clotting time in seconds: MCU/ml = 60/t. Separate assay procedures have been used for the determination of reduced (RMCU) and nonreduced (NRMCU) milk clot units. RMCU values are determined using enzyme and milk substrates, both of which are treated with cysteine-versene solution prior to clotting; NRMCU values are determined in the same manner except that the cysteine-versene treatment is omitted. NRMCU values are a measure of active enzyme only, whereas RMCU values measure both active and reversibly inactive enzyme activity. Examples of three typical assays are given below:

	pH	RMCU	NRMCU
Freshly prepared bromelain*	4.2	220	125
Aged bromelain**	9.0	186	17
Tetrathionate inactivated	9.0	220	0.5

* Only water added and analysis made 10 minutes later.
** pH adjusted to 9.0 and solution held at 15°C for 24 hours.

The data shows that freshly prepared bromelain contains a high level of active enzyme (NRMCU), and that aging at an alkaline pH reversibly inactivates a large portion of the active enzyme. However, the data also shows that aging caused a loss in total activity (RMCU), i.e., part of the active enzyme was irreversibly inactivated. In contrast, the tetrathionate-inactivated enzyme solution lost none of its original activity.

Example 2: In a typical inactivation procedure, 2.5 mmol of inactivator, 8 g of enzyme powder, and 8 g of glycerine were mixed to the consistency of a smooth paste. Separately, a 4 g quantity of tetrasodium ethylenediaminetetraacetic acid (EDTA) was dissolved in approximately 175 ml of ice cold distilled water containing enough 5N NaOH (predetermined) so that after mixing with the enzyme paste the solution will equilibrate to the desired pH. At time zero, the EDTA-NaOH solution was poured into the ice cold enzyme slurry, and stirred. The resulting mixture was diluted to 200 ml, bringing the enzyme powder concentration to 4%. When assayed at intervals, along with solutions of the same enzyme concentration without EDTA, values were obtained as shown in the following table.

NRMCU Values of 4% Bromelain Solutions After Treatment in 2.5 mM Hydroxyethyl Disulfide Solution for Various Time Intervals

	Minutes			
	15	**30**	**60**	**120**
pH 5				
With EDTA	12	12	12	12
Without EDTA	12	12	12	12
pH 6				
With EDTA	12	12	12	12
Without EDTA	12	12	12	12
pH 7				
With EDTA	12	12	9	–
Without EDTA	12	12	12	12
pH 8				
With EDTA	12	10	7	–
Without EDTA	12	12	10	6
pH 9				
With EDTA	3	<1	<1	<1
Without EDTA	9	6	4	3
pH 10				
With EDTA	2	<1	<1	<1
Without EDTA	5	4	3	2

The data illustrates the importance of proper pH and the presence of chelating agents in the efficient inactivation of the active enzyme fraction. Disulfide treatment at an acidic pH or without the presence of a chelating agent results in either no inactivation or very slow inactivation. On the other hand, when the enzyme solution contains a chelating agent, and the pH has been adjusted to from 7 to 10, reversible inactivation by the disulfide reagent takes place quite rapidly.

Example 3: Groups of six sheep per sample were injected with various enzyme preparations at 40 RMCU/lb. They were dispatched by bleeding twenty minutes after injection. Five organs, namely, heart, gall bladder, mediastinal lymph nodes, kidneys and larnyx were rated on a 0 to 10 scale as to severity of reaction. Then the organ rating scores for each animal were totaled and divided by the number of organs to get an average organ score. Another method of evaluating reactions is to calculate the percent of organs in a given group which were rated 5 or above. The average results were as follows.

Treatment of Bromelain	Average Organ Score	Percent of Organs Rated 5 or Above
Noninjection (no enzyme)	1.1	3
Inactivated by aging at pH 9.0	4.3	46
Inactivated with oxidized mercaptoethanol	2.8	24
Inactivated with oxidized thioglycerol	2.0	19
Inactivated with sodium tetrathionate	1.9	10

In all cases the severity of the animal reaction was reduced by prior treatment of the enzyme with the noted disulfide inactivators. It should be remembered that the level used was substantially higher than required for tenderization.

Selective Heat Inactivation

Meat pieces are used in commercial recipe formulations including gravies and vegetables to prepare products which are ultimately heat sterilized or frozen for distribution and consumer use. When either of these procedures is used, problems arise as to the tenderness of the meat in the product. Heating, which is primarily intended to sterilize the product, also cooks it, and this prolonged cooking tends to over-soften the meat. Frozen meat products such as beef pot pies tend to be insufficiently tender and require careful selection of meat or careful handling to provide sufficient tenderness of the meat prior to freezing.

Recently, high-temperature short-time canning procedures have been devised which substantially reduce the period of time needed to sterilize a canned product. As a consequence, the time which the product is subjected to cooking is considerably reduced. This factor gives rise to a distinct problem with regard to the tenderness of the meat component of products such as beef stew, beef stroganoff, beef goulash and beef pot pie. Because of the new shorter heating time, the meat is not adequately tenderized, and if the meat is not tenderized prior to canning, it will be too tough to appeal to the consumer.

However, if the meat pieces are fully cooked prior to canning, there is a problem of the meat pieces breaking up during handling when the can is subjected to agitating-type retorting or during high-temperature short-time sterilization. Additionally, if the meat pieces are completely pretenderized, normal handling can agitate the meat sufficiently to cause the outer layer of the meat to continually shred off.

It is also desirable to develop a procedure for preparing meat pieces to be used in canned meat products which are canned under high-temperature short-time canning methods, wherein the products are subjected to pumping action during processing. Meat pieces should possess properties such that they will be tender enough for consumer acceptance of the product, yet will be durable and stable enough to retain their shape during processing and in the can until ultimate consumption.

W.R. Schack and F.G. Connick; U.S. Patent 3,533,803; October 13, 1970; assigned to Swift & Company developed an enzyme process which gives meat pieces a relatively firm outer surface and a tenderized core. The method involves selective deactivation of the enzyme in the outer layers and activation of the enzyme in the interior. In the process, a proteolytic enzyme is introduced into a primal meat cut, either ante- or post-slaughter using methods such as syringe injection, gas propellant injection, puncturing the surface of the meat and placing the meat cut in contact with an enzyme solution so that the enzyme will permeate into the meat, or any other convenient procedure.

The amount of enzyme to be introduced may vary within a broad range, depending upon the degree of tenderization desired, the type of meat being treated, the initial toughness or tenderness of the meat, and other like considerations. However, it has been found that the amount of enzyme introduced into the meat should be at least about 0.001 mg of enzyme per pound of meat. Once the enzyme has been introduced, an amount of time is allowed to elapse in order to attain distribution of the enzyme throughout the meat. The holding period in this step is variable, and is dependent upon many factors such as the amount of enzyme introduced into the meat, the size of the primal meat cut when the enzyme is injected post-mortem and the degree of dispersion of the enzyme desired. Perhaps the most important factor in determining the holding time when post-mortem techniques are used is the temperature of the meat at the time the enzyme is introduced therein. It has been determined that when the meat temperature is about 80°F the holding time to allow for dispersion of the enzyme may be as short as about one hour or less, but at lower temperatures the holding time is considerably longer. For example, at a meat temperature of about 45°F, the holding time should be about 4 hours, and at temperatures of between about 30° and 40°F the holding time should be about 24 hours.

Notwithstanding the longer holding times required when the meat is at these lower temperatures, it is preferred to work at lower temperatures as a matter of economy since the cost of reducing the temperature of the meat to its optimal temperature for cutting into the desired shaped pieces, i.e., between about 26° and 32°F, will be considerably diminished.

After the enzyme has been allowed to disperse throughout the meat, the meat is cut into pieces. The meat pieces may be any shape desired, such as cubes, chunks, strips, slices, slabs, etc. It has been determined that the dimensions of these cut pieces should be at least ¼ inch per side. As above mentioned, the cutting operation is preferably performed when the temperature of the meat is reduced to between about 26° and 32°F. The next step is to selectively deactivate the enzyme in the outer layer of the meat piece to a sufficient depth so that this outer layer will act as a protective cover for the meat piece.

This deactivation can be accomplished by adjusting the temperature of the meat pieces to a temperature such that the enzyme in the outer layer of meat will be deactivated. This temperature adjustment may be made by any convenient surface heating technique such as by immersing the meat pieces in boiling water or in a gravy. The generally accepted temperatures for proteolytic enzyme activity are from approximately 98° to 185°F with varying specific optimal temperature ranges for specific enzymes, e.g., the optimal temperature range for papain activity is from about 150° to 185°F, the optimal temperature range for bromelain activity is from about 140° to 160°F, the optimum temperature range for ficin is from about 140° to 170°F, and the optimum temperature range for trypsin activity is from about 85° to 115°F. Temperatures in excess of these optimum ranges will cause the enzyme to become deactivated.

It has been found that for the meat pieces to exhibit the required physical characteristics, the protective outer layer of the meat piece in which the enzyme is deactivated without tenderizing the meat should be to a depth of at least about 1/16 inch on each side. Therefore, the temperature to which the meat pieces are subjected should be sufficient to raise the temperature of the entire outer portion of the meat pieces to a depth of at least about 1/16 inch to a degree which will deactivate the enzyme, i.e., a temperature in excess of the optimal temperature for enzyme activity.

The processed meat pieces having dispersed throughout the mass an unactivated proteolytic enzyme, the entire outer surface of each meat piece having enzyme deactivated, may be stored in a frozen condition and subsequently used in food formulations. Ultimately, the enzyme in the interior of the meat pieces is to be activated in order to tenderize the meat pieces. This activation of the interior enzyme may be performed by adjusting the temperature at the core of the meat pieces to the optimal temperature for the tenderizing activity of the particular enzyme. The temperature at the core of the meat piece should be held at this optimal level until sufficient tenderization has been accomplished. The interior of the meat piece should be tenderized to a level of at least about 6 on a tenderization scale wherein 1 represents tough meat and 10 represents mushy meat.

As soon as the interior of the meat has been tenderized to a sufficient degree, the enzyme is deactivated as rapidly as possible without causing a deleterious effect to the surface of the meat. This deactivation may be by any conventional heating technique including microwave heating. Alternatively, the selective deactivation of the enzyme at the surface of the meat and/or the deactivation of the enzyme at the core of the meat piece after the meat piece has been tenderized may be performed by techniques other than heating. Such possible deactivation techniques are controlled treatment of the meat with a chemical inhibitor of enzyme activity such as an oxidizing agent, e.g., hydrogen peroxide, or another suitable chemical inhibitor. Another alternative procedure for deactivating the enzyme is to adjust the pH of the meat surfaces to a level below pH 2.5 which will deactivate the enzyme.

Example 1: Papain (1.81 mg per pound of meat) was added to fresh (utility grade) beef clods by the nitrogen gas propellent method. The clods were held for 20 hours. The temperature of the meat was then adjusted to between 24° to 30°F and the meat diced into approximately 1-inch cubes. These cubes were placed into a boiling water bath for 4 minutes in order to heat the surface (1/16 to 1/8) inch deep to 210°F and also to raise the center temperature of the cube into the temperature range of 150° to 180°F.

The cubes were then removed from the boiling water and placed into 180°F water and held there for 1¼ hours. After this treatment, the cubes were mixed with other particulate food materials in gravy. The mixture was heated by a rapid heating to above 210°F and held for in excess of 15 minutes to deactivate substantially all the enzyme. The product was then chilled to eating temperature of about 130°F. Tenderness scores for the meat pieces indicated about an 8 to 9 rating on a 10 point scale, with 1 being tough and 10 mushy. The outside layer (about the outer 1/16 inch layer) had excellent bite.

Example 2: By the procedure of Example 1, approximately 1-inch beef cubes were prepared. The temperature of the cubes was then adjusted to between 36° to 40°F and the cubes were dipped into a solution containing approximately 100 ppm of hydrogen peroxide to deactivate the enzyme by adding approximately 50 to 90 ppm in the outer layer, not less than 1/16 inch deep into the cube. The dipped cubes were then placed into 180°F water and processed according to the procedure of Example 1.

Pasteurized Canned Meats Method

Bromelin is normally used for tenderizing meats, but it is difficult to control the treatment to avoid overtenderizing and softening of the product. Use of this enzyme for a pork product, namely pasteurized refrigerated canned hams is described by *F.G. Connick and H.F. Bernholdt; U.S. Patent 4,066,790; January 3, 1978; assigned to Swift & Company.*

A proteolytic enzyme that is functional at temperatures not exceeding 150°F and which may be inactivated by exposure to temperatures in the minimum range necessary for destruction of pathogens, is first introduced into meat items. This is done by combining the enzyme with a liquid curing pickle formula which is used to treat the meat as by injection. The meat is sealed in a container and gradually heat processed to an internal pasteurizing temperature of 150° to 185°F for a period of time sufficient to inactivate the enzyme.

The heat processing is carried out in a manner so as to gradually increase the meat temperature over a time period that allows the enzyme to tenderize the meat fibers, and the maximum temperature is maintained for a sufficient time to both destroy pathogenic organisms and to inactivate all of the added enzyme present in the meat. The product is cooled and moved into trade channels for delivery to consumers. The consumers in turn will normally remove the meat from the container and heat and recook it to individual taste and preference conditions.

While papain is the most widely used enzyme for meat tenderization and is sometimes used in combinations with bromelin, it is inappropriate for the present purpose of controlled tenderization as it is not destroyed or inactivated at temperatures normally associated with the processing and reheating or cooking of pasteurized meat items. Papain is active at temperatures of about 120° to 170°F and is temperature inactivated only by prolonged exposure to boiling or higher temperatures.

Thus where papain (and other enzymes having similar functional characteristics, temperature-wise) is added to meat that is canned and pasteurized it will tenderize the meat during the canning process and may remain available and further tenderize the meat during preparation by the consumer. Thus the tenderizing feature cannot be completely controlled by the processor using papain.

The process involves the addition of very small amounts of bromelin to virtually any normal curing pickle formula, such amounts being devised so as to introduce about 0.2 to 4.0 parts per million (ppm) bromelin enzyme in the meat tissue (0.09 mg/lb to 1.8 mg/lb of meat). To illustrate, where the pickle is added to the meat in the amount of 15% of the uncured weight, bromelin may be added in amounts of 2 ppm to about 20 ppm of the pickle formula. An optional step to further improve the uniformity of distribution of curing pickle and enzyme throughout the meat, and thereby enhance uniformity of tenderness, is to massage or agitate the meat after pickle introduction and before canning. This step can be performed on individual meat pieces or, more economically, on several pieces simultaneously.

Virtually any mechanical means capable of working and kneading meat pieces may be utilized; and it is permissible to perform the step with or without the presence of a liquid medium such as additional quantities of curing pickle. For instance, a churn type of device or tumbling equipment having slowly oscillating or rotating paddles may be used to agitate meat pieces for a period of two to three hours. This action is believed to assist distribution of the curing pickle solution uniformly throughout the meat tissue and to possibly mascerate or abrade or otherwise weaken the surfaces of meat tissues (both externally and internally of the meat pieces). At any rate it has been found that such massaging and agitation contribute to the development of tenderness in the final product and has an effect that is additive to the presence of enzyme. Following the curing steps (and massaging if desired) the meat, such as hams, are placed in containers, such as pear-shaped cans, and vacuum-sealed using standard canning practices. The canned products are then heat-processed through normal temperature and time conditions which are usually devised to raise the internal temperature of the product to about 150° to 160°F. For instance 3 lb canned hams may be processed through tanks of water held at 170°F for a period of 120 minutes; whereas 10 lb canned hams require about 285 minutes in 170°F water.

The bromelin distributed throughout the meat will gradually pass through a stage of activity when it will act to tenderize the meat fibers as the temperature of the meat increases through the enzyme activity range. Furthermore, virtually all of the enzyme will become inactivated as the various portions of the meat reach the maximum processing temperature. It will be noted that a temperature gradient will be established as heat penetrates inwardly from the outer portion with the center of the meat being relatively cooler than the outer portion.

Thus if any enzyme is left inactivated in the meat it will be limited to a very small amount confined to a small area in the center. Even this is not likely, however, since the minimum center temperature for USDA requirement is 150°F. The product is cooled and refrigerated and distributed to the consumer for cooking and consumption at which time no significant added enzyme will be present to soften the meat.

Example 1 through 6: Several cases of paired, as to left and right, hams from the same animal were prepared. Three-pound canned hams were processed under identical conditions, excepting the amount of bromelin added to the curing pickle. The basic ratio of pickle formula ingredients was 66.6% saturated brine, 21.25% water, 8.765% liquid dextrose; 3.0% sodium tripolyphosphate, 0.2% sodium erythorbate; 0.125% sodium nitrite and smaller quantities of bromelin. All hams were injected or stitch-pumped to 115% of the uncured weight so as to have the following amounts of bromelin present.

	. . . . . . Bromelin, ppm	. . . .
Sample	**Pickle**	**Meat**
1	30	4-6
2	20	3-4
3	10	1-2
4	5	0.5-1.0
5	2	0.2-0.5
6	0	0

Each ham was canned in three-pound pear-shaped cans and 1 oz liquid gelatin was added to each can after which the cans were closed with a minimum of 20 inches of vacuum, preferably 25 inches. The closed cans were placed in 170°F circulating water cook tank and processed until a minimum of 150°F was reached after approximately 2 hours. The cooked product was chilled in water at 40° to 45°F for approximately 1 hour, or until the internal temperature of the product reached about 80°F or less. The finished product was held in refrigerated storage 35° to 40°F until tested.

Sample hams of each lot were taken from storage for sensory panel testing by a panel of ten persons. All sample hams were heated in ovens at 325°F until an internal temperature of 125°F was reached. The hams were then sliced and examined and eaten by the panel members.

A summary of the rating scale which was used is as follows: (1) hams when sliced fell apart and were considered too soft and tender upon eating; (2) hams when sliced fell apart slightly and were considered somewhat soft and overtender upon eating; (3) hams when sliced were quite firm and considered acceptable for tenderness upon eating; (4) hams when sliced were firm, tender and considered acceptable when eaten; (5) hams when sliced were firm, fairly tender and considered acceptable when eaten; and (6) hams when sliced were firm, relatively tough and only slightly acceptable when eaten.

Panel ratings showed that the hams of lots (3) and (4) containing 0.5 to 2.0 ppm bromelin, were noticeably superior and preferred whereas the hams of lot (5) (0.2 to 0.5 ppm bromelin) were improved over the control lot [lot (6), no bromelin]. The hams of lot (2) (3.0 to 4.0 ppm bromelin), while beginning to show characteristics of overtenderization, were regarded as superior to the control lot. Those hams of lot (1) (4.0 to 6.0 ppm bromelin) were regarded as undesirably overtendered. Thus the examples show improved results at 0.2 to 4.0 ppm bromelin.

Amylose Free/Low pH System

Papain, the proteolytic enzyme meat tenderizer, is most commonly derived commercially from the unripe fruit of the tropical papaw tree (*Carica papaya*). In this form, it contains impurities which have not been considered significant. However, it is important to remove substantially all of the amylase prior to incorporation of the papain into a low pH carrier.

Papain has been used as a meat tenderizer both in dry granular form and in aqueous solutions. However, it has long been recognized that papain gradually loses its proteolytic activity during storage and particularly where the papain is in aqueous solution. It has been found that this loss of proteolytic activity in compositions containing papain is reduced by incorporating buffering agents capable of maintaining the pH level in the range from about pH 5.5 to pH 5.8. While the proteolytic activity of aqueous papain solutions buffered to pH levels in the range above pH 5.5 is satisfactory for some purposes of meat tenderizing, the use of buffered aqueous solutions of this type is subject to serious disadvantages where the object is to provide the ultimate consumer with a product which is acceptable for simultaneously imparting flavor and tenderness to meat products.

These disadvantages have been overcome in the process of *F.E. Metz, F.O. Boyce and J.L. Segmiller; U.S. Patent 3,930,030; December 30, 1975; assigned to H.J. Heinz Company* in which papain is purified to be amylose free. The enzyme is incorporated in a carrier at below pH 5.0 where it is stabilized against loss of proteolytic activity.

The process adapted to a meat tenderizing barbecue sauce consists of the following steps: (a) providing an aqueous mixture of spices and preservative at about 160°F, then, (b) adding tomato paste, sugar, gum-oil and salt to the mixture with agitation, while maintaining the temperature between 175° to 190°F for about 15 minutes, then, (c) milling the mixture in a Rietz mill, then, (d) deaerating and cooling the mixture to about 120°F, and then, (e) adding vinegar, oils and garnishes and amylase-free papain to the mixture with agitation, while maintaining the temperature of the mixture at about 120°F whereby the pH of the mixture is brought to about pH 3.5.

It is thought that the maintenance of the potency and proteolytic activity of the papain results from the substantial absence of amylase and the low pH environment in which the papain is kept in the barbecue sauce. Furthermore, it is thought that upon application to meat, which has a pH in the range of 6.0 to 7.0, the papain migrates from the barbecue sauce at the surface interface with the meat whereupon it enters a higher pH environment in which it can effectively act as a tenderizing agent.

While the exact manner in which the amylase-free papain penetrates into the meat and then uniformly tenderizes the same throughout is not completely understood at this time, it is thought that osmotic pressure and the initial low pH of the carrier sauce and its consistency are the primary contributing factors to such penetration, and that the rise in pH from pH 3.5 to pH 6.5 or higher after application to the meat accounts for the onset of the tenderizing action.

Example: A barbecue sauce with tenderizer was prepared in the following manner. First, the various prebatch operations were conducted. A barbecue mix was prepared by dissolving 950 lb of sugar with 87.5 lb of salt in 250 lb of water. A gum-oil mix was prepared by dispersing 17.5 lb of guar gum in 25 lb of vegetable oil and mixing until the dispersion was free of lumps. A preservative-spice slurry was prepared by dissolving 6.5 lb of preservatives in 250 lb of water and then adding 50 lb of spices and mixing until uniform.

A garnish mix was prepared by adding 25 lb of minced, dehydrated onions to 225 lb of water and boiling for 1 minute. A papain solution was prepared by dissolving 3 lb of papain in 50 lb of water at a temperature of about 60°F. The oil mix used in this example consisted simply of 4 lb of vegetable oil. In the first phase of the batch process, the sweetener mix, preservative-spice slurry and the gum-oil mix were pumped into a first-phase cook kettle and mixed with 100 lb of water and 625 lb of tomato paste. In the process of deaerating the mixture, the temperature was adjusted to about 120°F.

In the second phase of the batch process, 1,175 lb of the first-phase mixture was delivered into a second-phase cook kettle, followed by addition of 275 lb of 100-grain vinegar, 25 g of the oil mix and 250 lb of the garnish mix. This second-phase mixture was then gauged to a volume of 1,775 lb by addition of water and the temperature was adjusted to below 130°F. The papain solution prepared previously was then added and the final gauging and consistency adjustments were made so that 200 gal of the barbecue sauce was obtained. The barbecue sauce product which was thus prepared was then deaerated and filled at between 90° to 95°F into suitable containers.

The papain activity of the barbecue sauce product with tenderizer was then analyzed by preparing several types of meat and meat products, both with and without the barbecue sauce containing tenderizer. Comparisons were made, including taste panel organoleptic evaluation and shear-press texture measurements, in order to determine the tenderness of the meat and meat products. The effectiveness of the barbecue sauce of the process in tenderizing meat, or like protein-containing foods, was clearly observed.

Cured Meats Application

The conventional procedure for preparing corned meat products has included the following steps. First a curing or pickling solution containing salt, sugar, nitrate and nitrite has been pumped into the arterial system of a meat cut or injected deep into a meat cut at many points. After the meat cut has thus been pumped, it has been submerged or covered with a pickling solution. The pumped meat cut has been held in this cover pickle solution for about 7 to 15 days in order to completely cure the product. The resultant corned meat product has usually had a salt content of between 3 and 5.5% before cooking.

In cooking conventionally prepared corned meat products, it has been necessary to stand by to add more water, adjust the heat to keep the water simmering and not boiling. The product prepared by this long simmering cook has not exhibited the true flavor of the corned meat product since flavor has been lost in the cooking water. Furthermore, the long simmering cook has caused considerable dimensional shrink of the product and therefore, reduced the slice yield of the product.

It is desirable to shorten the curing time and to use lower quantities of salt in the pickling solution. Also, it is desirable to eliminate the simmer cooking of the corned meat product to obtain sufficient tenderness. *H.F. Bernholdt; U.S. Patent 3,549,385; December 22, 1970; assigned to Swift & Company* developed a process utilizing active proteolytic enzymes in the brine solution. Improvements are obtained in all conditions and characteristics discussed above. In fact, it is possible to dry roast instead of water cooking the pickled product.

The process requires a brine solution or pumping pickle containing from about 4 to 5% salt, plus sugar, nitrate and nitrite. The brine solution is then mixed with an enzyme solution in a manner such that the resultant brine solution will contain from about 0.0004 to 0.0015% active proteolytic enzyme (based on the total volume of the brine solution). Proteolytic enzymes which have been found to be most effective are papain and bromelain. The enzyme solution is prepared by forming a slurry of equal parts of enzyme and glycerin and then adding salt and dissolving the slurry in water.

Following mixing of the pumping pickle solution and the enzyme solution, the solution is introduced into the meat cut by spray pump injection. Arterial pumping should be avoided since this procedure results in undesirable pocketing of the enzyme resulting in lack of uniformity of tenderness and areas of mushiness. The meat cut should be pumped to a weight gain of from about 15 to 30% with a preferred weight gain of about 20%. With a weight gain greater than 30%, the roasted product will exhibit a tendency toward texture breakdown and hence mushiness.

Example: An enzyme solution having the following formulation of ingredients was prepared.

Ingredients	Percent
Papain*	0.0495
Bromelain**	0.0265
Glycerin	0.0760
Salt	2.0000
Water	97.8480

*Contributes 75% of total enzyme activity.
**Contributes 25% of total enzyme activity.

Ten ounces of the above-mentioned enzyme solution were added to 100 gal of a 25° salometer pumping pickle solution (4.71% salt). The pumping pickle had the following proportions of ingredients.

Ingredients	Percent
Sodium nitrate	0.08
Sodium nitrite	0.08
Raw sugar	2.14
Salt	4.71
Water	92.99

Following mixing of the enzyme solution and the pumping pickle solution, the resulting solution was spray pumped into several meat cuts (tip and plate ends of a beef brisket). The meat cuts were each pumped to approximately a 20% gain in weight. The pumped meat cuts were then drained and placed in a 25° salometer cover pickle solution (4.71% salt). The meat cuts were held in the cover pickle for a period of 7 days. The products were then removed from the cover pickle solution and drained. The resulting products were placed on racks in shallow open pans in an oven and dry roasted at a temperature of 325°F until the products were fork tender (about 3 hours).

For the purpose of comparison, conventionally prepared corned beef briskets were roasted under identical conditions. The roasted corned beef briskets were then randomly allocated to four panel evaluations of tenderness. The first two panels included tip samples, the second two, plate end samples. Both right and left sides of briskets were represented. The results (based on a 1 to 10 scoring scale where 1 equals very tough and 10 equals very tender) are set forth in the following table as mean panel scores.

	Corned Beef	
	Conventional	Enzyme Treated
Tip brisket	6.86	9.05
Plate brisket	6.38	8.09

VEGETABLE PROTEIN ENZYME APPLICATION

Vegetable protein fibers are used to fabricate so-called artificial meats. These vegetable proteins may be divided into two broad classes, i.e., fibrous protein foods and textured protein foods. The products in the former class have a texture like natural meat but require a complex manufacturing process which lowers the nutritive value and are obtained in an inferior yield which renders them costly, while those in the latter class do not have the texture of natural meat.

For making a nonfibrous protein fibrous, there has been adopted a method of extruding a solution of the protein into an acid solution by the aid of a special spinning device. With such method, however, the operation is complicated and the fibers produced are so uniform in shape and in properties that the product does not have the texture of natural meat when tasted. As a result of the extensive studies, it has been found that the treatment of a solution of a nonfibrous protein, suitably adjusted in pH, with a gelating agent and a protease produces a fibrous protein product. It has also been found that the stickiness and toughness of such fibrous protein product are made stable to heat by treatment with a polysaccharide.

K.Y. Kim and N. Yagi; U.S. Patent 3,840,671; October 8, 1974; assigned to Minaminihon Rabuno Kyodo Kabushiki Kaisha, Japan developed this process for the preparation of fibrous protein from a nonfibrous protein starting material. The starting nonfibrous protein is first dissolved in an aqueous medium and adjusted to a pH around 7. When the pH is higher than 9, the protein is unfavorably denatured. When the pH is lower than 6, the protein is dissolved with difficulty so that the gelating agent cannot act uniformly. The concentration of the protein in the solution is normally from 15 to 20% by weight.

The protein is usually dissolved in an aqueous solution of an alkaline substance such as sodium hydroxide, sodium phosphate or ammonia. The resultant dispersion is then treated with a gelating agent and a protease simultaneously or stepwise, usually at a temperature of 50° to 70°C. The gelating agent is normally used in 3 to 10 parts by weight, to 1 part by weight of the protein. The amount of the protease is dependent upon the enzymatic activity.

Ordinarily, the solution is first treated with the gelating agent to make a colloidal solution and the latter is then treated with the protease to make a gel, which releases water to give a fibrous product. Preparation of the colloidal solution is effected under stirring. Then, the protease is added. The formation of the fibrous product is normally completed in 10 to 30 minutes. A longer time of treatment with the protease is not favorable, because decomposition of the fibrous product commences once it is produced. The fibrous product is then subjected to a physical treatment such as stretching whereby orientation is imparted. The temperature under which the physical treatment is carried out is normally around 25° to 30°C. The physical treatment is desirable for accomplishing efficiently the fixation in the subsequent step.

The oriented fibrous product contains the protease and, when allowed to stand as such, undergoes gelatinization due to the enzymatic activity which will destroy the fibrous structure. In order to prevent gelatinization, the inactivation of the remaining protease is necessary. It may be also necessary to modify the side chain of the protein molecule. Thus, the fibrous product is fixed by a conventional procedure usually adopted for fixation of proteins. For instance, the fibrous product is treated with acetic anhydride, usually at a temperature from 0° to 30°C for 3 to 10 minutes whereby the resultant protein product can maintain a stable fibrous structure even when heated.

The use of any acidic solution in place of acetic anhydride is effective for the inactivation of the protease but is not effective for modification of the side chain of the protein molecule. Thus, the stabilization of the fibrous structure to heat cannot be attained. The fibrous protein product is, usually after neutralization of the fixing agent (e.g., acetic anhydride), washed with water and dried. The dried product can be stored stably for a long duration.

The fixed fibrous product is sticky and tough and may be used as such for the production of an artificial meat product. However, the stickiness and toughness are sometimes still unsatisfactory and, particularly when mixtures of the fibrous product and any natural fish or meat material are heated, such properties may be quickly lost. In order to avoid such disadvantages, the fibrous product may be treated with a polysaccharide capable of chemically or physically bonding with proteins.

This treatment for improving the stickiness and toughness on heating is effected by treating the fibrous product with an aqueous solution of a polysaccharide such as carboxymethylcellulose, sodium alginate, carrageenan or guar gum, normally at a temperature from 10° to 30°C for 5 to 30 minutes. The concentration of the polysaccharide in the solution may be usually from 0.5 to 5% by weight. The aqueous solution of the polysaccharide is favored to be absorbed in the fibrous product so as to make the water content of the resultant product to be 50 to 80% by weight. The aqueous solution of the polysaccharide may also include a coloring agent, a seasoning agent, a flavoring agent so that the product which absorbs such solution has an appearance, feeling and taste like natural meat.

A product having thick fibers or a coarse texture can be obtained as desired by using different kinds of nonfibrous proteins and gelating agents and using proteases under different conditions. In addition, by changing the method of cutting off the product, the product may be manufactured in the form of chunks, grains or flakes. Drying will produce a dry product which can readily be converted to a fibrous structure by treatment with hot water at about 60° to 80°C.

Example 1: A solution of casein (70 g) in water (500 ml) is heated at about 50°C and the pH is adjusted to around 7.0 with an aqueous sodium hydroxide solution while stirring. By the addition of calcium chloride (2 g) to the resulting solution under stirring, a white turbid solution is obtained. After the addition of protease (100 mg), the resulting mixture is kept at about 50°C while stirring. Gelation of the protein begins in about 5 minutes, gradually rendering the material fibrous with the release of water. By stretching the resultant material, a fibrous protein is obtained.

The fibrous protein (water content, 70%, 230 g) is treated with a 5% aqueous solution of acetic anhydride (300 ml) at 15°C for 5 minutes to modify the side chain and inactivate the enzyme. The product is then neutralized with an aqueous calcium hydroxide solution, washed with water and squeezed whereby about 200 g of a fibrous protein product (water content, 65%) is obtained.

Example 2: A suspension of casein (70 g) and glycinin (30 g) in water (500 ml) is heated at about 60°C, and the pH is adjusted to about 7.0 with ammonia water while stirring. Protease (150 mg) is added thereto. After 2 or 3 minutes, calcium chloride (5 g) is further added to the resultant mixture whereby the protein begins to gel and finally becomes fibrous. The resultant material is subjected to stretching. The obtained fibrous protein (water content, 70%, 333 g) is treated with a 5% aqueous solution of acetic anhydride (400 ml) as in Example 1. The product is then neutralized, washed with water and squeezed to give about 300 g of a fibrous protein product (water content 67%). In an aqueous suspension of beef flavor, beef paste and a coloring agent, the fibrous protein product is immersed and adjusted to a pH of around 6.6 with a phosphate whereby a beef-like product is obtained.

COLLAGEN PROCESSES

Meat Snack Application

An application of the enzyme treatment of collagen for improved wet burst strength is given by *R.F Autry; U.S. Patent 3,664,849; May 23, 1972; assigned to Devro, Inc.* for the production of a dried sausage snack-type product. The major problems associated with the production of this product are: (1) to develop a previously unavailable edible film which would serve some of the same functions as the casing now does for smoked and dried sausage meat products and (2) to develop a process which enables the processor to mass produce without the high level of hand labor needed for the sausage-type products now on the market, e.g., the variety of dried pepperoni-type sausages now sold in individual wrapped packages for use as snacks.

Initially an effort was made to make a meat snack product from a sausage meat emulsion without any edible film, but the following problems were encountered. (a) rolling the meat emulsion, without any film, caused the meat to stick to the rollers preventing a uniform layer of meat from being extruded and rolled in sheet form; (b) handling a sheet of meat, without any film for structural support, would not be possible on a commercial basis since while one could hand-carry small samples to the smokehouse for processing, any larger quantity fell apart; (c) processing the meat emulsion, without any film, caused the unsupported meat layer to drip through the holding screens, thereby losing all fat and moisture, and yielding a crust-like product which appeared quite unpalatable; (d) cutting this unsupported meat layer tended to cause splitting and fracturing which would create a problem in present types of commercial cutters.

To fill this need, the newly developed edible collagen film was extruded and dried in sheet form. The gel or dispersion used to make this extrusion mixture is made from the corium layer of fresh hide which was treated with a particular type of proteolytic enzyme to reduce the viscosity of the gel in order to give a more uniform extrusion and also to reduce the toughness of the resulting film. The dispersion used also contains a plasticizer and optionally contains one or more of an edible cellulose, a modified soy protein, a preservative, etc. The dispersion is dried to form a collagen film, which preferably is treated with gaseous ammonia to increase its wet burst strength. The clear collagen film resulting from this procedure retained all of the properties needed for a smoked and dried meat product.

The extruded meat emulsion is placed on a moving belt on which the edible film has first been placed. A second roll of the edible film is continually placed over this emulsion and rolling pressure is applied. This results in a continual laminate or blanket or belt of meat emulsion with the edible film adhered to both the top and bottom. Once the blanket or belt or laminate formed of meat emulsion placed between layers of edible film comes off this moving belt, it then goes into the smokehouse.

The process from this point on will be similar to that of any smoked and dried product, the laminates being cured, dried, and smoked in this form. In place of natural smoke, the meat snack product may be smoked using liquid smoke. The process is especially well adapted to be used with continuous driers, which will replace the present smokehouses. Such driers would likely use liquid smoke, if smoke taste is desired.

Once out of the smokehouse or drier, these laminates can be mechanically cut into candy-bar-size meat snacks and packaged by any standard candy bar overwrapper. After cutting, the meat bars will have the collagen film adhering to and bonded to both the top and bottom portions, but there will be no film on the sides. Tests have confirmed the results of greatly reducing hand labor in going from the sausage-type operation to this mechanical one. A second advantage is that because of the lessened thickness of the product (which is not critical and can vary as desired, with from between ⅛" to ½" thicknesses giving the most ease of handling) and the resultant large surface area, this product can pick up the required level of smoke as well as dry to the desired level necessary for shelf stability in a reduced time over a dry sausage-type operation. Present practice is to reduce the moisture content of dried sausage products below 20% by weight of moisture content to have a 90-day shelf life.

One typical emulsion made for use in the meat bar snack product would be a pepperoni type. This is prepared as follows. Chop two parts chuck tenders to one part navels. While chopping, the spices are blended in. If starter culture is desired to give the tang to the end product, it should be added at this time. The temperature of this emulsion should be kept cold (30° to 40°F) to prevent excess smearing of the fat. A typical formulation for such product is 120 lb chuck tenders, 60 lb navels, 1,700 g salt, 1,000 g dextrose, 250 g black pepper, 100 g red pepper, 90 g mustard, 90 g coriander, 70 g nutmeg, 50 g garlic, 25 g sodium erythorbate, and 100 g cure and 100 g starter culture.

The procedure on a bench-top preparation scale is as follows. Take the above meat emulsion and place it on a sheet of edible collagen film of the type previously discussed, which is about 1 mil thick and about 6 inches wide. Place a second sheet of the edible film over the emulsion and roll with a rolling pin to a given thickness such as ¼ inch. This laminated sheet can then be handled like a blanket. The sheet is then placed in a smokehouse or dried on wire sheets. If a starter culture is used, the initial part of the drying cycle is at 105°F with high humidity. This allows the culture to grow and produce the acid necessary for the tangy taste associated with a pepperoni stick.

If no starter culture is needed, higher temperatures and lower humidity conditions can be used which result in higher drying rates. At some point during the drying or smoking cycle, a temperature high enough to kill the naturally occurring organisms present in the meat emulsion should be reached. A temperature (internal meat temperature) of about 150°F for 30 minutes has been found operative for this purpose. With a product containing a starter culture and which is about ⅜-inch thick, a 23-hour drying cycle was used, in which temperatures varied from 100° to 160°F. With no starter culture used, a laminate 3/16-inch thick was dried in about an 8-hour drying cycle, with most of the drying cycle being at low humidity and higher temperatures.

The process utilizing the edible collagen film made by the enzyme process gives the ability of handling a thin sheet of ground meat as easily as a blanket. In addition, this sheet allows drying to take place and prevents any hard protein shell from forming on the surface with the undesired associated discoloration. The edible collagen film used is also permeable to smoke, permitting a smoked product if this is desired. The resultant meat bar snack product has a much greater shelf life than a similar product without the film on either side would have due to the fact that the film prevents the fat from migrating out and also acts as a barrier for mold growth.

Collagen Protease Treatment

Collagen found in the corium layer of animal skins can be converted by mechanical disintegration and by the swelling action of acids into a plastic mass or gel or dispersion, which usually is extruded in a tubular shape and treated in various ways to become the familiar collagen casings used for sausages; and it is also known that the plastic mass, gel, or dispersion, instead of being extruded in tubular shape, can be cast and dried to form a transparent film. These dispersions, however, are necessarily of very low solids content because of their high intrinsic viscosities. Thus, in order to obtain dry films of any workable thickness, a very tedious process of repetitively casting and drying layer upon layer of wet film is required. Thus the acid-swollen collagen dispersions of the type used to make casings, which usually have a solids content of about 3.5% collagen, are not practical for making continuous strong films.

A.T. Miller; U.S. Patent 3,664,844; May 23, 1972; assigned to Devro, Inc. uses high solids, gels or dispersions of acid-swollen collagen, treated with fungal protease enzymes derived from *Aspergillus flavus-oryzae* or *Aspergillus oryzae* to prepare films with improved wet strength.

Example 1: Preparation of Collagen Dispersion and Film – The corium of fresh steer hides is ground in a meat grinder until it will pass a ¼-inch screen. To 1,600 g of the ground hide corium (dry hide solids) is added 11,379 g water, 16 g Rhozyme 41 fungal protease (Rohm & Haas), 3.2 g sodium benzoate and 1.6 g sodium propionate. The slurry of collagen in the aqueous solution is finely ground in a Mikro-cut grinder. After two passes through the Mikro-cut grinder the slurry is mixed with 7,000 g of aqueous solution containing 480 g glycerin and 44 g citric acid.

The mixture is blended by passing it through a meat grinder fitted with a plate having ¼-inch diameter holes. A second pass is made through this grinder using a plate or head having 5/64-inch diameter holes. The swollen mixture is incubated at 32°C for 24 hours. At the end of this 24-hour period, the dispersion or gel is homogenized under a pressure of from about 2,000 to 3,000 psi and deaerated. It is then filtered through a 4-mil slotted filter screen to remove any unswollen particles and foreign material. The homogenized, deareated and filtered dispersion is metered through a metering pump which maintains a constant feed of dispersion under pressure to the extruder inlet.

As the resulting formed sheet of dispersion leaves the extruder, it is immediately cast onto an endless belt. The formed sheet of dispersion is conveyed into a warm air drier. The drier used is one built in two sections, an upper plenum for delivery of heated air and a lower drying chamber directly below the plenum, which is the part through which the endless belt actually passes. As the formed sheet of dispersion moves through the drying chamber and becomes partially dehydrated, it begins to take on the properties of a continuous flexible film. At the tail end of the drying chamber the dehydrated dried film has a moisture content of from 7 to 10%. At this point the film has enough strength and continuity to be released from the casting surface.

Extruded, formed sheet dispersions of approximately 8.0 mil thickness and containing 7.5 to 8.0% hide corium solids will generally dehydrate to yield films of approximately 1.0 mil thickness. The films are clear and transparent.

The dry tensile strength of the films is at least about 10,000 to 15,000 psi and the elongation about 13 to 17%. The dry burst strength is about 6 to 10 psi.

Example 2: Preparation of Soy Protein Film – To 1,600 g ground hide corium (dry solids) is added 10,280 g water, 100 g modified soy protein (Gunther D-100, Gunther Products, Inc.), 15.0 g Rhozyme P-11 concentrate, 3.2 g sodium benzoate, and 1.6 g sodium propionate. After two passes through the Mikro-cut grinder the slurry is mixed with 8,000 g of aqueous solution containing 480 g glycerin and 44 g citric acid. The mixture is blended through the meat grinder, one pass using a plate having ¼ inch diameter holes, and a second pass using a plate having 5⁄64 inch diameter holes. The swollen mixture is incubated at 32°C for 24 hours and then homogenized under 3,000 psi pressure.

After deaeration and filtering, the dispersion is extruded onto the casting belt and conveyed to the drier. One-half of the total quantity of the above dispersion was used to make a film which was dried and wound on a core as described in Example 1 without any ammonia treatment. The other half of the total quantity of the above dispersion was used to make a film which was treated with ammonia vapor. The following data illustrates the dramatic improvement in wet strength noted with soy protein containing collagen films of this example treated with gaseous ammonia over similar films of this example which were not treated with ammonia.

	Untreated	Ammonia Treated
pH	5.0	6.5
Wet burst strength	1.52	20.3

The ammonia-treated films survived very adverse handling conditions in preparation of meat emulsion products. Similar treatment with the untreated film resulted in breakage and rupturing after contact with the wet emulsion. Nevertheless both the untreated and ammonia-treated films of this example were found to possess improved moisture resistance and wet burst strength over films prepared from dispersions which did not contain modified soy protein. In addition superior adhesion of the film to the meat emulsion was demonstrated when films of this example were used to make meat bar snack products, and compared with similar meat bar snack products made with the film of the other example.

Tenderization Process

In a similar process *T. Tsuzuki and E.R. Lieberman; U.S. Patent 3,681,093; August 1, 1972; assigned to Devro, Inc.* use enzymes extracted from the fermentation products of *Aspergillus niger* to partially solubilize the collagen for improved dispersions before drying, stronger and more tender films after drying.

The process for preparing an extrudable dispersion of collagen fibrils includes the first steps of grinding and comminuting the slurry in a comminuting machine, adding acid and allowing the fibrils to swell during an incubation period. The procedures for swelling the collagen by adding an acid mixture of the slurry is modified in that a proteolytic enzyme is added to the collagen before it is swelled by addition of the acid mixture. The dispersion is homogenized and deaerated during the incubation period and filtered before being extruded. The enzyme-treated collagen mass is extruded into a coagulant. This may be a coagulation bath which may be an aqueous ammonium sulfate solution (from 37 to 42% weight/volume) having a pH of between 7.5 and 8.0, the contact time of the casing in the bath being about 8 seconds. Following extrusion, the casing is soaked in a prewash solution and then in a fixing solution (which is an alkaline medium having a pH of about 9.0 or higher) which hardens the casing and inactivates the enzyme. The fixed casing is then washed with chilled water and plasticized in a plasticizing solution. It is then dried and cut into convenient sections, shirred into the desired lengths, and heat-cured.

The enzyme-treated collagen mass may also be extruded using a gas-containing ammonia as the coagulant. When this procedure is used, the ammonia gas coagulant also serves as the alkaline medium having a pH of about 9.0 or higher which inactivates the enzyme as the coagulated casing remains in contact with ammonia gas. The casings made by this process are very similar to those made by the nonenzyme process in terms of their appearance and cooking properties in sausage products but have the advantage of increased tenderness.

In-Process Breaking Strength – After each wet processing step, sections of wet casing are cut from the continuous tube and stressed until they break. The maximum load (in grams) at the breaking point is recorded, the average of the five determinations being recorded as the In Process Breaking Strength.

Hot Acid Breaking Strength – The Hot Acid Breaking Strength is the load (in grams) that is applied to a casing soaked in a hot acid solution at the moment of breaking.

The following examples are illustrative of the invention and should not be regarded as suggesting that the process could not be carried out in other ways. Throughout the specification and examples that follow, all quantities are expressed in parts by weight unless otherwise indicated.

Example 1: Preparation of Control Without Enzyme – The corium of fresh acid-dehaired steer hides is ground in a meat grinder until it will pass a ¼ inch screen. To 22.8 parts of the ground hide corium is added 277.2 parts of water. The mixture is then passed through a comminuting machine to produce a uniform aqueous slurry. A control dispersion is prepared by mixing 100 parts of this aqueous collagen slurry with 100 parts of a dilute acid solution consisting of 1.52 parts of Kraft sulfite process alpha-cellulose, 2.28 parts of lactic acid, and 96.2 parts of water. The control collagen dispersion so obtained has the following composition.

Ingredient	Percent
Hide solids	3.8
Cellulose	0.76
Lactic acid	1.14

This extrusion composition is held at a temperature of 22°C for 20 hours, homogenized at a pressure of 3,000 psi, and deaerated. The fluid mass is then stored at 22°C for an additional 20 hours, filtered through 6 x 8 mil openings, and extruded through an annular orifice into a coagulation bath maintained at 21°C. The coagulation bath is a 40% weight/volume aqueous ammonium sulfate solution (100 liters of solution contains 40 kg of ammonium sulfate) the pH of which is adjusted to 7.5 to 8.0 by the addition of ammonium hydroxide. The extruded casing remains in the coagulation bath for 8 seconds and is passed through a 13% weight/volume prewash solution of ammonium sulfate, the pH of which is adjusted to 7.0 to 7.5 with ammonium hydroxide. The dwell time of the extruded casing in the prewash solution is 17 minutes. The casing then passes through a fixing solution of aqueous ammonium hydroxide maintained at pH 9.0 and 22°C. The dwell time of the casing in the aqueous ammonium hydroxide solution is also 17 minutes. The casing is next washed with water at 10°C for 40 minutes, plasticized in an aqueous solution which contains 4.5% weight/volume glycerin and 1.0% weight/volume carboxymethylcellulose. The plasticizing solution is maintained at 22°C, and the dwell time of the casing in this solution is 8 minutes.

The casing after removal from the plasticizing bath is dried, cut into 30-foot sections, and shirred. Some of the shirred sections of the casing so obtained are heat cured for 24 hours at 80°C. Other shirred sections of this casing are heat cured for 24 hours at 90°C.

Example 2: Test Samples with Enzyme – An extrusion mass is prepared by mixing 100 parts of the aqueous collagen slurry described in Example 1 with 100 parts of a dilute acid solution consisting of 1.52 parts of Kraft sulfite alpha-cellulose, 2.28 parts lactic acid, 0.000152 part of an enzyme produced by *Aspergillus niger* (Proctase), and 96.2 parts of water. The collagen dispersion so obtained has the following composition.

Ingredient	Percent
Hide solids	3.8
Cellulose	0.76
Lactic acid	1.14
Enzyme (*Aspergillus niger*)	0.000076

This extrusion composition is stored at a temperature of 22°C for 20 hours, homogenized at a pressure of 3,000 psi, and deaerated. The fluid mass is then stored at 22°C for an additional 20 hours, filtered through 6 x 8 mil openings, and extruded into a sheet. All samples of the product so obtained are then equilibrated at 70% relative humidity prior to testing. The extrusion characteristics and physical properties of the casing with different levels of enzyme are summarized in the following table. (Sample 1 is the control.)

	Sample 1	Sample 2	Sample 3
Proctase, ppm (based on hide solids)	0	20	100
Extruder pressure, psi	63-80	73	60
In-process wet breaking strength, g, after:			
Coagulant treatment	506	718	614
Prewash	552	628	616
NH_3 hardening	484	514	522
Water-washing	360	394	490
Plasticizer treatment	388	456	580
Cure temperature, °C	80-90	80-90	80-90
Hot acid breaking strength, g	604-874	602-873	332-872

FISH ENZYME DIGESTION

A special process was developed by *M. Rutman; U.S. Patent 3,561,973; February 9, 1971; assigned to Instituto de Fomento Pisquero, Chile* for the production of

FPC (fish protein concentrate). Autolysis can be considered as an enzymatic method, since it allows the enzymes present in the fish to digest the proteins under specific conditions. However, the autolytic process has taken too long (as long as 48 hours) because it was desired to completely break down protein and remove oil, and because of this microbial contamination became a serious problem. It was found that by use of proteolytic enzymes the time for digestion can be shortened and thus rapid multiplication of microorganisms can be prevented. In addition, selected and rapid enzymatic hydrolysis gives a final product that is odorless and has only a slightly bitter off-taste.

The method introduces the lipids in an early stage of the process in order to facilitate the dispersion of the protein concentrate, to form a lipid-protein complex that can be easily dried by conventional methods (i.e., spray-drying, drum-drying, etc.) and to avoid the production of an excessively hygroscopic powder which would be unstable during storage. Because of the lipids, a gross agglomeration of the protein particles is avoided and stability towards decantation is accomplished. The addition of lipids has a further advantage. It is possible to add antioxidants to a stable fat and when this mixture is homogenized with the meat proteins that carry a small fraction of the original unstable fish fat, total protection against oxidation is obtained.

The process is applicable to any type of lean fish. The fresh fish should be eviscerated and cleaned with heads and entrails removed. The fillets are passed through a meat separator machine to separate the meat from the scales, skin, and a large percentage of the undesirable lipids. The comminuted fish meat is then washed with water. Washing accomplishes several important features. It removes the blood and most bacteria, and it reduces the concentration of other compounds which can lead to an off-color and off-flavor product (desired color is white and desired flavor is neutral and, specifically, not fishy). Further, washing removes a significant fraction of the total lipids contained in the fish.

Such lipids are reactive (i.e., lead to spoilage). After washing, the comminuted fish meat is white, odorless and tasteless. This meat slurry is conveyed through a fish bone separator machine and then it is centrifuged to eliminate part of the water. The preparation of a fat-water emulsion can be carried out in parallel with the washing step. There are several reasons for adding suitable fat to the mixture. It is desirable from a dietetic point of view, as it provides a high-energy component to the final product. However, the fat also has several other functions. The added fat and natural lipids remaining in the fish are mutually soluble, but are insoluble in water. By adding an antioxidant to the added fat the entire fat content is rendered stable.

Further, without the fat, it would not be possible to produce a nonhygroscopic material. In order to incorporate the lipids in the reaction mixture it is necessary to emulsify them in water by conventional methods such as colloid mills, with the addition of edible surfactants.

The next step of the process consists in combining the meat slurry with the oil-in-water emulsion. This operation can be done in a colloid mill or in any other machine of equal efficiency which could also mince the fibers. The proportion of volume of emulsion to weight of meat depends, of course, on the percentage of proteins that are contained in the meat and on the percentage of lipids in the emulsion. But one also has to take into consideration the desired percentage of proteins in the final dried powder. In general, if the meat slurry is prepared in the way described above, it will contain 9 to 15% protein. In essence, the method is flexible and in order to calculate the proportion of emulsion volume to meat weight it is enough to specify the relation in weight of lipids/proteins of the final powder. The process has been applied to ratios in weight of lipids/proteins of 1/5 to 1/1 in the final dried product. When dissolved in water to form an emulsion, the product looks and tastes like milk, but is much more nutritious.

Example: Fresh samples of *Merluccius gayi gayi* were eviscerated, beheaded and cleaned with water. The resulting fillets were conveyed through a meat separator machine. The comminuted meat without the skin, scales and bones was washed with water and in this way blood and other materials were eliminated. The clean, white, odorless meat pulp was then passed through a fish bones separator machine and centrifuged to eliminate a fraction of the water. An approximate analysis of this meat pulp gave the following results.

Ingredient	Percent
Water	88
Protein	11
Ash	0.2
Fat	0.4

Ten parts of hydrogenated fish fat was melted at approximately 65°C, a mixture of antioxidants (BHA, BHT and dodecyl gallate) was added thereto, 0.15 part of sorbitan monostearate was dissolved in the melted fat, the whole mixture was stirred for 15 minutes and then poured into 90 parts of hot water that contained 0.35 part of polyoxyethylene (20) sorbitan monostearate. The resulting hot mixture (60°C) was then homogenized in a blender and left overnight. Next day the stable oil-in-water emulsion was separated from the cream of excess fat. Ten parts of bromelain of low activity was mixed with 100 parts of buffer, pH 6.0 (citric acid + sodium citrate, 0.1 M in water), gently stirred and then centrifuged at approximately 700 *g* for 10 minutes in order to settle the undissolved particles.

The bromelain solution was kept at 5°C. One hundred parts of meat pulp, 80 parts of emulsion and 20 parts of water were homogenized in a blender at room temperature for 5 minutes. To the resulting paste, 6 parts of the bromelain solution was added with slow agitation. The temperature of the reaction mixture was slowly raised as shown in the table below using slow agitation (the reaction kettle was kept at 57°C).

Temperature C°	Time, minutes
28	0
36	2
44	4
49	6
52	8
54	10
54	12
54	14

After 14 minutes, the already liquefied mixture was rapidly screened and heated to about 80°C using a heat exchanger and high-speed agitation. The cooking temperature was maintained for 12 minutes which rendered the material substantially sterile and the bromelain inactive. The hot liquid was homogenized in a high-speed blender (5 minutes) and then rapidly cooled to about 5°C (in about 7 minutes). Drying of the material was accomplished in a spray-drier. The product was white in color, stable, nonhygroscopic, odorless and it imparted a very slight bitter taste that could easily be masked. The powder was readily soluble in water and gave an emulsion that did not decant in several days. The final (dry) product had a PER (protein efficiency ratio) of 3.5, casein was 2.6, and had the following approximate composition.

Ingredient	Percent
Protein	65
Fat	27
Moisture	4
Ash	2

SHELLFISH ENZYME METHOD

While some mechanical methods have been attempted for the removal of edible flesh from shellfish and crustacea, usually hand labor is used for flesh removal. Individaul problems which are associated with the particular type of marine creatures must also be given special treatment, for instance shells which are produced from any manual method of shucking retain, to some degree, pieces of flesh attached to the shell and, hence, in a short time, develop an offensive odor due to the actions of microorganisms, necessitating these shells being further treated or rapidly discarded to avoid this problem.

In the case of shrimp, the large intestine often referred to as the vein of the shrimp, must be removed by hand operation to produce a commercial product. This vein removal not only also is again time-consuming but frequently results in a loss of edible flesh as it accompanies the discarded vein.

G.B. Fehmerling; U.S. Patent 3,513,071; May 19, 1970 discusses a synergistic combination of enzymes that attack the connective tissue holding edible flesh to the shell. The connecting tissue between the edible flesh and the shell of shellfish, such as clams, is composed, at least in part, of proteinaceous compounds which include collagen and form a narrow layer of cells between the shell and the true muscles of the shellfish. In the clam, the edible flesh includes the gills, the tongue and the adductor muscles but should not include the belly which is normally attached to the tongue.

It was naturally believed that proteolytic enzymes could alone be utilized to convert the proteins to less complex compounds, which include amino acids and peptides. However, the proteolytic enzymes were found to be wanting in desirable effect due to the complexities of the connecting tissue. It was then discovered that the action of the proteolytic enzymes would be synergistically improved, in addition to which other portions of the connecting tissue would be hydrolyzed, effecting a substantial loosening of the edible flesh in the shell if the proteolytic enzymes were combined with diastatic enzymes.

It is believed that the diasatic enzymes convert the polysaccharides such as chitin or starch into less complex compounds such as dextrin or dextrose. The syner gistic combination of proteolytic enzymes with diastatic enzymes has been further found to be particularly effective when used with crustacea such as shrimp. Shrimp, like other crustacea, have a shell chemically unlike the shell of the shellfish such as the clam or oyster in that the shrimp shell is composed of substantial amounts of chitin and it has been found that the diastatic enzymes hydrolyze the chitin readily so as to substantially loosen the edible portion of the shrimp. At the same time, the effect of the proteolytic enzymes upon the protein of the shell of the shrimp is enhanced to effect a complete breakdown of the shell to expose the edible flesh.

After the enzymes have been in contact with the marine creatures, it will be found that, in the case of clams or other shellfish, the hinge is weakened and partially dissolved along with the thin layer of connective tissues between the shell and the adductor muscles. It is believed that proteolytic enzymes are active for instance in hydrolyzing the collagen present in these connective tissues to substantially loosen the edible portion of the clam which includes the tongue, gill and the adductor muscle so that it may be subsequently removed from the shell.

In the instance that shrimp or other crustacea are subjected to the enzymatic treatment, there is a selective action of the proteolytic enzymes upon the chitin and connective tissue by which the edible flesh is attached to the chitin-containing shell. In this instance, the proteolytic enzymes also attack the protein, as well as the nitrogen atoms contained in the chitin molecule to enhance synergistically the action of the diastatic enzymes in their attack upon the polysaccharide chitin. At the same time, if cellulase is included in the enzymatic treating bath, the cellulose contained in the shell is degradated and hydrolyzed to lower molecular weight saccharides thereby effecting along with the proteolytic and diastatic enzymes a complete collapse and at least a partial distintegration of the shell.

Example: Five pounds of frozen, headed, unpeeled shrimp were thawed in tap water at 58°F. The thawed shrimp were immersed in a solution consisting of the following: 2 gallons water, 7.6 g proteolytic enzyme ficin, 15.2 g diastatic enzyme beta-amylose and 7 g of sodium bicarbonate; the pH of the solution was 7.8 and the temperature was 125°F.

The shrimp were added to a stainless steel vessel equipped with a discharge tube at the bottom of which was attached a stainless steel pump designed to introduce the added enzyme solution in the top and remove it from the bottom to provide circulation of the solution. The vessel containing the shrimp and solution was enclosed in a thermostatically controlled heating oven which maintained the temperature of the solution at 125°F. A vacuum equivalent to 25" of mercury was applied to the oven and maintained at this degree of evacuation for approximately 30 minutes to displace air or gases in the gut cavity of the shrimp and other tissues with the warm enzyme solution.

The vacuum was broken gradually by allowing air to enter the oven through a port regulated by a valve so that, at the end of 120 minutes, normal atmospheric pressure was reestablished. The shrimp were then drained of the enzyme solution and placed in a wire mesh container and jets of water and air, under pressure, were directed against the shrimp. The shells were blown through the wire mesh leaving the whole shrimp tails free of chitinized shell and having the gut vein removed. A final rinse was applied to the shrimp before they were put on a movable conveyor for inspection. The drained, deveined, peeled shrimp were weighed and the weight was 3 lb 15 oz. Samples of the treated shrimp were compared with manually peeled and deveined shrimp by applying a commercial batter and breading material to both and frying in deep fat at 350°F. The color, flavor and odor of both treated and conventional shrimp were judged to be identical by a trained taste panel.

Example 2: 90 lb of live clams were placed in a tank equipped with a flight conveyor which transported the clams through warm water (165°F) in 55 seconds. The partially anesthetized clams were collected in a stainless steel jacketed kettle. The clams were covered with a solution composed of the following: 15 gallons water, 114.0 g (0.25 lb) proteolytic enzyme solution of ficin, 28.0 g (0.062 lb) diastatic enzyme alpha-amylose, 12.5 g (0.027 lb) cellulolytic enzyme system of fungal derivation—cellulase, and 85.0 g citric acid.

Temperature of the solution surrounding the clams was maintained at 130±2°F by circulating thermostatically controlled warm water through the outer jacket of the kettle. The solution was gently circulated around and through the clams by a stainless steel pump which withdrew the liquid from the bottom of the kettle and discharged it through a stainless steel pipe back onto the top surface of the vessel. After 165 minutes the solution was drained from the kettle and the clams were emptied by gravity into a rod reel washer.

The clams were conveyed under sprays of water which washed the contents of the belly free from the remainder of the clam flesh as the flesh was floated and washed free of the shells. The flesh passed through the rods of the washer into a collection basin below the washer and by gravity, over a dewatering grid onto a conveyor belt for final inspection. The shells, free of clam flesh, were conveyed to the end of the washer and discharged onto a disposal conveyor.

The drained clam flesh was weighed and the weight was 13 pounds. This weight is equivalent to the average expected by conventional shucking. The flesh was minced and canned in the conventional manner followed by thermal processing. Comparison was made of the canned product with that from clams shucked manually by experts in this field. The comparison showed that the two products were identical in color, flavor and aroma.

FLAVORINGS

MEAT FLAVORS

Poultry Meat-Egg-Sulfur Digestion

A. Van Delft and C. Giacino; U.S. Patent 4,076,852; February 28, 1978; assigned to International Flavors & Fragrances Inc. use a mixture of enzymatically treated meat powder, poultry egg powder and a sulfur-containing compound to produce a meat- or poultry-like flavor. The relative ratio of powdered meat digest and egg product digest determines the resulting flavor as to meat or chicken.

The enzymatic digest of meat such as enzymatic digest of turkey meat is produced by mixing meat in the form of a powder in aqueous media with an enzyme which has the ability to effect selective cleavage of the peptide bonds located in the molecular structure of the meat protein. The enzyme used in digesting the egg protein are the same as those enzymes used in digesting the meat protein. Examples of such enzymes are as follows: papain, Pilzprotease S (Roehm GmbH), Rhozyme P 11 (Rohm & Haas GmbH), bromelin, pepsin, trypsin, and Protease 62 (Rohm & Haas Inc.), etc. When producing the enzymatic digest the enzyme is mixed with the meat material at a temperature of from 40° up to 75°C in aqueous media for a period of time up to 150 hours.

In producing the enzymatic digest, it is important that proper pH adjustments be made on the enzyme-protein material-water mix, prior to incubation. Suitable pH ranges vary from 3 up to 6 and constant pHs are obtainable by using appropriate buffers, e.g., sodium citrate and sodium diacid phosphate of pH 4.6.

The term "sulfur-containing compounds" is intended to mean sulfur-containing amino acids which have the ability to liberate hydrogen sulfide when reacted, or lower alkyl mercaptans, or lower alkyl sulfides, or lower alkyl disulfides, or amino alkane sulfonic acids or salts, e.g., the sodium salt of amino ethane sulfonic acid (taurine). The term "sulfur-containing compounds" may also include an inorganic sulfur compound or hydrogen sulfide itself, each of which is adapted to liberate a flavor under heating. Such inorganic sulfur compounds are designated as having the formula MS_x wherein M is selected from the group consisting of an alkali metal (e.g., sodium or potassium), an alkaline earth metal (e.g., calcium or magnesium) and ammonium and S_x is selected from the group consisting of sulfide, polysulfide and sulfhydrate.

Cysteine and methionine and their acid salts, such as cysteine hydrochloride, are especially preferred sulfur-containing amino acids. Products resembling natural flavor and having the natural texture and mouthfeel attributable to such natural flavor are produced by means of first producing a papain enzymatic digest of a poultry egg yolk and powdered meat (such as chicken meat), the weight ratio of the enzymatic digest of egg to powdered meat being approximately 2:1, and then heating the mixture of such enzymatic digest and powdered meat with cysteine hydrochloride.

Additional flavor nuances may be introduced into the reaction product by incorporating into the reaction mass prior to or during the heat reaction thiazole derivatives, such as, for example, thiamine hydrochloride and/or 4-methyl-5-(β-hydroxyethyl) thiazole. In addition, free amino acids, such as alanine, leucine, valine, glutamic acid and/or various polypeptides including dipeptides and tripeptides may also be added to the reaction mass prior to or during the heat reaction process. In addition, quantities of nucleotides may be added prior to, during or subsequent to the reaction between the sulfur-containing compound and meat powder or enzymatic digest of egg component.

When producing specific poultry meat flavors, lower alkyl carbonylic materials, such as ketones and alkyl aldehydes can be mixed into the reaction system during or subsequent to the heating step. It is generally preferred to add these materials after heating the reaction mass since they are generally volatile, and losses may occur during the heating. The ketonic materials are lower alkanones and hydroxy-substituted alkanones containing from about 4 to 8 carbon atoms such as diacetyl, acetyl-methylcarbinol and acetylpropionyl. The alkyl aldehydes are n-pentanal, n-hexanal, n-heptanal, 2,4-nonadienal, 2,4-decadienal, and 2,4-undecadienal. These aldehydes and/or ketones are used at levels ranging from 0.01 up to 0.1 times the combined weight of the meat powder or enzymatic digest and egg component or enzymatic digest.

The reaction between the sulfur-containing compound, the meat powder or enzymatic digest and the egg component or enzymatic digest is carried out at reflux conditions (usually over 100°C) at pressures of between 1 and 10 atmospheres. It is to be emphasized that, generally, temperatures above 400°C make the reaction difficult to control and may well produce uncharacteristic burnt notes even with very short reaction times (e.g., of the order of 10 to 20 min). Accordingly, it is preferred to carry out the reaction at temperatures of from 100° up to 350°C, until the desired flavor development is achieved.

Example 1: 1.5 kg of chicken egg yolk powder is dissolved in 5.15 kg of water and the resulting solution is mixed with 7.5 g papain for a period of 2 hours at a temperature of 65°C with stirring. At the end of the two-hour period, the resulting enzymatic digest is mixed with 44 g of cysteine hydrochloride, 70 g of dry dark chicken meat granules and 22 g of thiamine hydrochloride. The resulting mixture is then heated at reflux for a period of 4 hours. To 5 kg of the resulting product 2.22 kg of a 45% aqueous gum arabic solution is added. The resulting mixture is spray-dried to produce a chicken flavor. A chicken noodle soup is prepared by mixing together:

Ingredient	Grams
Salt	5.00
The spray-dried flavor of Example 1	6.00
Gelatin (180 bloom)	1.00
Monosodium glutamate	0.40
Caramel color	0.40
Garlic powder	0.10
White pepper, ground	0.06
Mixed vegetable pieces	36.00

Three cups of water are added to the above mixture; and the mix is brought to a full boil, simmered for 5 minutes, and is then ready to serve.

Example 2: In a pressurized scraped wall reaction vessel the following materials are introduced: 12 kg chicken egg yolk powder, 41.2 kg water and 0.06 kg papain. The reactor is operated for a period of two hours at 65°C after which 0.704 kg of cysteine hydrochloride, 0.704 kg of dry chicken meat granules and 0.352 kg of thiamine hydrochloride is added. The reaction vessel is operated for a period of 30 minutes at a temperature of 130°C and at a pressure in the range of from 2 up to 3 atmospheres. The reaction mass is then cooled yielding a product having a chicken flavor.

Poultry Meat-Sulfur Digestion

A similar process is given also by *L.C. Chhuy and E.A. Day; U.S. Patent 4,081,565; March 28, 1978; assigned to International Flavors & Fragrances Inc.* The process is carried out in the absence of any egg product as required in the previous process (U.S. Patent 4,076,852). The essential ingredients for the process are (1) a meat enzymatic digest, (2) a sulfur compound such as cysteine, taurine, thiamine and/or thiazole, and (3) a monosaccharide and/or disaccharide.

Example 1: Into a 300 ml flask equipped with stirrer, thermometer and heating mantle, the following materials are placed: 50 g ground raw chicken meat, 100 ml water and 0.10 g Mylase enzyme. The pH of the contents of the flask is adjusted to exactly 5.0 by addition thereto of phosphoric acid. The contents of the flask are prepasteurized for 30 minutes at 60°C. The contents of the flask are then incubated at 50°C using agitation (250 rpm) for a period of 120 hours. The resulting enzymatic chicken digest is used in the reaction flavors in the several examples following.

Example 2: The procedure of Example 1 is carried out using the following quantities of chicken meat paste and enzymes except that the pH of each of the mixtures is first adjusted to 4.6 using a buffer which is a one molar solution of sodium citrate and sodium diacid phosphate:

Quantity of Chicken Meat Paste, Grams	Enzyme, (Quantity Used—0.1 g)
100	Prolase
100	Bromelin

(continued)

Quantity of Chicken Meat Paste, Grams	Enzyme, (Quantity Used—0.1 g)
100	Papain
100	Pilzprotease S
75	Rhozyme P 11
75	Bromelin
50	Pepsin
50	Trypsin
50	J-25
50	Protease 62
40	HT proteolytic 200
25	Fungal protease
25	Prolease MT-7820
25	Molsin
20	Mylase
20	Bakterien

Each of the resulting enzymatic chicken digests is used separately in the reaction flavors.

Example 3: 68 g of cysteine hydrochloride, 500 g of water, 30 g of enzymatic chicken meat digest according to the process of Example 1, and 34 g of thiamine hydrochloride are intimately admixed. The resulting slurry is then heated to reflux and maintained at reflux with stirring for a period of four hours. The resulting slurry is filtered yielding 0.51 kg of filtrate. An aliquot of the filtrate is withdrawn and dried. Based on the weight of the dry product obtained, sufficient gum arabic is added to the filtrate to provide a composition containing one part by weight of flavor solids and one part by weight of gum arabic. The composition is then spray-dried to produce a dry product with an intense chicken flavor having a very close resemblance to natural chicken flavor insofar as flavor nuances, mouthfeel and texture are concerned.

Example 4: 44 g of cysteine hydrochloride, 600 g of water, 70 g of enzymatic chicken meat digest produced according to the process of Example 1, and 22 g of thiamine hydrochloride are intimately admixed. The resulting mixture is then heated at reflux for a period of four hours. To a ½ kg of the resulting product 2.22 kg of a 45% aqueous gum arabic solution is added. The resulting mixture is spray-dried to produce a chicken flavor insofar as flavor nuances, mouthfeel and texture are concerned.

Example 5: A chicken noodle soup is prepared by mixing together:

Ingredient	Grams
Salt	5.00
The spray-dried flavor of Example 4	6.00
Gelatin (180 bloom)	1.00
Monosodium glutamate	0.40
Caramel color	0.40
Garlic powder	0.10
White pepper, ground	0.06
Mixed vegetable pieces	36.00

Three cups of water are added to the above mixtures; and the mix is brought to a full boil, simmered for five minutes, and is then ready to serve.

Example 6: 12.75 g of cysteine hydrochloride, 800 g Neobee M-5, 20 g of enzymatic chicken meat digest according to the process of Example 2, and 6.375 g of thiamine hydrochloride are intimately admixed. The resulting mixture is refluxed for a period of four hours with stirring. 0.8 kg of the resulting flavor is then mixed with 0.356 kg of a 45% aqueous gum arabic solution and the resulting mixture is spray-dried yielding 315 g of a product having a flavor closely resembling natural chicken flavor insofar as flavor nuances, mouthfeel and texture are concerned.

Example 7: A chicken noodle soup is prepared by mixing together:

Ingredient	Grams
Salt	5.00
The spray-dried flavor of Example 6	1.00
Gelatin (180 bloom)	1.00
Monosodium glutamate	0.40
Caramel color	0.40
Garlic powder	0.10
White pepper, ground	0.06
Mixed vegetable pieces	36.00

Three cups of water are added to the above mixture; and the mix is brought to a full boil, simmered for five minutes, and is then ready to serve.

Yeast Autolysate-Beef Extract

The natural flavor of beef can be produced as a food flavoring in the process of *R.J. Gasser; U.S. Patent 3,645,753; February 29, 1972; assigned to Societe d'Assistance Technique Pour Produits Nestle, SA, Switzerland.* The process consists of successive steps: (1) heating a mixture of yeast autolysate containing a minimum amount of reducing sugars, an amount of amino nitrogen, hydrolyzed vegetable protein, enzymatically hydrolyzed meat extract and lactic acid; (2) approximately 25 parts by weight of water is added and the mixture is heated to 90°C for 1 to 3 hours; and (3) the pH is adjusted to 6.5 to 7.0 and a 5'-nucleotide is added and heating is continued for 1 to 3 hours.

It has been found that commercial extracts of beef, produced by aqueous extraction, are particularly satisfactory sources of extract and are readily available. The extract may be hydrolyzed by treatment with proteolytic enzymes, for example, papain. For example, when papain is used the extract, diluted to about 15 wt % of solids content, may be heated with papain, in an amount of 5 wt % of the amino nitrogen in the extract, for about 1 hour at about 65°C, followed by heating at 100°C for 5 to 10 minutes to inactivate the enzymes, and evaporation to a dry matter content of about 80 wt %.

After the pH has been adjusted to 6.5 to 7.0, one to two parts by weight of a 5'-nucleotide are added while the temperature is maintained within the range 60° to 90°C. The nucleotide selected should contain at least about 50 wt % of inosinic acid or inosinate, and it may be added as a mixture of equal quantities of sodium inosinate and sodium guanylate. The pH adjustments during the reaction may be made by adding alkali such as sodium hydroxide. The reaction mixture obtained by the process is in the form of a paste containing about 30 to 40 wt % of water.

The yeast autolysate contains from 1 to 2.5 wt % of reducing substances. The term "reducing substances" means substances which reduce Fehling's solution, such as ribose or glucose. Any commercially available yeast autolysate meeting the above specifications may be used; it is used as a paste containing about 80 wt % of dry matter. The hydrolyzed plant protein may be obtained by conventional acid hydrolysis of any plant protein. The aqueous product obtained by hydrolysis is decolorized after filtration by treatment with active charcoal and it may be evaporated to a paste containing about 80 wt % of dry matter.

Example 1: 55 parts of yeast autolysate (about 80% dry matter, 1.8% reducing substances, 7.2% total nitrogen and 3.5% amino nitrogen) are mixed with 45 parts hydrolyzed peanut protein containing about 80% dry matter which has been decolorized with two parts of active charcoal, and 20 parts of enzymatically hydrolyzed beef extract containing about 80% dry matter. The mixture is heated to 90°C for 4 hours, cooled to ambient temperature, 2 parts of lactic acid (containing 90% of free lactic acid) are added and the mixture is stored at ambient temperature for 10 days.

The water content is adjusted to 30%, the pH to 6.3 by addition of sodium hydroxide, and the mixture is heated to 90°C and held at this temperature for 3 hours. The pH is then adjusted to 6.8 (NaOH) and the mixture is again held at 90°C for 3 further hours before being evaporated to a paste having a moisture content of about 16 to 17%. The paste is finally dried in a vacuum oven at 80°C to a final moisture content of about 3%. Alternatively, the final mixture may be diluted to 50% solids and spray-dried.

The dry product is brown in color, and it may be ground to a powder having any desired particle size. It is soluble in hot or cold water, giving a solution having a flavor close to that of a broth prepared from roasted beef. The product also has an enhancing effect on the flavor of various foodstuffs, such as vegetables and mushrooms. The dry product can be stored for at least 1½ years at 20°C without any significant alteration in flavor.

Example 2: The procedure of Example 1 is repeated, except that after adjusting the pH to 6.8, 2 parts of a blend (50/50) of sodium inosinate and sodium guanylate are added while the mixture is maintained at 80°C. The final product is recovered as a paste containing 15% moisture. It may be stored at ambient temperature for at least one year without deterioration.

Milk Protein Digest

K. Jaeggi, V. Krasnobajew, P. Weber and J. Wild; U.S. Patent 4,001,437; Jan. 4, 1977; assigned to Givaudan Corporation describe a pretreatment of a liquid milk product with proteolytic enzymes and/or lactic acid fermentation to degrade the protein and carbohydrate to furnish amino acids and sugars for meat flavor development. These treatments provide the starting materials for a heating reaction which develops meat flavor.

The liquid product obtained after the pretreatment is denoted, for simplicity, as BTMP which is an abbreviation for biologically treated milk product. In the table shown on the following page the method of pretreatment is given for the principle milk products.

Milk Product	Pretreatment
Full-cream milk	Proteolysis, Lactic acid fermentation
Skim milk	Proteolysis, Lactic acid fermentation
Whey	Proteolysis, Lactic acid fermentation
Yogurt	Proteolysis
Buttermilk	Proteolysis, Lactic acid fermentation

The enzymatic proteolysis can be carried out using the usual proteolytically active enzymes or by the action of the enzyme system of proteolytically active microorganisms. The microorganisms should have a certain activity so that the proteolysis is completed after a relatively short time (e.g., after about 7 days). Suitable proteases which can be used are proteases of vegetable origin (e.g., ficin, bromelain and papain), of animal origin (e.g., trypsin and chymotrypsin) or of microbial origin which are capable of clarifying aqueous suspensions of milk products in a short time; that is, which are capable of degrading the milk proteins relatively quickly to peptides and amino acids.

The enzyme concentration can be determined by the conditions given in the following in the time not to exceed 5 to 24 hours. The determination itself is carried out according to methods which are common in the dairy industry; for example, by measurement of the optical density. For this purpose, samples are withdrawn from the mixture at regular intervals, mixed with 15% trichloroacetic acid, the insoluble material centrifuged off and the optical density of the solution measured at 280 mμ. If this has reached a value of about 20/ml (after subtraction of the zero value), this is equated with clarification. The substrate concentrations can be quite high; e.g., in the case of whey it lies at about 5 to about 20 wt %, in yogurt it lies at about 20 to about 50 wt % and in skim milk it lies at about 8 to about 20 wt %. The pH value of the medium lies at about 7, but it can also be carried out in weakly acidic media (e.g., at pH 4) or weakly alkaline media (e.g., at pH 7.5).

The suitable incubation temperature lies in the approximate range of from 30° to 60°C. When commercially available proteases are used, the proteolysis is generally completed after a few hours (e.g., 1 to 20 hours). The mixture is then neutralized (e.g., by means of sodium hydroxide), subjected to heat sterilization (e.g., at 120°C, steam) and worked up (e.g., by spray-drying or lyophilization).

If the proteolysis is carried out by the action of the enzyme system of microorganisms, there can be used the proteolytically highly active, apathogenic species of *Bacillus* (e.g., *B. cereus, B. subtilis, B. licheniformis, B. megaterium,* etc.), *Pseudomonas* (e.g., *Ps. aeruginosa*, etc.), *Streptomyces* (e.g., *S. narensis, S. griseus, S. faecalis,* etc.), *Aspergillus* (e.g., *A. oryzae, A. soyae, A. melleus,* etc.), or *Basidiomycetes* or yeasts (e.g., *Candida lipolytica, Rhodotorula glutinis,* etc.). The lactic acid fermentation can be carried out using any of the bacteria which are suitable in the food industry for lactic acid fermentation. There can thus be used *Lactobacilli* (e.g., *L. casei, L. bulgaricus,* etc.) and *Streptococci* (e.g., *S. lactis,* etc.).

When proteolysis using microorganisms and lactic acid fermentation are carried out simultaneously, measures are taken for the anaerobic lactic acid bacteria to be added to the medium after the proteolysis has first been carried out and fermentation is then carried out without aeration.

Carbohydrate-containing milk products, yogurt, skim milk and whey, are suitable for the manufacture of the aroma substances. Yogurt powder is especially suitable since it is inexpensive and readily obtainable. Being a product of an already completed lactic acid fermentation, yogurt only needs to be subjected to proteolysis. This is carried out by means of proteases. In addition, a yogurt concentrate of up to 50 wt % can be used. Whey is the least suitable insofar as it is a protein-deficient product which first has to be enriched with other proteins (e.g., casein) or a vegetable protein (e.g., soy protein) up to the content of milk or even higher (e.g., to 3 to 6 wt %).

The BTMP is normally a yellowish to brownish powder in the dry form. It can be heated in the form of such a powder or in the form of a concentrate (e.g., a 50% concentrate in water). The desired aroma is formed when the BTMP is heated at the temperatures customarily adopted in the known nonenzymatic browning reaction. For example, the heating can be carried out at temperatures of about 110° to 150°C. The heating time is normally inversely proportional to the temperature.

Cysteine can be used as such or in the form of derivatives or salts which form the free amino acid under the conditions of the process. Apart from cysteine, other amino acids can be added to the BTMP; e.g., small amounts of the well-known amino acids, peptides or related compounds present in meat such as alanine, arginine, cystine, proline, glycine, histidine, isoleucine, leucine, lysine, serine, threonine, valine and anserine.

Example 1: Manufacture of Pretreated Milk Products — (A) 10 liters of 10% sterilized skim milk (10% solids content) are incubated with a 24-hour culture of *Bacillus cereus* (isolated from milk) at 30°C for 24 hours in a shaking culture. After this time, the pH has risen from an original value of 6.3 to 7.5 and clarification has occurred; OD (280 mμ) = 20/ml. Then 1% (v/v) of a 48-hour *Lactobacillus casei* culture is added in order to ferment the lactose. The fermentation is carried out at 31°C without aeration and stirring and takes 3 days. The pH value of the solution now amounts to 4.5. The medium is subsequently brought to pH 7 by the addition of 6 N sodium hydroxide, sterilized at 120°C and lyophilized or spray-dried.

(B) Instead of 10 liters of skim milk, 1 kg of skim milk powder can be added to to 10 liters of water.

(C) In the incubation of the skim milk or suspension of reconstituted skim milk powder with *Bacillus licheniformis,* the time necessary for clarification amounts to 7 days.

Example 2: (A) 1 liter of an enzyme solution containing 13 g of Rhozyme 41 concentrate is added to 130 liters of sterilized or pasteurized 10% skim milk. The mixture is incubated with stirring at 31°C for 20 hours, after which time clarification of the milk occurs. Then, for the lactose fermentation, the mixture is fermented with 1.25 liters of a 48-hour old *Lactobacillus casei* preculture for 3 days at 31°C. At the end of the fermentation, the suspension is neutralized with concentrated sodium hydroxide, sterilized and spray-dried.

(B) Proteolysis and lactic acid fermentation can be carried out simultaneously, which gives a time saving of about 2 days. In this case, protease (Rhozyme) and

Lactobacillus casei are added simultaneously to the milk suspension and incubated at 31°C with stirring for 3 days. The working-up is carried out as described earlier.

Controlled Sugar-Protein Reaction

Reaction products of sugars and protein hydrolysates, classified as nonenzymatic browning reactions, usually contain high amounts of sodium chloride. Such products are undesirable because they cannot be used in salt-free diets. A salt-free, sugar-based protein hydrolyzate flavor enhancer is described by *G.V. Rao and O.B. Gerrish; U.S. Patent 3,952,109; April 20, 1976; assigned to Far-Mar-Co., Inc.* In this process reducing sugars which are formed by enzyme action on liquefied starch, are reacted immediately with hydrolyzed protein to form the hydrolyzate flavor enhancer.

Processes prepare protein hydrolyzates by hydrolyzing protein material with about 6 N HCl, neutralizing the hydrolyzed amino acids with sodium carbonate and concentrating the resulting hydrolyzate. The salt content in a hydrolyzate prepared in this fashion varies from 45 to 55%. In the process, the salt content is eliminated from the hydrolyzate by in situ enzymatic production of reducing sugars from starch followed by relatively low temperature reaction of the in situ formed sugars with the amino acids of the hydrolyzed protein.

The process consists of three steps: (1) preparation of amino acids from protein; (2) liquefaction of starch; and (3) amino acid-liquefied starch reaction under carefully controlled conditions in the presence of amyloglucosidase to form salt-free, sugar-based protein hydrolyzates. The resulting hydrolyzate has enhanced sweetness and flavor, improved keeping qualities and thus increased storage life.

Preparation of amino acids from protein is accomplished by conventional acid hydrolysis using 2 N to 12 N hydrochloric acid, or its equivalent. The normality of the acid for hydrolysis is 4 N to 6 N. Typically, 6 N hydrochloric acid is heated to 110° to 120°C, in a steam jacketed, glass-lined reaction kettle equipped with an agitator. Protein material in an amount, by weight, about 50% greater than the hydrochloric acid is next added to the hot hydrochloric acid and heating is continued under reflux with continuous stirring for 5 to 6 hours. The resulting hydrolyzed protein may be filtered to remove insolubles, primarily humin, and the filtered material discarded.

The solid black humin is formed during acid hydrolysis due to reaction between amino acids and carbohydrates. Thus, an excess of humin formation indicates a loss of amino acids. To suppress excess humin formation, it is recommended and desirable to use a protein component having a low carbohydrate content and a high percentage protein, such as wheat gluten. In addition to minimizing humin formation it is the object of the protein hydrolysis to also minimize chloride ion concentration and polypeptide formation while maximizing free amino nitrogen. It has been found that these objects are best accomplished when the hydrolysis is continued at least until about 80% of the hydrolyzed protein is present as free amino acid.

The hydrolyzed protein pH is very low, i.e., less than about 1.0. Since the subsequent reaction with liquefied starch to produce the hydrolyzates is accomplished at pHs in the range of 3.5 to 5.0, it is desirable to increase the pH of the hydrolyzed protein by removing excess acid. This can be most readily accomplished by pumping the unfiltered hydrolyzed protein, or, if the humin has been removed, the hydrolyzed protein filtrate resulting from the humin filtration, into a glass-lined vacuum kettle and distilling it under vacuum to about one-tenth of its volume, after which the concentrated hydrolyzed protein is diluted with distilled water and the vacuum concentration step is repeated to recover any excess hydrochloric acid in the reaction mixture.

Liquefaction of starch is accomplished using α-amylase to break down the starch. Virtually any starch or combination of starches can be used. For example, readily available starches include wheat, corn, and potato. The starch is slurried in about three times its weight of distilled water and sufficient calcium chloride is added to produce a calcium ion concentration in the slurry ranging from 0.01 to 0.05 molar. Since the calcium ion serves to enhance the activity of the enzyme, and a high concentration of calcium in the system is undesirable, 0.01 molar appears to be the optimum calcium ion level in the slurry. At this point the slurry pH is adjusted to the range 5 to 8 and the enzyme α-amylase added to the slurry.

It has been found that the presence of excess enzyme during liquefaction contributes to undesirable side reactions and darkening of the liquefied starch, and therefore is to be avoided.

The reaction between the amino acid and the liquefied starch occurs only in the presence of an enzyme such as amyloglucosidase. It has been found that the enzyme operates effectively in the pH range from 3.5 to 5.0, but has optimal activity in the pH range 3.5 to 4.0. To raise the pH of the amino acid-liquefied starch mixture to at least 3.5 requires the addition of an alkali, such as sodium hydroxide. Since the added alkali reacts with any residual uncombined HCl to form NaCl, it follows that the higher the pH of the mixture the lower the salt content of the final hydrolyzate product. Where sufficient alkali is added to raise the pH from about 2.8 to 2.9 to about 3.5 to 3.8, the resulting hydrolyzate product contains about 4 to 5% salt.

If it is desired to reduce the salt content of the final product, the protein hydrolyzate following vacuum concentration and following decolorization can be passed through an ion-exchange bed to remove excess HCl and to raise the pH to about 3.0 to 3.2. When the pH 3.0 to 3.2 hydrolyzed protein is mixed with the liquefied starch, the diluting effect of the starch is sufficient to raise the pH into the 3.5 to 4.0 optimum activity range for the amyloglucosidase. Where ion exchange is used to this extent, alkali additions are unnecessary to adjust the pH and the only salt present in the final product comes from the reaction of HCl in the hydrolyzed protein with sodium ion contamination present on the ion exchange bed.

The amyloglucosidase converts the liquefied starch to a reducing sugar in situ, at a relatively slow rate. As the sugar is produced, as long as the temperatures are maintained relatively low, i.e., in the range of 40° to 75°C, the browning reaction of the sugar with the amino acid predominates to form salt-free, sugar-based protein hydrolyzate. However, at higher temperatures, competing polymerization and caramelization reactions involving the reducing sugars tend to become of increased significance with the result that side reaction products having an adverse effect upon hydrolyzate flavor are produced. Thus, it is essential that temperatures be maintained below 75°C. At these low temperatures, reaction time can vary from 30 to 40 hours where the reaction takes place under preferred pH and temperature conditions.

The advantage of in situ production of the reducing sugars is that the relatively slow production rate permits the low temperature favored browning reaction to totally dominate the competing side reactions, such as high temperature favored polymerization and caramelization, in the competition for the produced sugars. If, on the other hand, the reducing sugars were added in massive quantities to the amino acids, even at low temperatures, there would be sufficient sugar present to fuel both the browning and the competing reactions, with the result that the protein hydrolyzate produced by browning would be contaminated by the polymerization and caramelization reaction products which have been found to adversely influence the hydrolyzate flavor.

Example 1: Regular Process – 250 g of 6 N hydrochloric acid was heated in a round bottomed flask to 120°C. 100 g of wheat gluten (74.5% protein) was added and the gluten and acid were heated for 5 hours at 120°C with constant stirring. After completing hydrolysis, the humin formed was filtered. The filtrate (hydrolyzate) was then concentrated to one-third of its volume. After dilution with distilled water the hydrolyzate was again concentrated. The concentrated hydrolyzate was diluted to 250 g with distilled water and decolorized with 1 g of activated carbon by heating at 80°C for 1 hour. The decolorized filtrate was neutralized to pH 5.2 with sodium carbonate. The neutralized wheat gluten hydrolyzate was concentrated and dried to obtain a powdered protein hydrolyzate. Analysis: 9.55 wt % humin formed; 51.41 wt % sodium chloride in the powdered protein hydrolyzate; and 3.65 wt % amino nitrogen in the protein hydrolyzate.

Example 2: New Process – Step 1: 250 g of 6 N hydrochloric acid was heated in a round bottomed flask to 120°C. 100 g of wheat gluten (74.5% protein) was added to the hot acid and the gluten and acid were heated for 5 hours at 120°C with stirring. After hydrolysis, the humin formed was filtered. The filtrate (hydrolyzate) containing in excess of 80% free amino acids based upon the starting material protein was concentrated to one-third its volume to recover excess hydrochloric acid. The resulting concentrate was diluted with distilled water and concentration was repeated. This second concentration step recovered excess hydrochloric acid. The concentrated hydrolyzed protein was diluted with water to 250 g (resulting hydrolyzate pH 1.75) and decolorized with 2 g of activated carbon by heating at 80°C for 1 hour.

Step 2: 111 g of starch was made a slurry in 333 g of distilled water. 0.2 g of calcium chloride was added to the slurry and the pH adjusted to 6.8. 0.25 ml of α-amylase was added and the slurry heated to 85°C with vigorous agitation. The temperature of 85°C and the stirring was maintained and continued for 40 minutes to obtain complete liquefaction of the starch.

Step 3: 200 g of the pH 1.75, decolorized, hydrolyzed wheat gluten from Step 1 was mixed with 200 g of liquefied starch from Step 2 and the pH of the mixture was adjusted to 3.5 by addition of concentrated sodium hydroxide solution. 0.5 ml of amyloglucosidase was added and the gluten-starch-enzyme mixture incubated at 60°C for 72 hours. The resulting protein hydrolyzate was filtered and concentrated at low temperature under vacuum to about 68 wt % solids. Analysis: 0.92 wt % amino nitrogen in the protein hydrolyzate and 8.9 wt % sodium chloride in the protein hydrolyzate. The final product protein hydrolyzate obtained was brown in color and had a beefy flavor with a little sweetness. No flavor or storage problems were observed after storing the product for 2 months at room temperature.

Glutaminase Process

T. Yokotsuka, T. Iwaasa and M. Fujii; U.S. Patent 3,852,479; December 3, 1974; assigned to Kikkoman Shoyu Co., Ltd., Japan show a process for producing a protein hydrolyzate in a one-step operation having a high glutamic acid content. The process uses a glutaminase in combination with a proteolytic enzyme. This enzyme combination prevents the liberation of glutamine by proteolysis by conversion to pyroglutamic acid. The amino acid solution obtained can be used as a seasoning for soups, sauces and protein-containing products.

When the protein is hydrolyzed by the proteolytic enzymes, unnecessary excessive decomposition can be avoided, whereby a product having a good quality and nutritive value can be obtained. However, in that case, the protein is hydrolyzed to amino acids in poor efficiency, and as a result, the glutamic acid content of the hydrolyzate is considerably smaller, as compared with that obtained by the hydrochloric acid hydrolysis. This is a disadvantage of the enzymatic hydrolysis process.

According to one of the hydrolysis methods utilizing an enzyme, a protein-containing raw material to be hydrolyzed is subjected to a denaturation treatment such as heat treatment, etc., as in the conventional process for producing soy sauce, miso (fermented soybean paste), etc. and then inoculated with a microorganism capable of producing a proteolytic enzyme, such as Koji mold, and then mixed with a saline water or water to complete the protein hydrolysis. According to another method propagation of Koji mold mycelia is positively carried out in the hydrolyzate of the raw material, as seen in a paste-like, salt-free moromi (mash).

However, according to the fermentation method usually carried out, that is, the method of denaturing a raw material and adding to the denatured raw material a commercially available proteolytic enzyme preparation (a filtrate of submerged culture of Koji mold or an enzyme preparation prepared by separating the enzyme from the filtrate) to effect the protein hydrolysis, the glutamic acid content in the hydrolyzate, or yield of the glutamic acid, is usually inferior, i.e., the glutamic acid content per total nitrogen of the protein-containing raw material usually has the values shown in the table on the following page.

The glutamic acid content is smaller in Method 2 than in the Method 1, because the glutamic acid in the raw material is liberated in the hydrolyzate as glutamine and almost all of the glutamine is converted to pyroglutamic acid by nonenzymatic reaction, or because the glutaminase, though present in Koji, has no salt resistance and its activity is inhibited, and as a result the glutamine is converted not to glutamic acid, but to pyroglutamic acid.

Glutamic Acid/Total Nitrogen in Protein Hydrolyzate

Hydrolysis Method	Defatted Soybean	Wheat Gluten
1. Hydrochloric acid	1.14	2.36
2. Fermented soy sauce	0.71-0.89	0.86
3. Proteolytic enzyme	0.56-0.69	0.47

The yield of glutamic acid is smaller in Method 3 than in Method 2, because the Koji mold mycelia propagated in the raw material accumulate glutamic acid within the mycelia and the accumulated glutamic acid is later released out of the mycelia by autolysis as free glutamic acid in Method 2, whereby the glutamic acid content is enhanced in the hydrolyzate.

That is, the proteolytic enzyme preparation contains no glutaminase or a very small amount of glutaminase, and consequently the glutamine liberated is converted not to glutamic acid, but to pyroglutamic acid. For example, when the production of proteolytic enzyme is carried out by a Koji culture and a Koji extract solution is used as an enzyme preparation, only 2 to 5% of total glutaminase activity present in the Koji is extracted.

It is possible to convert glutamine liberated from the raw material by hydrolysis to glutamic acid without any conversion to pyroglutamic acid, by adding a glutaminase of a microorganism, which is extracted from the cells or is brought into a state workable upon the glutamine at least outside the cells, to the proteolytic enzyme. That is, a process is provided for hydrolysis of protein, characterized by adding glutaminase together with a proteolytic enzyme to a protein-containing raw material and hydrolyzing the raw material by parallel action of the proteolytic enzyme and the glutaminase. The ratio of glutamic acid to the total nitrogen in the protein hydrolyzate is higher than those of Methods 2 and 3, as shown below in the table.

Hydrolysis Method	Defatted Soybean	Wheat Gluten
4. Adding glutaminase to Koji in fermented soy sauce method	(A) 0.89 (B) 1.15	(A) 0.85 (B) 1.44
5. Adding glutaminase to proteolytic enzyme preparation	(A) 1.23 (B) 1.28	(A) 2.15 (B) 2.11

Note: (A) = Koji mold glutaminase was used.
(B) = *Escherichia coli* glutaminase was used.

Example 1: To 10 g of defatted soybean was added 13 ml of potable water, and after the heat treatment, 500 mg of proteolytic enzyme preparation (a commercially available product of enzyme produced by *Aspergillus oryzae*), 1 g of a glutaminase preparation (acetone powders of mycelia obtained by culturing *Aspergillus sojae* IAM 2665 in a Czapek's medium containing 1% L-glutamine and 0.5% glucose as a source of carbon at 30°C by shaking) and 50 ml of potable water were added. Further, a small amount of toluene was added, and then the hydrolysis was carried out at 37°C for 7 days with shaking. As a control, the hydrolysis was carried out in the same manner as above, using only the proteolytic enzyme without the addition of the glutaminase. The ratio of free glutamic acid to soluble total nitrogen of the hydrolyzate was 1.23 for the case that the glutaminase was added, and 0.69 for the case that no glutaminase was added. The amino acid solution thus obtained could be used as a seasoning or a raw material for others.

Example 2: To 1 g of milk casein were added 50 mg of Pronase as a proteolytic enzyme, 200 mg of acetone-dried powders of cells obtained by culturing *Pseudomonas schuylkilliensis* IAM 1154 (NRRL B-6) in a medium containing meat extract and peptone, as a glutaminase, 5 ml of water and a small amount of toluene. Hydrolysis was carried out at 37°C with shaking for 4 days. The ratio of free glutamic acid to soluble total nitrogen in the hydrolyzate was 1.57 for the case that the glutaminase was added, and 1.01 for the case that no glutaminase was added.

Example 3: To 100 g of wheat gluten were added 500 ml of potable water and 300 mg of a proteolytic enzyme (dried precipitate obtained by culturing *Bacillus badius* FERM-P No. 543 and adding 3 volumes of ethyl alcohol to one volume of the culture filtrate), and hydrolysis was carried out at 65°C with shaking for 5 hours. The amount of amino acid-state nitrogen in the hydrolyzate thus obtained was very small, and it was presumed that the most of the hydrolyzate was in a peptide state. To the hydrolyzate was added 1 g of acetone-dried cells obtained by culturing *Chryptococcus albidus* K-3566 (ATCC 20293), and reaction was carried out at 65°C by stirring for 1 hour. Then, the resulting solution was heated at 100°C for 5 minutes, cooled and filtered.

To the filtrate was added 300 mg of a proteolytic enzyme of the Koji mold (dried precipitate obtained by adding 3 volumes of ethyl alcohol to 1 volume of water extract of Koji culture of *Aspergillus sojae*), and the hydrolysis was carried out at 50°C for 12 hours. As a control, the treatment and hydrolysis were carried out in the same manner as above without the addition of acetone-dried cells of *Chryptococcus albidus* K-3566 (ATCC 20293). The ratio of free glutamic acid to the total nitrogen of the hydrolyzate was 1.58 for the case that the acetone-dried cells of *Chryptococcus albidus* were added, and 0.46 for the control. The amino acid solution thus obtained could be used as a seasoning or a raw material for others.

Special Peptidoglutaminase Process

In a similar process a special peptidoglutaminase (PGase) which has an activity to react with glutaminyl-peptides is used by *M. Kikuchi, K. Sakaguchi and E. Nakano; U.S. Patent 3,857,967; December 31, 1974; assigned to Kikkoman Shoyu Co., Ltd. and Noda Institute for Scientific Research, Japan* to enhance the flavor of protein foods and beverages.

Glutaminase is an enzyme which is able to hydrolyze glutamine to glutamic acid and ammonia. It is well known that the glutaminase is obtained from animals, plants and microorganisms. This enzyme, however, specifically reacts with only free glutamine and is unable to deamidate γ-amide group of glutamine where the glutamine is bonded to peptides. In the production of glutamic acid using glutaminase, therefore, the protein is first decomposed using such enzymes as protease or peptidase to liberate glutamine and then the liberated glutamine is allowed to react with glutaminase to deamidate γ-amide group of the liberated glutamine, thereby to yield the desired glutamic acid.

However, when protein is first hydrolyzed, by aid of the action of protease or peptidase, a free glutamine or a glutamine being bonded to N-terminal of peptide is produced. Such glutamines are liable to form tasteless pyroglutamic acid because γ- and α-positions of glutamine tend to bond cyclically to each other.

The glutamine being bonded to N-terminal of peptide is not subject to deamidation even when glutaminase is present, and forms pyroglutamyl peptides in a state of being bonded to peptide, and hence cannot be converted to glutamic acid. It is possible to use a transglutaminase which reacts with peptide-bonded glutamine, but this enzyme is generally expensive and has weak deamination activity.

Studies conducted on the new enzyme PGase have resulted in the finding that one form of this enzyme reacts with the glutamine located at the carboxyl terminal (C-terminal) of peptide, but does not react with the glutamine located at the amino terminal (N-terminal) or at the internal position of the peptides. Another form of the enzyme does not react with the glutamine located at the carboxyl terminal (C-terminal) of peptide, but reacts with only glutamine located at the amino terminal (N-terminal) of peptide or at the internal position of the peptides. The former enzyme is named peptidoglutaminase-I (PGase-I) and the latter peptidoglutaminase-II (PGase-II).

Example 1: 5 kg of defatted soybean, which had been subjected to water spray at a rate of 125%, was cooked under steam pressure of 1 kg/cm^2 for 40 minutes. The cooked soybean was mixed with 5 kg of roasted and powdered wheat, and soy sauce seed koji was sprayed on the mixture. The culture was carried out for 4 days to obtain 12 kg of koji of excellent quality. The obtained koji was charged in 12 liters of a 24% saline water and the saline koji broth was subjected to fermentation and ripening at about 30°C for 3 months with aeration from time to time.

Separately, strain No. 466 of *Bacillus circulans* was cultured in a liquid medium (pH 7.2) prepared by adding 0.3% of yeast extracts, 1% of peptone, 0.5% of lactose, 0.025% of magnesium sulfate, 0.001% of ferrous sulfate and 0.17% of dipotassium hydrogen phosphate to 400 liters of water, at 30°C, for 16 hours with aeration-agitation. The bacterial cells, in which peptidoglutaminase-I and peptidoglutaminase-II had been accumulated, were ground by means of a sonic wave disintegrator and centrifuged to collect a supernatant solution, i.e., a crude enzyme solution.

To the crude enzyme solution was added gradually a solution containing streptomycin and a sediment of nucleic acid was removed. The solution was further saturated with ammonium sulfate to a degree of 38 to 54% and a sediment fraction was collected. The sediment thus obtained was lyophilized to obtain 4 g of powdery crude enzyme (a mixture of PGase-I and PGase-II). This crude enzyme contained 4,300 units of PGase-I and 3,650 units of PGase-II. The powdery peptidoglutaminase was added in an amount in terms of 2,000 units of PGase-I and 1,800 units of PGase-II to the above-prepared saline koji broth just before being subjected to fermentation and ripening.

After the 3-month ripening at 30°C, the ripened mash was pressed to obtain 23 liters of a soy sauce containing the total nitrogen content of 1.7%, of which the amount of glutamic acid was 18.9 mg/ml. This glutamic acid content represents 10.56% of the product soy sauce taken as the proportion of the resultant glutamic acid nitrogen to the total nitrogen content. In a control run a soy sauce, which had been prepared in the same manner as above but without addition of the peptidoglutaminase, contained glutamic acid nitrogen at 7.69% to the total nitrogen content.

Example 2: 100 g of wheat gluten was charged with tap water to total up to 380 ml, stirred and cooked under a pressure of 1 kg/cm² for 30 minutes. Thereafter, the cooked gluten was charged with 5 g of a proteolytic enzyme (an enzyme commercially available, which is produced from *Aspergillus oryzae* and is partially refined) and 2.4 g of a peptidoglutaminase preparation (the ammonium sulfate precipitated powder obtained in Example 1). The mixture was subjected to hydrolysis at 30°C for 3 days in the presence of toluene, where the pH was 5.0.

After the hydrolysis was over, the reaction mixture was pressed to give 360 ml of a gluten hydrolyzate which was excellent in palatability. The liquid seasoning thus obtained contained 30 mg/ml of glutamic acid, whereas the same seasoning but without addition of the PGase preparation, contained only 8 mg/ml of the same. This gluten hydrolyzate may be usable for the preparation of various seasonings and soups, either as such or after being concentrated or, if necessary, dried and powdered.

Yeast Cell Lytic Enzyme

Yeast extract has previously been prepared from cultured yeast as a raw material by (1) an extraction method, (2) a protease decomposition method, (3) an acid decomposition method, and (4) an autolysis method, but none of these conventional methods is completely satisfactory because in methods (1) and (2), the yield is low and a bitter taste develops; in method (3) an undesirable smell of acid decomposition is generated and useful components may be broken down; and in method (4) the control of the process is difficult and the method cannot be applied to heat-dried yeast cells. A strong cell lytic enzyme is used by *K. Ishida, M. Kawai and N. Mukai; U.S. Patent 3,809,780; May 7, 1974; assigned to Kyowa Hakko Kogyo Co., Ltd., Japan* to produce yeast extracts without the deficient characteristics discussed above.

The cell lytic enzymes can be obtained by culturing in a solid medium or a liquid medium strains, such as *Coprinus macrorhizus* f. *microsporus, Coprinus radians, Coprinus micaceus,* etc. or strains such as *Daedaleopsis styracina, Daedaleopsis confragosa, Daedaleopsis* tricolor, and *Daedaleopsis nipponica.*

The raw material yeast used includes all available yeasts such as bakers' yeast cultured according to conventional methods, sulfite pulp yeast cultured in a sulfite pulp waste liquor, the so-called petroleum yeast cultured in a hydrocarbon medium such as normal paraffin, alcohol waste liquor yeast separated from the alcohol fermentation waste liquor, etc. These yeasts can be used in a dried state after culturing and separation, or the culture liquor can be used without further treatment. The decomposition efficiency of the cell lytic enzyme is generally increased when cells killed by heat treatment are used, but the living cells can also be used without any problem.

The yeast concentration, amount of enzyme added, reaction temperature, reaction time, reaction pH, etc., for preparing the yeast extract depend upon the kind of enzyme used and a sampling method of an enzyme source.

When the enzyme obtained from the strains belonging to the genus *Daedaleopsis* or *Irpex* such as *Daedaleopsis styracina, Irpex lacteus,* etc., is used, 0.02 to 2% of enzyme is added to the suspension of yeast cells at 5 to 25% concentration and enzymatic decomposition is carried out at a pH of 2.5 to 3.0 and 45° to 50°C. These decomposition reactions are not generally high, but ethyl acetate, alcohols, etc., can be added and used as preservatives, if necessary.

By the decomposition reaction, the cell walls of yeast cells, polysaccharides and proteins are decomposed, and such sugars as glucose, mannose, galactose, ribose, etc., various palatable amino acids such as glutamic acid and alanine principally, and lower and medium peptides are formed. On the other hand, various vitamins, amino acids, nucleic acid-related substances, organic acids and phosphoric acid compounds contained in the yeast cells or culture liquor are utilized as soluble components. When the yeast is decomposed by *Coprinus macrorhizus* f. *microsporus*, more than 90% of the yeast cells is solubilized and recovered as yeast extract. Further, in the decomposition products, no bitter-tasting substance is formed. It is possible to increase the amount of amino acids formed, by adding a commercially available protease preparation after decomposition by cell lytic enzyme.

Example 1: Coprinus macrorhizus f. *microsporus* (ATCC 20120) was cultured under aeration with stirring at 28°C for 48 hours in 20 liters of a medium consisting of 3% sucrose, 3% of distillers' solubles, 0.5% of yeast cells, 0.02% of magnesium sulfate and 0.5% of potassium primary phosphate (pH 6.0), the cells were filtered off, and a filtrate was obtained. Precipitates obtained by adding acetone to the filtrate to give 75% acetone (v/v) were freeze-dried, whereby 170 g of crude enzyme was obtained.

In a separate vessel, 80 liters of a suspension of sulfite pulp waste liquor yeast, *Candida utilis*, having a 10% concentration and a pH adjusted to 8.5 was heat-treated at 100°C for 10 minutes, and then cooled and adjusted to 55°C. A quantity of 17 g of crude cell lytic enzyme was added and decomposition was conducted at 55°C with stirring for 10 hours. The cells were decomposed, and sugars, amino acids, peptides, etc., were liberated.

After the decomposition liquor was adjusted to a pH of 6.0 with hydrochloric acid, 62 liters of separated liquor was obtained by centrifugal separation. Total solid content of the separated liquor was 118 g/l, and thus more than 90% of the raw material yeast was deemed to be solubilized. The liquor was heated at 60°C for 2 hours, then concentrated to give 8.8 kg of yeast extract in a paste state. General analytical values of the seasoning agent were 7.9% total nitrogen, 4.4% sodium glutamate, 8.2% total sugars, 14.3% ash, 2.8% sodium chloride and 16% water.

A palatability test of a 2% aqueous solution of the yeast extract was conducted by a staff panel of five members. No bitter taste was observed, but a thick, good taste was felt, a meat-like taste and flavor and a good body were recognized.

Example 2: Daedaleopsis styracina (ATCC 20188) was cultured under aeration with stirring for 72 hours in 20 liters of a medium consisting of 3% of sucrose, 3% of distillers' solubles, 0.3% of yeast extract, 0.04% of magnesium sulfate and 0.5% of potassium primary phosphate (pH 4.5), the cells were filtered off and subjected to acetone dehydration according to the conventional method, whereby 400 g of dry cells were obtained.

In a separate vessel, 50 liters of 8% yeast suspension (dry basis) was adjusted to pH 3.0, heat-treated at 100°C for 10 minutes and then cooled. The entire amount of the so-obtained dry cells of *Daedaleopsis styracina* was added to the suspension and the decomposition was carried out at 45°C for 5 hours to dissolve the yeast cells. The pH was then adjusted to 7.2 with caustic soda, and then 8 g of protease enzyme Prozyme was added and decomposition was carried out at 48°C for three hours to form a large quantity of amino acids. Then, the pH was adjusted to 6.0 with hydrochloric acid, and 38 liters of separated liquor was obtained by centrifugal separation. The total solids present in the product liquor was 95 g/l and 90% of the yeast cells was deemed to be solubilized. From this liquor, 20 liters of a liquid seasoning agent was obtained by concentrating under reduced pressure.

ONION FLAVOR–GLUTAMYL TRANSPEPTIDASE

S. Schwimmer; U.S. Patent 3,725,085; April 3, 1973; assigned to the U.S. Secretary of Agriculture uses γ-glutamyl transpeptidase (GGT) to enhance the flavor of allium (onion and garlic) products. The enzyme may be obtained from various sources such as kidney beans, asparagus, mammalian kidney, sprouted onions, etc. It should be noted that the enzyme is present only in sprouted onions, i.e., onions which have just begun to grow. Mature onions do not contain the enzyme in question.

To enhance the flavor of dehydrated onion flakes or powder, the dehydrated material is rehydrated by contact with water and GGT is added to the rehydrated material. The rehydrated onion enriched in flavor can then be used directly for flavoring any desired food product, or, alternatively, it may be preserved for future use by such conventional procedures as dehydration or freezing.

Another plan is to incorporate dehydrated onion with GGT in dry form (e.g., prepared by freeze-drying an isolate of the enzyme from sprouted onions). The resulting composition can be held indefinitely in such condition. It should be realized that as long as the composition is maintained in the dry state, the flavor enhancement is potential rather than actual. However, when the composition is rehydrated for ultimate use, action of GGT occurs and the flavor enhancement is realized.

To enhance the flavor of fresh onion, the bulbs are sliced, chopped, crushed, or otherwise reduced to particles and the resulting mass contacted with GGT. Here the flavor enhancement takes place directly because of the moist conditions. The resulting material may be used as such for flavoring a desired food, or, alternatively, it may be preserved for future use by such conventional procedures as dehydration or freezing.

The mode of action of GGT in enhancing flavor is explained as follows. The characteristic flavor of onion is due to the formation of volatile organic sulfur-containing compounds. These compounds are formed by the action of a particular enzyme on substrates which may be designated as a primary flavor precursor. Both the enzyme and the precursor are naturally present in the onion tissue. The enzyme is a lyase and the primary flavor precursors are S-alkenyl-L-cysteine sulfoxides. The development of flavor may be illustrated by the equation shown on the following page.

$$\text{S-alkenyl-L-cysteine sulfoxide} + H_2O \xrightarrow{\text{lyase}} \text{S-containing flavor compounds} + NH_3 + \text{pyruvic acid}$$

It may be noted that lyase is present not only in fresh allium products but also in dehydrated products because the conditions used in drying do not substantially affect this enzyme and the step of blanching conventionally used prior to dehydration with many fruits and vegetables is not used with allium products.

There also exist in the onion tissue certain substrates which may be termed secondary flavor precursors. More particularly, these are γ-L-glutamyl-S-alkenyl-cysteine sulfoxides. The lyase which is normally present in the tissue is unable to convert the secondary flavor precursors into flavor components so that these precursors are not utilized. It was found, however, that GGT has the ability to convert the secondary flavor precursors into primary flavor precursors. The net effect then is that GGT makes available the primary precursors which can then be converted by lyase into the desired flavor components. The effect of GGT may be illustrated by the following equation:

$$\gamma\text{-L-glutamyl-S-alkenyl-cysteine sulfoxide} + H_2O \xrightarrow{\text{GGT}} \text{L-glutamic acid} + \text{S-alkenyl-L-cysteine sulfoxide}$$

The activity of GGT preparations is expressed in terms of milliunits. One milliunit (mu) of GGT activity is the amount of enzyme which will liberate one nanomol (10^{-9}) mol of p-nitroaniline from L-γ-glutamyl-p-nitroanilide in one minute under standard assay conditions. For example, 230 mu would liberate 230×10^{-9} mol of p-nitroaniline.

The products of various experiments were assayed for flavor by measurement of pyruvic acid. This acid is not itself a flavor component but provides a measure of flavor because when the primary flavor precursors are converted by lyase into flavor components, pyruvic acid is released in chemically equivalent amount. In sum, the determination of pyruvic acid content constitutes a more accurate measure of flavor than could be done by organoleptic appraisals.

Example 1: Isolation of γ-Glutamyl Transpeptidase from Sprouted Onion – Step 1: 1,100 g of chilled sprouted onion bulbs, divested of skin, roots, most of the stems, lateral buds and bladeless leaves were sliced into wedges small enough to pass through the hopper (3 x 5 cm) of a vegetable juicer fitted with a strip of Whatman No. 1 filter paper to cover the perforations in the side of the cylindrical rotatable basket. The combined grinding, centrifugation, and filtration accomplished by this device resulted, after washing the pulp with 100 ml of water, in 890 ml of liquid.

Step 2: The slightly turbid extract was centrifuged at 27,000 *g* for 30 minutes at 0°C. The supernatant liquid was discarded; the precipitate was washed with 179 ml of water and recentrifuged.

Step 3: To this precipitate was added 90 ml of 0.5 M sodium chloride. One hour after thorough mixing to insure complete dispersal, the resulting suspension was centrifiged, and the supernatant was discarded. The precipitate was dispersed in 90 ml of 2 M sodium chloride. After one hour the suspension was again centrifuged and the precipitate was discarded to yield a supernatant liquid.

Step 4: After 10 days at 4°C, the precipitate which developed in this liquid was removed after centrifugation and the supernatant fraction was retained.

Step 5: This fraction, 90 ml, was concentrated at 0°C by means of pressure dialysis for 16 hours against 4 liters of 0.005 M tris-hydrogen chloride buffer (pH 8.0) and centrifuged to remove precipitate to yield 11.5 ml of supernatant liquid.

Step 6: To this liquid was added, at 4°C, 3.4 g ammonium sulfate to achieve 0.5 saturation. After centrifugation at 34,000 *g* for 10 minutes, 4.6 g of ammonium sulfate (0.99 saturation) was added to the supernatant. Centrifugation at 34,000 *g* for 20 minutes and filtration gave the desired γ-glutamyl transpeptidase.

Example 2: Release of Flavor Components by Addition of GGT to Rehydrated Allium Powders – A sample of dehydrated garlic powder was rehydrated and 100 mg was mixed with 230 mu of GGT isolated from sprouted onions, and the mixture was held for 3 hours at room temperature. After this period, 1 ml of 2 N hydrochloric acid was added to inactivate the enzymes. The resulting suspension was diluted to 100 ml with water and the increase in flavor assayed by determination of the increase in pyruvic acid content.

In a second experiment the same procedure as above was followed with the exception that the pH of the mixture of rehydrated garlic and GGT was adjusted to pH 8.1 prior to starting the 3-hour holding period. The pH adjustment was made by adding 0.08 M aqueous tris(hydroxymethyl)aminomethane hydrochloride.

The two experiments described above were repeated, but using different starting materials, namely: recently-prepared dehydrated onion powder (designated onion A) and dehydrated onion powder which had been stored for several months (designated onion B) and dehydrated onion powders prepared from particular portions of the bulb, i.e., inner bulb and outer bulb. The results are summarized in the following table.

	Increase in Pyruvic Acid			
	Buffer		No Buffer	
	Without GGT	With GGT	Without GGT	With GGT
Sample	(μmol/g)			
Garlic	80	192	112	164
Onion A	27	67	23	52
Onion B	13	28	11	20
Onion inner bulb	20	91	*	52
Onion outer bulb	42	53	28	33

*Not determined.

SOY SAUCE PROCESSES

Fermented Whey Extrusion

Saccharomyces cerevisiae in combination with lactase enzyme is used to ferment whey to obtain a fermented type soy sauce as described by *A.J. Luksas; U.S. Patent 3,558,328; January 26, 1971; assigned to Beatrice Foods Co.* In this process a lactase producing organism such as *Saccharomyces fragilis* or *Saccharomyces lactis* can be used. The fluid whey is normally pasteurized before adding the enzyme. Pasteurization can be accomplished under conventional pasteurization conditions, e.g., 145°F for 30 minutes or 161°F for 15 seconds.

Following pasteurization there is added to the fluid whey a lactase enzyme such as Pectinol 42E concentrate. The Pectinol 42E concentrate is added as a powder, 0.1%, of the liquid system. After adding the enzyme the system is inoculated with *Saccharomyces cerevisiae* and the mixture aerated and agitated throughout the fermentation. Fermentation is carried out at 26°C but this can be varied, e.g., from 16° to 40°C. Stirring and agitation are continued for 12 to 24 hours. It is critical that *Saccharomyces cerevisiae* be used since *Saccharomyces fragilis, Saccharomyces lactis* and *Saccharomyces unisporum* by themselves do not produce the characteristic flavor.

As intense flavor develops as a result of the growth of the *Saccharomyces cerevisiae* and the lactase enzyme on the whey. This intense flavor composition can be combined with a nonfermented soy sauce or a hydrolyzed vegetable protein to obtain a fermented oriental type soy sauce flavor. Thus, there can be used 2 parts of fermented whey to 1 part of soy sauce by volume, or 1 part of fermented whey to 1 part of soy sauce by volume or 5 parts of fermented whey to 1 part of soy sauce by volume. As little as 0.5 part of fermented whey can be used with 1.5 parts by volume of soy sauce. There can be added extra salt as desired, e.g., 5% salt on a dry basis. There can be 18% salt by weight in the final liquid product to act as a preservative. The ferment acts as an extender for fermented soy sauce.

Example: Whey was pasteurized at 161°F for 15 seconds. Then there was added 0.1 wt % of Pectinol 42E concentrate (a lactase) and the system was inoculated with *Saccharomyces cerevisiae*. The mixture was continuously agitated strongly for 18 hours and at the same time air was passed into the mixture constantly. An intense flavor developed. To 2 parts of the fermented whey prepared in the example there was added 1 part of soy sauce by volume to obtain a fermented oriental type soy sauce flavor. To the product there was added 18 wt % salt as a preservative.

Fermented Whey-Lactase Combination

A similar process without *Saccharomyces cerevisiae* is described also by *A.J. Luksas; U.S. Patent 3,552,981; January 5, 1971; assigned to Beatrice Foods Co.* This limitation of the process was necessary since when the lactase or lactase-producing organism was used alone, without *Saccharomyces cerevisiae*, the resulting fermented composition exhibited a very disagreeable odor, which was not acceptable for commercial utilization of the produced flavor.

The process involves inoculating whey with a lactase-producing organism such as *Saccharomyces fragilis, Saccharomyces lactis* and *Saccharomyces unisporum*, fermenting the inoculated whey with aeration and stirring to produce the soy sauce flavor (a very disagreeable odor also develops), mixing the fermented whey with the soy sauce flavor developed with hydrolyzed vegetable protein and salt and heating to at least 150°F, whereby the disagreeable odor is completely removed, while the soy sauce flavor is retained.

While the disagreeable odor will be removed almost immediately after reaching 150°F, in order to insure that the odors are substantially completely removed, the mixture should dwell at that temperature, with aeration and vigorous stirring, for up to 10 minutes. However, no dwell time at all is necessary when the mixture is spray dried. The spray drying is simply accomplished in a conventional spray drying tower. When spray drying is used, the dry powder may be reconstituted to the original volume or some other volume, as desired, by mixing with any edible liquid, e.g., water, so that the reconstituted flavor may be mixed in the liquid state with other ingredients.

Example 1: Whey was pasteurized at 161°F for 15 seconds. Then there was added 10,000 organisms per cc of *Sacchromyces fragilis.* The mixture was continuously agitated strongly for 18 hours and at the same time air was passed into the mixture constantly. The mixture had an intense flavor but with a disagreeable odor.

To 60 parts of the mixture was added 35 parts of hydrolyzed soy protein and 5 parts of salt (all parts by weight). The resulting mixture was heated to 160°F with vigorous stirring while passing air therethrough. After dwelling at 160°F for 5 minutes the mixture was cooled. An intense flavor remained without the previous disagreeable odors. To 2 parts of the fermented whey prepared there was added 1 part of soy sauce by volume to obtain a fermented oriental type of soy sauce flavor.

Example 2: Example 1 was repeated except that the mixture was spray dried in a conventional spray dryer with an inlet temperature of 160°F. The dried ferment was reconstituted with water to the volume before spray drying. Two parts of the reconstituted flavor was mixed with 1 part of soy sauce and the same fermented oriental type soy sauce as in Example 1 was obtained. Instead of hydrolyzed soy protein, hydrolyzed corn, maize, wheat, rye, barley, oat, etc., protein may be used.

CITRUS BITTERNESS REDUCTION

Limonoate:NAD Oxidoreductase

It is possible to treat citrus juices and citrus by-products with enzymes to prevent or at least reduce the development of bitterness after juicing or processing. Background approaches and applications of the enzyme process are given by *S. Hasegawa, L.C. Brewster and V.P. Maier; U.S. Patents 3,920,851; November 18, 1975 and 3,917,512; November 4, 1975; both assigned to the U.S. Secretary of Agriculture*.

Limonin is the compound responsible for bitterness in citrus juices. It is formed from the nonbitter precursor, limonoate A-ring lactone. In the following description, the application to navel orange juice is stressed. It is applicable to products and by-products from all kinds of citrus fruits, including oranges, lemons, grapefruit, tangerines, mandarins, limes, tangelos, and citrus hybrids. Such products and by-products include juices, concentrates, purees, pulps, and seed material. Despite the abundance of high-quality navel oranges each year, very little of the fruit is consumed in the form of unblended navel orange juice. This is due to the fact that the juice from navel oranges becomes bitter soon after it is extracted from the fruit.

Early investigators observed that juice from late-season navel oranges tended to have less bitterness than juice from early-season fruit. Unfortunately, the low bitterness levels were reached very late in the harvest season, after most of the crop had been harvested. Other investigators attempted to simulate this natural debittering process by storing early-season navel oranges in warm, moist rooms. Although some debittering was achieved during prolonged storage, this approach had a number of serious drawbacks which prevented its commercialization. These disadvantages included the growth of molds and other microorganisms, the large amount of time required, the development of off-flavors, and the necessity for special storage rooms.

In this process, navel orange juice (or other citrus juice) is treated with an enzyme named limonoate:NAD(P) oxidoreductase. For the sake of brevity, this enzyme is referred to as LOR-P. The mechanism by which bitterness is developed in navel orange juice and the action of LOR-P on this mechanism are explained as follows. The compound responsible for the bitter flavor in citrus juice is limonin. This bitter principle is produced, after the juice is extracted, in a manner represented by the following scheme:

$$\underset{\text{(nonbitter)}}{\text{Limonoate A-Ring Lactone}} \xrightarrow[\text{acids}]{\text{juice}} \underset{\text{(bitter principle)}}{\text{Limonin}}$$

The nonbitter substance, limonoate A-ring lactone, occurs naturally in the fruit tissues where it is stable and remains nonbitter. However, when the fruit tissues are ruptured, as in juice extraction, this normally nonbitter substance is attacked by the juice acids and is converted into limonin. Because of its intense bitterness, only a very small amount of limonin is needed to render the juice unpalatable.

In the process LOR-P acts upon the limonin precursor (limonate A-ring lactone), producing 17-dehydrolimonoate A-ring lactone as the end product. This compound is not bitter and is not convertible to limonin by any of the substances (chemical or biochemical) present in citrus juices. The net result is that the limonin precursor is effectively removed from the juice, whereby development of bitterness is prevented or at least substantially reduced.

LOR-P exhibits the specific ability of forming 17-dehydro derivatives from limonoids. For example, it converts limonoate A-ring lactone into 17-dehydrolimonoate A-ring lactone. It also converts limonoic acid, or its salts, into 17-dehydrolimonoate. The enzyme has optimum activity at a pH of about 8.0.

LOR-P requires for its action the presence of nicotinamide-adenine nucleotide (NAD) or nicotinamide-adenine dinucleotide phosphate (NADP). That is, either NAD or NADP must be present in the reaction mixture as a hydrogen acceptor in order for the enzyme to exert its dehydrogenating action. Consequently, this enzyme is classified as a limonoate:NAD(P) oxidoreductase. It should be noted that it is not essential to add NAD or NADP when orange juice or other citrus products or by-products (except products from seeds) are treated with the enzyme in order to prevent the development of bitterness. There is a sufficient amount of NAD and NADP naturally present in these citrus materials so that the enzyme can act on the bitterness precursor (LARL). However, if desired, either NAD or NADP can be added to expedite the desired reaction. LOR-P exhibits twice as much activity in the presence of NADP as in the presence of NAD.

Another method for increasing the rate of debittering can be explained as follows. Orange juice contains the reduced forms of NAD and NADP, referred to as NADH and NADPH, naturally and as a result of the debittering reaction. These latter materials can be converted enzymatically to NAD and NADP to form a cycle for reuse of the cofactors. Therefore, to expedite debittering one simply adds the proper enzyme, i.e., one which oxidizes NADH and NADPH to NAD and NADP, respectively. One such enzyme is diaphorase.

It has been found that zinc ion increases the activity of LOR-P. Since zinc ion is present in the medium used to prepare the enzyme, this ion is incorporated into the enzyme and it is not required to add any further zinc ion when the enzyme is used to prevent the development of bitterness in citrus juices and the like. However, addition of zinc ion might be used to increase the rate of debittering. LOR-P is further characterized by its stability in acidic solutions. For example, the enzyme retains 90% of its original activity at 23°C when held for 5 minutes at pH 3.5 and 31% of its original activity at 33°C when held for 5 minutes at pH 3.5.

LOR-P is prepared by culturing *Pseudomonas* species 321-18 on a nutrient medium containing limonoate ion. The critical ingredient of the medium is the limonoate ion, and this may be supplied by limonoic acid or its alkali metal salts, the sodium salt being preferred. The other ingredients of the medium are those conventionally used in culturing bacteria.

The enzyme is applied to the juice without delay after its extraction from the fruit. The point is that LOR-P is only effective on the limonin precursor (limonoate A-ring lactone) which has an open D-ring, rather than on limonin itself, in which both A- and D-rings are closed. In freshly-prepared juices, most of the limonin precursor is present as such and thus available for dehydrogenation by the enzyme. Where it is planned to subject the juice to a heat treatment (as in canning or concentration or drying), the process should be applied to the juice prior to such heating operation.

In conducting the debittering treatment the juice may be at its natural pH, or, for faster results, it may be adjusted to a pH of about 5.0 to 8.0. The pH adjustment may be accomplished by adding food-grade sodium hydroxide or other food-grade alkali. Next, LOR-P is added in a quantity sufficient to reduce the limonin precursor content to the level desired. If a completely nonbitter juice is desired, the precursor level should be reduced to about 5 to 8 ppm. In the case of a juice with a high concentration of LARL or where rapid reduction of LARL is desired, more units of LOR-P may be used.

Example: Freshly-prepared navel orange juice was divided into a series of lots. Each lot was treated as follows. No pH adjustment was made; thus each remained at the natural pH of the juice (3.5). A measured quantity of LOR-P was added to each lot together with measured amounts of NADP and NAD. Each mixture was then held at room temperature for a specific time period.

After the above-described enzyme treatment, the various lots and untreated samples (controls) were acidified and boiled for 15 minutes to convert all the limonin precursor to limonin. The limonin content of each was determined by specific thin-layer chromatographic assay. Taking into account the limonin contents of the treated samples and that of the controls, the effectiveness of each treatment was calculated. One unit of LOR-P is defined as the amount which catalyzes the production of 1 μmol of 17-dehydrolimonoate per minute.

Limonin Loss vs Time

Lot No.	pH	Time (hr)	LOR-P Added (u/100 g juice)	NADP Added (μmol/100 g juice)	NAD Added (μmol/100 g juice)	Effectiveness (% Limonin Loss)
Control	3.5	0	–	–	–	0
1	3.5	0.25	6.0	100	100	23
2	3.5	0.50	6.0	100	100	37
3	3.5	0.67	6.0	100	100	42
4	3.5	2.0	6.0	100	100	57

Limonin Loss vs Enzyme and Cofactor Conc.

Lot No.	pH	Time (hr)	LOR-P Added (u/100 g juice)	NADP Added (μmol/100 g juice)	NAD Added (μmol/100 g juice)	Effectiveness (% Limonin Loss)
Control	3.5	–	–	–	–	0
5	3.5	0.25	6.0	0	0	14
6	3.5	0.25	6.0	100	0	23
7	3.5	0.25	6.0	100	100	23
8	3.5	0.25	12.0	200	200	57

Limonoate Dehydrogenase

In another process *S. Hasegawa and L.C. Brewster; U.S. Patent 3,911,103; October 7, 1975; assigned to the U.S. Secretary of Agriculture* use limonoate dehydrogenase (LD) prepared by culturing *Arthrobacter globiformis* on a medium containing sodium limonoate.

Limonin precursor (limonoate A-ring lactone) produces 17-dehydrolimonoate A-ring lactone as the end product. This compound is not bitter and is not convertible to limonin by any of the substances (chemical or biochemical) present in citrus juices. The net result is that the limonin precursor is effectively removed from the juice, whereby development of bitterness is prevented or at least substantially reduced.

It has been found that zinc ion increases the activity of limonoate dehydrogenase. Since zinc ion is present in the medium used to prepare the enzyme, this ion is incorporated into the enzyme and it is not required to add any further zinc ion when the enzyme is used to prevent the development of bitterness in citrus juices and the like. However, addition of zinc ion might be used to increase the rate of debittering.

Limonoate dehydrogenase is further characterized by its greater stability in neutral and alkaline solutions than in acidic solutions. For example, the enzyme retains 50% of its original activity at 46°C when held for 5 minutes at pH 7, whereas no activity is retained at pH 3.5 under these conditions and only 6% of the activity is retained at 23°C at pH 3.5. Limonoate dehydrogenase is prepared by culturing *Arthrobacter globiformis* on a nutrient medium containing limonoate ion. An aqueous nutrient medium containing about from 0.1 to 2% of sodium limonoate is inoculated with a culture of *A. globiformis*. The inoculated medium is then cultured under aerobic conditions for a period of 24 to 96 hours, depending on such factors as temperature, concentration of nutrients, etc. The cells containing the enzyme can be collected by centrifugation, washed, and frozen until used. For further separation the cell cake may be allowed to stand at room temperature to cause autolysis of the cellular material, or the cells can be disrupted by physical means.

On centrifuging the autolysate, the limonoate dehydrogenase activity will then be in the liquid fraction. This liquid fraction may be further purified by conventional techniques used in the purification of enzymes, for example, dialysis, precipitation with salts, column chromatography, etc. The utilization of limonoate dehydrogenase is demonstrated by the following illustrative example.

Assay Method: Limonoate dehydrogenase activity was assayed by following the increase in absorbancy at 340 mμ due to formation of dihydronicotinamide adenine dinucleotide (NADH) from nicotinamide adenine dinucleotide (NAD) added to the reaction system as a hydrogen acceptor. In particular, activity was assayed in 1 ml of a reaction mixture consisting of 10^{-2} M of Na-limonoate, 0.07 M Tris buffer at pH 9.5, 5 x 10^{-4} M of NAD and 1 to 4 mμ of enzyme. The reaction was carried out at 23°C in a standard silica gel cuvette with 1 cm light path. One unit of limonoate dehydrogenase is defined as the amount which catalyzes the production of 1 μmol of 17-dehydrolimonoate per minute under the above conditions.

Example: Debittering of Citrus Juices – Citrus juice is treated with limonoate dehydrogenase (referred to as LD) in order to reduce or eliminate the formation of bitterness in the juice. The enzyme is applied to the juice without delay after its extraction from the fruit. The point is that LD is only effective on the limonin precursor (limonoate A-ring lactone) which has an open D-ring, rather than on limonin itself, in which both A- and D-rings are closed. In freshly prepared juices, most of limonin precursor is present as such and thus available for dehydrogenation by the enzyme. Where it is planned to subject the juice to a heat treatment (as in canning or concentration or drying), the process should be applied to the juice prior to such treating operation.

In conducting the debittering treatment of the process the juice may be at its natural pH, or, for faster results, it may be adjusted to a pH of about 5.5 to 9.5. The pH adjustment may be accomplished by adding food-grade sodium hydroxide or other food-grade alkali.

More complete utilization of LD can be realized by the addition of nicotinamide adenine dinucleotide (NAD) in a quantity equivalent to or some fraction of the molar amounts of limonin precursor in the juice. The addition of NAD is, however, not essential and the debittering proceeds well without its addition. The time required for the debittering treatment depends on such factors as the pH, the amount of LD used and the temperature. More time is required at the lower temperatures. Usually for convenience, the reaction is conducted at room temperature (about 25°C).

In the case of citrus products and by-products already containing limonin (such as ground seed products or heated or stored juice, concentrate, pulp, or peel products) it is necessary to first treat the product with alkali to achieve pH 7 or higher in order to hydrolyze the D-ring of limonin before or simultaneously with the LD debittering treatment.

After the limonin precursor content has been reduced to the level desired (and when the pH of the juice has been initially increased), the juice is acidified to its original pH with an appropriate amount of a food-grade acid, such as citric, acetic, malic, fumaric, hydrochloric, and phosphoric. The use of organic acids, especially citric, is preferred as they do not result in saltiness as do the inorganic acids. It may be desired to balance the salt formed with added sugar.

ENZYME DEACTIVATED MUSTARD

Mustard flour produced from yellow mustard seed sells for a higher price than the mustard flour produced from oriental mustard seed, because of the lower yield (and resulting higher cost per pound) of yellow mustard seed per acre. However, the flour produced from yellow mustard seed, when used as a mayonnaise or processed meat ingredient, is milder in flavor. When used in mayonnaise, a higher viscosity product is obtained. Accordingly, manufacturers of these products primarily use the higher priced flour made from yellow mustard seed. Brown mustard seed is also used to produce mustard flour and prepared mustard products.

When water is added to dry ground mustard flour, an enzyme, myrosin, catalyzes a glucoside, e.g., sinigrin to produce a volatile, pungent oil, principally allyl isothiocyanate. If the enzyme is destroyed before the flour is hydrated, a bland product is produced, which has found use in the food processing industry as a binder and emulsifier, for instance in frankfurters.

Spray dried mustard flour with controlled pungency is prepared in the process of *M.P. Haak; U.S. Patent 4,062,979; December 13, 1977; assigned to McCormick & Company, Inc.* and described as follows. Mustard seed, dehulled or not and fresh or aged, is mixed with water to produce a slurry which is wet milled and spray dried to produce mustard flour. When hot water is used to produce the slurry, an enzyme is deactivated, and the flour contains a characteristic glucoside. When cold water is used to produce the slurry, the enzyme splits the glucoside and produces a pungent volatile oil that is driven off in the spray drying step, but the enzyme survives in the mustard flour. Both mustard flours, alone, are mild in flavor. When the two are blended and mixed with water, the enzyme in one splits the glucoside in the other producing the pungent flavoring oil. Thus, the degree of mildness versus pungency of a mustard product can be predetermined by blending the two flours.

Spray dried mustard flour is an excellent emulsifier for mayonnaise, and may be increased in proportion and egg yolk content reduced, with a wholly satisfactory emulsification, viscosity and flavor being obtained. Alternatively, the proportion of mustard flour may be reduced and egg yolk content maintained when spray dried mustard flour is used, again with wholly satisfactory emulsification, viscosity and flavor being obtained.

Raw mustard seed is withdrawn from storage **10** (Figure 7.1) and blended at **12**. The seed is dehulled at **14** and the dehulled seed and hulls are separated from one another at **16**. The dehulled seed is stored at **18** until it is forwarded to mixing and blending tanks at **20** where water is added to create a slurry. The slurry is pumped to a wet mill **22** where wet grinding takes place. The slurry of wet ground seed is pumped to a homogenizer **24** and, following homogenization, to a spray drier **26**. The dried mustard flour is cooled and sifted at **28**, then dry blended and packaged at **30, 32**.

Figure 7.1: Preparation of Spray Dried Mustard Flour

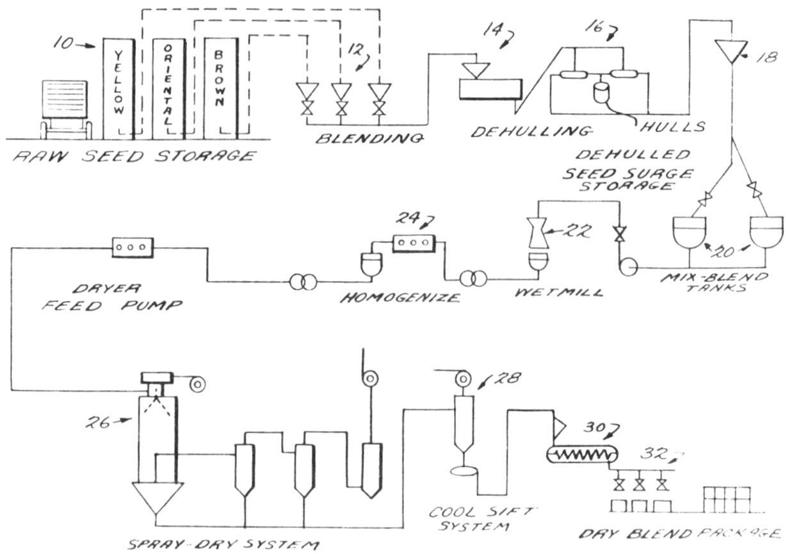

Source: U.S. Patent 4,062,979

A typical system, run continuously, can process a seed from storage at **10** into spray dried flour and to packaging at **32** in ten minutes. The flow chart depicts several instances where blending may occur. For instance, the blending step at **12** is used to blend fresh seed with aged seed, and/or to blend yellow, brown and/or oriental seed together in any desired ratio. If only one kind of mustard seed is being used, this step is not needed. If desired, different kinds of seed may be withdrawn from storage at **10** and dehulled one kind at a time, with blending of the dehulled seed occurring at **20**.

Similarly, although one processing line is shown extending from the slurry formation at **20** to the dry blending step at **30**, the portion from **20** to **26** or **28** may be duplicated to provide separate lines for hot water slurry processing and cold water slurry processing. In one such variation, the output of the two wet grinders **22** and homogenizers **24** may feed the sprayers of the same spray drying tower. In such a case the blending step shown conducted at **30** is not needed, since blending is actually conducted at **26** as the particles intermix in the dryer. A similar result is achieved if the parallel slurry feed lines combine through a proportioning valve just before the spray dryer.

Example: Three samples of spray dried mustard flour and one sample of conventionally dried ground mustard flour were manufactured and used at the same concentration (weight percent) in a standard mayonnaise recipe. Viscosity of the mayonnaise was tested with a Brookfield viscometer immediately following mixing, four weeks after then three months after mixing. Samples of the same mustard flours were also tested for volatile oil content. The test results are tabulated as follows:

Mustard Seed Blend	Process	Volatile Oil (%)	Mayonnaise Viscosity (in 1,000 cp) Immed.	4 Weeks	3 Months
60% Oriental 40% yellow (dehulled seeds)	85% water/15% seed; both heated to 200°F; wet-ground; homogenized; spray-dried	0.45	303	292	285
70% Oriental 30% yellow (whole seeds)	80% water/20% seed; oriental heated; wet-ground; spray-dried; yellow processed cold	0.67	284	230	196
30% Oriental 70% yellow (whole seeds)	80% water/20% seed; both heated to 200°F; wet-ground; homogenized; spray-dried	0.07	291	257	238
50% Oriental 50% yellow (dehulled seeds)	Roller ground control. This is regular mustard flour of commerce.	0.49	237	148*	130

*5 weeks.

FRUIT AND VEGETABLE PROCESSING

CITRUS JUICE PROCESSING

Cloud Stability Pectin Treatment

Preservation of cloud in citrus juices is the object in the process of *R.A. Baker and J.H. Bruemmer; U.S. Patent 3,754,932; August 28, 1973; assigned to The U.S. Secretary of Agriculture.* It was found that enzymatic removal of water-soluble pectins and insoluble pectins will protect the cloud in juice without heating. Previously, it was found that a heat treatment of citrus juice is necessary not only for the destruction of spoilage organisms, but also for the inactivation of enzymes which bring about clarification of the juice. Without the heat treatment the loss of cloud lowers the quality of the juice before spoilage occurs.

The term "cloud" refers to an ill-defined aggregation of particulate colloidal materials in citrus juices which impart a cloudy appearance which one normally associates with fresh juice. If this cloud settles out the remaining clear, watery serum is visually unappealing and relatively flavorless. It is felt that the enzyme pectinesterase is responsible for the events leading to the loss of the cloud. It is presumed that this enzyme converts the soluble pectins of the juice to pectic acids, which are precipitated as insoluble pectates. In precipitating, these pectic acid salts entrain the finely divided particles of the juice and coprecipitate them.

At the pH of citrus juices a more severe temperature treatment is required to inactivate the pectinesterase than is necessary for the prevention of microbial deterioration. In addition there is no quick practical way to measure the degree of cloud stability being produced. Thus, there is a tendency to exceed the minimum temperature requirements to remove any possibility of clarification. The result of excessive heat treatment is a partial or complete loss of the fresh natural flavor and the development to some degree of a cooked off-flavor.

No attempt is made to inhibit pectinesterase activity. Instead, polygalacturonase, an enzyme which destroys demethylated pectins by hydrolysis is added. Thus, the formation of a calcium pectate precipitate is prevented by the removal, by hydrolysis, of demethylated pectins before it can combine with calcium. For example, in three experiments in which a commercially available pectinase containing a high level of polygalacturonase was added to fresh juice at the rate of one part in 2,000 and stored 24 hours at 80°F, the levels of water-soluble pectins, calcium pectates, and protopectin were reduced by 36, 53, and 75% respectively.

Since pure preparations of polygalacturonase are not easily obtainable, commercially available pectinases which are mixtures of pectic enzymes must be used. These commonly contain pectinesterase (PE), polygalacturonase (PG), and polymethylgalacturonase (PMG). It is desirable to use a preparation having a high ratio of PG activity to PE activity. In addition it was found that the presence of substantial PMG activity is detrimental to the function of the pectinase as a cloud stabilizer. Examination of a number of commercial pectinases revealed that those having the highest ratio of PG activity to PMG activity were the most effective in stabilizing cloud. Several of these pectinases have provided good control of cloud loss at a concentration of one part to 500 parts of juice.

A modification which can be used is the addition of a proteolytic enzyme such as ficin, papain, or bromelin in addition to the polygalacturonase. The addition of a protease enhances the breakdown of calcium pectate by polygalacturonase. Juice treated thus has a stable cloud of greater turbidity than the cloud of freshly extracted juice.

Example: Freshly extracted orange juice was treated with Pectinol 41-P at a rate of one part enzyme to 5,000 parts juice. In addition ficin, a proteolytic enzyme derived from the sap of certain trees of the genus *Ficus,* was added to the juice at a rate of one part ficin to 10,000 parts juice. The containers were then placed in 40°F storage.

After centrifuging for 10 minutes at 470 *g* to remove pulp, the freshly extracted juice had a light transmission of 51.9%, when measured with a Lumetron colorimeter equipped with a 650 mμ filter. After 38 days storage at 40°F, a sample of this juice with no enzyme treatment had a light transmission of 92.8%, indicating almost complete clarification. The juice treated with Pectinol 41-P and ficin had a light transmission of 54.6% after 38 days, demonstrating the retention of a stable juice cloud.

Specific Cloud Enzyme Process

Orange base syrup suitable for manufacture of bottled orange drinks is prepared in the process of *C.D. Atkins, J.A. Attaway and M.D. Maraulja; U.S. Patent 3,711,294; January 16, 1973; assigned to State of Florida, Department of Citrus.* Orange juice contains the naturally occurring colloid stabilizer known as pectin which gives the juice a viscosity or consistency termed "body" along with suspended fatty materials and phospholipids which are partly responsible for the cloud present in citrus juices. Attempts to remove insoluble pulp solids from citrus juices have resulted in also detrimentally affecting the body and cloud of the citrus juice.

It has been found that highly colored, smooth, low viscosity orange base, which can be concentrated into a 100% orange juice syrup suitable for production of bottled or canned beverages and meeting all the requirements of a bottle of syrup and, in addition, having satisfactory cloud and body while being free of suspended solids can be obtained from fresh or reconstituted orange juice, by hydrolyzing the pectic substances present in the orange juice specifically without affecting or removing phospholipids present in the orange juice and then separating the insoluble fatty materials from the juice.

It has been found that by using enzymes which are specific for pectic substances, it is possible to effect the hydrolysis of pectic substances in orange juice and thereby permit the removal of fatty solids without also removing phospholipids which are responsible for the desirable optical properties (i.e., cloud) and body in natural orange juice.

Freshly extracted orange juice or reconstituted bulk orange juice concentrate is first heated to a temperature of about 165°F, in order to inactivate the naturally present pectic enzymes in the orange juice such as pectin-sterase. Subsequent to this heating, the juice is cooled to a temperature about 80°F, and sufficient pectic enzyme added to hydrolyze a substantial portion of the pectins present in the juice.

The pectic enzyme used (which may contain traces of adulterating nonspecific enzymes) should, however, not be permitted to remain in contact with the citrus juice substrate for sufficient time to affect the substances responsible for maintaining the cloud, color and body of the juice. The amount of enzyme actually used will depend on a number of factors such as time of treatment until the enzyme is deactivated, enzyme concentration and temperature. Generally, these parameters can be varied as convenient to effect a reduction of the pectic fraction to about 0.01 to 0.15 weight percent pectins and about 0.002 to 0.01 weight percent H_2O insoluble solids. Typically, for example, about 4 to 8 volume percent based on the volume of single-strength juice of enzyme is sufficient when the juice containing the enzyme is allowed to stand for about 18-30 hours at about 70°-120°F before raising the temperature to deactivate the enzyme.

Concentrates of the natural pectinesterase enzyme can be prepared, for example, where processing of citrus fruits is in progress. Insoluble solids from juice extraction procedures in the form of pulp, juice sacs, and rag are washed with distilled water to remove sugars and acids present in the mixture.

The citrus pulp solids are rapidly screened and pressed to remove excess water. The dewatered residue is then comminuted to pass a 100 Tyler mesh sieve. The slurry is tested to determine the potency of the native pectinesterase extract at the pH of about 7.0 to 7.5. The insoluble solids slurry containing the pectinesterase is added to 80° Brix clarified citrus concentrate at a natural pH. This addition should not lower the natural sugar concentration to below 65% concentration. The enzyme contained in the 65% natural sugar concentrate is then subjected to temperatures below 0°F. Dehydration of the yeast and fungi bodies effectively destroys the ability of these microorganisms to produce additional enzymes destructive to the color and cloud of citrus juices. When this citric enzyme mixture is used for clarification, the pulp or insoluble solids content of the juice to be clarified should not be in excess of 12% insoluble solids content.

The polygalacturonase activity of a pectinol preparation, representing an alcohol precipitate from an extract of mold cultures was compared to the activity of the pectinesterase prepared from citrus residues of pulp, juice sacs and rag. The crude enzyme preparation of pectinesterase, without the presence of NaCl (salt normally used in pectinesterase extraction) retained sufficient activity for successful clarification, and gave more evidence that there are at least two enzymes, pectinesterase and pectinase, capable of demethylating pectin, one occurring in higher plants (citrus) and the other in commercial pectinase preparations made from mold cultures. These commercially prepared enzyme solutions performed, when tested on citrus juices, in a similar manner to the pectinesterase enzyme concentrated from citrus pulp or insoluble solids normally present in extracted unheated citrus juices.

Pectinesterase and polygalacturonase or mixtures of the two were used to clarify citrus juices. The pectinesterase enzyme catalyzes the cleavage of methoxyl groups from the pectic chain. These groups in the presence of calcium form low methoxyl gel particles that lack solubility in water or citrus juices and are readily removed by centrifugation of the citrus juice substrate. Polygalacturonase rapidly breaks the main pectin chain to form insoluble pectinates that can be centrifuged or filtered from the juice substrate.

Example 1: Freshly extracted single-strength orange juice was heated to 165°F and cooled to approximately 80°F. One half pint of a suitable specific pectic enzyme was added to each 100 gallons of single-strength juice. The juice was allowed to stand for 25 hours. The treated juice was then heated to about 200°F to completely inactivate the enzymes and destroy any microorganisms present. The excess insoluble solids of the juice were decanted and the remaining juice centrifuged to yield a colored single-strength juice that represented that present in the cell sac of the fruit prior to normal juice extraction.

Following this preparation, the juice was then concentrated under vacuum to yield a sparkling colored syrup of 65° Brix concentration. The analyses of the base and a control concentrate was as follows:

Concentrate	Apparent Viscosity (cp)	Insoluble Solids (vol %)	Hunter Citrus Colorimeter
New base, 65° Brix	70	trace	38
Standard (control), 65° Brix	3,000	9	38

Example 2: Carbonated bottled orange juice was prepared using a deeply colored Valencia orange base having a ratio of 21.6, a Brix of 64.83° and an apparent viscosity of 70 centipoises. The base was first thawed and filtered through rice papers to remove remaining pulp. Citric acid was added (64.4 grams in 242 ml of solution) to 3 liters or 3,954 grams of the base to adjust the ratio to 14.0 and the Brix to 62.5°. Cold-pressed peel oil (2.30 ml) was then added to obtain an 0.015% level in the reconstituted juice for flavor enhancement. The fold was calculated based on a 14.0° Brix reconstituted juice as follows:

$$\frac{\text{lb/gal } 62.5^\circ \text{ Brix}}{\text{lb/gal } 14.0^\circ \text{ Brix}} = \frac{6.782}{1.231} = 5.51 \text{ fold}$$

To obtain 14.0° Brix, one part by volume of the adjusted base must be reconstituted with 4.51 parts of carbonated water. Therefore, to each 6 fluid ounce bottle of juice (178 ml), 32 ml of the orange base was added.

A solution of sodium benzoate was prepared containing 8.54 grams per 48 ml of solution and one ml was added per bottle to obtain 1/10 of 1% benzoate content.

The bottles were prechilled to 0°C prior to addition of the concentrate base and the sodium benzoate and for one hour afterward in order to solidify the contents to prevent gushing or foaming when filling with carbonated water. The water temperature during carbonation and filling of the bottles was 32°F. After the bottles and contents had reached approximately 40°F, they were inverted repeatedly until reconstitution was complete and then placed in 40°F storage.

Analysis of Bottled Product

Corrected refract	14.20° Brix
Acid	1.01%
Ratio	14.06
pH	3.5
Pressure-bottled juice	12-15 psi
Pressure-bottled water	20-23 psi

Flow diagrams for the manufacture of the citrus juice bases using natural and added enzymes for stable citrus drinks are given in the patent.

Naringinase Debittering

Increased utilization of citrus fruit has been proposed for whole fruit puree, however the process either incorporates the seeds and rag into the finished puree, or else an additional separation or coring step is required. Incorporation of seeds and rag into whole fruit puree has been shown to impart a bitter and/or a "mealy" texture to the puree, which, when further processed or utilized, particularly in the preparation of beverages, gives an off-flavor, as well as an undesirable physical effect in the mouth and lips. Separate removal of seeds adds costs to the finished product in the form of additional processing time and labor, plus losses of desirable portions of the fruit, particularly if coring is used.

R.R. Cruse and B.J. Lime; U.S. Patent 3,749,585; July 31, 1973; assigned to the U.S. Secretary of Agriculture prepare a citrus puree in a process in which flavonoids present in the citrus pulp and peel are treated enzymatically to destroy their bitter taste properties.

It was found that whole citrus fruit can be blanched in boiling water at atmospheric pressure to soften the tissues; partially comminuted in a Fitzpatrick comminuting machine operating at slow speed; further comminuted with concommitant separation of seeds, tough scaly peel, and tough rag in a paddle finisher; and finally pureed in a homogenizer to reduce the solid pulp particles.

The bitter principles present are hydrolyzed after homogenization with an enzyme such as naringinase, which will specifically hydrolyze and thus neutralize such bitter principles. The resulting puree is suitable as a base for citrus beverages, glazes, sauces, and other food applications utilizing citrus flavor. The puree lends itself to many applications in the baking and dairy industries. The puree may be preserved by frozen storage, or pasteurization and subsequent canning while hot.

Example 1: A field box of redblush grapefruit (about 90 lb of fruit) was blanched 10 minutes in boiling water; the pulp in the center of the fruit had an average temperature, after blanching, of 150°-155°F. The blanched fruit were coarsely cut up while still hot, and passed through a screw finisher with a conical screen having 0.125" openings and a central exit for rejecting seeds and other portions of the pulp not expressed through the screen. The pulp was well-mixed and passed through a 2-stage laboratory homogenizer. The homogenized puree weighed 80 pounds, which represents a utilization of 88.9% of the fruit.

Example 2: A field box of redblush grapefruit was treated as in Example 1, except that the finisher screen had 0.062" openings. During the final homogenization, 0.05% naringinase, based on the weight of the fruit taken, was added in portions to the pulp in the feed hopper of the homogenizer. The homogenized puree was placed in a kettle supported in a hot water bath and maintained at 115°-125°F until the bitterness had reached an acceptable level. This required 3 hours. The debittered puree was packaged and stored at -10°F. A beverage was prepared from the above pulp as follows:

	Parts by Weight
Grapefruit puree	100
Sugar	125
Six-fold lemon concentrate	25
Water to 1,000	

Analysis of this beverage gave the following results:

Brix	14.3°
Acid	0.75%
pH	2.9
Suspended solids	9%
Cloud	40.3
Recoverable oil	0.0104

Example 3: Eighty pounds of field run redblush grapefruit were blanched 10 minutes in boiling water at 212°F, in two batches. The fruit were passed through a Fitzpatrick Model D comminuting machine operating at 625 rpm and fitted with a screen having ½" openings. This combination of speed and screen openings rapidly comminuted the fruit without significantly cracking or attritioning the seeds. The combined ground grapefruit pulp was passed through a laboratory model paddle finisher, fitted with 0.060" screens, and set to operate at 1,725 revolutions per minute. This operation separated seeds and tough and scaly portions of peel and rag continuously with further comminution. The seeds were rejected intact, and the rejected peel and rag (total rejected weight 11 to 15% of the starting weight of fruit) was in pieces small enough to dry and use for livestock feed without further milling or comminution.

The screened fine pulp was passed through a laboratory homogenizer, first stage setting, 1,500 psig; second stage, 3,000 psig. The homogenized puree was treated 3 hours at 45°-50°C with 0.05% naringinase (Kumitanase) prior to pasteurization. The puree was heated to 195°F using a tubular pasteurizer, and hot canned. Storage at 40°F showed satisfactory storage stability over a 9 month period.

Protein Citrus Drink

A mixture of milk protein, citrus juice and a partially hydrolyzed starch is used by *M.D. Maraulja, J.A. Attaway and C.D. Atkins; U.S. Patent 3,917,852; Nov. 4, 1975; assigned to State of Florida Department of Citrus* to prepare a high protein citrus juice beverage which is equivalent to a milk-citrus beverage. An enzyme method is used to produce insoluble amylose which complexes with milk protein.

The solubility or partial hydrolysis of the starch into the soluble amylopectin and relatively insoluble amylose fractions and the complexing of these fractions with the milk protein is effected by boiling the starch in sufficient water for about 10-15 minutes at a temperature of about 210°-260°F. The necessary water for the complete dissolution of the starch can either be added to the starch-protein mixture and boiling carried out prior to the addition of the citrus juice, or the citrus juice, which will generally contain sufficient water for reaction with the starch, is added to the starch-protein first and the entire mixture boiled. Preferably, however, the water is added to the starch-protein mixture and boiling carried out prior to the addition of the citrus juice since the risk of impairing the flavor of the juice is not thereby encountered.

When relatively high protein and starch levels are used, hydrolysis of the starch can be facilitated by heating of the mixture at a temperature below 140°F in the presence of about 0.10 to 0.15% by weight, based on the starch, of suitable enzymes to effect additional solubilization of the relatively insoluble amylose fraction to soluble maltose. While this partial hydrolysis of the amylose fraction takes place rapidly below 140°F, above this temperature inactivation of the enzyme occurs and the hydrolyses stops. Below starch levels of about 2% by weight, the use of enzymes is not necessary. Suitable enzymes which can be used to further solubilize the amylose fraction is α-amylose (α-1,4-glucan-4-glucanohydrolase). This heating in the presence of an enzyme can be carried out either as an alternative to or subsequent to heating in water at 210° to 260°F which was described above.

It is believed that the soluble amylopectin starch fraction acts primarily to form a stable complex with the soluble protein, and thereby remains stable when mixed with acid citrus juices. This action of the amylopectin is not adversely affected by the use of the α-amylose enzyme since this enzyme is specific for the amylose during the short reaction time of the heating procedure (about 10-15 minutes).

The various food stabilizers which can be used include hydrophilic colloidal stabilizers such as gum arabic, gelatin, and pectin, however, particularly suitable stabilizers have been found to be anionic polymers derived from cellulose, which are water-soluble and tolerant of low pHs encountered in citrus foods, and, in particular, sodium carboxymethylcellulose polymers having a low viscosity and a degree of carboxymethyl substitution of about 0.65-0.85 DS.

Either before or subsequent to mixing and heating of the starch various sweeteners can be added in appropriate amounts to assure a final product having a desired flavor.

Example 1: 12 quarts of distilled water were added to a steam jacketed kettle and while stirring, the following ingredients slowly added: 1,230 grams of milk protein material containing about 89% glycoprotein and soluble in the pH range 2-3.2, 205 grams of starch and 105 grams of low viscosity sodium carboxymethylcellulose having a degree of carboxymethyl substitution of 0.65-0.85 DS.

The mixture was slowly heated using steam pressures of 25 to 28 psi until the mixture had been brought to a boil. Boiling was then maintained for 15 min to insure the complete hydrolysis of the starch and complete solution of the sodium carboxymethylcellulose polymer. During this period of time a temperature of 210° to 225°F was maintained. The mixture was then transferred to a larger tank and while stirring, a mixture of 2 quarts of distilled water and 1,281 grams of corn syrup added, followed by 3 quarts of 68° Brix concentrate and 1.45 quarts of water. Finally, 2.5 ml of Valencia orange essence oil, 2.5 ml of Valencia orange cold-pressed oil and 3.0 ml of vanilla extract were added for flavor adjustment. The product temperature was below 120°F at this point. The beverage was then passed through a heat exchange unit and the orange juice constituent heat stabilized at 185° to 210°F, filled into 12 oz cans, cooled under a water spray and stored at approximately 40°F.

Ingredients	Grams	Percent by Weight
18.45 quarts of 14.0° Brix orange juice*	18,472.0	86.74
Milk protein material		
Milk protein	1,095	5.14
Ash and moisture	135	0.63
Starch	205	0.96
Sodium carboxymethylcellulose	105	0.49
Corn syrup	1,281.0	6.01
Orange essence oil (0.009%)	2	–
Cold-pressed orange oil (0.009%)	2	0.03
Vanilla extract (0.013%)	3	–

*3 quarts 68° Brix concentrate plus 15.45 quarts water.

Example 2: The orange juice-high protein food product of Example 1 was prepared except that to the mixture of distilled water, milk protein material, starch, and sodium carboxymethylcellulose, 0.25 gram of α-amylose was added to the unheated mixture when 205 grams of starch was used. The mixture was then slowly heated using steam pressure of 25 to 28 psi until a temperature of 140°F was reached. During this time, approximately 12 to 15 minutes, the enzyme α-amylase, had partially hydrolyzed the starch fraction amylose to maltose. The mixture was then brought to boil and boiling maintained for 15 minutes at 210° to 225°F to deactivate the enzyme and insure the complete solution of the sodium carboxymethylcellulose polymer. The remainder of the processing procedures were then continued as given in Example 1.

FRUIT AND VEGETABLE JUICE PROCESSING

Monitored Depectinization

Stable fruit and vegetable juices with controlled clarity are prepared in the process of *J.-P. Richard; U.S. Patent 3,795,521; March 5, 1974* by use of depectinizing enzymes. The conventional enzyme process is as follows: suppression of the coarser elements, such as cellular debris, either by sieving, centrifuging or by decanting continuously; depectinization, which has as its object the cutting of the pectin molecules in such a way as to favor subsequent filtration, and which utilizes pectinolytic enzymes; fining, since depectinization has not been sufficient to ensure complete stabilization of the juice, to prevent subsequent turbidity from appearing, a setting operation is carried out using an excess of positively charged fillers either in the form of gelatine or albumen, the most common examples, or in the form of tannin; and filtration, after depectinization and fining, which generally yields a perfectly clear and bright juice.

However, in spite of depectinization and fining, it is found that after several weeks of storage in tanks, turbidity again occurs in the juice necessitating supplementary filtering before the final bottling. This means a continuous production line from fruit to bottles cannot be envisaged.

The usual conditions of operation of pectinolytic enzymes employ slow enzymatic reaction speeds. Generally the pectinolytic enzymes are introduced into the juice to be treated in a proportion of from 5 to 50 grams of pectinolytic enzymes per hectoliter and the duration of depectinization varies from 1 to 8 hours, depending upon the temperature at which this operation is carried out. No monitoring of this process has previously been practiced and the absence of a check upon the depectinization operation means that the action of the pectinolytic enzymes may be inadequate, thus making filtration difficult, as a result of clogging of the filter or it may be too drastic, leading to the existence within the juice treated of small negatively charged particles which associate themselves progressively with positively charged particles, the aggregation of these together leading to insolubility.

In theory, the addition of an excess of positively charged particles in the form of gelatin or tannin for fining serves to carry away the entirety of the fillers as a mass and thereby eliminate the appearance of subsequent turbidity. However an excess of fining leads to absorption of the aromatic principles contained in the juice; and, although such absorption does not present a major difficulty in the case of juices which are sufficiently acid, it is unacceptable in the case of juices which are sparingly acid; in these it can alter or even completely suppress the taste, especially in the case of juices extracted from golden apples. In addition, although the gelatin is eliminated at the end of the fining operation, it nevertheless leaves in the juice a heavier taste, which is not the taste of juices which have not been fined.

Finally, if the pectinolytic enzymes have not been completely destroyed during the depectinization operation and the pasteurization operation which follows it, they continue to act and cause a new disequilibrium of the fillers, leading to turbidity in the storage tanks. By control of the important variables, mainly the enzymatic reaction, it is possible to use a continuous process for production of stable fruit or vegetable juice products.

Figure 8.1 is a schematic diagram of a continuous fruit juice production line. A production line for fruit or vegetable juices has a supply system **1** for fruits or vegetables, such as a feed screw for example, a grater **2**, where applicable (for those cases where it is intended to obtain juices from fruits or vegetables

Figure 8.1: Continuous Fruit Juice Production Line

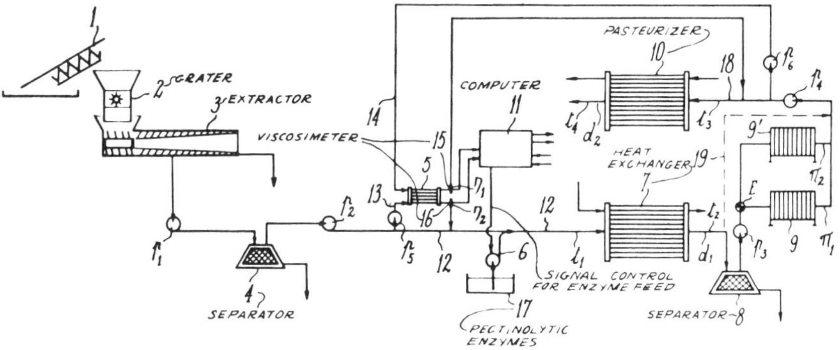

Source: U.S. Patent 3,795,521

which require preliminary grating), and an extractor **3** in which the juice is extracted from the fruits or vegetables, thus favoring the rupturing of the cells of which the fruits or vegetables are composed, the juice obtained being full of cellular debris containing in particular cellulose fractions and pectin fractions; following the extractor **3** is an apparatus **4** suitable for separating the juice from the cellulose fractions; this apparatus for example may be a centrifuge, a sieve or a continuous decanter. The juice from the apparatus **4** is then passed to a heated container through which it continuously passes, in this case a heat exchanger **7** in which a depectinization operation is carried out.

The depectinization operation is made necessary by the fact that the pectin fractions contained by the juice prevent the filtering of the latter, as they are the principal cause of the clogging of the filters. A portion of the juice is taken off on a bypass line **13** by pump p_5 and passed through a heat exchanger **5** before being returned to the line **12**.

A pump **6** under the control of a servo control unit **11** pumps a controlled quantity of pectinolytic enzymes from a reservoir **17** into the juice in line **12**. The depectinized juice obtained at the outlet from the heat exchanger **7** is continuously passed through separator **8** and is then filtered on a plate or cloth filter **9–9′**. The filtered juice obtained is conveyed on line **18** for pasteurization which is carried out continuously in a heat exchanger **10**. At the outlet from the heat exchanger, the juice needs no further treatment and is transmitted to a conditioning unit, or for storage under sterile conditions, or for immediate bottling or canning.

Pasteurization should be carried out under time and temperature conditions which are also defined to avoid the appearance of subsequent turbidity. If it is desired to obtain a juice having controlled and stable turbidity rather than clarity, it is sufficient to modify the conditions for concentrating with enzymes and the depectinization time/temperature combination so that coagulation in particular, but not solely coagulation, takes place not during heating in the exchanger **7** but during the pasteurization or cooling of the juice. A portion of the filtered juice in line **18** is extracted on line **14** by a pump $\mathbf{p}_6$ and passed through the heat exchanger **5** to heat the untreated juice in line **13** before being returned to line **18**. The treated and untreated juices leaving the heat exchanger **5** then have approximately the same temperatures and their respective viscosities are measured by two viscosimeter probes **15, 16,** the viscosity measurements being supplied to the controller **11** and the difference in these measurements being used as a control parameter.

The difference in viscosity $\eta_2 - \eta_1$ (η_2 is viscosity before depectinization and η_1 is viscosity after depectinization and filtering) is used to control the addition of enzymes by controlling the pump **6** so as to maintain a predetermined desired constant viscosity difference.

Pectin Transeliminase Method

T. Yokotsuka and S. Ishii; U.S. Patent 3,666,487; May 30, 1972; assigned to Kikkoman Shoyu Co., Ltd., Japan describes a process in which pectin transeliminase is added to mashed fruits or fruit juices to degrade pectin substances and obtain improvements in yields and clarity.

It has been known that pectic substances present in fruit juice maintain the stability of the colloid system present in juice, thereby making the juice turbid, or because of being extremely rich in viscosity the pectic substances make it difficult to carry out the treatments required in a process of manufacturing product fruit juices, such as steps of pressing and filtering fruits or mashed products or the steps of filtering the pressed juices from fruits, thereby prolonging the operation time or lowering the yields of product fruit juices. Moreover, the pectic substances act frequently as contributory causes for a jellification during the course of operating the concentration step.

Two kinds of enzymes having activities entirely different from each other, i.e., esterase and glycosidase, are necessary for enzymatically degrading pectin present in fruit juices. That is, by treating with pectin esterase (PE) the methyl ester bond in the pectin molecule is first cleaved to yield pectic acid, and subsequently, by treating with polygalacturonase (PG) the α-1,4-galacturonide bond of the resultant pectic acid is cleaved. Thus, not only the degradation of pectin with PE and PG independently is hardly possible but also is entirely impossible in the case where the order of treatment with PE and PG is reversed, i.e., when PG is first allowed to react and subsequently, followed by PE.

The pectinase being used in clarifying fruit juices as well as other treatments is composed principally of PE and PG. It is necessary that both PE and PG be allowed to react in a well balanced state existing between the two. For instance, the proportion of methyl esterification of pectin greatly varies depending on the kind and species of fruit used or on the method of preservation of the fruit.

It is therefore necessary to enhance the activity of PE when applied to a fruit having a high degree of esterification. In case, however, PE has a higher activity than needed, a large amount of undecomposed pectic acid is left, which tends readily to coagulate owing to the presence of an acid or metal ion such as calcium ion, magnesium ion or the like, whereby the subsequent treatments become extremely difficult to operate. For that reason, when PE and PG are used in treating fruit juices, the determination of the activity ratio of PE to PG according to the kind and condition of fruit must receive extensive consideration.

Pectin transeliminase referred to as PTE, which is entirely different in action and function from PE and PG, will act independently and satisfactorily upon the pectin, whereby the treatment of fruit juice becomes easy, the juicing step is improved, the yield of juice pressed out of fruit is increased and the clarification of product fruit juice can be accomplished. This is ascribable to the fact that PTE reacts with an α-1,4-galacturonide bond, thereby effectively degrading the pectin.

Usually, PTE is active over a widespread pH ranging from 3 to 8, and it has in most cases an optimum pH ranging from 3.5 to 6.0. The temperature at which PTE is active is from 30° to 70°C, but an optimum temperature is from 45° to 65°C.

The improved method for the treatment of fruit juice consists of adding PTE to a material to be treated, such as mashed product of fruit or juice pressed from fruit, or, if necessary, crude filtrate at any stage of mashing, pressing and filtering, and of treating such material with the PTE under the conditions of pH value of from 3 to 6, treatment temperature of from 40° to 60°C and thereby degrading pectin contained in the resulting fruit juice.

After having been treated with PTE, the resulting fruit juice is heated to inactivate the enzymatic activity of the PTE, and then solid matters contained are removed by centrifugal separation or filtration to give a desired clear product fruit juice.

PTE obtained from microorganism includes, for example, a PTE-containing cultured product obtained by culturing a PTE-producing microorganism, such as mold, yeast, or actinomyces, in a liquid or solid culture medium; a crude suspension of enzyme obtained by treating the cultured product by means of centrifugal separation, extraction, etc.; an enzyme-containing precipitate obtained with ammonium sulfate; and a dried crude powder of enzyme obtained by drying.

Industrial supply sources of PTE include PTE-producing fungi which belong to genus *Aspergillus* among molds, for example, yellow *Aspergillus* such as *Aspergillus oryzae* 1773, *Aspergillus sojae* 48, ATCC 20235 and *Aspergillus sojae* 22 and black *Aspergillus,* for example, *Aspergillus niger* 1742, *Aspergillus saitoi* 1540, and *Aspergillus inui.*

Examples of fruit juices that can be treated include, for example, pressed out or water-extracted juices of fruits, such as apples, grapes, citrus fruits, apricots, quinces, peaches, pears and plums and in addition, pressed out juices or water-extracts, such as vegetable juice of pectin-containing vegetable products.

Example: A cultured product (Koji) obtained by solid-culturing *Aspergillus sojae* 48 ATCC 20235 in wheat bran for two or three days was extracted with 5 times as much water, and the extract was then centrifuged to give a transparent crude enzyme suspension. A precipitate formed by addition to the enzyme suspension of ammonium sulfate in an amount necessary for attaining a 40% saturation was removed. Subsequently, a precipitate formed by addition of ammonium sulfate in an amount necessary for attaining a 75% saturation to a supernatant liquid of the enzyme suspension was collected by centrifugal separation. The precipitate thus collected was dissolved in a small amount of distilled water and the solution was passed through a column (2 x 140 cm) packed with Sephadex G-25 to remove the ammonium sulfate, and then an eluted fraction of enzyme protein was lyophilized to enzyme powder.

The enzyme powder was dissolved in 0.01 M acetate buffer solution adjusted to pH 4.0, and adsorption was effected by means of a column packed with CM-cellulose, the column being equilibrated with the buffer solution. PTE absorbed to the column was eluted with 0.1 M acetate buffer solution adjusted to pH 5.0. After repeating this operation, the eluted fractions of PTE were collected and concentrated. Subsequently, the gel formed was filtered off using a column packed with Sephadex G-100. In this manner, there was obtained an ultracentrifugally and electrophoretically homogeneous purified PTE that is absolutely free from other enzymes than PTE.

One liter of apple (species: Jonathan) juice was treated with 150 units (protein content: 0.5 mg) of the purified PTE obtained according to the above method at 40°C for 60 minutes. The enzyme reaction was stopped by heat treatment. The reaction product was cooled and then centrifuged at 3,000 rpm for 5 min to give an extremely clarified apple juice having a light transmission value of 97.00 at 660 mμ. On the other hand, a control treated, without addition of the enzyme, in the same manner as above failed completely to give a clear apple juice.

Juice/Wine Enzyme Process

The method of treating grape pulp for the production of wines, cognac and nonalcoholic beverages consists in separation of a must from the pulp by means of mechanical or physical disintegration of the berry tissue cells of a fresh or partially prealcoholized pulp or a pulp preliminarily fermented completely or partially fermented with a subsequent alcoholization to a definite content of alcohol for the given grade of wine.

The pulp treated in such a way provides a limited yield of a high-quality fraction must (free run must, must of the first and second pressings of low-pressure grape presses), which comes to 50 to 60 dal per ton of grapes depending on the grade of grapes and the ecological conditions of their growth, while the total yield of the must accounts for about 75 dal per ton of grapes. The yield of the must may be increased by increasing the pressure. However, in this case the content of suspended solids in the must, colloids, tanning and nitrous compounds, is much increased. This material considerably deteriorates the quality of the wines, particularly white and dessert wines, champagne and cognac wine materials.

When producing white table wines and champagne wine materials by using a mechanically worked (pressed) pulp, the process includes the following basic operation: immediate extraction of the must, supplying it to settling tanks, while simultaneously sulfating the must flow, clarification of the must, decantation of the clarified must and its fermentation. Such a process provides for a yield of a high-quality fraction within 50 to 60 dal per ton of grapes. The clarification of the must during the settling lasts about 24 hours while the yield of clarified must accounts for not more than 60% of the yield of the high-quality fraction of the must. The wines made from such a must, particularly high-grade wines, have a flat aroma and an inadequately soft taste.

During the production of red table wines with the use of the pulp treated by mechanically or physically disintegrating the cells of the berry tissue, the process is effected according to three basic procedures:

(1) Fermentation of the pulp up to complete fermentation of the sugar and separation of the fermented must from the pulp;

(2) Heat treatment of the fresh pulp at a temperature of 60°-65°C for removing the coloring matter, cooling of the pulp, extraction of the must and fermentation of the must;

(3) Extraction of the coloring matter from the fresh pulp by means of a prefermented dry wine and further fermentation of the sugar extracted from the pulp.

V.I. Zinchenko, L.S. Salmanova and F.L. Minchuk; U.S. Patent 3,737,324; June 5, 1973 developed a process which treats grape pulp with a purified cytolytic ferment or enzyme preparation of a culture of fungus *Trichothecium roseum.* This enzyme preparation contains a mixture of cellulase, hemicellulase, and cellobiase-destroying humin substances. These enzymes together with pectolytic enzyme attack the cell wall constituents for maximum extraction of juice and compounds affecting quality of wines.

The addition of the enzyme preparation increases the yield of a high-quality fraction of the must from European grapes from 6 to 10 dal and the total yield of the must from 3 to 7 dal per ton of grapes depending on the grade of grapes, pH of the medium, dose of the preparation and the infusion conditions (temperature, time). For the hybrid grapes, particularly of the Isabella group, the yield increase of high-quality fractions accounts for 10 to 15 dal and the total increased yield comes to 5 to 10 dal. There is also an increase of the sugar content in the free run and first pressing must from 0.5 to 3.0% together with an increase of the content of aromatics and other compounds which provide a fine tasting wine.

The grapes from a transport means are unloaded into a receiving bunker **1** (Figure 8.2) and fed by a feed screw **2** to a machine **3** for crushing and stemming. The pulp produced after the crushing is forced by a pulp pump **4** into a tank **5** for infusion. Continuously added into the pulp flow by a doser **6** is a working solution of purified cytolytic enzyme preparation of a culture of fungus *Trichothecium roseum* in the amount of 0.001 to 0.05% by weight with the activity of 2,500 units and added by a sulfodoser **7** is sulfur dioxide for sulfating the pulp taken in the amount of 100 mg/l of the pulp.

The pulp is infused in the tank **5** during a period of not less than 3 hours for white table wines and not more than 6 hours for grape juice at a temperature of 15°-20°C. Then the pulp by means of a pulp pump **8** is fed to a flow feeder, for example a screw feeder, for separation of the free run must and, after that, the pulp is fed into a press **10**, for example a screw press, for producing must of the first and second pressings.

Figure 8.2: Enzyme Treatment of Grape Pulp for Wine Production

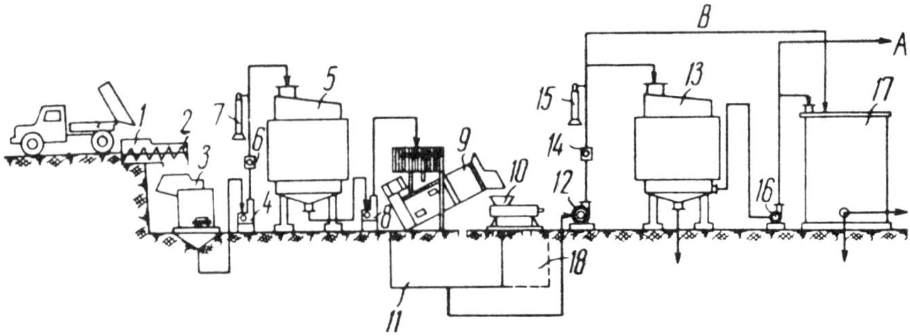

Source: U.S. Patent 3,737,324

The produced free run must and the first and second pressings' must are fed into a receiving tank **11** and therefrom through a pump **12** are fed into a tank **13** for settling, bentonite being added into the flow of must through a doser **14** in the amount at least 2 g/dal.

Simultaneously with addition of bentonite, by means of a sulfodoser **15**, there is added sulfur dioxide in the amount which does not exceed its total content in the must or 100 to 200 mg/l depending on the must temperature. The infusion is effected during 12 to 15 hours, then, the clarified must is decanted and fed through a pump **16** into a tank **17** for fermentation. The fermentation is effected at a temperature not higher than 28°C. After the fermentation and clarification the young wine is removed from the yeast sediment and is transported for further treatment and holding.

In the process of pressing the pulp there can be produced must of the third fraction which is collected in a tank **18**. This fraction is used for making strong wines or for distillation of crude grape alcohol. The yeast sediment, after removing the wine material, is removed for utilization.

For making high-quality grape juice the infusion of the must is effected without sulfation at a temperature of 10° to 12°C. The clarified must through the pump **16** is fed for cooling and stabilization (the movement of the juice is shown by an arrow **A**).

On producing red table wines from grapes having a sufficient content of coloring matter, the process is carried out as described above until the must is fed into the tank **5**. In this tank the must is fermented on the pulp at a temperature not exceeding 30°C. After the complete fermentation of the sugar the fermented pulp is supplied into the flow feeder **9** and, after that, the process is effected as described above. The produced young wine material of a fine fraction through the pump **12** is fed for further treatment and storage (the movement of the young wine material is shown by arrow **B**).

During the production of red wines the maximum amount of the cytolytic enzyme preparation is 0.02% by weight of the pulp while transportation of the young wine material is effected without treating this material with bentonite.

When producing white and rose high-quality dessert wines, the process is carried out similarly to the process of production of white table wines until the pulp is fed into the tank **5**.

In this tank the pulp is infused without prealcoholization at least 24 hours at a temperature of 15°-20°C. After the infusion the pulp through the pulp pump **8** is fed into the flow feeder **9** for separation of the free run must and for further pressing of the pulp in the press **10**. The must of a fine fraction through the pump **12** is fed into the tank **13** for fermentation and alcoholization to a required content of sugar and alcohol. Then the produced wine material is transported by the pump **16** into the tank **17** for clarification.

After the clarification, the wine material is removed from the yeast sediment and fed for further holding and treatment. Cytolytic enzyme preparation is added in the amount of 0.005 to 0.015% by weight of the pulp. When the must is fed for fermentation, it is used without sulfation and bentonite treatment.

FRUIT AND VEGETABLE PROCESSING

Stone Cells Removal

Peptidase is added to crude fruit puree containing stone cells and the stone cells are separated without causing a decrease in viscosity of the crude puree. The process is discussed by *A. Yamane; U.S. Patent 3,860,726; January 14, 1975*. The stone cells which contain lignin, tannin and cellulose tend to incur the following problems.

(a) They tend to cause a color change to yellow due to the alkali used for peeling fruit.

(b) They tend to impair the color tone of the drinks, producing a color change to light black due to iron ions.

(c) They produce difficulties in the operation of the exfoliating device and of the homogenizer, because the stone cells are hard grains, around which are tightly entangled pulp-like materials. The exfoliating device is used to separate and remove the pulp-like material and the homogenizer is used to homogenize the crude puree.

In the production of fruit base drinks from starting materials such as pears which contain stone cells, it is mandatory to remove the stone cells to a satisfactory degree and to maintain the viscosity of the crude puree, because the crude puree from pears, particularly from Japanese pears, is lower in viscosity and the amount of the flesh pulp is less.

Two processes to avoid such difficulties are as follows. One process consists of crushing the pears and straining to obtain crude puree; adding water to the crude puree (in an amount of about twice the crude puree for Japanese pears, and over twice the crude puree for European pears); and subjecting the crude puree thus treated to agitation to separate the stone cells by sedimentation. The second process involves treating the crude puree in a super-micronizer and placing the puree thus treated in a bath under moderate agitation to separate the stone cells by sedimentation. However, these two earlier processes still suffer from disadvantages in that: the former has proven not practicable, because the separation of the stone cells is insufficient due to the mere use of water dilution and because there arises a certain limitation for the amount of water to be added in the case of the production of nectar; and in that the latter has been found to fail in maintaining the viscosity of the crude puree during the treatments by a super-micronizer or a homogenizer, although the separation of the stone cells is satisfactory.

This process involves the following: alkali-treatment of the starting fruit which has been selected and washed, peeling, removing the core and subjecting the fruit flesh thus treated to crushing and straining to obtain a crude puree. The crushing treatment may be heat-crushing in the presence of an antioxidant to prevent color change and odors followed by straining. To the crude puree is added a peptidase, followed by rapid cooling to room temperature where it is allowed to stand for a predetermined period of time. Then, the crude puree thus treated is subjected to an exfoliating treatment to remove from the stone cells the elongated pulp-like material which retains fine pieces of fruit flesh. The separating treatment which follows separates the stone cells from the crude puree.

Illustrative of suitable peptidases are dipeptidase, aminopeptidase, carboxypeptidase, prolinase and, dehydropeptidase. For purpose of comparison there was used, in the same manner as the peptidase, pectinase, cellulase or protease having similar properties to those of peptidase. The results of the tests are shown in the following table which gives the viscosity versus treatment duration in the separating of the pulp-like material from stone cells. The conditions of the treatment are as follows: B-type viscosimeter, rotor No. 3, 20 rpm, 20°C; peripheral speed (m/min) 1,640; and crude puree → enzyme treatment (2 hr at 20°C) → deactivation treatment (10 min at 90°C).

Enzyme	Concentration of Enzyme (%)	Viscosity Preparatory to Separation (cp)	Viscosity Subsequent to Separation (cp)	Time Required for Completion (min)	Remarks
Peptidase	0.1	1,600	990	15	*
Pectinase	0.1	1,685	675	13	
Cellulase	0.1	435	590	1	
Protease	0.1	1,635	750	13	**
Peptidase	0.3	1,400	1,300	8	*
Pectinase	0.3	1,550	960	8	
Cellulase	0.3	400	510	0.5	
Protease	0.3	1,485	1,350	8	**
Control		1,950	700	18	

*Aminopeptidase was used as the peptidase.
**Color tone changed to dark brown.

It has been found that peptidase has an accelerating effect on the separation of the pulp-like matter from the stone cells and in the prevention of marked changes in the viscosity of the crude puree. There was no appreciable difference observed in viscosity when protease was used, although the color changed to dark brown.

While the viscosity with cellulase increased, the viscosity was too low to be suitable as a starting material for nectar. As can be seen from the table, the peptidase is most effective. The percentage of peptidase should be in the range from 0.05 to 1.0, and the time for treatment should fall in the range from 1 to 5 hours at room temperature.

Candied Fruit Enzyme Process

K. Mochizuki, K. Isobe and Y. Sawada; U.S. Patent 3,615,687; October 26, 1971; assigned to Takeda Chemical Industries, Ltd., Japan developed a process which has a shorter time for candying fruit. Enzymes produced by *Aspergillus niger* are used in the process.

The greater part of the enzymatic composition produced is accumulated in the culture medium coming out of the cells of *Aspergillus niger.* When the incubation is carried out with the use of a liquid medium it is necessary to filter or centrifuge the whole culture broth and, if required, to subject the resultant cleared broth to further recovery procedures. When the incubation is carried out with the use of a solid medium, it is possible to extract the culture broth with water and, if required, to subject the resultant extract to further recovery procedures.

The enzymatic composition is composed of many kinds of enzymes, e.g., pectinase, protease, cellulase, hemicellulase, peptidase, glucanase, RNA depolymerase, sucrase, maltase, lactase, xylanase, inulase, dextranase, mannase, α-amylase, β-amylase, lipase and cellobiase.

The enzymatic composition is used after being purified at least to make its pectinase activity higher than about 50 units per gram; its cellulase activity higher than about 10,000 units per gram and protease activity higher than about 5,000 units per gram.

The time for the candying is shortened by increasing the rate of the impregnation of the fruits with sugar without causing the shrinking of the fruits by treating sap fruits as starting material with the enzymatic composition at a state not later than the candying in the processes for producing candied fruits. The effect of shortening the period of the candying process is found in all cases for sap fruits usable as starting material for producing candied fruits.

In the case of cherries as starting material, the period of time required in order to have cherries impregnated with sugar to make the sugar concentration of cherries approximately 72% by weight is shortened to about 4 or 5 days without causing the shrinking of cherries, while at least 3 weeks are required for the same purpose which does not include the treatment with the enzymatic composition.

Example: 10 kg of cherries are preserved in 10 liters of aqueous brine containing 3% of sodium hydrogen sulfite and 0.1% of calcium chloride, adjusting the concentration of sulfur dioxide in the brine to above 0.7% by the addition of sodium hydrogen sulfite. After about 3 months storage, the cherries are pitted and stemmed, and then immersed in about 10 liters of 1% calcium chloride solution to harden the tissues of the cherries. The cherries are heated to boiling with incessant changing of water to reduce the sulfur dioxide below 500 ppm.

On the other hand, 20 grams of wheat bran and 20 milliliters of water are mixed in a 500 milliliter flask. The mixture is sterilized by heating at 120°C under pressure for 30 minutes. *Aspergillus niger* (ATCC No. 10254) is inoculated in the above prepared medium and then incubated for 5 days at 28°C. After incubation, the cultured broth is extracted with water, and the aqueous layer is separated by filtration.

To the filtrate there is added acetone of twice the volume of the filtrate to give precipitates. The precipitates are separated by filtration, and dried to give a white powdery enzymatic composition which has 320 units/gram of pectinase activity, 750,000 units/gram of cellulase activity and 5,500 units/gram of protease activity, respectively. 22 grams of thus-obtained enzymatic composition is dissolved in 10 liters of water, and then the solution is adjusted to pH 3.0 by the addition of citric acid.

The cherries are immersed in the enzymatic solution for 24 hours at 28°C. After the enzymatic treatment, the cherries are transferred into 10 liters of water. 250 ml of dye solution containing 3.5 grams of erythrosine is added to the mixture, while keeping the temperature at 80°C. The mixture is kept at the same temperature for 40 minutes, and then 100 ml of 1% aqueous tartaric acid solution is added to the mixture. The mixture is further heated at 80°C for 40 min, whereby the cherries are dyed and the enzymes in the cherries are inactivated. The cherries are immersed in 10 liters of 40% aqueous sucrose syrup.

The mixture is allowed to stand at about 50°C for 24 hours, then the whole portion of syrup is drained from the immersion tank and the sucrose concentration of the syrup is made up to 60% by the addition of sucrose. The cherries are again immersed in the 60% sucrose syrup and then the mixture is allowed to stand at about 50°C. After 24 hour immersion, the syrup is drained from the immersion tank and the sucrose concentration of the syrup is made up to 70% by the addition of sucrose. After 24 hour immersion in the 70% syrup, the syrup is increased to 80% of sucrose in the manner described above. After 24 hour immersion, the syrup is readjusted to 80% of sucrose by the addition of sucrose, and then the immersion is further continued for 24 hours.

By candying for 5 days, the concentration of sucrose in the cherries has reached approximately 72% without shrinking the cherries. The candied cherries are taken out of the syrup and dried by a conventional method.

In comparative tests when the candying is carried out without previously performing the enzymatic treatment, 21 days are required in order to have cherries impregnated with sucrose until sucrose concentration of the cherries reaches approximately 72% without causing shrinking of the cherries.

Pea Cook-Time Reduction

Dehydrated peas can be cooked and prepared with less boiling time by pretreatment of the peas in a slurry with pectolytic enzymes. The process is given by *W. Hitze, K. Legler, R. Stute and H.-U. Woelk; U.S. Patent 3,867,553; February 18, 1975; assigned to CPC International Inc.*

Treating peas with pectolytic enzymes causes the pectins, which can be regarded as the intercellular cementing substances for the palisade cells forming the husk, to decompose to an extent which makes the husks sufficiently and irreversibly permeable to water and provides for faster rehydration and softening of the husks. As a result the boiling time necessary to cook the peas which normally is about 25 minutes is reduced to about 10 to 15 minutes. The overall structure of the peas is not unfavorably changed by the treatment with pectolytic enzymes. The table shows figures reflecting the accelerated uptake of water by enzymatically treated dehydrated peas.

Time (min)	% Water Uptake Untreated Peas	% Water Uptake Treated Peas
5	44-46	51-53
10	50-56	63-65

The final state of water absorption is reached about 10 minutes earlier in the peas treated with pectolytic enzymes than in the untreated peas.

The use of pectolytic enzymes precludes any adverse effect on the pea proteins which play a key role from a nutritional and physiological point of view. The process permits the treatment of the peas with pectolytic enzymes to be combined with the soaking process (preparation) usually preceding drying. By choosing a proper enzyme concentration it is even possible to omit the adjustment of the pH by adding acids or alkali which would impair the tastes of the peas treated.

Change of the husks may also be caused when using enzymes capable of decomposing cellulose and/or hemicellulose. But it is difficult to control the decomposition process in a way insuring that no serious changes are produced in the texture of the coat and interior of the seed. A method of husking soy beans by treating them with a solution which is obtained by cultivating cellulase producing microorganisms and may contain pectinase thus proves to be unsuitable for reducing the boiling time of dehydrated peas. This treatment of peas with cellulytic enzymes provides maximum husking when the pH is adjusted to an optimal level of 4.0 by means of addition of glacial acetic acid. Under these conditions extensive changes in the shape, structure and notably the taste of the peas occur.

Example: 50 kg of green peas are added to 125 liters of soaking water, 3°-7° hardness (German standard) which contains 0.5% (referred to the quantity of peas) of the enzyme preparation Pektinol B 20 which decomposes pectin, and are left to soak for 20 hours at room temperature. The water is then drained off and the moist peas are cooked in streaming steam at about 110°C, whereupon they are exposed to air for cooking and subsequently placed on a belt-type dryer to be dried in an air current at between 50° and 90°C to a moisture content of 9.5%.

Organoleptic evaluation showed the boiling time of the peas so treated to be 12 minutes. Peas which had been treated in the same manner but in the absence of enzymes required 25 minutes of boiling to reach the same degree of softness.

Potato Snack Amylase Method

Potato chips usually contain about 40% oil and 60% potato solids. The caloric content of potato chips is about 600 cal/100 grams. Because of the high caloric content there has been a considerable amount of interest in a chip with a reduced caloric content. If a potato chip is fabricated from dry ingredients it becomes possible to replace some of the nutritional ingredients with non-nutritive ingredients such as edible cellulose or to replace starch with proteins by using for instance soy flour or soy isolates. The oil content that accounts for the major part of the caloric content of a potato chip must also be reduced.

With unmodified potato flour to which was added 33⅓% rice flour it was found that the practical limit of the solids content was around 23%. A higher solids content would produce a viscosity that is not suitable for aeration and casting. Drying of the cast film is carried out on a steel belt to a moisture content of about 30%. To produce one pound of the film containing 30% moisture it is necessary to remove 2.04 pounds of water during the drying step. By increasing the solids content of the batter to 35% it is only necessary to evaporate 1.00 pound of water to produce one pound of film with 30% moisture remaining. Drying requirements have been reduced by 50%.

A further advantage of an increased solids content is the fact that it is possible to reduce the casting thickness while retaining the same amount of solids per square inch. Thus it was found that whereas the batter with 23% solids should be cast at a thickness of 60 mils to produce a chip with acceptable thickness and strength it is possible to reduce the casting thickness to around 40 mils when batter of 35% solids is used. Since drying rate relates to the square of the thickness, the reduction in thickness from 60 to 40 mils would theoretically produce an increase in drying rate of 125%.

C. Kortschot and P.F. Adams; U.S. Patent 3,840,673; October 8, 1974; assigned to Corporate Foods Limited, Canada uses α-amylase in the batter of potato solids in the production of a snack food. Use of this enzyme gives a snack with lower oil content and more crisp texture. These improvements result primarily from increased solids content and reduced viscosity of the batter produced from potato flour when the amylase is added to the batter. Reduced viscosity is accomplished by the α-amylase enzyme hydrolyzing at least part of the starch molecules.

Using this process, it is possible to reduce the viscosity so that a fluid batter can be produced containing as much as 40% solids, approximately 30% potato flour and 10% rice flour. The resulting product does not have any unpleasant flavor and possesses a good potato flavor after cooking, e.g., by frying. The enzymatic reaction produces a material which has good film-forming properties and improved drying rates. This improvement in properties probably results from the formation of dextrins from the starch by the action of the enzyme.

A critical requirement of the process is a short reaction time of the enzyme in the potato solids suspension, otherwise an off-flavor described as chalky and astringent results in the finished dry snack. The steps of the process are: mixing a foodstuff containing potato solids with an α-amylase enzyme to form a premix, dispersing this premix in water at a temperature at which the enzyme has good activity, heating the mixture until the enzyme is deactivated, shaping the resulting mixture to the described thickness, and drying the shaped mixture to a moisture content so that it will be of a golden color when cooked.

Example: Preparation of Potato-Rice Chip —

(1) A premix was prepared by blending together the following dry ingredients: 75 parts potato flour, 25 parts rice flour, and 0.05 part amylase enzyme (fungal diastase).

(2) 150 parts of water were heated to 80°C.

(3) The dry premix from (1) was added to the hot water and because the premix was at ambient temperature, the temperature of the mix dropped to the 60°C range where the enzyme has good activity. The addition of the premix was regulated so that the viscosity of the mixture remained low enough to allow proper agitation. The addition took 3 to 4 minutes.

(4) After the addition of the premix was completed the temperature of the mix was raised to about 80°C and this temperature was maintained for ten minutes to completely deactivate the enzyme.

(5) The next step was to aerate the material. The objective of this step is to create a number of very small air bubbles. The aeration is not essential and a chip with good eating qualities can be made without it but aeration improves drying rates and by means of aeration it is possible to control tenderness of the finished chip. Aeration alters the fragility of the product by creating thinner layers of solids in the product.

(6) The resulting aerated, hot suspension of potato and rice flours from step (5) was then cast into a sheet with a thickness of about 45 mils. The preferred method of casting and drying is to use a so-called Microflake dehydrator (American Machine and Foundry Company).

(7) The resulting film was dried to a moisture content of about 30% and flash-fried in oil (or shortening) at about 375°F for a period of 10-15 seconds. Alternatively the film can be preserved by drying to lower moisture or by freezing and kept as a nonperishable semifabricated product.

L-TARTARIC ACID PROCESS

Y. Miura, K. Yutani and Y. Izumi; U.S. Patent 4,010,072; March 1, 1977; assigned to Tokuyama Soda Kabushiki Kaisha, Japan and Y. Miura, K. Yutani, H. Takesue and K. Fujii; U.S. Patent 4,017,362; April 12, 1977; assigned to Tokuyama Soda Kubushiki Kaisha, Japan report that the microorganism *Nocardia tartaricans* (ATCC 31191) produces an enzyme, cis-epoxysuccinate hydrolase which hydrolyzes cis-epoxysuccinate to produce L(+)-tartaric acid. L-tartaric acid is naturally present in grapes and other fruits, and is widely used in formulated fruit flavored juices, jellies and fillings.

Cis-epoxysuccinic acid or its derivatives can be easily obtained, for example, by epoxidizing maleic anhydride, maleic acid or a derivative of any of these with hydrogen peroxide. Examples of the derivatives of cis-epoxysuccinic acid are inorganic salts such as sodium, potassium, calcium and ammonium salts, organic salts such as primary, secondary and tertiary aliphatic amine, aromatic amine, pyridine and quinoline salts and esters or amides or aromatic alcohols or primary or secondary aliphatic or aromatic amines. Any derivatives can be used which are soluble in the aqueous medium and do not adversely affect the ability of the cis-epoxysuccinate hydrolase to hydrolyze the epoxy ring of cis-epoxysuccinic acid. The cis-epoxysuccinic acid or its derivatives may be in the form of crystals or the reaction mixture as obtained by a synthetic process. However, the presence of impurities in amounts which will retard the reaction, such as too much hydrogen peroxide, is not desirable.

When cis-epoxysuccinic acid or its derivatives are present at the time of growing microbial cells in culture media, the growth of the cells tends to be inhibited by too high a concentration of cis-epoxysuccinic acid. Therefore, the concentration of the cis-epoxysuccinic acid or its derivatives should be adjusted, for example, to not more than 5% by weight. Higher concentrations can be used when the microorganism or an L-tartaric acid-forming enzyme isolated is contacted with cis-epoxysuccinic acid or its derivatives in a system for hydrolyzing the cis-epoxysuccinic acid or its derivatives which is free from the growth of microbial cells.

Nocardia vaccinii, Nocardia lutea, Nocardia corallina, Nocardia salmonicolor, Nocardia rubra, and Nocardia opaca can be cited as species analogous to the strains of *Nocardia tartaricans* (ATCC 31191). The ES1-T strain and ES3-T strain differ from *Nocardia vaccinii* and *Nocardia lutea* in regard to the growth characteristics on a nutrient agar plate, the formation of mycelium, acid-fastness, the pH range for growth, and the hydrolyzability of deoxyribonucleic acid (DNA). Furthermore, these strains differ from *Nocardia corallina* in regard to the growth characteristics on a nutrient agar plate, the optimum growth temperature, tolerance to NaCl, the formation of acid from sucrose, and the utilization of phenol and m-cresol; and from *Nocardia salmonicolor* in regard to the growth characteristics on a nutrient agar plate, the growth characteristics on a glycerol-asparagine agar plate, the optimum growth temperature, the growth pH, tolerance to NaCl, the formation of acid from dextrin, and the utilization of p-cresol.

The contacting of the microorganism having the ability to produce cis-epoxysuccinate hydrolase or an enzyme isolated with cis-epoxysuccinic acid or its derivatives can be effected in various modes. For example, the contacting can be carried out while the microorganism is in the form of growing culture, culture broth, living cells, dried cells, pulverized cells or cell extract. It can also be in the form of purified enzyme.

cis-Epoxysuccinate hydrolase isolated from *Nocardia tartaricans* (ATCC 31190) and purified by the procedure described above has the following properties.

Catalytic action: hydrolyzes cis-epoxysuccinate, producing L(+)-tartaric acid
Molecular weight: 60,000-95,000 by the Andrews method
Substrate specificity: does not hydrolyze d- or l-transepoxysuccinate
Optimum pH: 8-9
Optimum temperature: 50°C in 20-80 minutes' incubation, and 45°C in several hours' incubation
pH stability: stable at pH 7.0, unstable at pH below 5.0 and pH over 10.0
Temperature stability: stable at temperatures below 40°C in 10 minute incubation at pH 7.0; lost activity at temperatures higher than 60°C in 10 minute incubation at pH 7.0

Example: Each 500 ml shaking flask was charged with 50 ml of a culture medium (pH 7.0) containing 1% of meat extract, 1% of peptone and 0.3% of sodium chloride, and the culture medium was sterilized at 121°C for 15 minutes. Sterilized sodium cis-epoxysuccinate was added in a concentration of 1% to the culture medium, and *Nocardia tartaricans* (ATCC 31191) was inoculated, and cultivated with shaking at 30°C for 24 hours. The cell concentration was 3.5 g/l. The cells were separated centrifugally from the resulting culture broth, washed once with a physiological saline solution and further centrifuged to collect the resulting cells.

The collected cells, 25 mg as a dry weight, were added to 25 ml of a 1 mol/l aqueous solution of sodium cis-epoxysuccinate (pH 7.0), and the reaction was carried out at 30°C. The results are shown below:

Time (hr)	Amount of Tartaric Acid Formed (g/l)	Yield of Tartaric Acid (%)
24	39.5	26.3
48	90.0	60.0
72	139.1	92.7
90	145.7	97.1

The maximum rate of tartaric acid formation was 2.04 g/l-hr which was 2.04 g/g of cells per hour. This corresponded to 47.6 units as the enzymatic activity of 1 ml of the original culture broth.

SOLUBLE TEA PRODUCTION

GREEN TEA/BLACK TEA TANNASE PROCESS

G.W. Sanderson and P. Coggon; U.S. Patent 3,812,266; May 21, 1974; assigned to Thomas J. Lipton Inc. describe an enzymatic process to convert green tea to black tea by treatment with tannase and natural green tea enzymes.

The term "green tea," includes freshly gathered tea leaves, that are dried immediately after gathering, fresh green tea leaves that have been heat treated before drying to inactivate their enzymes and aqueous extracts of these leaves. It also, in its broader sense, includes fresh green tea leaf material, the latter being defined as any tea leaf material containing the enzymes naturally present in the tissues in an active form.

Fresh green tea leaf includes the entire tea shoot tip and leaves from other portions of the tea plant, *Camellia sinensis*, as well as the stem material to which the leaves are attached. Thus, the term "fresh green tea leaf" includes freshly harvested tea leaf, withered tea leaf and fresh frozen tea leaf. In each instance, the green tea described above has undergone substantially no fermentation to the black tea state.

Fresh green tea leaf is comminuted to ensure that a major portion of the soluble tea leaf constituents are brought into the liquid phase of the tea conversion system. Comminution may be carried out by any convenient means which will not destroy the activity of the tea leaf enzymes, the latter being required for the subsequent step of converting the green tea to black. Suitable comminution techniques include cryogenic milling, i.e., milling the tea leaf while maintaining the tea leaf in a frozen state, and homogenizing the tea leaf in an apparatus, such as a Waring blender. The comminuted tea leaf is then suspended in water to form a fluid tea leaf homogenate slurry.

The tea leaf homogenate may be maintained under anaerobic conditions for the time it is prepared so as to prevent the tea conversion process, which is an aerobic process, from taking place until the preconversion tannase treatment is completed. Maintenance of anaerobic conditions may be accomplished by placing the homogenate in a closed vessel and sparging the homogenate with nitrogen gas to remove entrained oxygen. Conditions sufficiently anaerobic for satisfactory tannase preconversion treatment are obtained by placing a tea leaf homogenate in a container open to the atmosphere and by providing gentle mechanical agitation, where required, with a minimum of aeration.

The tea leaf homogenate is next given a preconversion tannase treatment by contacting the homogenate with a tannase enzyme preparation. The tannase enzyme is added to the homogenate in the amount of about eight enzyme units per gram dry weight of tea leaf tissue. The treatment is carried out for about 15 minutes to about 2 hours at about 45°C under anaerobic conditions with gentle continuous stirring.

Following completion of the preconversion tannase treatment, the tea conversion process is effected by adjusting the temperature of the tea leaf homogenate to about 25°C and sparging the homogenate with air. When the rate of aeration is adjusted so as to maintain about 25% oxygen saturation in the homogenate, a satisfactory conversion is effected in about 80 minutes. Hydrogen peroxide may be added to the tea leaf homogenate in controlled amounts to shorten the conversion period.

The end of the tea conversion process is signaled when the degree of oxygen saturation in the tea homogenate rises abruptly. Aeration of the homogenate is continued for about 10 minutes after this rise in oxygen saturation if hydrogen peroxide is used, at which time the conversion process is terminated by heating the homogenate at about 85°C.

While the exact mechanism by which the process operates is not understood with certainty, it is believed that tannase catalyzes a reaction which, in turn, leads to the formation of cold-water-soluble tea pigments during the tea conversion process. This is in contrast to the poorly soluble tea pigments which are formed in the conversion process when there is no preconversion tannase treatment. It is thought that, in the preconversion treatment, tannase catalyzes hydrolysis of the ester linkages between galloyl groups and various compounds known to be present in unconverted tea leaves. The reaction involved is believed to be a deesterification which may be written as follows:

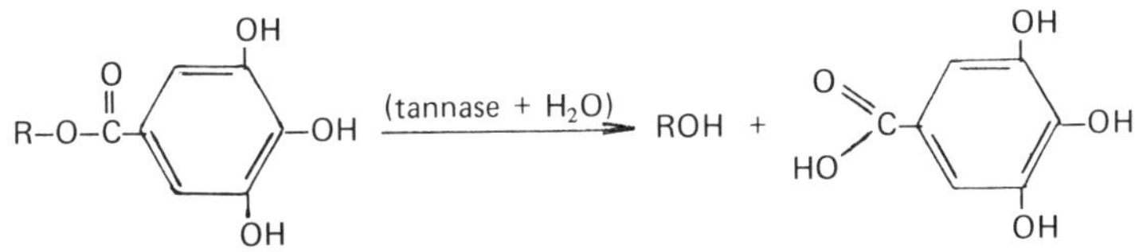

Gallic acid

where ROH is any of several different compounds present in green tea leaf, including epicatechin and epigallo-catechin.

It has been established that the effect of this deesterification mediated by tannase is to enhance the natural level of gallic acid and epicatechin in noncoverted green tea leaf material. This appears to favor the formation of large amounts of epitheaflavic acid during the tea conversion process. This reaction may be written as seen on the following page.

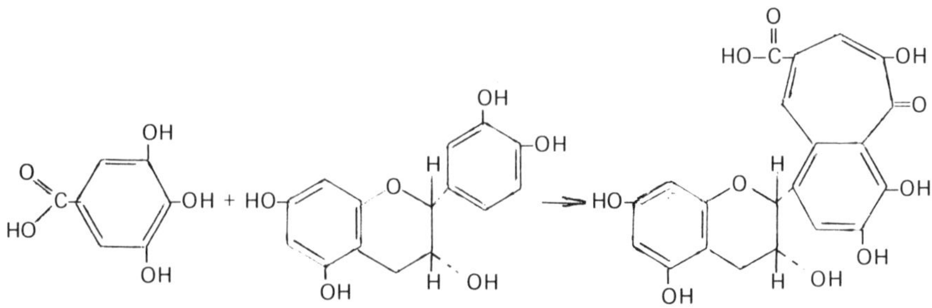

Gallic acid Epicatechin Epitheaflavic acid

Epitheaflavic acid has been shown to be formed in unusually large amounts during the tea conversion process in tea leaf materials which have undergone a preconversion tannase treatment. Further, epitheaflavic acid has been shown to have a desirable bright reddish black tea-like color, and it has very good cold-water-solubility properties. Accordingly, one of the major benefits derived from a preconversion tannase treatment is the setting up of conditions which lead to the presence of large amounts of epitheaflavic acid in an instant tea product.

In addition to the above, deesterification of the green tea leaf constituents may prevent the formation of any gallated tea oxidation products by eliminating the precursors of these compounds which are normally present in black tea infusions. The gallated forms of black tea constituents appear to be essential components of the cold-water-insoluble portion of black tea extracts. Therefore, elimination of the possibility of forming such poorly soluble compounds is probably important in producing instant teas with good color and solubility in good yield when the tea conversion process is carried out after a preconversion tannase treatment.

Example: Fresh green tea leaf was frozen to -40°C and cryogenically milled to a uniform particle size of about 0.5 to 1.0 mm^2 in a hammer mill. A 171 g sample of the milled fresh tea leaf, having a total solids content of about 25%, was slurried with 600 ml of water in the reaction vessel and nitrogen gas was flushed through the system while the temperature was equilibrated to 25°C by circulating controlled temperature water through the jacket.

Preconversion tannase treatment was carried out by adding 16 mg tannase enzyme, containing 320 units of activity to the tea leaf homogenate and holding the homogenate under anaerobic conditions with continuous nitrogen sparging for 2 hours at 25°C. By the end of the tannase treatment period, the pH of the green tea leaf homogenate had dropped to pH 4.7, and so it was necessary to adjust the pH of the homogenate back to the initial pH 5.6 with 2 N KOH.

The enzymic conversion of the green tea was then carried out in the following way: The fermentation was commenced by changing the sparge gas from nitrogen to air. The rate of aeration was adjusted to maintain the dissolved oxygen level in the homogenate at 25% saturation as determined by a dissolved oxygen meter. The temperature was held at 25°C throughout the ensuing conversion process. After 15 minutes of aeration, a 2 M hydrogen peroxide solution was added dropwise over a period of 20 minutes so that the total volume of hydrogen peroxide solution added was 16.0 ml. Conversion continued for about another 10 minutes until the dissolved oxygen level began to increase rapidly from 25 to about 40% saturation in the homogenate. The aeration was continued for an additional 10 minutes, after which the conversion was stopped by changing the sparge gas from air to nitrogen, and the pH was again adjusted with 2 N KOH back to pH 5.6.

The converted homogenate was heated to 85°C and held at that temperature for 5 minutes. The tea extract was then separated from the spent tea leaf by filtering the heated homogenate through Miracloth. Calcium chloride (1% on a tea solids basis) was added to the filtered tea extract, and the solution was held at 25°C for 45 minutes. Finally, the extract was cooled to 4°C and polished by centrifuging at 6,900 *g* for 10 minutes and freeze-dried to produce a cold-water-soluble instant tea powder. This high yield instant tea product, when made into a beverage with cold water at the normal solids concentration of about 0.35%, was tea-like, having excellent color and a milk reaction typical of beverage strength infusions of conventionally prepared black tea leaf.

TEA CREAM SOLUBILIZATION

Y. Takino; U.S. Patent 3,959,497; May 25, 1976; assigned to The Coca-Cola Co. gives the general process for manufacture of cold-water-soluble tea, and describes how tannase can be used to prepare high quality extracts.

Black tea is generally made from green tea leaves by subjecting tea leaves to a series of processing conditions including (1) withering, (2) rolling or otherwise disintegrating, (3) fermenting and (4) firing. Rolling or other disintegration initiates the enzymic reactions of fermentation, during which the characteristic color, flavor and aroma of black tea are developed. When the fermentation has progressed sufficiently, the tea leaves are fired (i.e., dried). This arrests the enzymic reactions and the resulting product possesses the characteristic appearance of black tea.

Black tea extracts or infusions are normally produced by a hot water countercurrent extraction process. Black tea extracts, and particularly dried black tea extracts, when made to beverage concentration, generally become turbid, if the beverage or extract is allowed to cool, for example, to room temperature or lower. The turbidity is caused by precipitated tea material, which in the trade is normally referred to as cream. Clouding or creaming has been a serious obstacle to the preparation of a stable commercial tea concentrate and to its acceptability by the consumer of soluble tea powders for use in cool beverages.

It has been proposed to eliminate the objectionable clouding or creaming by incorporating certain agents in the concentrate which are said to have the property of maintaining the insoluble ingredients in solution. Suggested suitable agents for the purpose include carbohydrates and various chemicals used to adjust the pH. In other instances it has been proposed to extract the tea with cold, rather than with hot water to avoid the presence of the less soluble ingredients which cause the clouding. However, the problems of clouding or creaming have not been completely overcome by these methods.

In order to produce a turbidity-free, water-soluble instant tea or tea extract, and particularly one that is soluble in cold water, it is generally necessary to separate some or all of the cream from the extract. This is usually accomplished by cooling the extract to cause the cream to precipitate. The cream is then removed, for example, by centrifuging. The cream solids, normally removed by centrifuging, contain a substantial proportion of the desirable tea color and typical tea astringency, and thus the supernatant remaining after centrifuging no longer contains these proportions of the desirable tea color and tea astringency.

If desired, the removed cream may be chemically solubilized such as by means of oxidation at elevated pH and temperature. Suitable oxidants include oxygen and hydrogen peroxide. pH is elevated using a basic material such as, for example, sodium hydroxide or potassium hydroxide. After oxidation, pH may be lowered using edible acids such as phosphoric or acetic acid. The supernatant fraction and the solubilized cream fraction may then be recombined, concentrated, and dried.

Alternatively, all or a portion of the cream may be discarded. The drying may be accomplished, for example, by spray drying, vacuum drying or freeze drying. However, if the chemical solubilization procedures on the cream are followed, much of the desirable tea color properties and typical tea astringency are destroyed or significantly reduced and a substantial portion of the natural color and flavor properties of the tea solids is lost.

The amount of cream removed by centrifuging varies with the leaf tea used in the extraction, the concentration of the tea extract, the temperature of the extract, and the gravitational force supplied by the centrifuge. The amount of cream removed, for example, at 7°C may be in the range between 20 and 35% of the extracted tea solids if the extract is at high concentrations (5 to 10% solids concentration) and as low as 4 to 8% if the extract is at lower concentrations (for example, 0.5 to 1% solids concentration).

The method of solubilizing tea cream by enzyme treatment utilizes tannase-producing molds, for example, *Aspergillus oryzae, Aspergillus niger* or *Aspergillus flavus.* The whole tea liquor may be treated with the tannase preparation prior to any centrifuging step. In the alternative, the whole tea liquor may be centrifuged prior to treatment with the tannase preparation. The isolated tea cream may be treated, or the isolated tea cream may be resuspended, for example, to a 4% solids suspension, prior to treatment with the tannase preparation. This latter method, however, entails an additional centrifuging. Any remaining cream may be discarded or treated, for example, by a chemical solubilization method, if desired.

Example 1: 1 g of black tea was extracted with 50 ml of boiling water for five minutes. 4 ml of the filtered extract was then mixed with 1 ml of enzyme solution prepared by dissolving 5 mg of partially purified tannase (23 units/mg activity) from *Aspergillus oryzae* in 50 ml of McIlvaine's buffer at a pH of 5.5. The solution was allowed to stand at 30°C for 60 minutes. Another 4 ml of the above tea extract was mixed with 1 ml of McIlvaine's buffer only. After incubation, the solutions were held at various temperatures for 30-minute periods, and the turbidity of the solution was measured in the turbidimeter. The results obtained are shown on the following page.

Standing Temperature for 30 Minutes	.... Turbidity (%)	
	Control	Enzyme Treated
20°C	15	5
15°C	35	7
10°C	60	7.5
5°C	80	8

The enzyme treated samples were further stored at about 5°C for two days without any increase in the turbidity.

Example 2: Approximately 1,000 ml of about 6% black tea extract was incubated with 0.6% tannase prepration (*Aspergillus oryzae,* 10 units per mg activity) for one hour at 50°C with intermittent stirring. The pH of the extract was adjusted from 4.5 to the normal pH of 5.0 using a sodium hydroxide solution, and was then cooled and centrifuged at 12,000 rpm for 10 minutes at 7°C in a refrigerated international centrifuge. The solids of the whole liquor and supernatant were determined, and the amount of 7°C insolubles was calculated to be 7.9%. An untreated sample (1,000 ml) of the same tea extract had 27.0% of 7°C insolubles.

The supernatant from the tannase treated extract, containing 92.1% of its tea solids, was evaporated on a rotary evaporator to approximately 40% solids, and was then dried in a laboratory Niro spray dryer. The resulting powder, when added to 7°C tap water at beverage strength (0.435%) produced a very clear solution of tea solids of pleasant reddish tea color and tea astringency. The supernatant from the untreated sample, containing 77% of its tea solids was also evaporated on a rotary evaporator, and dried in a laboratory Niro spray dryer. Powder made from the untreated supernatant, when added to 7°C tap water at beverage strength (0.435%) also produced a clear solution, but the beverage was very pale yellow and had no astringency.

TREATMENT WITH PECTINASE

In another enzyme application in the production of instant tea, *G.W. Sanderson and W.S. Simpson; U.S. Patent 3,787,582; January 22, 1974; assigned to Thomas J. Lipton, Inc.* use a pectinase which is added to tea extracts to remove those carbohydrates which cause foaming upon reconstitution with cold water. Also, the pectinase treatment enables a more efficient removal of suspended particles which affect clarity and color.

Prior to the addition of the pectinase preparation, an aqueous extract is prepared from tea leaves and the aroma may be stripped off and concentrated if desired. The extract is then decreamed by cooling it to a temperature at which heavy clouding (or creaming) or separation of a part of the extracted solid material from the solution occurs. The cream portion of the extract can be separated from the main portion of the extract by centrifugation and the separated tea cream can then be slurried with water, treated with KOH and oxygen sparged. The cream is then bleached by adding H_2O_2 and brought to a gentle boil. After cooling the mixture, the treated cream is added back to the decreamed extract. $CaCl_2 \cdot H_2O$ is then added to the cream add-back extract, which is held for a period of time and thereafter polished by centrifugation.

Following the decreaming and polishing steps, a pectinase enzyme preparation is added to the tea extract containing solubilized tea cream at a temperature of about 90°F. The enzyme-containing extract is then held for about 30 minutes after which the extract is subjected to a final polishing by centrifugation which removes the precipitated enzyme and other insolubles. The polished extract can then be concentrated, filled with corn syrup solids and spray-dried to a high bulk density, i.e., about 0.35 g/cc or higher or freeze-dried. The resulting instant tea powder is red in color and more brilliant in clarity than instant teas produced by the same procedure but without the pectinase treatment.

The use of the pectinase preparation produces no marked flavor taint and effectively eliminates the tendency of the instant tea powder to form foam when reconstituted. This is especially advantageous when the reconstitution is effected by machine dispensing. Moreover, by treating the cream add-back tea extract with a pectinase preparation, the stability of instant tea beverages toward hazing on standing is greatly improved.

The percent of the pectinase enzyme preparation added to the tea extract is critical because too little unduly prolongs the reaction, while the use of excessive amounts creates a haze problem of its own. Preferably about 0.2 to 3%, based on the weight of solids in the aqueous tea extract, is used. The temperature maintained during the enzyme treatment is from about 80° to 90°F. This range is critical since temperatures which are too low unduly prolong the treatment period and temperatures which are too high inactivate the enzyme. The properties of the particular pectinase preparation used should be taken into account in determining the optimum temperatures to be used.

The amount of time that the enzyme is held in the extract is determined by the amount of enzyme used and the temperatures at which the treatment is carried out. For example, as the amount of enzyme used approaches the lower limits of the above range a longer contact time between the enzyme and extract is required, and the converse of this is also true. Furthermore, as the temperature at which the reaction is carried out increases, the rate of the reaction also increases.

The pectinase enzyme preparation used in this work is Klerzyme 200 powder, Wallerstein Co. Klerzyme 200 powder was found to possess 640 units of pectinase activity per gram.

Example 1: 26 pounds of a blend of black teas were extracted in a 2-cell countercurrent extraction system. A water-to-leaf ratio of 8 to 1 was used in this work. The extract was concentrated under reduced pressure to 6.08% solids, cooled to 40°F, and the tea cream which formed was removed using a centrifuge. The cream, which amounted to 24.9% of the extract solids, was solubilized by alkalinizing with 13% (tea solids basis) KOH, air sparged for 10 minutes, and finally bleached with 13% H_2O_2. The solubilized tea cream was added back to the decreamed extract and the temperature was adjusted to 85°F. Then 2% Klerzyme 200 was added to the extract which was held between 80° and 90°F, and stirred gently for 20 minutes.

At this time, the extract was polished by adding 1.8% $CaCl_2$, total tea solids basis, cooling to 50°F and centrifuging. The polished extract was concentrated and filled by by adding corn syrup solids equal in weight to the tea solids. Different extracts prepared by this procedure were spray-dried to bulk densities of 0.294 g/cc, 0.321 g/cc, 0.432 g/cc, and 0.450 g/cc, the variations being obtained by introduction of varying amounts of a gas into the spray-dried feed according to methods known in the art.

Based upon visual observation, all of these instant tea products were much clearer and redder in color than a control product made by the same process but without the Klerzyme treatment. Further, the three products with the highest bulk densities had an appreciably reduced tendency to form foam on reconstitution with cold water.

Analytical data obtained on two of the instant tea products, i.e., products A and B which were treated with pectinase are summarized below. Analytical data for a control instant tea product, i.e., product C, which was prepared by the same procedure except that the pectinase treatment was omitted is also summarized in the table. The tristimulus color determinations are measurements of the transmitted light. The L value is the lightness factor, a value of 100 indicates no color, whereas a value of 0 indicates a very dark color. The a value is a measure of the red-blue range of colors whereas a b is a measure of the green-yellow line.

Product	Powder Bulk Density (g/cc)	Reconstituted Beverage Color and Clarity Tristimulus Color L	a	b	Subjective Description	Haze (Hunter units)	. . . Powder Characteristics. . . Rate of Solution	Reconstituted Beverage Foam
A	0.432	25.0	31.8	17.3	clear, bright, reddish	12.9	excellent	fair amount, active, clean
B	0.294	23.6	31.2	16.5	clear, dark red	14.8	good to excellent	fair amount, not clean
C*	0.104	24.4	25.4	17.4	slight hazy, brownish yellow	64.8	fair to good	fair amount, not active, dirty

*Control - instant tea product made without pectinase treatment.

Example 2: 10,100 pounds of an extract of tea solids that contained tea cream solids which had been solubilized by an alkaline/air sparge/hydrogen peroxide oxidation process similar to that described in Example 1 above was treated with pectinase as follows: 12.6 pounds of Klerzyme 200 was added to the tank in which the tea extract was being held at 83°F. The extract was stirred for 20 minutes after which 1.8% calcium chloride, based on total tea solids was added, and the extract was polished at 60°F by centrifugation, according to Example 1. The polished extract was concentrated, filled with an equal weight of corn syrup solids, and spray-dried. A control run, identical to that described above was made in which the pectinase treatment was eliminated. Analytical data obtained on these products are summarized in the table below.

Instant Tea	Tapped Bulk Density (g/cc)	Tristimulus . . . . Color L	a	b	Haze (Hunter units)	Foam on Reconstitution
Pectinase treatment	0.54	20.8	26.0	14.4	49.8	virtually none
No pectinase treatment (control)	0.52	27.8	27.9	19.2	74.2	some foam

These results show that the pectinase-treated instant tea product was redder in color (it had a larger a minus L value), clearer (the haze value was lower), and it formed far less foam than the control product.

BEER AND ALCOHOL PRODUCTION

BREWER'S WORT PROCESSES

Enzyme Malt Replacement

The production of beer, ales and lagers normally involves the initial formation of a wort in a mashing process followed by a fermentation process in which fermentable sugars such as maltose present in the wort are converted into alcohol and carbon dioxide. In the brewing of beer, the wort is produced by mashing a slurry of barley malt and adjuncts such as prepared cereals, unmalted raw cereal grains such as corn and rice, or some other carbohydrate source. Unmodified starch-bearing materials such as raw corn grits, must be precooked in a separate cooker before being added to the mash. This is generally done by mixing them with water and finely ground malt, and then boiling the mixture. The malt liquefies the starch material, thereby permitting the subsequent conversion of starch to sugar during the mashing operation.

In the mashing operation itself, the malt, by virtue of enzymes present, plays an important role. Thus, α-amylases liquefy the starch material of the grain producing mainly nonfermentable sugars like dextrin, while β-amylases saccharify the liquid starch to fermentable sugars, principally maltose. Further, proteolytic enzymes present break down the high molecular weight proteins to form lower molecular weight peptides and also significant amounts of amino acids and other intermediate fractions. These decomposition products of proteins not only provide nutrients for the subsequent yeast growth, but also contribute toward characteristic properties of the beer, for example, foam and haze stability, and a smooth mellow palate.

This reliance upon malt is attended by several significant disadvantages. For instance, the material is relatively expensive because of the high cost of barley of malting quality, the time and cost of converting barley into barley malt, and especially because of the investment, in both plant equipment and labor associated with its production. Moreover, malt contains husks (8 to 12%), and typically about 2 to 3% of a viscous, fatty liquid which tends to impart an inferior color and a bitter taste to the wort. For some time, the brewing art has recognized these factors, and proposals have been made to lessen the importance of malt in the manufacture of a nitrogenous wort.

Several processes were developed with the objective of process and product improvement. These processes vary slightly in specific details on the enzyme mixtures used, the composition and proportion of the cereal adjunct, the conditions for the mash cycle, etc. Three of these processes are given below in U.S. Patents 3,712,820, 3,716,365 and 3,713,840.

In the process of *M.F. Walmsley and J.V. Cross; U.S. Patent 3,712,820; Jan. 23, 1973; assigned to John Labatt Limited, Canada* unmalted barley or corn may be used to supply all the carbohydrates needed for fermentation instead of the more costly malted grains. The enzymatic process gives high concentration of fermentable sugars and high formol nitrogen content.

Referring to Figure 10.1, **11** indicates a stirred tank reactor in which the discrete protease and α-amylase mixture is added to an aqueous slurry of ground barley grains. A corn cooker **12** is connected to the tank reactor **11**. The pumped fluid is transported to a heat exchanger **20**. An outlet end of this heat exchanger is connected to a stirred tank reactor **22** which leads to a conventional filter press or continuous centrifuge **24**. This filter press is connected to an evaporator **26**.

The slurry in the stirred tank reactor **11** is heated by direct steam injection to an initial temperature of between about 40° and 50°C and with continuous and vigorous agitation, held at this temperature for between about 40 to 60 minutes. With proteolysis substantially complete, the liquefied corn adjunct from the corn cooker **12** is dropped into the reactor. The liquefied adjunct is thoroughly dispersed in the slurry by stirring. The temperature of the combined mass is then gradually raised to between about 75° and 85°C, again by direct steam injection, and this temperature held for a period of between about 10 and 40 minutes. At the end of this period, the liquefying process is completed, and the medium is no longer susceptible to gelatinization on cooling. The liquefied mass is continuously passed from the reactor **11** to the heat exchanger **20** in which it is cooled.

From the heat exchanger **20**, the medium, at or about the appropriate temperature for saccharification passes to the stirred tank reactor **22** to which the ground barley malt is added and dispersed by vigorous stirring. After saccharification is complete, the medium from the reactor is passed to the filter press **24**. The filtrate may be used directly in the manufacture of beer or passed to the evaporator **26** in which the wort is concentrated to a syrup ready for storage.

Example: Mash Cycle – The enzymic conversion was effected in the apparatus assembly shown in Figure 10.1. 24 liters of water were added to the stirred tank reactor **11** and the salt-stabilized enzyme-containing broth slowly stirred in at an enzymic level of 115 α-amylase units and 0.6 protease units per gram of barley subsequently to be incorporated. Then 6.205 kg ground barley were added and the slurry was vigorously stirred. The temperature of the aqueous slurry was raised to 44.5°C by direct steam injection. This temperature was held for 45 minutes, during which time the slurry was continuously stirred.

Figure 10.1: Production of Brewer's Wort Beer

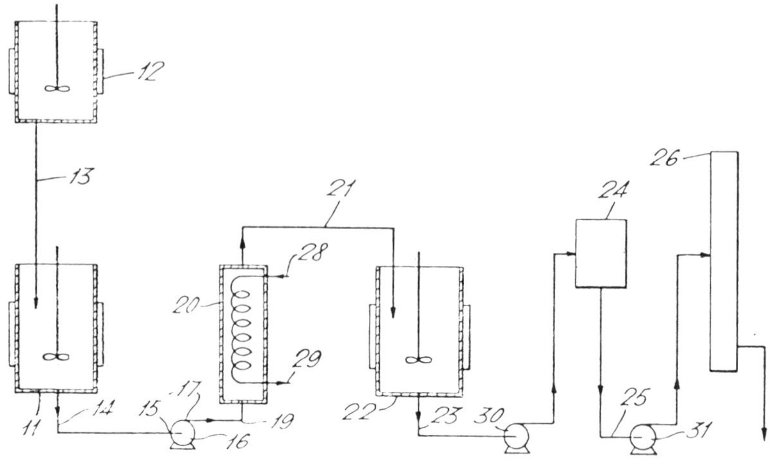

Source: U.S. Patent 3,712,820

Simultaneously the raw corn grits were liquefied. 12 liters of water were added to the corn cooker **12** and the salt-stabilized enzyme-containing broth slowly stirred in at an enzymic level of 14 α-amylase units per gram of raw corn grits subsequently to be incorporated. 3.458 kg raw corn grits were then added, and the slurry vigorously stirred. The slurry was first heated to 71°C and this temperature held for 16 minutes. The corn mash was brought to the boil (come-up time = 2 minutes), and held at boiling point for 2 minutes. The liquefied mash was then dropped into the barley mash in the stirred tank reactor **11** at the end of the 45 minutes proteolytic reaction time. The combined mass was vigorously stirred to disperse the liquefied corn mash.

The temperature of the combined mass was then gradually raised to 78°C (come-up time = 15 minutes), again by direct steam injection. This temperature was maintained for 25 minutes during which period the α-amylase enzyme liquefied (solubilized) the barley grains with the production of oligosaccharides and branched limit dextrins. Throughout this entire step, the mass was vigorously stirred.

The liquefied mass was then conveyed from the reactor **11**, through the heat exchanger **20** in which it is cooled. The residence time in the heat exchanger was around 20 minutes, at the end of which period the temperature of the medium had fallen to about 56°C. From this heat exchanger, the medium was passed to a second stirred tank reactor **22** to which 0.610 kg ground barley malt was added rapidly and without interruption, and dispersed by stirring. The residence time in this reactor was 1 hour, and during this period the temperature was maintained, by direct steam injection, at 56°C and the medium was continuously stirred. At the end of this period, the saccharified mash was pumped to a conventional filter press **24** in which the mash was allowed to settle for about 10 minutes. It was noted that the clarity of the filtrate (wort) collected was good. The run-off time was about 25 minutes. The wort was light-colored, and had a properly balanced amino acid and carbohydrate spectrum. This wort was then boiled in the kettle. Before the boil was started, 42.8 grams of hops were added to the wort in the kettle. The wort was boiled for 90 minutes. 30 minutes before the end of the boil, an additional 28.8 grams of hops and 2.9 grams Irish moss were added followed, 5 minutes before the end of the boil, with a further 14.4 grams of hops.

During the open boil the volume fell by evaporation from 61 to about 56 to 59 liters. At the conclusion of the boil the wort was placed in the hop jack where it remained for 10 minutes. The wort was then slowly run into the pan where it was allowed to settle for 30 minutes. An analysis showing respective wort properties (after this boil) is given in the table below, which also includes for comparative purposes, an analysis of a typical conventional wort suitable for the commercial manufacture of beer.

Property	Barley Enzyme Wort	Conventional Wort
Extract value, °P	11.7	11.9
Total nitrogen, mg/l	962	875
Formol nitrogen, mg/l	298	293
pH	5.2	5.1
Apparent attenuation, %	78	76

The amino acid spectrum of the wort when compared with the amino acid spectrum of a conventional malt wort showed no essential difference. The wort can be converted directly into beer by the following procedure:

Fermentation – A lager yeast *(Saccharomyces carlsbergensis)* was added at a rate of 100 grams pressed yeast per 40 liters wort. Wort was placed in a glass carboy and oxygen bubbled through to give dissolved oxygen level of 20 ppm. The yeast was then pitched and well-mixed with the wort. Fermentation was continued for 7 days at 14.4°C. At the end of the fermentation, the natural sulfur dioxide content of the primary storage beer was adjusted to 15 ppm by the addition of sodium metabisulfite.

Aging – A stainless steel keg was used for aging. The beer was dropped into the keg and 0.068 ml of the chill proofing enzyme Protesal was added and the keg put under 20 psig carbon dioxide pressure. Primary aging was effected at 1°C for 14 days, after which the beer was passed to filtration.

Filtration – The primary filtration was made through a sparkler filter containing 34 grams of a diatomaceous earth filter aid. The beer was filtered into another keg to which was added 7 grams Clearfil (a synthetic silicate filter aid). The beer was then filtered again through a sparkler filter containing 23 grams of a diatomaceous earth filter aid into another keg. Following secondary filtration the beer was carbonated to 2.8 to 3.0 volumes.

Bottling – The beer was bottled with the keg under 15 psig CO_2; before capping the bottle was tapped to release dissolved oxygen.

Pasteurization – The beer was pasteurized by heating at 60°C for about 2 minutes. The total time in the pasteurizing tunnel was 26.6 minutes, with an exit temperature of about 27°C.

Low Malt System

Additional details can be found also in the process descriptions given by *M.F. Walmsley and J.V. Cross; U.S. Patent 3,716,365; February 13, 1973; assigned to John Labatt Limited, Canada* which specifies low malt content, in the 8 to 12% range, cereal adjuncts and an enzyme mixture. Descriptions are given for the ingredients in the wort composition.

Aqueous Slurry: The cereal grain barley is present in the slurry at a concentration of between 28 to 33 grams per 100 cc water (ratio = 1:3.0 to 1:3.5). The hardness of the slurry water is between 20 and 35 equivalent parts by weight of Ca and Mg carbonates per 1,000,000 parts by weight of water; if the hardness is less than about 20 ppm then calcium chloride or some other calcium salt may be added to increase the hardness. The addition of calcium ions, in the form of a salt, at this stage for the purpose (aside from increasing the hardness) of enhancing the heat stability of the enzymes offers a convenient alternative to their incorporation elsewhere during the process, e.g., in the preparation of the enzyme mixture. The pH of the water is adjusted to between about 5.2 and 5.8. Usually the pH remains essentially the same throughout the process.

Proteolysis: In this step, the aqueous slurry containing the ground barley and at least the protease enzyme is heated at about 44° to 48°C for the required period of between 40 to 60 minutes with stirring. Heating within this temperature range for this period permits both proteolysis of the grain protein by the proteases and digestion of the barley grain by barley enzyme systems.

The proteolytic reaction is directly reflected in the total nitrogen content as well as the α-amino acid content (formol nitrogen) of the wort. Typically, in conventional worts, a total nitrogen content of around 800 to 950 mg/l and a formol nitrogen content of around 200 to 250 mg/l are considered satisfactory. With a protease level of at least 0.9 unit per gram and with the temperature held at between 44° and 48°C, these levels can usually be attained in the relatively short time of around 45 minutes.

Cereal Adjunct: At, or toward the end of the proteolytic reaction period (when proteolysis is substantially complete), a cereal adjunct is introduced into the main mass. The use of a cereal adjunct permits substantial cost savings and, at the same time, is considered to give a paler colored beer with a better shelf life.

The cereal adjunct may be derived from raw or unprepared starch-containing cereal grains such, for example, as corn grits, corn meal, wheat flour, barley flour, rice, rice flour, and sorghum corn. Alternatively, pregelatinized starch-bearing cereal grains may be used. The cereal grains should be used in an amount of between about 42 and 55%, by weight relative to the weight of cereal substrate. The prepared cereal grains may be introduced directly into the main mass for such materials are accessible to the enzymes in the liquefaction and saccharification stages.

The precooking operation may be performed by mixing the raw cereal grains, for instance, corn grits, with water and either finely ground barley malt or a suitable discrete α-amylase enzyme. The mixture is heated at about 70° to 80°C, for about 10 to 30 minutes to gelatinize the starch and liquefy it by the action of α-amylases derived from the malt or the discrete enzyme, and then boiled.

Solubilization and Saccharification: Following any addition of cereal adjunct, the discrete α-amylase enzyme and/or the malt are introduced into the mash if not incorporated during the initial step involving formation of the aqueous slurry of cereal grains. In the event the enzyme complex is used to provide the necessary α-amylase, some additional protease is also included at this stage.

To effect solubilization and saccharification, the temperature of the mash is raised to between about 60° and 80°C and held at a temperature within this range for the required period of about 30 to 120 minutes. During this period, the amylases are highly active in digesting the starch by acting upon, and breaking down, amylose and amylopectin polymers of which starch is composed.

Wort Separation: At the conclusion of the previous step, the temperature of the mash is raised for a brief period, for instance, 2 to 5 minutes, to over 80°C to inactivate the enzymes. Then, it is run off into a conventional brewery lauter tun or mash filter so as to separate the wort from the spent grains. Other separation methods, for example, centrifuging, or a combination of methods, such as lautering and centrifuging, may be used. The mash is filtered without cooling, but, if desired, may be cooled to ambient temperature before filtration. The filtered digest is then sparged and brought up to the desired volume.

The wort so-obtained may be used directly in making beer by the conventional process steps, so serving as a full replacement for a conventionally produced wort which simplifies the plant required and results in other economies. Alternatively, the wort may be evaporated to a syrup using, for example, a vacuum film evaporator. This syrup may then be stored until required, i.e., to increase the throughput of a conventional process at peak times.

Example: Various commercially available enzymes were blended together in defined proportions to provide different enzyme systems each of which was then used to convert a given grain bill into a brewer's wort using laboratory mash units (fitted with cooker and mash mixer cups).

Part A Grain Bill —

	Grams	Percent
Barley	66.4	100
Corn	29.2	44
Malt	5.3	8

Part A Enzyme System —

Amylase (Tenase)	
Amylase activity (u/ml)	14,100
Amylase level (u/g barley)	100
Volume (percent of barley) v/w	0.7
Protease (Ficin)	
Protease activity (u/g)	543.6
Protease level (u/g barley)	1.1
Weight (percent of barley) w/w	0.2
Amylase:protease	91:1

Part B Grain Bill –

	Grams	Percent
Barley	64.4	100
Corn	30.9	48
Malt	5.1	8

Part B Enzyme System –

Amylase (HT 1000)	
Amylase activity (u/g)	55,400
Amylase level (u/g barley)	140
Volume (percent of barley) v/w	0.25
Protease (Ficin)	
Protease activity (u/g)	543.6
Protease level (u/g barley)	0.9
Volume (percent of barley) v/w	0.16
Amylase:protease	155:1

Combined Discrete Enzyme System

Specific details are given by *R.C. Quittenton; U.S. Patent 3,713,840; January 30, 1973; assigned to John Labatt Limited, Canada* for the discrete enzyme system used in the production of brewer's wort in which raw starch-containing materials (unmalted raw cereal grains) are used as partial replacement for barley malt.

The expression "discrete enzyme" as used in relation to the protease, α-amylase, amyloglucosidase and β-amylase enzymes refers to an enzyme derived from a plant, bacterial or fungal source, and which has been extracted and purified on an industrial scale, and which manifests a significant protease, α-amylase, amyloglucosidase and/or β-amylase activity. Other enzymes, aside from these particular four enzymes may also be present such, for example, as cellulases, hemicellulases and pectinases. Further a discrete enzyme mixture containing significant amounts of two (especially the protease and amylase), three or all four enzymes may be used in the process.

Protease and α-Amylase Mixture: With a suitable bacterial species, and using appropriate fermentation conditions, an enzyme mixture comprising a complex of extracellular protease and α-amylase enzymes may be isolated using a bacterium of the genus *Bacillus,* conveniently a strain of *Bacillus subtilis* designated ATCC 21556.

Amyloglucosidase: The amyloglucosidase enzyme, also known as glucoamylase or limit dextrinase, hydrolyzes and splits α-D-(1→4) and α-D-(1→6) linkages in the starch oligosaccharides with the removal of successive glucose units from the nonreducing ends of the oligosaccharide chains. The primary effect of the amyloglucosidase enzyme is to convert carbohydrate material into the fermentable sugar dextrose. The specific action of this enzyme in hydrolyzing (1→4) and (1→6) linkages with the production of the fermentable sugar dextrose reinforces the action of the amylases, especially the saccharifying amylase, usually a β-amylase, in producing such sugars from the oligosaccharides. This effect of the amyloglucosidase enzyme is reflected in a higher fermentable sugar content, i.e., higher apparent attenuation, in the wort and, accordingly, a higher alcohol content, in the beer. The discrete amyloglucosidase enzyme may be derived from a fungal source such as, for example: *Aspergillus niger; Aspergillus awamori; Aspergillus oryzae; Rhizopus javanicus, Rhizopus niyveus;* and *Endomyces* sp.

The β-amylase enzyme, under suitable conditions, attacks at the nonreducing ends of the amylose and amylopectin chains in the starch, and proceeds by stepwise removal of maltose units. An inversion of the D-glucosidic linkage occurs, and the maltose liberated is of the β-configuration. Amylose with an even number of D-glucose units is converted completely to maltose while amylose with an odd number of units is converted to maltose and maltotriose which contains the reducing D-glucose unit of the original amylose chain. Amylopectin is hydrolyzed like amylose beginning at the nonreducing ends of the the other chains, though (1→6) linkages present in the amylopectin are not attacked by the amylase and remain as residual or branched-limit dextrins. The main effect of the β-amylase attack is to produce reducing sugars, principally maltose, which are available for subsequent conversion, in the fermentation process, to alcohol.

In this procedure (see Figure 10.2), all of the enzymes are added, in an initial step, to the aqueous slurry of barley grains at a pH of between 5.0 and 6.5.

Figure 10.2: Process for Making a Brewer's Wort

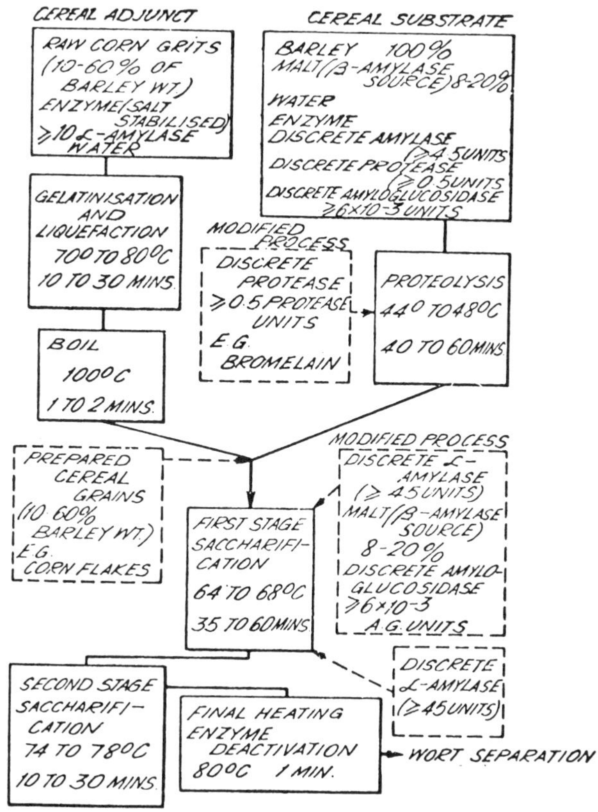

Source: U.S. Patent 3,713,840

The enzymes may be added simultaneously, in the form of an enzyme blend, or consecutively. Thereafter, the enzymes are thoroughly dispersed in the slurry by vigorous agitating. The enzyme-containing slurry is then heated at about 40° to 55°C, preferably at about 44° to 50°C, for a period of between about 30 and 240 minutes, usually between about 30 and 120 minutes and preferably between about 40 and 60 minutes. While heating, it is desirable to agitate the slurry vigorously, as by stirring, to ensure intimate contact between the barley substrate and the enzyme(s). Heating within this temperature range for this period permits both proteolysis of the grain protein by the proteases, and digestion of the barley grain by barley enzyme systems.

Single Protease Method

Another process which permits reduction of malt content by using raw cereal grains and an enzyme complex is described by *M.W. Brenner; U.S. Patent 3,795,745; March 5, 1974; assigned to Schwarz Services International Limited.* A proteolytic enzyme alone, not in combination with α-amylase, is employed in this case. This proteolytic enzyme liberates the bound and latent enzyme systems naturally present in the unmalted grain and permits the raw grain to be used in conventional brewing procedures and techniques without having to carry it through the costly and lengthy malting process.

The mash bill includes three components: a raw, enzymically rich cereal grain, typically barley, malted barley and a starchy, low-enzymic cereal adjunct, such as corn grits. Many successful experiments have been performed using a mash bill consisting of 35 parts by weight ground barley, 25 parts by weight malted barley and 40 parts by weight corn. It will be seen that this sum of 60 parts of barley and malt in relation to 40 parts of corn closely approaches the conventional ratio for the production of beers in which malt and adjunct are used.

Process: The barley used in the mash bill is crushed, ground or rolled to the desired size. Prior or subsequently to the barley grains being ground, granulated or comminuted they may be steeped or soaked in water, in the ratio of 1:1, at a temperature of from about 20° to 60°C. Such a pretreatment conditions the barley grains and permits more rapid penetration of the proteolytic enzyme into the interior of the grain during the mashing process. Alternatively, the barley grains may be exposed to an atmosphere of water vapor or steam to increase the moisture content of the hull without wetting the endosperm.

A slurry is prepared in which the barley is present in an amount of between about 20 and 40 grams per 100 cc water. The slurry water has a calcium hardness equivalent to not less than about 100 ppm by weight of $CaCl_2$. In cases where the calcium level or the water supply is below the desired level, adjustment, for instance, by addition of calcium chloride, may be made. The discrete proteolytic enzyme, papain, is added to this aqueous slurry in combination with an activator such as sodium metabisulfite. The quantity of enzyme required is relatively small, but the actual amount depends upon the activity level of the particular proteolytic enzyme used.

For instance, a commercially available papain enzyme has an activity level of between about 30 and 300 Azocoll units per mg and this particular enzyme is used in an amount of between about 50 parts and about 150 parts of 100,000 parts unmalted cereal grains by weight. This aqueous slurry is held at a temperature around 40°C, for between 10 and 60 minutes. The cereal adjunct is mashed separately in a cooker in a conventional manner using a small amount of bacterial α-amylase to help keep the viscosity of the boiling adjunct-water mixture sufficiently low so that it may be pumped or handled through the usual pumps, pipelines, and other brewhouse equipment.

The prepared adjunct mash is transferred from the cooker to the main barley mash and the combined mash thoroughly stirred. The temperature of the combined mash during this mashing should not exceed 70°C, so as to conserve enzymic activity. The small amount of malt, i.e., 10 to 30 parts by weight, is usually also incorporated in the mash at this stage.

After the mixture of the liquefied nonbarley adjunct with the main mash, the latter also containing the major portion, or all of the malt used in the particular mash bill, the combined mash is allowed to rest at temperatures between about 60° and 75°C, a predetermined period of time whereupon the enzymes, principally β-amylase, liberated from the barley by the action of the proteolytic enzyme act autolytically and upon the starch of the cereal adjunct to produce, along with the saccharifying enzymes present in the malt, the desired content of fermentable sugars, without the need to add glucoamylase or similar enzymes to achieve normal saccharification or fermentability. This saccharification may take anywhere from 1½ to 6 hours. If saccharification is permitted to continue for too long, the physicochemical and organoleptic properties of the beer manufactured from the wort may be detrimentally affected.

At the conclusion of the saccharification step, the mash may be briefly heated at a higher temperature, i.e., 72° to 78°C. This heating tends to destroy the enzyme systems and therefore to stabilize the composition of the wort. After the heating, the mash is transferred into, for example, a conventional brewery lauter tun, mash filter or a centrifuge where the wort is separated from the spent grains.

When barley is used as the cereal substrate in the foregoing procedures, a properly balanced, light-colored wort with satisfactory starch and protein breakdown may be obtained using simply a discrete proteolytic enzyme. Moreover, such a wort normally has comparable fermentable sugar and nitrogen contents, as indicated by apparent attenuations of around 75% or more, total nitrogen levels of over 900 ppm or more and formol nitrogen levels of around 250 to 350 mg/l, to the worts derived from more complex enzyme preparations such as mixtures of a proteolytic and an α-amylase enzyme using the same mash bills or the worts derived from a conventional 65:35 malt:adjunct mash bill.

The wort may be used directly in making beer by the standard process steps, so serving as a full replacement for a conventionally produced wort which simplifies the plant required and results in other economies.

Enzyme Viscosity Reduction

A mixture of enzymes is used by *D.H. Clayton and J. Karkalas; U.S. Patent 3,679,431; July 25, 1972* in wort production for beer manufacture to reduce the viscosity which in turn facilitates the action of starch liquefying enzymes. The enzyme mixture contains a commercial bacterial α-amylase derived, for example, from *Bacillus subtilis;* β-amylase is added as the saccharifying enzyme and a bacterial protease as the proteolytic enzyme. The wort is treated with the proteolytic enzyme and the saccharifying enzyme simultaneously at a temperature about 55°C, the medium being cooled to the required temperature following the liquefaction step; the pH is preferably 5.5. By cooling the liquefied product to a temperature within the temperature range and adjusting the pH value to between 5 and 6, an optimum effect of both the proteolytic enzymes and the saccharifying enzymes is obtained.

Following the treatment with proteolytic enzyme and saccharifying enzyme the resultant wort is, for example, filtered and passed directly to normal subsequent processing to produce beer. Alternatively, for example, the wort is filtered and evaporated to a syrup which is stored and subsequently diluted to provide a wort for normal subsequent processing.

If malt is used as a source of the saccharifying enzyme or if starch is used as an enzyme substrate it is preferred to heat the medium treated with proteolytic and saccharifying enzymes to a temperature of between 70° and 80°C before further processing. In this manner the nongelatinized starch contained in the malt is converted into soluble carbohydrates and the yield of soluble carbohydrates in the wort produced is consequently increased. Furthermore, the filtration of the mash is considerably improved. A flow diagram for the process is shown in Figure 10.3.

Figure 10.3: Wort Production

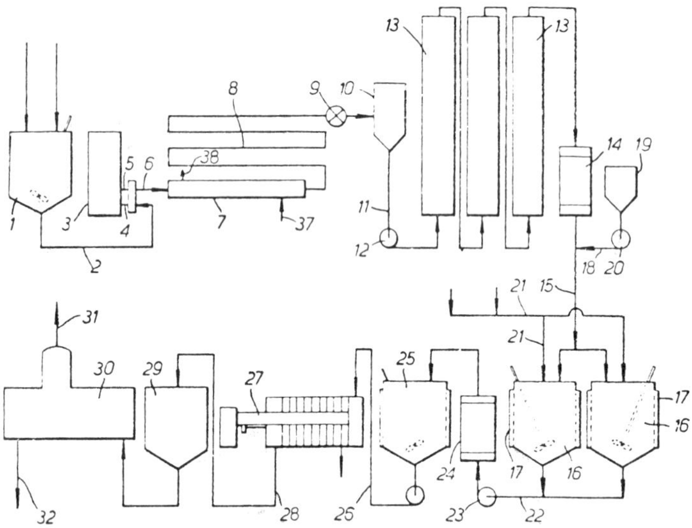

Source: U.S. Patent 3,679,431

In the figure, **1** indicates a vessel in which commercial bacterial α-amylase is mixed with a slurry of finely ground unmalted barley; the barley may be wet-ground or ground in a dry condition. Connections are provided to introduce malt or other saccharifying enzymes and proteinase into the saccharification tank **16**. The slurry from the vessel **1**, which is adjusted to a pH value of about 6.5, is pumped by means of a diaphragm pump through the heat exchanger **7**, and the tubular reactor **8** in a continuous rhythmically pulsating flow movement. During the passage of the medium through the heat exchanger **7**, it is gradually heated to the selected liquefying temperature, which is about 86°C and the medium is maintained at this temperature while passing through the tubular reactor **8**. The length of the tubular reactor **8** is such that when the medium has passed through the tubular reactor **8**, the liquefying process has proceeded to a point where the medium is no longer gelatinizable by cooling. The viscosity of the medium entering the tubular reactor **8** is reduced as a result of the enzymatic liquefaction conditions in the heat exchanger and reactor.

The medium is collected in the cyclone collector **10** and is subsequently pumped through the continuous reactor vessels **13** which are maintained at a temperature of about 86°C. The reactor vessels **13** are constructed in such a manner that no back mixing takes place. The total residence time of the medium passing through the reactor vessels is between 2 and 5 hours, depending on the amount of liquefying enzyme added. When the medium leaves the reactor vessels **13** the dextrose equivalent value is for example about 15%. In the cooler **14** the liquefied medium is cooled to a temperature of about 55°C and subsequently the pH value is adjusted to about 5.5 by introducing acid such as hydrochloric acid from the reservoir **19**. The acidified medium is then passed to one of the tanks **16**.

Malt and proteinase are introduced into the tanks through the appropriate pipe **21** and the liquefied medium which is maintained at a temperature of about 55°C is subjected to the influence of the saccharifying enzymes and the proteinase. The malt which is introduced into the tank **16** contains some nongelatinized starch and in order to convert the starch into soluble carbohydrates the medium is heated to a temperature of about 75°C in the heater **24**, and then passed to the tank **25** in which it is maintained at a temperature of 75°C for a time sufficient to convert the starch introduced in the form of malt to soluble carbohydrates.

The medium is then pumped through the filter press **27** in which the insoluble materials are separated in the form of a filter cake. The filtrate passes through the pipe **28** to the filtrate tank **29** and from this tank to the evaporator **30** in which it is concentrated to a total solids content of about 80%.

Example: 48 kg barley flour (43 kg dry substance) were mixed with 120 kg of water. The pH value was adjusted to 6.5 with Na_2CO_3 and 66 grams $CaCl_2 \cdot 2H_2O$, (calcium ions were found to enhance the stability of the α-amylase at high temperature) were added followed by 430 grams of commercial α-amylase preparation. The resulting slurry was pumped through the heat exchanger **7**, the reactor **8** and the vessels **13** of the apparatus at a rate of 128 l/hr. The total liquefaction time was 4 hours 43 minutes at 86°C.

The temperature of the product was lowered to 55°C in the cooler **14** and the pH adjusted to 5.5 by means of HCl. Then 800 grams malt and 400 grams commercial proteinase were added. After 16 hours the product was ready for filtration in the filter press **27**. The filtrate had a clear amber color and a pleasant flavor. It was concentrated to about 80% by weight total solids in the evaporator **30**. The resulting wort concentrate is stable for long periods of time and can be diluted to any desired concentration before fermentation. The wort composition is in good agreement with the composition of wort prepared by the traditional method where malt is the starting material.

Chit Malt Enzyme Process

Another process for substitution of regular barley malting using special enzyme systems is shown by *A. Delhaye, M.M. Moll and J.R. Pollock; U.S. Patent 4,004,034; January 18, 1977; assigned to Grandes Malteries Modernes and Groupement D'Interet Economique Tepral, both of France and Pollock International Limited, Bermuda.* The treatment cycle requires steeping barley in water; then the barley is separated from the steeping water and exposed to air inside a chamber in which the temperature is kept at between about 10° and 40°C. A temperature much below 10°C, would, in effect, markedly slow down the process and a temperature much above 40°C would denature the barley proteins.

The duration of this exposure must be sufficient to cause the barley to germinate, that is to say to cause the appearance of the white tip of the shoot emerging from the husk of the grain. Experience shows that the minimum duration necessary for the appearance of this shoot is about 5 hours. This treatment, which results in the chitting of the barley, is generally stopped before clearly distinguished rootlets appear.

The chit malt is ground with a mill to reduce it to particles, the diameter of which is sufficiently small to make a good subsequent extraction possible. It is then extracted with a mixture of hot water, malt and an enzymatic agent, especially an amylase protease and/or β-glucanase. Once the malt and the enzyme or enzymes have acted on the chit malt, the liquid phase, which forms the wort, is separated from the insoluble parts forming the draff.

For storage, it is convenient to subject the chit malt to a treatment to stop the development of the biological processes occurring in it, while retaining its enzymatic potential. Such a treatment can consist of storing it at a low temperature, of the order of 0°C, or of drying it, for example, with air so that the temperature of the grain does not exceed about 45°C. When a dehydration is used, it is continued until the water content of the grain is less than 15%.

The chit malt may be rolled (i.e., the grain is flattened, possibly with an alteration in its internal structure). If this rolling is carried out before the dehydration treatment, this dehydration treatment, as well as the subsequent mixing and grinding operations, is facilitated. If it is carried out after the dehydration, only the subsequent mixings and grinding are facilitated.

The rest of the operations for producing the chit malt and the subsequent treatments to which the latter is subjected, are carried out under conditions to assist the appearance and retention of a maximum enzymatic potential. To achieve these conditions, it is essential to take care that the grain has sufficient oxygen and is not asphyxiated, for example by an accumulation of CO_2; it is also necessary to prevent any deterioration caused by abnormal heating.

The enzymatic agent, which may be used in the process may contain at least one amylolytic enzyme, for example, α-amylase, and/or one proteolytic enzyme. The enzymatic mixture has an activity of the amyloglucosidase type, the β-amylase type, the pullulanase type and other types. To improve the yield of the process, it can be advantageous to remove the husk of the barley, at least partially, before subjecting it to the various stages of the treatment. Furthermore, in order to accelerate the germination and to improve the qualities of the wort, it can be valuable to add to the barley, at or after the first stage of steeping, an amount of gibberellic acid sufficient to give a concentration as much as 5 ppm (based on the dry weight of the grain).

To accelerate the start of the germination, substances such as hydrogen peroxide, H_2O_2, can be added to the steeping water. Furthermore, the addition of lime makes it possible to dissolve certain undesirable constituents of the husk of the grain, such as resins and tannins.

Example: Some barley is treated by a mechanical process to remove a part of its husks (for example 5 to 6% of the total dry weight of the barley). The barley is then immersed in water for 1 hour and 30 minutes at 25°C. The water is poured off and the barley is left uncovered for 6 hours, aerating it continuously or intermittently to retain a temperature of 25°C in the mass of grains. At the end of this period, the moisture content of the grain is about 30%.

The grain is then immersed for a second time, for 1 hour, in water to which 0.5% of hydrogen peroxide has previously been added, and the water is then poured off and the grain is left uncovered again for 15 hours and 30 minutes, aerating it continuously to retain a temperature of 25°C.

The chit malt is then rolled by passing it between two rollers, so as to alter its shape slightly and to alter its internal structure. The grain is then moistened with a solution containing amylolytic and/or proteolytic and/or α-glycanasic enzymes. After this addition, the grain is dried by a stream of hot air, the temperature in the grain remaining below 40°C; this drying is stopped, for example, when the moisture content reaches 12%. The chit malt can be transported like ordinary malt and can be stored for 2 weeks or more before being used.

Before use, the stored chit malt is ground in a mill, for example, a roll mill of the conventional type used in brewing. The malt which, in the rest of the treatment, is mixed with the chit malt, for example, in a proportion of 20%, can be mixed with the chit malt before the grinding, or it can be ground separately and then mixed. This mixture of the various constituents is used to manufacture a wort intended to be fermented for the purpose of producing beer. A conventional brewing method can be used.

For example, the mash may be made from: 30% of maize in the form of grits (maize semolina), or maize starch, or glucose syrup; 20% of malt which has been well broken up and which has a high enzymatic potential; and 50% of chit malt, treated as described above. The maize is made into a paste (doughed in) separately and it can be worked, by a conventional brewing process, with part of the malt or with an enzyme which has a high α-amylase activity.

The mash (malt + chit malt) is made into a paste at 52°C and this temperature is maintained for 45 minutes. The soaked maize is then combined with the mash and the brewing is continued in a conventional brewing manner. The enzymatic agent added while the mash is made into a paste is, e.g., a preparation of amyloglucosidase, used in a proportion of 0.05 to 0.08% (based on the dry weight of the mash).

BEER CHILLPROOFING PROCEDURES

Decoupling Protein-Tannin Enzyme

I.M. Stone; U.S. Patent 3,749,582; July 31, 1973; assigned to Baxter Laboratories, Inc. uses a decomplexing enzyme for the protein-tannin complex in beer to prevent the development of haze and turbidity (chillproofing) in fermented malt beverages.

The chillproofing of fermented malt beverages is a necessary step in the brewing process. The beer-drinking consumer expects to receive a beverage that is sparklingly brilliant in clarity and one that will retain this clarity on long storage at low temperatures. The slightest haze appearing in the cold beer could be a cause for consumer complaint and rejection. The step in the brewing process that insures this crystal clarity under varied storage conditions and counteracts the deleterious effects of time and temperature is known as chillproofing.

The procedure widely used in the industry for chillproofing of beer involves the addition of a trace of proteolytic enzyme to the cold beer contained in tanks. On further cold storage and filtration a chillproofed beverage results. The proteolytic enzyme most generally used for chillproofing is papain, although ficin, bromelain or other plant proteases can also be used. The chillproofing action is an enzymatic breakdown of the traces of high molecular weight proteins in the beer into smaller fragments which are more soluble and less precipitatable.

Another enzymatic chillproofing process utilizes a mixture of chillproofing enzymes obtained as the extract or the concentrate of the extract of the water-soluble growth products of certain mold microorganisms such as *Aspergillus niger* and other mold microorganisms. A product containing a mixture of chillproofing enzymes can be prepared by extracting the mold growth with water which can be used after inactivating the limit dextrinase.

The specific mode of the chillproofing action of the mixture of enzymes contained in the enzyme complex preparation is not completely known. It is known, however, that the chillproofing activity is not entirely proteolytic, as the proteolytic activity in the enzyme complex preparation is not sufficient for this purpose, being much less than that of commercial proteolytic enzymatic chillproofing preparations. A larger part of its action is due to its strong decomplexing action on the protein-tannin complex in the beer. The appearance of a chill haze in beer is generally regarded as due to the formation of highly insoluble protein-tannin combinations. The enzymes contained in the enzyme complex preparation prevent the formation of these protein-tannin complexes and tend to decomplex any protein-tannin combinations formed before the enzyme complex preparation is added.

In the chillproofing of American-type beers prepared with about 70% malt and 30% adjuncts such as corn or rice, about 15 to 25 ppm of the enzyme complex preparation are required to produce a satisfactorily chillproofed beverage containing only the tannin bodies derived from the malt and hops. It has been found, however, that if these beers are given a preliminary treatment with tannic acid, a synergism exists between the tannic acid treatment and the enzyme complex preparation such that there is an approximately tenfold reduction in the amount of the enzyme complex preparation required to eliminate haze development and produce a highly stable beer. With a pretreatment of from about 25 to 80 ppm tannic acid, only from about 1 to 3 ppm of the enzyme complex preparation is required instead of the usual 15 to 25 ppm.

It is known that tannic acid or tannins have been proposed for the chillproofing of beer for a long time. However, the tannic acid chillproofing of beer has not found wide acceptance by the industry because its use is beset with many practical difficulties.

A significant drawback of tannic acid chillproofing is the necessity for adding the precise stoichiometric amount of tannic acid to exactly precipitate the protein-type compounds in the beer. If too much or too little tannic acid is added, very poor chillproofing results are obtained and the resulting beers are unstable and become hazy on storage. Because of batch-to-batch variations in beers it has been a practical necessity to titrate each tank of beer before tannin addition to determine the precise stoichiometric amount of tannic acid required for that particular tank.

The use of the enzyme complex preparation in combination with a tannic acid pretreatment of beer eliminates the need for this exact control of the levels of tannic acid additions. Levels of tannic acid below or above the stoichiometric amount can be used without experiencing the development of hazes and turbidities in the finished packaged beer.

In the example below the enzyme complex preparation is obtained as follows: A wheat bran mash is prepared from wheat bran, wheat flour and water, and is sterilized. 125 grams of the sterile mash, contained in a Fernbach flask, is inoculated with an actively growing culture of *Aspergillus niger.* After suitable growth for about 5 days at 30°C, the resultant bran-growth product is dried at low temperature until ready for extraction. 10 parts by weight of water is then mixed with the bran and stirred or shaken continually for a period of an hour. The liquor is expressed from the wet bran by pressing (centrifugation or filtration can also be used). The liquor thus obtained is then concentrated by low temperature vacuum evaporation to a specific gravity of approximately 12°Bé. The chillproofing materials are precipitated from this concentrated liquor by adding ammonium sulfate to give salt concentration of about 25 to 33% by weight. This mixture is permitted to stand to allow the precipitate to form and settle.

Some filter aid is added and the precipitate is centrifuged (filtration can also be used) to obtain a relatively dry cake. The cake is then redissolved in about 1.5 parts by weight of water at about 25°C and the pH is adjusted from its original value of 4.5 to 5.0 to a value of 1.8 with hydrochloric acid. The pH is held at this value for about 1 hour at this temperature and is then readjusted back to its original pH value with sodium hydroxide solution. This liquor is then refiltered and the clear filtrate is dried by low-temperature vacuum drying (spray drying or freeze drying can also be used) to give a friable powder which is retained as the enzyme complex preparation.

Example: End fermented beer is decanted from the settled yeast and transferred into 3 ruh storage tanks. 32 ppm of tannic acid is added to the beer of each of two tanks during the transfer. To the beer in the third tank is added 20 ppm of a commercial proteolytic enzyme (papain) chillproofing preparation during the transfer. The beer is held at approximately -2°C overnight and 2 ppm of the enzyme complex preparation is added to one tank of the tannic acid treated beer and thoroughly mixed. Three chillproofing treatments have thus been prepared in this example: (1) tannic acid alone, (2) tannic acid plus the enzyme complex preparation, and (3) commercial proteolytic enzyme (papain) treatment.

The beer is bottled off and pasteurized. Some finished bottles from these 3 batches are subjected to storage stability tests by keeping the bottles at 30°C and each week the bottles are placed for 18 to 20 hours in a 0°C bath. After this cold treatment the amount of haze developed is measured with a Radiometer Haze meter. The bottles are then returned to storage at 30°C for another week and the cycle repeated. The table below shows the chill haze developed after various storage periods. The haze measurements are reported in Formazin Turbidity Units (FTU). The higher the FTU figure the more the haze. A reading in FTU of about 75 to 100 is a noticeable but allowable haze while about 150 and higher is an objectional haze.

	Chill Haze Measurements After Storage					
	.... Formazin Turbidity Units*.........					
		4	8	12	16	24
Chillproofing Treatment	**Initial**	.... (weeks)...........				
Tannic acid, 32 ppm	36	65	135	190	270	400
Tannic acid, 32 ppm, plus enzyme complex preparation, 22 ppm	28	50	45	40	45	60
Double-strength commercial proteolytic enzyme preparation, 20 ppm	64	80	130	160	234	234

*Each result is the average of measurements on three bottles.

Tannic Acid Elimination

In the refinement and modification of the process for chillproofing beer with decomplexing enzymes *I.M. Stone; U.S. Patent 3,770,454; November 6, 1973; assigned to Baxter Laboratories, Inc.* uses alkali metal sulfite salts or alkali metal or alkaline earth metal ascorbate salts as additives for stabilization against the effects of oxidation. These additives eliminate the need to use tannic acid in the enzyme decomplexing step.

Chillproofing can result from the combined use of tannic acid and the enzyme complex preparation but relatively few U.S. breweries use tannic acid chillproofing because there is a reluctance on the part of many of the other breweries to convert to a process which requires the use of tannic acid.

It has been found, however, that similar synergistic chillproofing results are obtained when the enzyme complex preparation is used in combination with at least one substance selected from the group consisting of the alkali metal sulfite salts and the alkali metal and alkaline earth metal ascorbate salts. When the beer is treated with a total of from about 10 to 100 ppm of one or more of these substances, only from about 1 to 5 ppm of the enzyme complex preparation is required instead of the usual 15 to 25 ppm.

Example: A purified enzyme complex is prepared as described in the previous patent (p 339). The enzyme complex preparation is mixed with sodium metabisulfite and sodium ascorbate in the following proportions: 3.3 parts by weight enzyme complex preparation; 37.4 parts by weight sodium metabisulfite; and 59.3 parts by weight sodium ascorbate.

The mixture is then added to finished beer before placement in cold storage at a concentration of 40 ppm (equivalent to 1 pound per 100 barrels). Excellent chillproofing of the beer is obtained in extended cold storage of the beer.

Mucor pusillus (Milk Clotting) Enzyme

Another approach in the application of enzymes to inhibit haze and turbidity in beer, or chillproofing, is the use of milk clotting enzymes as described by *J.H. Nelson and P.R. Witt; U.S. Patent 3,740,233; June 19, 1973; assigned to Dairyland Food Laboratories, Inc.* Usually chillproofing is accomplished by enzyme systems which are strongly proteolytic. It is proposed that a mild, highly specific proteolytic enzyme be used for effective chillproofing.

The enzyme system used is classified as a milk clotting enzyme. Milk clotting enzymes, which are used to curdle milk in cheese making are characterized by their specific action on the kappa casein fraction of milk. Milk clotting enzymes catalyze the hydrolysis of the kappa casein fraction but have little or no effect on the other casein fractions or on the milk serum proteins. The enzyme is obtained via pure culture fermentation by *Mucor pusillus* Lindt (MPE).

There are significant differences between the enzyme characteristics of MPE and conventional chillproofing enzymes. One such difference is illustrated by the relative enzyme activities of MPE and a conventional papain chillproofing enzyme preparation assayed by two procedures: the milk clotting enzyme assay and by a casein digestion assay normally used to standardize chillproofing preparations. Data obtained on one such comparison follows:

Enzyme Preparation	Milk Clotting Activity*	Casein Digestion Activity**
MPE	94.3	217
Commercial papain chillproofing	69.9	2,455

*Percent of normal or standard reference enzyme
**OD units

The proteolytic activity patterns exhibited by MPE and the commercial chillproofing preparation are markedly different. MPE has 35% more milk clotting activity than commercial chillproofing yet only 8.8% as much casein digestion activity.

A second observation on the relative proteolytic characteristics of MPE and commercial papain chillproofing enzymes is a test utilizing the gelatin emulsion on commercial photographic film.

The test consists of suspending pieces of exposed photographic film in the beer to which either MPE or commercial chillproofing enzyme has been added under standard conditions of time and temperature. The film pieces are removed from the beer and the remaining photographic emulsion is removed by the solvent and measured spectrophotometrically. The proteolytic activity of the enzyme is inversely proportionate to the amount of gelatin emulsion remaining after the film has been submerged in the beer sample and transmittance values are, therefore, directly proportional to enzyme activity.

Solutions of MPE or commercial papain were prepared by diluting 0.35 ml of enzyme preparation to 100 ml with freshly decarbonated beer. 10 ml of these diluted solutions were added to 12 ounces (350 ml) of primary filtered, unpasteurized beer. Spectrophotometric analyses per the procedure described above yielded the following results:

Enzyme System	% Transmittance
Control (no enzyme added)	22.0
MPE	25.5
Commercial papain enzyme	100.0

From this data it is apparent that MPE does not attack the gelatin while conventional papain chillproofing enzyme readily attacks gelatin. A third difference between MPE and conventional chillproofing enzymes is characterized by the presence (or absence) of milk clotting activity in unpasteurized beers to which MPE or commercial chillproofing enzyme has been added.

Example 1: A series of sub-lots of primary filtered beer were treated with MPE or a commercial chillproofing enzyme preparation. Addition levels ranged from 0.2 to 1.0 pound of enzyme preparation per 100 barrels of beer; within usual commercial practice. A control sub-lot containing no chillproofing enzyme was also included.

After a 5 day storage period at 32°F, the beer was given the second or final filtration, carbonated, bottled and pasteurized. The beer was incubated at 110°F for 2 weeks, chilled at 32°F for 5 days and examined at 32°F and also after warming to 70°F. It was found that MPE is superior to regular chillproofing in fresh beer and is comparable or superior in beer as aging occurs.

Example 2: The effectiveness of MPE treatment before primary filtration is demonstrated by adding MPE at low levels to ruh beer before primary filtrations. After primary filtration either additional MPE or commercial chillproofing enzyme preparations are added in the conventional manner. Improvement was noted in clarity stability resulting from the conventional addition of either MPE or commercial chillproofing enzyme to beer which was first treated with 0.25 pound of MPE per 100 barrels at the ruh stage, before primary filtration.

Experimentation was also conducted in which commercial chillproofing enzyme preparation was used to treat the ruh beer. Although some improvement in final chill stability was accomplished, the marked, undesirable deterioration of foam stability resulted.

Insoluble Polymer-Enzyme Process

B.S. Wildi and D.C. Boyce; U.S. Patent 3,597,219; August 3, 1971; assigned to Monsanto Company use an insoluble polymer-enzyme system for chillproofing beer. The process permits removal of the enzyme for reuse. A particularly effective source of mixed enzymes which can be used as starting material for the enzyme-polymer products is a mutated *Bacillus subtilis* organism. A culture of this *Bacillus subtilis* (species AM) organism has been assigned No. NRRL B-3411. The enzymatically active material produced by this organism has been found to consist of two proteases, approximately 65 to 75% neutral protease (activity at a pH of 7.0 to 7.5) and about 25 to 35% alkaline protease (activity at a pH of 9 to 10). A significant amount of amylase is also present.

There are generally about 700,000 to about 1.2 million units of neutral protease activity per gram and about 250,000 to about 400,000 units of alkaline protease activity per gram. There are generally about 300,000 to 350,000 units of amylase activity per gram.

The relative proportions of protease to amylase will vary depending on the exact conditions of growth of the microorganism, but it was found that the neutral and alkaline protease and the amylase will always be produced, in at least some amounts, almost regardless of changes in the culture medium and other conditions of growth of the microorganism. The ratio of the activity of the alkaline protease to the activity of the neutral protease in the starting materials and in the polymer-enzyme product is no greater than about 0.25-1.2 to 1.

Another source of mixed enzymes which can be used as starting material is *B. subtilis* strain NRRL 644, *B. subtilis* strain NRRL 941, and *B. subtilis* strain IAM 1523 (Japanese Culture Collection).

The polymer to which the enzyme is coupled for use contains carboxyl or anhydride linkages, especially where the enzyme contains an amino, hydroxyl, or sulfhydryl group not essential for its enzymatic activity. Where the enzyme contains a carboxyl group not essential for activity, the polymer can contain hydroxyl or amine groups for reaction therewith. The polymer may be EMA (ethylene-maleic anhydride) or an EMA-type polymer, or be any of those types for coupling reactions with an enzyme, and is adapted to couple or react with the enzyme to effect covalent bonding and production of the desired enzyme-polymer product.

The polymers are selected from the group consisting of: ethylene-maleic anhydride copolymer (EMA); styrene-maleic anhydride copolymer; vinyl methyl ether-maleic anhydride copolymer; vinyl acetate-maleic anhydride copolymer; divinyl ether-maleic anhydride cyclocopolymer; polymaleic anhydride and polyacrylic anhydride.

The enzymes are selected from the group consisting of: papain (including zinc papain); neutral protease; acid protease and ficin.

Experimental Preparation of Polymer-Enzyme Products: The general procedure used consisted of allowing cold solutions of enzymes in appropriate buffers to react overnight at 4°C with cold, homogenized polymer, e.g., EMA, suspensions. EMA-21 has a molecular weight of about 20,000 to 30,000. Other molecular weight polymers may also be used. For example, EMA-11, having a molecular weight of about 2,000 to 3,000, and EMA-31, having a molecular weight of about 60,000, may also be used. Separation of soluble and insoluble adducts, after reaction, was achieved by centrifugation in the cold centrifuge, about 10,000 rpm and 10 minutes centrifugation time. Insoluble adducts were washed and centrifuged, usually 10 times with cold buffer and 5 times with cold distilled water and then lyophilized. Addition of a small amount of a crosslinker, e.g., hexamethylenediamine, in 1 to 2% concentration by weight, based upon the amount of polymer, to the reaction mixture increases yields of desired insoluble product.

The reaction of the polymer with enzymes, as in some of the examples, can obviously be carried out stepwise, one enzyme at a time, with or without intermediate isolation, or with all enzymes at once. The latter procedure is considered best for reasons of time, convenience, and economy.

Preparation 1: Water-Insoluble EMA-Papain — A solution of 0.5 gram of crystalline papain is suspended in 55 ml of 0.05 M acetate buffer, pH 4.5, until a clear solution is obtained. The papain solution is added with stirring to a cold (0° to 5°C) homogenized suspension of 2.5 grams of EMA-21 copolymer in 250 ml of 0.1 M phosphate buffer at a pH of 7.6. Hexamethylenediamine (1.25 grams) is added to crosslink the polymer-enzyme complex as it is formed. The reaction mixture is stirred overnight at 4°C. The insoluble EMA-papain derivative is separated by centrifugation and washed thoroughly with 0.1 M phosphate buffer, pH 7.5, and 0.1 M NaCl until the washings are free from enzymatic activity. Finally the product is washed with water to remove the salt and the product lyophilized yielding 3.2 grams of EMA-papain retaining 67% of the esterase activity found in the initial crystalline papain.

Preparation 2: Water-Insoluble EMA-Zn Papain — 0.1 gram crystalline papain is dissolved in 60 ml of a solution containing 0.005 M cysteine and 0.002 M EDTA (ethylenediaminetetraacetic acid) at a pH of 6, this solution being aged at 37°C for 15 minutes prior to addition of papain. 0.125 gram zinc chloride is dissolved in 10 ml of 0.005 M Tris [tris(hydroxymethyl)aminomethane] buffer at pH 8.0 and added to the papain solution. Excess zinc and chloride ions are then removed by exhaustive dialysis against 0.005 M Tris buffer and then against 0.005 molar phosphate buffer. This product is used without isolation to attach it to EMA-21. Lyophilization of an aliquot indicates a yield of Zn-papain of 77%.

The EMA-Zn papain is prepared by chilling a volume of the Zn papain suspension containing 12 mg of the zinc salt to 0°C and adding 60 mg of EMA (dissolved in 5 ml of dimethylsulfoxide) to the chilled Zn-papain with vigorous stirring. The crosslinking agent (1 ml of a 1% hexamethylenediamine aqueous solution) is then added and the mixture stirred overnight at 4°C. The insoluble EMA-Zn papain is then recovered by centrifugation followed by washing with water.

Prior to use in chillproofing of beverages the Zn can be removed by treating a suspension with a solution containing cysteine and EDTA. The product upon removal of the Zn possesses 78% of its initial esterase activity. The insoluble EMA-Zn papain can be used in treating beverages without removal of the Zn with a somewhat lower activity, possibly aided by partial removal of Zn by components of the beverage.

Example: Use of Insoluble EMA-Papain in Beverage Chillproofing (During Fermentation) — 100 barrels of wort is prepared using 60% malt and 40% corn grits. To the wort at 47°F is added the brewer's yeast and this is allowed to ferment for 24 hours. At this time, 75.7 grams of an insoluble EMA-papain polymer as prepared in Preparation 1 is added. This product has 300 tyrosine units per mg.

After fermentation is complete (an additional 96 hours), the beer with the EMA-papain product suspended as a gel is decanted from the settled yeast, stored for 7 days at 3°C, filtered, carbonated, stored for 4 to 5 additional days at low temperature, polished, filtered, bottled and pasteurized. The beer produced by this method has superior clarity, stability, and a decidedly improved taste as compared with beer made in the traditional manner by the addition of soluble papain in the cellar, essentially all enzymes having been removed by filtration removal of insolubles.

The EMA-papain is manually removed from the filter, washed with 1 liter of 0.1 M NaOH, filtered, followed by two washes with 1 liter of 0.2 M acetic acid and two washes with 1 liter of water. This washed, used polymer-enzyme product is then put into another 100 barrel batch of fermenting wort which is treated as described above with the same results.

Insoluble Polymer-Enzyme Variations

In a refinement and/or extension of the EMA and EMA-type polymers in beer chillproofing *B.W. Weinrich, J.H. Johnson, B.S. Wildi and D.C. Boyce; U.S. Patent 3,597,220; August 3, 1971; assigned to Monsanto Company* use insoluble basic polymer-enzyme products with basic or cationic groups as variables.

Preparation 1: Partial Dimethylaminopropylamine Imide of EMA — 30 grams EMA-21 (a copolymer of ethylene and maleic anhydride) was suspended in 500 ml xylene. The mixture was heated to reflux temperature under a nitrogen atmosphere. 14.58 grams dimethylaminopropylamine was added and the mixture held at reflux for 4.5 hours. Water, a by-product of imide formation, was separated in a Dean-Stark trap and reaction was judged complete when water was no longer being evolved. The product was precipitated by addition of hexane and dried in a vacuum oven at 105°C. It contained 8.86% N indicating an imide content of 54.3%.

Preparation 2: Water-Insoluble Basic EMA-Papain — A solution of 0.5 gram of crystalline papain is suspended in 55 ml of 0.05 M acetate buffer, pH 4.5, until a clear solution is obtained. The papain solution is added with stirring to a cold (0° to 5°C) homogenized suspension of 2.5 grams of basic EMA copolymer prepared in Preparation 1 in 250 ml of 0.1 M phosphate buffer at a pH of 7.6. Hexamethylenediamine (1.25 grams) is added to crosslink the polymer-enzyme complex as it is formed. The reaction mixture is stirred overnight at 4°C. The insoluble basic EMA-papain derivative is separated by centrifugation and washed thoroughly with 0.1 M phosphate buffer, pH 7.5, and 0.1 M NaCl until the washings are free from enzymatic activity. Finally the product is washed with water to remove the salt and the product is lyophilized to yield 3.4 grams of EMA-papain, retaining 62% of the esterease activity found in the initial crystalline papain.

Preparation 3: Water-Insoluble Basic EMA-Zn Papain — 0.1 gram crystalline papain is dissolved in 60 ml of a solution containing 0.005 M cysteine and 0.002 M EDTA at a pH of 6, this solution being aged at 37°C for 15 minutes prior to addition of papain. 0.125 gram zinc chloride is dissolved in 10 ml of 0.005 M Tris [tris(hydroxymethyl)aminomethane] buffer at pH 8.0 and added to the papain solution. Excess zinc and chloride ions are then removed by exhaustive dialysis against 0.005 M Tris buffer and then against 0.005 M phosphate buffer. This product is used without isolation to attach it to EMA-21. Lyophilization of an aliquot indicates a yield of Zn papain of 77%.

The basic EMA-Zn papain is prepared by chilling a volume of the Zn papain suspension containing 12 mg of the zinc salt to 0°C and adding 60 mg of the basic EMA of Preparation 1 (dissolved in 5 ml of dimethylsulfoxide) to the chilled Zn papain with vigorous stirring. The crosslinking agent (1 ml of a 1% hexamethylenediamine aqueous solution) is then added and the mixture stirred overnight at 4°C. The insoluble basic EMA-Zn papain is recovered by centrifugation followed by washing with water.

Example 1: Use of Insoluble Basic EMA-Papain in Beverage Chillproofing (During Fermentation) – 100 barrels of wort is prepared using 60% malt and 40% corn grits. To the wort at 47°F is added the brewer's yeast and this is allowed to ferment for 24 hours. At this time, 65.7 grams of an insoluble basic EMA-papain polymer (prepared as in Preparation 2) is added. This product has 320 tyrosine units per mg. After fermentation is complete (an additional 96 hours), the beer with the basic EMA-papain product suspended as a gel is decanted from the settled yeast, stored for 7 days at 3°C, filtered, carbonated, stored for 4 to 5 additional days at low temperature, polished, filtered, bottled and pasteurized. The beer produced by this method has superior clarity, stability, and a decidedly improved taste as compared with beer made in the traditional manner by the addition of soluble papain in the cellar, essentially all enzymes having been removed therefrom by filtration removal of insolubles.

The basic EMA-papain is manually removed from the filter, washed with 1 liter of 0.1 M NaOH, filtered, followed by two washes with 1 liter of 0.2 M acetic acid and two washes with 1 liter of water. This washed, used polymer enzyme product is then put into another 100 barrel batch of fermenting wort and treated as described above with the same results.

Example 2: Use of Insoluble Basic EMA-Papain in Beverage Chillproofing (in Ruh Stage with Pasteurization) – To 9.3 kg of filtered carbonated ruh beer in a 5-gallon stainless steel carbonated beverage can is added 52 mg of a basic EMA-papain insoluble polymer (prepared as in Preparation 3) having after removal of Zn ions an activity of 285 tyrosine u/mg. The can is sealed and mixed by tipping end to end several times. The CO_2 pressure is raised to 10 pounds and the can then stored in a cooler at 0°C for 7 days with daily mixing. At this time the beer is pasteurized in the can at 60°C for 15 minutes and then immediately cooled. The beer is filtered through a 0.22 μ Millipore filter, by which filtration the insoluble polymer enzyme product is removed, and then bottled and again pasteurized. Clarity, stability, and taste is superior to the traditional beer and the product maintains these favorable characteristics 6 to 12 months longer than conventionally chillproofed beer.

Preparations and examples are given for polymer-insoluble neutral protease, polymer-insoluble acid protease and polymer-insoluble bromelin.

Protease Clarifying Method

An enzyme protease is used in the clarification step in beer manufacture. In beer, turbidity is contributed by proteins and tannins which are not removed by physical methods, and often the turbidity is more intense on storage and on chilling. A special protease for this process is derived from *Streptomyces fradiae* as described by *R.S. Ferns; U.S. Patent 4,038,419; July 26, 1977; assigned to Instituto de Farmacologia Espanola, S.L. (Fundacion Marques de Urquijo), Spain.* This method is applicable to any protein-containing solution, particularly gelatin solutions, which are used by way of example, in addition to beer clarification.

Example: 15 grams of *Streptomyces fradiae* protease, in dry condition and with a unitary strength of 11,000 AU/mg, were dissolved, with shaking and at ambient temperature, for 1 hour, 4 grams of dry calcium chloride having previously been added to the buffers.

When shaking had been completed, the enzyme solution was filtered and subjected to suction through an ordinary filter paper covered in situ by a washed, dry diatomite cake of about 5 mm height.

When filtration had been completed, the cake, which had been drained in vacuo, was washed twice with 100 ml of buffers, the cake being drained after each washing and the discharging liquids being collected on the filtration mother liquors. The collected, transparent filtrate liquors were homogenized, and 100 ml of propylene glycol were added to these filtrate liquors, the pH value being adjusted to a value of 6.0 by an addition of sodium hydroxide solution.

The proteolytic activity of the enzyme solution was evaluated by Anson's method and from the analytical data obtained, it was possible to deduce the quantity of buffers which it was necessary to add to the filtered enzyme solution to adjust its enzyme strength to 250,000 AU/ml.

This solution was used for the clarification of beer, 3 ml being added per hl of beer to be clarified, following the American method, after the initial filtration, subsequent to the primary fermentation. The beer was stored for 6 weeks in the starting tank, to which the enzyme solution was added.

When the action (activity) phase had been completed, the second filtration took place, and then the clean beer was pasteurized and conditioned, the beer then passing to a bottling station and finally being sent to storage.

It will be appreciated that the clarification results achieved in the comparative test carried out are clearly favorable to the use of proteases derived from *Streptomyces fradiae,* both for the absolute initial brilliance and also for the maintenance of the clear solution, comparing the results with those obtained in the first manufacturing batch, in which the clarification was carried out with products commercially in use by the beer brewing industry.

The residual filtration turbidity for those beers treated with proteases derived from *Streptomyces* is about 18±5% lower than that of the same beers treated with proteases of vegetable origin; and also the turbidity developed during storage, at ambient temperature, increases more rapidly in the beers treated with vegetable proteases—up to 71% greater turbidity. Equally good comparative results are obtained with thermal shock.

Protease Cellulose Carrier

In the process of *B.S. Wildi and L.E. Weeks; U.S. Patent 4,013,514; March 22, 1977; assigned to Monsanto Company* the matrix or carrier for the enzyme is cellulose or a water-insoluble derivative. The specific derivative is dialdehyde cellulose prepared by oxidation of glycoside units to dialdehyde groups. The amino group of the enzyme is crosslinked with dialdehyde cellulose.

The degree of oxidation of the cellulose to provide dialdehyde groups can vary. A degree of oxidation higher than about 80% is undesirable because the structural integrity of the dialdehyde cellulose matrix appears to be impaired. About 10 to 50% of the glucoside units or rings usually are opened and oxidized to dialdehydes. Covalent bonding of the enzyme to dialdehyde cellulose occurs through an amino group of the enzyme which is not essential for biological activity.

A wide variety of enzymes can be coupled directly to the dialdehyde cotton carrier. Active enzymes suitable for coupling may be obtained from any suitable source, either vegetable, animal, or microbial. Many such enzymes are available commercially. Typical are the proteolytic enzymes such as the proteases, e.g., neutral and/or alkaline protease. In some instances another differently active enzyme such as amylase can be employed, along or admixed with proteases so as to maximize the operative enzyme activity of the composite. Still other enzymes such as a lipase may be used instead of or in addition to amylase. Other suitable enzymes are esterase, nuclease, or other types of hydrolase. A hydrase or oxidoreductase may also be utilized, depending upon the ultimate activity and intended application.

Examples of application for continuous enzyme reactors having fibrous dialdehyde cotton core and an enzyme covalently bonded are a papain-dialdehyde cotton reactor useful for the chillproofing of beer, a rennin- or pepsin-dialdehyde cotton reactor useful for curdling milk, an amylase-dialdehyde cotton reactor useful for making glucose from starch, a naringinase- or hesperidinase-dialdehyde cotton reactor for the removal of bitter flavors from citrus juices. Similarly, excess hydrogen peroxide used as a bactericide in milk treatment can be removed by passing the treated milk through a catalase-dialdehyde cotton reactor. Also, a lactase-dialdehyde cotton reactor can be used to hydrolyze lactose to galactose and glucose.

Enzymes can also be covalently bound to dialdehyde cotton to provide means for clinical analysis and assays, e.g., urease covalently bonded to dialdehyde cotton can be used for blood urea nitrogen determinations, also to provide means for resolving dl-amino acids, e.g., a racemic mixture of N-acylamino acid can be passed through an aminoacylase-dialdehyde cotton reactor to hydrolyze the l-form which can then be separated from the d-form.

Example 1: Attachment of Papain to Dialdehyde Cotton — A portion of fibrous dialdehyde cotton (2.5 grams), was packed in a stainless steel nipple (0.75" i.d. by 1.5" long). Foamed polyethylene discs were placed at each end of the packed nipple to retain the fibers and the nipple was closed at both ends with stainless steel caps provided with 1/16" inlet and outlet fittings, respectively.

Papain (0.5 gram, purified) was dissolved in water (50 ml) and the resulting solution adjusted to pH 8.0 with aqueous 0.2 N sodium hydroxide solution (1.6 ml). The papain solution was then recirculated through the prepared dialdehyde cotton reactor bed at a rate of 130 ml per minute for 50 minutes and until pH was stable at 8.0. During recirculation 2.0 ml of the 0.2 N sodium hydroxide solution were added to the recirculating papain solution to maintain pH at 8.0.

Recirculation of the papain solution was then terminated and the reactor bed washed by pumping through at 130 ml per minute 2 liters of water followed by 2 liters of aqueous 1 M sodium chloride solution.

Example 2: Chillproofing of Beer in a Dialdehyde Cotton Reactor – A cartridge comprising cotton yarn was wound to a depth of 0.75" on a roll made of stainless steel screen (1" o.d. by 4" long). Outside dimensions of the cartridge were 2.5" diameter by 4" long. The cartridge was placed in a stainless steel holder and 3.5 liters of aqueous 0.11 M periodic acid solution was recirculated through the cartridge for 4 days. Thereafter the cartridge was washed with 30 liters of water and a papain solution (16 grams papain in 1,000 ml of water, pH 8.0) was recycled through the periodic acid-treated cartridge. Then the excess papain was washed out with water, 1 M aqueous sodium chloride solution and more water.

For the treatment of beer the prepared cartridge was first conditioned by recycling draft beer containing about 0.2 to 0.3 weight percent protein, i.e., not chillproofed, for 24 hours at 5°C and at a rate of 1 liter per minute. Thereafter 5 kg of test beer (not chillproofed) was recycled through the cartridge at 5°C for 4 days at a rate of 1 liter per minute. Pressure drop across the cartridge was observed to be 2.25 psig. In another test run, beer was passed through the prepared cartridge at 5°C in a single pass at a rate of 400 ml per minute.

PROTEASE MICROCRYSTALLINE CELLULOSE CARRIER

Enzyme preparations, insoluble in water because of the chemical coupling of enzymes with derivatives of microcrystalline cellulose can be used in a wide range of product applications where production efficiency and heat stability have distinct advantages over conventional technology. A specific application with papain is described by *A.N. Emery; U.S. Patent 3,713,982; January 30, 1973; assigned to Ranks Hovis McDougall Limited, England.* The preparation consists of reacting, at 0° to 3°C, papain dissolved in a buffer within a pH range 6.5 to 8.5 and containing L-cysteine and diaminoethane tetraacetic acid with the p-diazophenoxy hydroxypropyl ether of cellulose.

Example 1: Samples (100 mg) of p-amino phenoxy hydroxypropyl cellulose ether were placed in a centrifuge tube. Aliquots (5 ml) of ice-cold, 1 N hydrochloric acid were added and the slurry stirred magnetically for 15 minutes at 0°C. Aliquots (4 ml) of ice-cold 2% sodium nitrite solution were added and after a further 15 minutes the tubes were centrifuged and the supernatant discarded. The solid was washed 3 times with aliquots (15 ml) of phosphate buffer (0.075 M, pH 6.5 to 8.5). After the final washings had been decanted, aliquots (1 ml) of a solution of papain (10 mg/ml activity, 0.148 unit/mg protein) in phosphate buffer (0.075 M, pH 6.5 to 8.5) containing L-cysteine (5 mmol) and EDTA (diaminoethane tetraacetic acid) (2 mmol), were added and the tubes stirred magnetically at 0° to 3°C for 48 hours. After coupling, 5 ml of an ice-cold 0.01% solution of β-naphthol in saturated aqueous sodium acetate were added. After a further 15 minutes stirring, the water-insoluble papain derivatives were subjected to 5 cycles of washing with phosphate buffer (0.03 M, pH 6.9, 15 ml) and sodium chloride solution (0.5 M) in the same buffer. The papain derivatives were finally washed twice with phosphate buffer (0.02 M, pH 6.9).

The activity of the water-insoluble papain was determined in the following manner. Samples were suspended (or dissolved in the case of free papain) in phosphate buffer (5 ml, 0.075 M, pH 6.5). 5 ml of activator solution, (L-cysteine 5 mmol and EDTA 2 mmol) were added and the mixture allowed to incubate with 10 ml of a prewarmed 2% casein solution, stirred magnetically at 600 rpm.

4 ml aliquots were removed at intervals and added to 2 ml of 10% trichloracetic acid in test tubes, agitated, and allowed to stand for 5 minutes. The optical density of the solutions was read at 280 mμ in a spectrophotometer using 1 cm silica cuvettes. The activities were calculated from a plot of the optical density change with time. The results are as follows.

Coupling (pH)	Bound Protein per Derivative (mg/100 mg)	Enzyme per Bound Protein (u/mg)	Enzyme Activity Retention (%)
6.5	1.23	0.051	34.5
7.5	1.30	0.048	32.5
8.5	1.09	0.049	33.1

Example 2: A sample of water-insoluble papain was prepared as in Example 1. The coupling was carried out at pH 6.9 and gave a derivative containing 0.809 mg protein/100 mg carrier and with a specific activity of 0.105 enzyme units per mg of protein.

One gram of the water-insoluble papain derivative was placed in a plastic 1 liter centrifuge pot. Two 1 pint bottles of beer (original gravity 1.035, fermented for 5 days, chilled and filtered) were opened and poured into the centrifuge pot. Potassium metabisulfite was added (50 mg/l) and the beer was then magnetically stirred at room temperature for 72 hours.

The pot was centrifuged for 15 minutes at 1,100 *g*, the beer decanted back into bottles, which were crowned and repasteurized. Further bottles were opened and treated in the same way for periods of 24 hours and 18 hours using the same solid.

After the final treatment the water-insoluble papain was washed in distilled water and resuspended in phosphate buffer (0.02 M, pH 6.9, 100 ml). An aliquot (5 ml) of the suspension was assayed as before.

The bottles of treated beer, together with similar but untreated controls were stored for alternate periods at 4°C and room temperature to accelerate the development of haze. Haze readings were taken at intervals. The results are shown below.

	Activity of Papain (OD units/min)	Activity Retention (%)	. . . Haze (EBC units) . . . . . Days of Incubation. . . 0	12	14	15
Preparation of Ex. 2						
Initially	8.9×10^{-2}	–				
After 10 days of treatment	4.7×10^{-2}	53				
Time of treatment						
72 hours			2.5	3.3	2.5	2.3
24 hours			2.5	6.3	8.8	9.3
18 hours			2.5	6.4	9.7	10.5
Control			2.5	9.8	13.0	>13

LOW CALORIE BEER

Two-Step Fermentation

Slow decomposition of dextrins in malt extracts occurs in beer manufacture. In some cases it is not possible to completely ferment the last amounts of sugars present.

It has been found that the enzymatic hydrolysis is inhibited by the high content of ethyl alcohol, mainly in the last stage of the fermentation. The fermentation may also be inhibited by amyl alcohol, which is contained in highly fermented beers in much larger amounts than in normal beers and has a toxicity that is nine times higher than that of ethyl alcohol.

The high alcohol content of the special beers intended for diabetics is undesired for another reason too. Diabetics sometimes have a weight problem, and their liver is more sensitive than that of a healthy person. For this reason, many physicians believe that a beer intended for diabetics should not have a higher alcohol content than a normal beer. Because the highly fermented beers are not full-bodied, they cannot be diluted with beers having a lower alcohol content in order to reduce the alcohol content.

F.-W. Schimpf and W. Rinke; U.S. Patent 3,852,495; December 3, 1974; assigned to AG fur Brauerei-Industrie, Switzerland developed a process for producing a dietary beer which has low alcohol and low carbohydrate content.

The dietary beer is made from a light-colored malt, which is mashed in the usual manner at the temperature which is optimum for the formation of maltose. The wort is boiled with hops and then strained and cooled. In a first main fermentation step, the cooled wort is fermented at a temperature of 4° to 8°C to an apparent degree of fermentation corresponding to 40 to 70% of the extract. This generally requires 2 to 5 days. The fermentation is then interrupted. The partly fermented wort is pumped into the brew kettle and boiled for 40 to 50 minutes, whereby about 1.5% of the about 2.5% ethyl alcohol which are present are stripped so that the finished beer has in spite of its high fermentation only an ethyl alcohol content which is about the same as that of a normal beer.

At the same time, approximately all of the higher aliphatic alcohols (fusel oils) contained in the beer are stripped so that the final fermented beer has only about 50% of the normal content of these substances. The boiling of the beer affords further advantages because the sulfur-containing substances as well as the diacetyl and acetoin are stripped almost completely whereas they are normally removed only during the storage and maturing. Besides, the formation of esters is promoted. As a result, the time for which the beer is stored to effect a physical or biochemical removal of these substances and to induce a formation of esters can be reduced. The partly fermented wort is usually boiled under atmospheric pressure for about 45 minutes. When the boiling has been terminated, the partly fermented beer is cooled and is filtered from the precipitates which have formed.

The subsequent second main fermentation step is carried out at a temperature of 12° to 15°C and after an addition of fresh yeast. The fermentation is initiated quickly by the maltose which is still present in the new beer (partly fermented wort). After about 12 hours, the beer is pumped from the sediment which has separated and a diastase solution which has a particularly high content of marginal dextrinases and consists of a malt extract made in a special manner is added in an amount of 3 to 7%. The new beer is then stored at 12° to 20°C for additional 7 to 10 days to complete the fermentation.

To produce the diastase solution which has an increased content of marginal dextrinases, grits of light-colored malt are mixed with water at 45°C and the resulting mash is stirred for 30 minutes. The operation in this temperature range results in the formation of an optimum amount of marginal dextrinases whereas the danger of a bacterial infection is eliminated to a large extent. After mashing for 30 minutes, the stirrer is stopped and the mash is allowed to stand for about 30 minutes for clarification. The supernatant and clarified malt extract is added to the new beer in the second fermentation step.

When the second main fermentation step has been terminated, the beer is cooled and is pumped into a storage tank, where the yeast is separated and in which the beer is stored for about one week to two weeks for a precipitation of the substances which cause turbidity. Hence, the total production time of the dietary beer is about 4 to 5 weeks.

Example: A pale malt is mashed at a temperature of 62° to 65°C which is optimum for the formation for maltose. When the mash has been strained, the wort is boiled for 45 minutes together with 50% of the total amount of hops to be added. The wort which is withdrawn contains 11.5% extract and is then cooled.

In the fermenting vessel, 0.5 liter yeast per hectoliter of the cooled wort is added to the latter at 5°C, then the wort is fermented at temperatures up to 8°C until an apparent degree of fermentation of about 55% has been reached after about 3 to 4 days. This degree of fermentation corresponds to an apparent residual extract content of about 5.1%. A calculation shows that about 5.2% real extract have been fermented so that the new beer (partly fermented wort) contains about 2.5% alcohol. The new beer is pumped into the brew kettle and boiled in the open kettle for 45 minutes after the remaining 50% of the total amount of hops has been added. This boiling reduces the alcohol content to about 1.0%. The evaporated liquid is then replaced by water, whereafter the partly fermented wort is cooled.

After the removal of alcohol, the new beer is filtered and 1 liter of yeast per hectoliter of this beer is added thereto. When the fermentation has been initiated after about 12 hours, the new beer is decanted from the sediment and about 6% malt extract prepared as described hereinbefore to have a high content of marginal dextrinases is added to the decanted new beer. The fermentation is continued at temperatures in the range of 12° to 20°C and is terminated after 7 days. The apparent residual extract amounts to about 0.3%.

The fermented beer is cooled, pumped from the yeast and stored for 12 days. When the substances causing turbidity have been separated, the beer is mature for consumption.

Protein Hydrolysate Process

Special toasted protein hydrolysate can be used as a flavor precursor in the production of beer. This is explained by *M.R. Sfat and B.J. Morton; U.S. Patent 3,689,277; September 5, 1972; assigned to Bio-Technical Resources, Inc.* in a process where protein hydrolysate and sugar are added to the fermentable wort in place of cereal grains. The unique feature of this process is low carbohydrate yield and high protein content in the finished beer. It is possible in this process to eliminate malt from the brewing process, completely or partially, without loss of malt-like flavor properties.

Divided grain or legume is mixed with water, and a proteolytic enzyme is added to the mixture for hydrolysis and solubilization of the protein. It was found that when more than an appreciable amount of amylolytic enzyme was added to a starch-containing grain or legume, the protein content of the product decreased and the carbohydrate content increased markedly. The protein hydrolysate product solids should contain at least about 40% protein by weight of the solids, thereby providing the desired concentration for independent control of the protein and carbohydrate components of the fermentation wort in beer making, and also providing low carbohydrate product for the heat treatment and other purposes.

The proteolytic enzyme is used as a concentrate or isolate from natural sources, and may actually constitute a mixture of proteolytic enzymes. Numerous products are available, and depending upon their source, they may include amylolytic enzyme. The products should have from low amylolytic enzyme content to no amylolytic enzyme content. They are also referred to in the trade as having relatively low diastatic activity.

Such products may have less than about 1,000 starch liquefying units and less than about 200 starch dextrinization units (SKB) per gram. (An enzyme with 1,000 starch liquefying units per gram will reduce the viscosity of 300 times its weight of potato starch by 90% in 10 minutes at 70°C and pH 6.7. The starch dextrinization units represent the number of grams of beta-amylase-treated soluble starch dextrinized by 1 gram of enzyme in one hour at 30°C and pH 4.85.)

The proteolytic enzyme treatment of the aqueous dispersion of the seed product hydrolyzes the seed protein to soluble hydrolysis products and also results in the extraction of a starchy carbohydrate fraction from the seed product when starch is present. The extracted starch would interfere with brewing a satisfactory alcoholic beverage, and it causes the formation of a viscous gel if the solution is concentrated beyond about 33% solids.

The starch content of the solution resulting from such proteolytic enzyme treatment is hydrolyzed by treating the solution with added amylolytic enzyme. It has been found that treatment of such solution in the presence of the undissolved residue results in further extraction of carbohydrate, reducing the protein concentration. Separation of the solution from undissolved residue followed by amylolytic enzyme addition and treatment of the separated solution avoids additional carbohydrate extraction and provides for rapid and complete hydrolysis of starch. The resulting solution or the solids content then may be used for brewing and it may be concentrated to a solids content in excess of about 40% by weight.

The amylolytic enzyme may be supplied by any of numerous enzyme-containing materials, including natural sources and isolates or concentrates. The presence of other enzymes is, in general, not a significant factor. Thus, enzyme products containing substantial proteolytic enzyme activity in addition to amylolytic activity may be used and the proteolytic enzymes may serve to provide additional hydrolysis of the soluble protein hydrolysis products of the proteolytic enzyme treatment.

The protein hydrolysate is intimately mixed with a sugar and sufficient water to provide a moist granular mixture. The water may be provided by a concentrated solution of the hydrolysate or by a sugar syrup, and/or may be added to relatively dry hydrolysate and sugar materials. The mixture is sufficiently moist to insure that the hydrolysate and the sugar adhere to each other in fine granules, requiring about 10 to 25% of water in mixture, as based upon the total wet weight of the mixture. The desired condition at treatment temperature is that of a soft pliable or spongy mass, resembling brown sugar at room temperature, not clinging to the finger yet not dry like granulated sugar (sucrose).

The mixture is heated to a bed temperature of about 80° to 85°C and maintained at such temperature to lightly toast the mixture. The bed temperature is held until a very faint or slight caramel flavor is imparted. During such heating, the moisture content of the mixture is lowered to about 2 to 10% by weight of the mixture.

The toasted product is rapidly cooled to about 38°C or lower by circulating air at ambient temperature. The toasted product is free-flowing, having the consistency of granular sugar. The product from cereal grain protein is cream-colored and has a taste resembling honey.

Example 1: A 2,000 gram quantity of hull-containing barley whole grain containing about 13.5% protein is ground providing a particle size distribution, percent retained on the U.S. Sieve Series sieve number indicated: No. 18, 0.2%; No. 30, 16.7%; No. 60, 59%; No. 100, 18.5%; through No. 100, 5.6%.

The ground grain is added to 10 liters of water, made up of 9 liters of tap water and 1 liter of brewing water containing 240 ppm calcium sulfate, 100 ppm sodium chloride, and 55 ppm magnesium sulfate. After mixing, the pH is adjusted to about 6.6 to 6.7 by the addition of about 15 ml of 15% sodium hydroxide.

As a source of proteolytic enzyme, 4 grams of standardized papain is added to the mixture, with 5 grams of potassium metabisulfite enzyme activator and the mixture is maintained at 45°C with stirring for 4 hours. The mixture is stored overnight at 0°C and decanted. The clear decanted solution is treated with a source of amylolytic enzyme.

The enzyme source is added in a quantity of 0.2 gram per liter, and the solution is maintained at 45°C with stirring for 2 hours. The resulting solution is starch-free by the starch-iodine test. It contains about 4.4 to 4.8% solids including about 2.2 to 2.4% protein. The protein recovery from the grain is about 75 to 85%.

Solutions may be concentrated in a circulating evaporator at temperatures in the range of 35° to 55°C and under a vacuum of 27" of mercury (3" pressure) to solids concentrations in the range of 50 to 55% by weight, and concentrations up to about 65% can be obtained if desired. The protein content of the solids in such solutions is about 45 to 50% by weight of the solids.

Alternatively, barley malt having an alpha-amylase activity of about 40 dextrinizing units (DU) per gram (ASBC Methods of Analysis) or higher may be used for amylolysis, at a rate of about 5 to 10% by weight of the solids content of the solution. The malt treatment is followed by filtration, to remove the malt residue.

Example 2: A 400 gram portion of the hydrolysate concentrate was blended by hand with 1,500 grams of finely granulated glucose, producing a granular mixture containing 16.8% water. The mixture was dried in a kiln with forced air at room temperature for about 15 minutes. The air then was heated at 82°C and the bottom of the bed reached 82°C in about 12 minutes. Circulation of air at 82°C continued for an additional 30 minutes, during the last 20 minutes of which the bed temperature was 80°C. The mixture had a slight caramel aroma at this time, and the toasted material was removed from the kiln, spread on trays, and cooled by exposure to air at room temperature.

Low carbohydrate beers were produced from the toasted product. The beers had greater beer-like quality than those from untoasted material, and there was no winey note. They were superior in flavor and aroma to prior commercial low carbohydrate beers.

Protein/Amylase Process

A modification of the previous process using protein hydrolysate which is starch-free is discussed by *M.R. Sfat and B.J. Morton; U.S. Patent 3,711,292; Jan. 16, 1973; assigned to Bio-Technical Resources, Inc.* A cereal grain is used in the process, and in the last stages of the process an amylolytic enzyme is added to produce a starch-free solution.

In the process, a starchy proteinaceous cereal grain product is treated or extracted in aqueous medium with added proteolytic enzyme in the absence of added amylolytic enzyme. The grains are barley, wheat, oats and rye. For availability and production reasons, it is preferred to use the whole grain. It has been noted in the case of barley that a beer of improved graininess is produced from the grain having the hull intact. Proteinaceous grain fractions may be used such as are obtained as by-products in milling operations. For example, barley pearlings are high protein by-product fractions which have been found to be good sources of protein concentrate. High protein wheat flour is another example of a good source of protein concentrate. The grain product is finely divided or ground for use in the process. The grain product is mixed with water and a proteolytic enzyme is added to the mixture for hydrolysis and solubilization of the grain protein.

The proteolytic enzyme treatment is conducted so as to obtain a hydrolysate or extract solution containing at least about 40% protein by weight of the solids content. This condition is achieved by appropriate selection of the grain or grain fraction, proteolytic enzyme and amount, absence of amylolytic enzyme, pH adjustment for the proteolytic enzyme, and treatment temperature and time. The yield of protein from the cereal grains, in general, ranges upwards from 60% by weight of the total grain protein.

The starch content of the solution resulting from proteolytic enzyme treatment is hydrolyzed by treating the solution with added amylolytic enzyme. It has been found that treatment of such solution in the presence of the undissolved residue results in further extraction of carbohydrate, reducing the protein concentration. Separation of the solution from undissolved residue followed by amylolytic enzyme addition and treatment of the separated solution avoids additional carbohydrate extraction and provides for rapid and complete hydrolysis of starch, as determined by the starch-iodine test. The resulting solution or the solids content then may be used for brewing, and it may be concentrated to a solids content of at least about 40% by weight for brewing or other purposes and to at least about 55% by weight for biological stability.

The product solution may be incorporated in a fermentation wort with a highly fermentable sugar to produce a beer having a carbohydrate caloric content, i.e., exclusive of alcohol and protein, of less than about 1.1 calories per ounce, as calculated on the basis of 1.19 calories per ounce for each 1% of real extract carbohydrate. Such caloric content is highly desirable to the ultimate consumer and in general, may provide as much as one-third less total calories than regular beers having the same alcohol and protein content.

When beer is made using both the product and malt, the product is added to a wort prepared from the malt in a conventional manner, following separation of the wort from spent grains. Adjunct is added to the resulting wort in a quantity sufficient to produce beer of the desired alcohol content.

After suitable dilution with brewing water, the wort is boiled with the addition of hops, and strained, cooled and again adjusted with brewing water, all in a conventional manner to provide a wort ready for fermentation. Thus, for example, the wort is brought to a boil and boiled for about 1½ to 2 hours during which hops are added at the normal rate, the hops are strained out, the wort is cooled to about 20°C and diluted with brewing water to the desired concentration and the wort is finally cooled to about 8° to 14°C. The wort then is ready for fermentation with brewer's yeast.

The wort is fermented in a conventional manner. The wort is pitched with brewer's yeast and fermented for about 5 to 14 days at a temperature of about 8° to 14°C. The beer product is finished in a conventional manner, as by transferring it to storage tanks for up to about 2 months, filtering, chillproofing, carbonating, bottling and pasteurizing.

Example: Barley whole grain was treated with papain proteolytic enzyme and thereafter with Takamine acid fungal protease amylolytic enzyme and the resulting protein hydrolysate solution was concentrated to 55.7% solids. The protein content of the solids was 44.8%.

Beers were comparatively brewed by the procedure described below, employing a standard wort from barley malt and common yellow corn grits and a wort one-half standard and one-half made up of the protein hydrolysate solution concentrate and liquid adjunct (Vel-Bru IX 43°Bé corn adjunct) combined in the kettle.

In preparing the standard wort, a grain bill consisting of approximately 70% malt and 30% yellow corn grits (as-is basis) was used. A two-roller laboratory mill was used to grind the malt and was adjusted so that 73 to 75% of the crushed malt remained on the No. 1 screen of the Pfungstadt series. Laboratory distilled water was made up to 240 ppm calcium sulfate, 100 ppm sodium chloride and 55 ppm magnesium sulfate for the brewing process (including sparge water). The procedure was as follows:

A cereal mash was made up by adding 190.5 grams of corn grits (dry basis) and 50 grams of malt (dry basis) to 1,060 grams of 45°C water and was then agitated continuously with a turbine-type impeller. The temperature was maintained at 45°C for 15 minutes and then raised to boiling within 35 minutes. After a 15-minute boil, the cereal mash was combined with a malt mash.

The malt mash was made up by adding 450 grams of malt (dry basis) to 1,340 grams of 38°C water. This mixture was agitated for 10 minutes with a blade-type mixer and then allowed to rest for 60 minutes at 38°C. After combining the cereal and malt mashes, the temperature was increased to 73°C conversion temperature within approximately 5 minutes, and was then held at 73°C for 30 minutes. The temperature was raised to 75°C and the mash was transferred to a lauter column.

Following a rest period of 15 minutes, run-off was initiated. The wort was run off until clear and the first portion was recycled. Sparging was begun when the liquid level was one inch above the grain bed, and was adjusted to approximate the run-off rate. A temperature of 75°C was maintained during the entire lautering period. Approximately 5,000 grams of wort, referred to as the standard wort, was collected in a two-gallon stainless steel container.

In producing a wort one-half standard and one-half protein hydrolysate and adjunct, one-half of a standard wort prepared in the foregoing manner was combined with 108 grams of the protein hydrolysate solution concentrate, 550 grams of the Vel-Bru adjunct, 16 drops of caramel coloring (about 65% solids), and brewing water to make 5 liters.

Each wort was then processed as follows: The wort was brought to boil and hops were added at the rate of 2 grams at the start of boiling, 2 grams one hour after the start of boiling and 5 grams 20 minutes before the kettle was knocked out at the completion of a 2-hour boiling period. The wort then was strained to remove the hops, cooled to 20°C and adjusted to 12.2 to 12.3% solids with brewing water. The wort was aerated and pitched with washed, centrifuged brewer's yeast at a rate of about one pound per barrel and then placed in a closed glass column fermenter. The wort was fermented for about 7 days at

12°C, until the apparent extract was within 0.3 degree Plato of that determined by the ASBC rapid fermentation test. Following fermentation, the beer was held at 1°C for 24 hours to inactivate and help settle the yeast from suspension. It was then transferred by means of carbon dioxide pressure to storage bottles and aged at 1°C for three weeks.

After aging, the beer was filtered through a one-plate powder filter into a two-gallon stainless steel carbonation tank, where it was carbonated to 2.6 volumes. The beer was then bottled by means of a hand filler, using carbon dioxide counter pressure, pasteurized, cooled and stored at 1°C. Approximately 16 six-ounce bottles of beer were obtained from each brew.

The analytical and flavor properties of the two brews were quite similar, with the beer made from one-half standard wort and one-half protein hydrolysate and adjunct being slightly more winey. Analyses for the brews are shown below.

	Standard Brew	50% Standard + 50% Hydrolysate Brew
Wort		
Color (after boil), °L	3.3	2.6
pH (after boil)	5.50	5.72
Protein, %	0.49	0.52
Beer		
Real extract, %	5.073	4.806
Alcohol, %	3.74	3.77
Real degree of fermentation	58.2	60.3
Color, °L	2.4	2.0
pH	4.35	4.20
Protein, %	0.32	0.38
Foam stability, sigma	127	113

KVASS WORT CONCENTRATE

A special process for production of concentrated kvass wort for cold beverages was developed by *D.A. Korolev, L.S. Salmanova, V.I. Bukanova, P.I. Bukovsky, L.S. Sheptun and E.N. Chipchina; U.S. Patent 3,594,179; July 20, 1971.* Kvass extract or grain extract is prepared by mixing concentrated kvass wort with sugar syrup, citric or lactic acid and other kvass ingredients and then evaporating under atmospheric pressure to the required concentration.

While drying the malt, the enzyme complexes are destroyed through thermal inactivation, which necessitates the production of concentrated wort or kvass only from a mixture of dry rye and barley malt and a small amount of flour. The potential for melanoidin formation during the process of drying the malt is very low due to the presence of insufficient amounts of the reducing sugars and amino acids reacting in the process.

In this process, to a mixture of malt, flour and water, a cytolytic enzyme preparation of a culture of the fungus *Trichothecium roseum* is added and enzymatic hydrolysis of the substances of the intercellular walls in flour and malt (cytolysis) is carried out, followed by enzymatic hydrolysis of the mash proteins (proteolysis). The formation of melanoidins imparting the bread and malt flavor, aroma and color takes place directly in the wort during the step of evaporating in vacuum.

Kvass wort is prepared from freshly germinated rye malt (30 to 55% by weight), rye flour (44 to 70% by weight) and a cytolytic preparation (0.1 to 0.4% by weight) by mixing the entire mass with water in a ratio of 1:3 to 1:4.

The method of producing concentrated kvass wort consists of the following steps: Cleaned and sorted rye is steeped in water by the air-water or air-spray method at a temperature of 10° to 12°C for 20 to 24 hours until the moisture content reaches 42 to 44% and sprouts emerge.

The steeped rye is germinated for malt so as to achieve maximum formation and accumulation of the required enzyme complexes (cytolytic, proteolytic, aminolytic and others). Germination is carried out at a temperature of 12° to 15°C for 4 days. During germination the grain is stirred, thus affording the access of atmospheric oxygen and removal of carbon dioxide. The grain is remoistened if necessary.

For better solution of the malt and accumulation of maximum amounts of proteolytic enzymes, the grain, before germination is completed, is held for 6 to 10 hours without access of air and without stirring, the temperature of the malt not exceeding 50°C. Freshly germinated rye malt is sometimes slightly sun dried and kibbled in water in a ratio of 1:3 to 1:4 utilizing hammer mills, thus converting it to malt milk which is used for making kvass wort.

Before mashing, rye flour, dry or precooked in water in pressure cookers and then cooled, is mixed with the malt milk and cytolytic enzyme preparation of a fungus culture or purified preparation of *Trichothecium roseum.* Precooking of the flour is necessary to promote rupturing of the starch grains of the flour, paste formation and partial enrichment of the mash with melanoidins.

The process of preparing (mashing) the first kvass wort with the object of achieving maximum enzymatic hydrolysis of the starch and proteins of the malt and flour, thus converting them into soluble sugars, dextrins and amino acids, which are brought into aqueous solution, consists of the following steps:

5% of rye flour, either cooked or uncooked, and 50% of freshly germinated rye malt in the form of malt milk (in terms of flour moisture content) are mixed with warm water (not warmer than 42°C) in a ratio of 1:3 to 1:4, depending on the preset concentration of the kvass wort. When the temperature of the mash (mixture) reaches 40°C, the abovementioned enzyme preparation is added in a proportion of 0.2% by weight of the malt and flour. The mash is prepared in such a way as to obtain an initial concentration of wort extractives of the order of 14° to 15° Balling. The kvass mash thus prepared must undergo the following steps:

(a) Cytolysis at a temperature of 40° to 42°C for 1 to 2 hours periodically stirring the mash for 5 minutes at 30-minute intervals; the object of this step is to achieve enzymatic hydrolysis (splitting) of the substances of the intercellular walls in the malt and flour (hemicelluloses, gummy substances, etc.).

(b) Deeply penetrating proteolysis at a temperature of 50° to 52°C for 1 to 1.5 hours periodically stirring for 5 minutes at 30-minute intervals as is done during cytolysis; the object of proteolysis is the deeper and more complete splitting of the proteins to soluble amino acids and proteins having medium molecular weights.

(c) Sugar formation in the mash at a temperature of 62° to 63°C for 30 minutes and a temperature of 73° to 75°C for 20 to 30 minutes; the object is enzymatic hydrolysis of the starch of the malt and flour to fermentable sugars.

(d) Coagulation of sparingly soluble proteins of the mash by boiling the latter for 2 hours; and

(e) Separation of the first kvass wort from the grounds and its clarification.

The first kvass wort is fed to a vacuum apparatus to carry out the process of melanoidin formation, i.e., the formation of melanoidins imparting the bread flavor, aroma and color, which determine the characteristic features and quality of grain kvass as a beverage.

This process consists in bringing the concentration of the extractives contained in the first kvass wort to 63° to 65° Balling at a temperature of 53° to 64°C, raising it to 80° to 105°C after reaching the specified concentration and holding the wort at this temperature for 30 to 35 minutes or longer, depending on the intended purpose for the concentrate and completion of the reaction of melanoidin formation at the stage ensuring the characteristic bread flavor, aroma and color of the concentrate and kvass made therefrom.

The concentrated kvass wort thus prepared is blended with all the ingredients of grain kvass in order to obtain concentrated kvass. When producing concentrated bottled grain kvasses, concentrated wort is mixed with sugar syrup to give a concentration of 63° to 68° Balling, the content of grain extractives being not lower than 27° to 28° Balling. To the concentrated wort is added citric or lactic acid to give an acidity in the finished grain kvass equivalent to 3 to 4 ml of normal alkaline solution per 10 ml of kvass as determined by titration. Aqueous or alcoholic infusions of aromatic herbs, roots and spices can be added to the concentrated wort, while to barreled concentrated kvass wort lactic acid bacteria and yeasts may be added.

When producing concentrated beer wort, instead of freshly germinated rye malt and rye flour, there are utilized 10 to 30% of freshly germinated barley malt, 69 to 90% of barley flour, 0.4 to 0.8% (by weight of malt and flour) of cytolytic enzyme preparation in the form of a culture of the fungus *Trichothecium roseum* and water in a ratio of 1:3 to 1:4. After mashing, the following processes are carried out: cytolysis, proteolysis, conversion, coagulation of mash proteins, separation of the wort from the grounds, clarification and vacuum-evaporation of the wort in order to produce malt melanoidins imparting the characteristic flavor, aroma and color.

ENCAPSULATED DIACETYL REDUCTASE

A special application of this enzyme in the manufacture of beer and ale is explained by *J. Shovers and W.E. Sandine; U.S. Patent 3,733,205; May 15, 1973; assigned to Pfizer Inc.* In the production of alcoholic beverages such as beer and ale, a phase of the manufacturing process involves fermentation of cereal extracts (often referred to as wort) with certain suitable yeasts. During the fermentation process, the yeast cells produce a variety of chemical compounds in addition to ethyl alcohol which contribute to the flavor of the finished product. Selection of the strains of yeast to be used in the commercial fermentation of beer or ale is based on the relative amounts of compounds they produce, as determined by actual chemical analyses and by flavor of the final product. Careful attention must be paid by the brewer to select and preserve the strain of yeast used for his fermentation. Rigid quality control standards must be employed to prevent the introduction of extraneous flavors which detract from the characteristic and highly desirable taste and flavor of the finished product.

Diacetyl (2,3-butanedione) is produced in varying amounts by the yeast during the carbohydrate fermentation. Beyond a threshold level of 0.20 to 0.30 ppm w/v, the presence of diacetyl adversely affects the flavor and taste of beer and ale. The enzyme diacetyl reductase reduces diacetyl in the presence of a cofactor, reduced nicotinamide adenine dinucleotide (NADH).

Microencapsulation of diacetyl reductase and reduced NADH was found to be effective in the latter stages of fermentation or during finishing steps of processing prior to bottling.

Baker's or brewer's yeast (containing diacetyl reductase and reduced nicotinamide adenine dinucleotide) is incorporated with diacetyl reductase and reduced nicotinamide adenine dinucleotide or with diacetyl reductase alone. The microencapsulating film prevents the baker's yeast *(Saccharomyces cerevisiae)* or the brewer's yeast *(Saccharomyces carlsbergensis)* from escaping into the beer where it could be considered a wild type yeast by the brewer, while maintaining cell viability. Furthermore, the protected yeast cells can operate at the naturally acidic pH of beer (3.9 to 4.4) without undergoing denaturation or inhibition of the enzyme systems contained within the cell structure of the yeast.

Diacetyl reductase is widely distributed in microbial, animal and plant life. For example, the enzyme is prepared from a strain of *Aerobacter aerogenes.* The diacetyl reductase cofactor, reduced nicotinamide adenine dinucleotide, may be prepared from microbial sources such as baker's yeast. A high-bloom (high molecular weight) gelatin is used for the microencapsulation process because of its insolubility in beer.

Gelatin is added to water and heated to about 40°C until the gelatin has dissolved. The solution is cooled to about 30°C and the pH adjusted to 6.0 with 1 M NaOH. A mixture, in stoichiometric ratio, of diacetyl reductase and reduced nicotinamide adenine dinucleotide is stirred in and the mixture spread uniformly as a thin film on sheets of polyethylene. After about 24 hours at room temperature, the translucent gelatin-enzyme-cofactor film is peeled from the polyethylene and cut into small pieces approximately 1.0 cm^2. The film pieces are stored in a tightly stoppered container at 2° to 5°C.

In an alternate process, the gelatin-enzyme mixture is chilled to about 2° to 10°C until the gelatin has set and the coated mass is macerated with a high shear mixer. The material is tray dried at about 2° to 10°C and then about 22° to 25°C to a final moisture content of about 7 to 11%, preferably about 8%.

The regeneration of cofactor (reduced nicotinamide adenine dinucleotide) by viable cells of baker's or brewer's yeast makes possible a mixture of whole yeast cells, diacetyl reductase and the reduction of cofactor to about 0.1 the normal amount or even its elimination entirely.

When mixtures of diacetyl reductase and reduced nicotinamide adenine dinucleotide containing whole yeast cells are encapsulated and recovered as ground or milled material, there is the possibility of sheared surfaces exposing nonencapsulated yeast cells. This is obviated by a coacervation process in which the yeast-containing material is dispersed in light mineral oil which is then added in a thin stream with vigorous stirring into an aqueous solution of high-bloom gelatin. The enzymes and yeast cells are dispersed in the oil medium and each particle is coated with a gelatin-water film. The mixture is chilled to about 10° to 20°C which hardens the gelatin around the yeast cells and enzyme particles with the formation of discrete beadlets. Several liters of cold ethanol are added to partially dehydrate the beadlets. The beadlets are removed by filtration, washed with cold hexane and tray dried or fluid-bed dried at about 5° to 10°C to a final moisture content of about 8%.

Example 1: Aerobacter aerogenes ATCC No. 8724 is grown from a 1% inoculum for 24 hours at 30°C in 40 liters of sterile broth of the following composition: 20 g/l glucose, 10 g/l yeast extract and 20 g/l peptone.

Following growth, the bacterial cells are harvested with the use of a continuous flow attachment for a refrigerated centrifuge at 12,000 *g* with a flow rate of about 300 ml per minute. The packed cells are resuspended in 0.1 M potassium phosphate buffer at pH 7.2. The cells are washed three times in buffer and then resuspended in buffer to a volume of about 50 ml.

A crude diacetyl reductase extract is prepared by disrupting the bacterial cells in a sonic oscillator for about 20 minutes. The cell debris is removed by centrifugation at 27,750 *g* for about 1.5 hours in the refrigerated centrifuge. The supernatant is placed in a cellophane bag and dialyzed at 0° to 5°C against three, 4-liter changes of distilled water, with each dialysis lasting about 8 hours. The material inside the cellophane bag is lyophilized and stored at -20°C until used.

Example 2: 10 grams of gelatin are added to 50 ml of distilled water, and the mixture heated to about 40°C until the gelatin is dissolved. The solution is cooled to 30°C and the pH adjusted to about 6.0 with 1 M NaOH. 400 mg of freeze-dried diacetyl reductase enzyme preparation and 160 mg of reduced nicotinamide adenine dinucleotide are then mixed in, and the mixture spread uniformly as a thin film on sheets of polyethylene. After about 24 hours at room temperature, the translucent film is peeled from the polyethylene sheets, cut into pieces approximately 1.0 cm^2 and stored in a tightly stoppered container at 2° to 5°C.

To 120 ml of beer containing 0.5 ppm w/v of diacetyl is added 100 mg of the gelatin coated enzyme system. After approximately 72 hours at 5° to 7°C, the diacetyl concentration is 0.05 ppm w/v.

ALCOHOL PRODUCTION

Low Temperature Mash Process

It has been a common practice that before fermenting cereals, the cereals are ground and mixed with water or a mixture of water and stillage to obtain mash, which is then heated to burst the starch cell and liquefy starch contained in cereals, followed by saccharification.

Before or after the above cooking, saccharification proceeds to some extent along with the liquefaction, due to the addition of acids or liquefying enzymes, for instance, germinated cereals or enzymes from microorganisms, thereby reducing the viscosity of the mash. The mash, in which starch has been liquefied and partially saccharified, is filtered, as required, and then cooled to an optimum fermentation temperature, such as 25° to 34°C, after which the saccharifying enzymes and alcohol fermentative yeasts are added for fermentation.

Alternatively, the mash is cooled to the temperature optimum for saccharifying enzymes, such as 50° to 65°C, following the cooking step, then saccharifying enzymes are added, and a mixture is held at the above temperature for a while for allowing further saccharification of starch. Then, the mixture is cooled to the fermenting temperature of 25° to 34°C for fermentation.

The best method known for obtaining alcohol at a high yield is to cook mash at a high temperature of nearly 150°C in the case of conventional cooking or of nearly 180°C in the case of continuous cooking under a high pressure when using maize or milomaize. The object of cooking the cereals at a high temperature is to rupture the structures of cereal grits to elute starch from the cereal grits as well as to sterilize cereals. Upon cooking, the corn exhibits the maximum viscosity in the temperature range of 90° to 95°C in the case of using maize and milomaize, although this temperature depends on the kinds of cereals.

An increase of a temperature more than the maximum viscosity temperature causes a decrease in the viscosity of the mash. A decrease in viscosity due to such an increased temperature facilitates handling of the mash as well as enhancing the liquefaction of starch and the action of saccharifying enzymes on starch. In addition, cooking at a high temperature has a sterilizing effect on undesirable-type microorganisms. Thus, it has been widely accepted that cooking at a lower temperature results in the failure to elute starch or in difficulty in agitation due to increased viscosity after gelatinization, which in turn brings about incomplete liquefaction and saccharification. This lowers the manufacturing yields, i.e., the amount of alcohol manufactured.

H. Yoshizumi, N. Matsumoto and O. Fukushi; U.S. Patent 4,092,434; May 30, 1978; assigned to Suntory Ltd., Japan developed a process in which a high-temperature, high-pressure cooking step is eliminated and thus reduces energy and equipment costs. The process includes the steps of:

(a) adding water or a mixture of water and stillage, to grits of maize or milomaize containing grits of over 30% through 20 mesh to obtain a slurry having 13 to 40% by weight of solids content;

(b) heating the mash at a temperature 5° to 15°C lower than the temperature presenting the maximum viscosity, i.e., from 75° to 85°C, for the mash but higher than the sterilization temperature of undesirable-type microorganisms, for a period up to 30 minutes under substantially atmospheric pressure, with or without liquefying enzymes being added;

(c) cooling the mash to 25° to 34°C and adding saccharifying enzymes and alcohol fermentative yeasts and distilling the fermented slurry thus obtained.

As required, after heating at 75° to 85°C, the slurry may be cooled to 50° to 65°C and then a saccharifying agent is added followed by standing temperature for up to 30 minutes and then cooling to 25° to 34°C for fermentation.

The ground maize or milomaize may include the cereals themselves and grits of the cereals excluding embryo and starch derived from the cereals. In addition, this may include a mixture of raw materials containing over 50% by weight of such ground cereals with other starchy materials. Included as cereals to be added to maize or milomaize are rice, barley, wheat, rye, barnyard millet, common millet and Italian millet. In this respect, the finer the size of grits, the better will be the results. However, grits of over 30% through 20 mesh may suffice for this purpose, and water or a mixture of water and stillage is added to such grits to obtain a slurry having 13 to 40% by weight, of solid content. The term stillage refers to a residue of the mash, from which alcoholic components have been distilled after the completion of fermentation. The stillage in amounts of up to about 40% may be added with or without filtration.

The temperature exhibiting the high viscosity due to the heating and the resulting gelatinization of mash of the ground cereals varies depending on the kinds of cereals, lots and mash concentrations. The heating temperature should fall between 75° and 85°C, which temperatures are also sterilization temperatures to prevent growth of microorganisms. Upon heating, a liquefying agent may be added to the slurry.

The use of the liquefying enzymes may include germinated cereals and liquefying enzyme preparations from microorganisms, for instance α-amylase from *Bacillus* sp., which are used alone or in combination.

The slurry is cooled to the fermenting temperature of 25° to 34°C and then malts and/or saccharifying enzymes and alcohol fermentative yeasts are added to the mash. Depending on the case, the mash is cooled to 50° to 65°C after heating, then saccharifying enzymes or malts are added and the mixture is held for up to 30 minutes for the progress of saccharification, after which the mixture may be cooled to 25° to 34°C, and then alcohol fermentative yeasts are added for fermentation.

The use of the saccharifying enzymes may include germinated cereals (malts) and the saccharifying enzyme (e.g., glucoamylase from fungi) preparations derived from microorganisms, which are used alone or in combination. The fermentation is normally carried out at 25° to 34°C for 4 to 5 days, after which, if required, the mash is filtered and then alcohol or alcoholic beverage may be obtained by distillation.

The structures of starch are partially transformed due to the heating at a temperature on the order of the sterilization temperature of undesirable-type microorganisms, that is, from 75° to 85°C. This dispenses with cooking and sterilization at a temperature higher than 115°C, yet presents satisfactory fermentation.

On the other hand, it is known that starch is saccharified under the action of enzymes at a temperature of about 55°C, without liquefying the starch. This is possible, only when a special kind of cereal or special kind of enzyme is used. However, the use of direct saccharification of cereals at the temperature of about 55°C necessitates the sterilization of cereals by using chemicals or x-rays, beforehand, or the addition of chemicals for interrupting the propagation of undesirable type microorganisms or a method for interrupting the propagation of undesirable type microorganisms during the step of saccharification.

Example: Water was added to 100 grams of various kinds of ground cereals to obtain a slurry containing about 18% by weight of solids content, after which 3 grams of malt powder as liquefying enzymes was added to the slurry and agitated, followed by standing at 80°C for 30 minutes. The slurry was cooled to 28°C and 17 grams of malt powder and 0.1 gram of glucoamylase (800 units/g JIS K 7001-1972) were added, followed by further addition of 25 ml of culture yeast *(Saccharomyces cerevisiae).* The slurry was fermented at 28°C for 72 hours and analyzed for comparison of fermentation results.

The results show that there may be obtained fermentation efficiency of over 80%, presenting the same results as those obtained from the prior art fermentation which uses heating at a temperature of over 120°C. The viscosity measurements of slurry obtained from the cooking condition were not different from those of slurry which has been subjected to cooking at a temperature of over 120°C.

Glucoamylase Fermentation

A special glucoamylase ferment is used on cooked mash prior to fermentation with yeast to produce grain alcohol. According to *A.A. Andreasen, A.J. Bronsky and W.L. Bruce; U.S. Patent 3,988,204; October 26, 1976; assigned to Joseph E. Seagram & Sons, Inc.* the glucoamylase ferment is prepared from a culture propagated in the same cooked mash dispersed in water. The glucoamylase ferment is obtained from the culture *Aspergillus awamori* NRRL 3112 which shows a 100% increase in enzyme activity in the process.

The number of glucoamylase units required for conversion of high grain content mashes is reduced by about 50%. A unit of glucoamylase activity is that quantity of glucoamylase necessary to form 1 gram of glucose from a 4% starch solution, in 1 hour, at 60°C. The smaller amount of glucoamylase ferment required for converting mashes provides an economic and operating advantage in the process of preparing grain alcohol.

The mash is prepared by slurrying an appropriate cereal grain such as corn in water. The slurry is then pressure cooked using conventional batch pressure cooking apparatus or continuous pressure cooking apparatus. In general, cooking is carried out at a relatively high temperature range of from about 300° to 350°F and under a pressure of from about 50 to 120 psig.

The pH of the mash being cooked is adjusted within the range of from about 5.0 to 5.6, prior to addition of the propagation ferment and inoculation with yeast. The pH of the mash may be adjusted by the addition of stillage or other acidic materials such as mineral or organic acids such as sulfuric acid or citric acid. The propagation ferment added for conversion of the pressure cooked mash is prepared by incubating the microorganism *A. awamori* NRRL 3112 on a substrate of high grain content mash, prepared by cooking a grain dispersion under relatively high temperature and pressure conditions. Incubation is carried out using conventional aeration apparatus and at a temperature within the range of from about 90° to 100°F. Incubation is carried out upon a mash medium having a pH about 4.8.

Example 1: (A) Starter Culture — Spores obtained from a stock culture of *A. awamori* NRRL 3112 (maintained on Czapek's agar) are transferred to a starter culture medium consisting of 5% corn and 0.5% yeast extract in water, adjusted initially to a pH of about 4.8. The inoculated starting material is shaken and maintained at about 72°F. Two subcultivations are made at 30 and 48 hours respectively.

(B) Propagator Inoculum — Soured yeast mash of any conventional proportions of grain and water and as conventionally prepared for the growth of yeast inoculum is charged into a stainless steel, unbaffled, 90 gallon propagator equipped for air admission through a tube sparger and for agitation with one turbine-type impeller rotating at 90 rpm. The mash is diluted with water to 6° Balling resulting in 70 to 90 gallons of mash which is sterilized at 250°F for one-half hour and cooled to inoculation temperature.

About 1% of the mash volume or 3 liters of second stage starter culture prepared according to (A) above is used to inoculate the propagator contents. The propagator is then held at 75° to 80°F under 6 to 10 psig of positive pressure and aerated at 1 to 2 ft^3/min for a period of 30 to 40 hours.

(C) Propagator Medium — A corn grain slurry containing sufficient stillage for pH adjustment to 5.0 at a 16 gallon of liquid per bushel of grain ratio is pressure cooked in a continuous cooking system at a temperature of 320°F and under a pressure of 80 psig for 3.2 minutes. The mash is liquefied and partially converted with 3% malt at 145°F. There is thus obtained a mash of 28 gallons per bushel of grain in the ratio of 97% corn and 3% malt.

(D) Propagation — To a stainless steel, baffled, 6,000 gallon propagator equipped for air admission through a ring sparger and for agitation with two turbine-type impellers rotating at 90 rpm there is charged 4,600 gallons of mash prepared according to (C) above. The propagator contents are sterilized for ½ hr at 248°F, cooled to 95°F and then inoculated with 1.5% by volume of the propagator inoculum from (B) above. The propagator is held at 94° to 96°F under 6 to 10 lb of positive pressure and aerated at about 300 ft^3/min (0.45 vol air/vol liquid/min) for 180 hours. At intervals during the incubation, samples of the ferment are taken and analyzed for pH and glucoamylase activity. The results are listed in Table 1.

Table 1

Incubation Time, hours	pH	Glucoamylase Activity, units/ml
0	5.0	—
85	3.8	6.1
108	3.6	8.3
132	3.6	10.8
156	3.5	14.6
180	3.5	16.0

Similarly, repeating the above procedure, a second run is obtained with the following results shown in Table 2.

Table 2

Incubation Time, hours	pH	Glucoamylase Activity, units/ml
0	5.0	–
72	3.9	6.0
108	3.6	10.3
132	3.6	12.0
156	3.5	14.4
180	3.5	15.4
204	3.4	16.3

As shown in Tables 1 and 2, after 180 hours, the glucoamylase activity is 16.0 units per milliliter for the first plant run (Table 1) and 15.4 units per milliliter for the second plant run (Table 2). These yields are nearly double the average obtained in runs carried out following the above procedure but using mash cooked at atmospheric pressures.

Example 2: Fermentation of Plant Mashes Converted with Glucoamylase to Produce Grain Alcohol – A series of fermenters are charged with mashes prepared by cooking slurries of corn for 3.2 minutes at a temperature of 320°F and under a pressure of 80 psig. The pressure cooked mash is then liquefied with 2 to 3% gibberellin-treated malt at 145°F and cooled to fermentation temperatures. The mash so obtained contains 97 to 98% corn, a pH of about 5.0 and a mash ratio of 28 gallons per bushel of grain.

The fermenters are then inoculated with a 1% (by volume) yeast culture (200 million yeast cells per ml) and varying quantities of the glucoamylase-rich ferment obtained in Example 1. The fermentation is carried out by maintaining the temperature at about 90°F. The amounts of malt and glucoamylase ferment employed and the alcohol yields obtained are listed in the table below.

Conversion Agent				
Malt	Glucoamylase		No. of	Alcohol Yield
(%)	(u/bu)	(ml/bu)	Fermenters	(gal/bu)
3	1,536	97	4	5.24
3	1,024	65	45	5.31
3	640	41	103	5.32
3	416	26	4	5.26
2	960	60	2	5.24

As shown, all levels of glucoamylase gave approximately the same alcohol yield. Therefore, 416 units per bushel or 26 milliliters per bushel of an enzyme ferment having 16 units of glucoamylase activity per milliliter was sufficient for conversion of pressure cooked mash.

DISTILLER'S YEAST PRODUCTION

Improvements in operating ranges and accompanying economic advantages in the production of yeast from distilling grain mashes are explained by *J.M. Van Lanen,*

M.B. Smith and W.F. Maisch; U.S. Patent 3,868,307; February 25, 1975; assigned to Hiram Walker & Sons, Inc. Rye, barley, and barley malt are the usual grains used because, under the mashing conditions these grains provide adequate amounts of the nutrients required to propagate distiller's yeasts. Without special treatment, corn and grain sorghum (milo) do not support either rapid yeast growth or rapid growth and acid production by the lactic acid organisms used to sour yeast mashes. Therefore, mashes with high percentages of corn and/or milo, as normally processed, cannot be used satisfactorily to propagate distiller's yeast.

Various additives to corn and milo mashes have been recommended to overcome these nutritional deficiencies, among them urea, inorganic nitrogen, protein concentrates, yeast cell preparations, vitamins, etc., but each of these adds to the cost of yeast propagation and often has undesirable physiological effects upon the yeast or flavor effects upon the alcoholic distillate being produced.

The method involves the steps of preparing a cooked mash of ground cereal grains, especially mashes with at least about 95% corn and/or grain sorghum (milo). The mash is cooked until the starch content is gelatinized, and a minor amount (i.e., no more than about 2%, preferably no more than about 1%) of malted cereal (e.g., barley malt) or microbial amylase may be added in order to liquefy and thin the cooked mash.

The resulting mash is then treated with glucamylase ferment and, if desired, with lactic acid-producing bacteria, and the resulting mixture is incubated for a time and at a temperature such that starch and protein of the mash are hydrolyzed (i.e., for about 4 to 6 hours at 120° to 140°F). The mash is then sterilized, cooled, inoculated with yeast, incubated and then cooled. The resulting incubated inoculum has higher than usual yeast cell counts and greatly extended cell viability such that the cells may be stored for 72 hours or longer before use.

Ferments, which are rich in enzymes, notably amylases, proteinase, phytase, etc., may, for example, be produced by cultivating selected strains of *Aspergilli* on cereal grain substrates, which normally contain 10 to 20% grain by weight, under submerged aerobic conditions with intensive aeration and agitation. The resulting ferment is called glucamylase ferment because it has been primarily used in the rapid and near-complete saccharification of grain starch to glucose in alcoholic fermentations.

It has been found that this type of ferment, which can be produced using several different species and strains of *Aspergilli,* has an exceptional capacity to modify cereal grain mashes for the growth and viability of distiller's yeast.

Glucamylase ferments are important for yeast propagation because they make available yeast nutrients from grains, especially corn and/or milo, thereby promoting yeast growth and viability. As a result, yeast cultures can be produced which possess exceptional numbers of active yeast and which maintain exceptional viability over long periods of time. Yet all of these features are attained more rapidly and at lower cost than by conventional methods.

Example: The improved growth and viability of distiller's yeast have been demonstrated as follows: Hiram Walker Distiller's Yeast No. 109 was inoculated into three different substrates, viz: (1) 90% rye and 10% barley malt, which is commonly used in distillery practice, (2) 90% rye, 10% barley malt plus glucamylase ferment at the rate of 1 pint per bushel of grain, and (3) 99% corn, 1% malt plus glucamylase ferment at the rate of about 1 pint per bushel of grain. These glucamylase ferments were produced with *A. awamori* NRRL 3112. All mashes were soured with the same lactic acid culture and were then sterilized and cooled. Yeast culture was added and the mashes were incubated at the same temperature until the Balling was reduced 4°. The mashes were then placed at 60°F and viable yeast numbers were measured during the next 3 days by the usual method of serial dilution and plating in a medium containing 1% yeast extract (Difco), 1% glucose and 1.5% agar adjusted to pH 6.8. Yeast colonies were counted after 48 hours incubation at 90°F.

The viable yeast count reached an appreciably higher figure in the 99% corn, 1% malt, glucamylase ferment mash than it did in either of the 90% rye, 10% malt mashes and these high viable counts persisted many hours longer. The glucamylase ferment improved the 90% rye, 10% malt mash insofar as viable yeast counts were concerned but did not improve the yeast viability.

PRESERVATION AND ANTIOXIDANT METHODS

BACTERIAL SPORE GERMINATING METHOD

Germination is described as an irreversible process in which a number of events take place shortly after exposure of spores to specific stimulants. During this period of time, referred to as the germination phase, the spore loses heat resistance, dipicolinic acid, impermeability to dyes, calcium refractility and optical density to visible light. The spore is now referred to as a germinated form or a germinated spore. However, all of the changes do not take place at the same rate, for example, the complete germination phase for a spore may take two hours, but it may take only one hour for the spore to become heat sensitive whereas it may take two hours to lose dipicolinic acid. Outgrowth is described as the period of development after the germination phase until the beginning of the first cell division.

During this period of time, referred to as the outgrowth phase, the spore core membrane becomes a cell wall of the emerging vegetative cell which then elongates and divides. Vegetative cells sporulate and then the life cycle is repeated. The vegetative cells from Bacillus or Clostridium spores are responsible for food spoilage including flat sour spoilage and thermophilic anaerobic spoilage. For example, various salad dressings, vegetables and canned milk are spoiled by *B. subtilis* vegetative cells; likewise, dairy foods are spoiled by *B. megaterium* and fruits and vegetables by *B. stearothermophilus*. In order to destroy the nongerminated form of the spores present in food, they must be heated at high temperatures for extended periods of time.

By destroying these spores under such severe conditions, generally flavor, texture and vitamin content of the food is adversely affected. One way of reducing these adverse effects is to first germinate the spores to make them heat sensitive so that these forms may be easily destroyed by heating at lower temperatures for short periods of time.

L.G. Scharpf, Jr.; U.S. Patent 3,672,956; June 27, 1972; assigned to Monsanto Company reports that a lipase enzyme may be used directly to induce germination of bacterial spores of Bacillus and Clostridium genera, such as *Bacillus subtilis, Bacillus megaterium, Bacillus stearothermophilus*, for example, with corresponding decreases in the time required for the germination phase.

Additionally, the spores become heat sensitive at the same rate or before many of the other mentioned characteristic changes of germination. This is a substantial advantage in food processing, for as soon as the spores become heat sensitive, they may be destroyed, which decreases the overall processing time required for food. Another advantage found was that a percentage of the spores undergoing germination were rendered not viable, i.e., not able to continue their development in the outgrowth phase. This is another advantage in food processes, as in some cases, further preservation steps may not be needed.

The microorganisms that may be used to supply the lipases may be of a fungal origin or bacterial origin. Those of fungal origin include Aspergillus and Candida and Rhizopus. Examples of particular species include *Candida cylindracea, Rhizopus delemar, Mycotorula lipolytica, Deotricum candidum, Aspergillus niger* and *Aspergillus oryzae*. Those of bacterial origin include Pseudomonas, with *Pseudomonas fragii* being a specific species.

The lipase enzyme is used in combination with a physiological germinant or supplemental enzymes or both. By combining lipase with supplementary enzymes to germinate spores, the time required for the germination phase is decreased. These include the well-known enzymes such as proteases, amylases, or phosphatases.

As mentioned previously, the spores of the Bacillus or Clostridium genera may be germinated in any environment by bringing the lipase in contact with the spores. In most instances, however, the enzyme is brought into contact with the spores in the presence of some moisture. The amount of moisture is not critical, for example, that found in ordinary vegetables such as carrots would be sufficient to bring about the germination. Likewise, when the spores are germinated by a lipase in combination with a physiological germinant or supplementary enzyme or both, the components may be added separately or may be added as one composition containing all of the components.

It was found, however, that by first pretreating the spores with one germinating agent for a period of time then adding another, better germination rates were obtained than by using either alone. Generally, the spores are treated with one component for a period of about 20 minutes to about 45 minutes at a temperature of about 35° to about 45°C. This procedure is very effective in treating *B. megaterium* spores. The spores are pretreated with lipase for about 30 minutes then with alkaline protease. When the spores are germinated in the presence of moisture, generally the pH of such aqueous medium is from about 7 to about 9. This pH can readily be maintained by using any of the normal buffering agents such as the alkali metal phosphates including sodium phosphate or potassium phosphate.

Example 1: *B. megaterium* spores are prepared in the following way: A culture of *B. megaterium* is rehydrated, then plated out on 1% casein agar and incubated at 37°C for 24 hours. A transfer from a colony on this plate is made to a trypticase soy agar slant containing 0.5% glucose. The culture from this slant is transferred into 150 ml of nutrient broth having the following formula:

Glucose, g	1.0
KH_2PO_4, g	4.0
$(NH_4)_2HPO_4$, g	1.0
$MnSO_4$, g	0.2
NaCl, g	1.0
$CaCl_2$, g	5.0
$MgSO_4 \cdot H_2O$, mg	7.0
$ZnSO_4$, mg	10.0
$FeSO_4$, mg	10.0
Glutamic acid, g	1.0
Yeast extract, g	0.5
Agar, g	20.0
Deionized H_2O, liter	1.0

A 500 ml Erlenmeyer flask containing the culture and the broth is placed on a rotary shaker at 300 rpm for 24 hours. Sterile nutrient agar 100 x 15 mm plates are prepared and inoculated with 0.5 to 1.0 ml of active inoculum from the flask and incubated at 5°C for 3 days during which time sporulation takes place.

After sporulation is complete, autolysis is effected by keeping the plates at 4°C for 3 days. The plates are flooded with chilled, sterile deionized water to suspend the spores and the spores are collected by centrifugation at 3°C. The spores are resuspended in sterile (millipore filtered) lysozyme solution containing 0.5 ml/ml of lysozyme. The temperature is maintained at 37°C and agitation is carried out for 2 hours on a magnetic stirrer. The pH of the suspension is raised to pH 10 to effect further solubilization of vegetative cells. After 5 minutes, the pH is readjusted back to pH 7. The spores are again collected and washed from 6 to 8 times with sterile deionized water.

Concentrated spore suspensions are stored in glass bottles at 3°C. Spore suspensions are free of nonspore fragments and other debris and contain less than 3% of germinated forms. The spore concentration is determined by a dilution of the heat activated suspension on nutrient agar. Colonies are counted after 24 hours at 30°C. The spore concentration is 10^8 spores/ml of suspension.

B. subtilis spores are prepared in the following manner: a slant culture of *B. subtilis* (ATCC 7953) is inoculated into 1 liter Roux bottles containing 250 ml of nutrient agar at a pH of 7.4 supplemented with 0.01 mg/ml of $MnSO_4 \cdot H_2O$ and 8 mg/ml of NaCl. These bottles were held at 30°C for 3 days. The spores are harvested according to the procedure set forth previously. Spore concentration is determined by plating a dilution of heat activated suspensions. Colonies are counted after 24 hours at 30°C. The spores are plated out on nutrient agar.

Example 2: In order to test the effectiveness of these enzymes against spore contamination in canned food, the following procedure is used. 25 2 oz cans of sliced mushrooms are treated by separately injecting 5 ml of enzyme lipase to provide about 20,000 units/can and 10 mg of glucose/can. Cans are sealed with solder and incubated along with a control, i.e., 25 cans containing no enzymes, sealed at 37°C for 3 hours on a rotary shaker. Cans are then subjected to a temperature of 105°C for 10 minutes in the autoclave to inactivate the enzyme.

Cans were allowed to cool and held at 55°C for 30 days after which time they were evaluated. Cans are then determined to see if they have buckled or if they are swollen but not buckled. Gas produced in these cans is the cause for swelling and even bursting of cans. When these cans are opened, a very sour odor is noted and the pH has usually decreased from about 6.2 to about 5.1. A direct microscopic observation of the liquor from the swollen cans reveals the presence of many long thin rods. These observations are characteristics of thermophilic anaerobic spoilage frequently encountered in canned mushrooms. At the end of the storage period, control and enzyme treated cans are opened. The enzymes had no observable effect on the mushrooms and the overall quality of the enzyme treated mushrooms appeared to be as high as controlled mushrooms.

The same procedure is used with the exception that 25 cans of mushrooms are injected with a combination of lipase and a supplementary enzyme and alkaline protease. The combination is injected so that there are present about 20,000 units of lipase and about 70,000 units of activity of alkaline protease/can.

CELL WALL-LYSING ENZYMES

Yeasts and chlorella can be produced from stock cultures on a large scale for use in foods and feeds. However, the cell walls of these materials are poorly digested and it has been proposed that treatment with cell wall-lysing enzymes be used to improve digestibility. Such treatment will facilitate the extraction of useful intracellular substances from the cultured microorganism. Moreover, the action of these enzymes on cell wall constituents provides a method for destruction of yeasts, molds and other spoilage bacteria.

R. Kobayashi, H. Sato, K. Takita and N. Toyama; U.S. Patents 3,969,189; July 13, 1976; and *4,032,663; June 28, 1977; both assigned to Kumiai Chemical Industry Co., Ltd., Japan* identified a specific group of enzymes which have excellent capability of lysing cells and cell walls of various organisms. Two strains of the genus *Pellicularia* are used for supplying these enzymes, namely *P. filamentosa* (ATCC 14701) and *P. sasakii* (ATCC 13289).

The complex enzyme which is lytic to the cell of various microorganisms is produced and accumulated in the culture medium during the cultivation of the Pellicularia strain. A supernate of the culture broth which is obtained by filtering off or centrifuging the solid matters such as mycelium and solid residue of the culture medium may be used as such, for lysing the cells of various microorganisms. Such a supernate may also be freeze dried to give a crude enzyme preparation. When the solid cultivation has been conducted, the resulting solid culture may be extracted with water or buffered water of pH 4.0 to 7.0 to give an aqueous extract containing the complex enzyme, which may be utilized similarly to the abovementioned filtrate.

It is considered that the cell wall-lysing complex enzyme produced by *Pellicularia sasakii* and the complex enzyme produced by *Pellicularia filamentosa* are each a mixture of cellulase, glucanase, chitinase, protease, hemi-cellulase and carboxymethyl cellulase in the form of a complex and exhibit their high activity of lysing the cells of various microorganisms through the synergistic effect of the actions of the particular component enzymes, though it is presumed that these complex enzymes produced by the different species of Pellicularia have compositions which are more or less different from each other, as it is observed that these complex enzymes show more or less different properties and potencies for the particular enzymatic activities.

The cell wall-lysing complex enzyme may be applied in the extraction of useful intracellular substances from the cells as well as in the production of a concentrated extract comprising nutrient substances or other useful substances present in the protoplasm in the cells of various microorganisms such as yeasts, chlorella, bacteria and fungi. For these purposes the cells are treated with the complex enzyme in water to lyse the cell wall and to release the protoplasm so as to give a cell solution which may then by dehydrated partially or completely to give a concentrate or a dry powder.

Furthermore, the cell wall-lysing complex enzyme may be applied for the purpose of preventing foodstuff or drinks from spoiling, by incorporating an effective amount of the complex enzyme and allowing it to dissolve the cells of bacteria or fungi which would possibly occur or be present in the foodstuff or drinks and to which the spoiling of the foodstuffs or drinks is attributable. Thus, the complex enzymes find wide applications in many fields such as the food industry, pharmaceutical industry, pesticide industry and feed-preparing industry.

Example 1: In a conical flask of a 300 ml capacity were placed 19 g of wheat bran, 1 g of dried yeast and 18 ml of water. The flask was then warmed and the content of the flask was agitated to give a uniform pasty mixture. This culture medium was steamed at 120°C for 20 minutes for sterilization. A stock culture of *Pellicularia sasakii* was inoculated to the sterile culture medium. Solid cultivation of the Pellicularia strain was carried out for 120 hours at a constant temperature of 28°C. The resulting culture was mixed with 80 ml of an acetate buffer solution, pH 4.0, and the mixture was agitated at ambient temperature for 1 hour to ensure that the enzyme was extracted into the liquid phase.

The mixture was then filtered under pressure and the filtrate obtained was centrifuged to give 70 ml of a clear solution containing the complex enzyme which was produced by *Pellicularia sasakii.* To this clear solution was added 210 ml of acetone, so that the complex enzyme was precipitated. The precipitate was filtered out and dried to give 700 mg of an enzyme preparation in the powder form. This enzyme preparation was found to be active to lyse the cell walls of a variety of microorganisms, including bacteria, fungi, yeasts, Basidiomycetes and chlorella.

Example 2: This example is to show that the complex enzyme is useful to prevent various foodstuffs and drinks such as tomato juice, potato juice and orange juice from spoiling. 10 ml portions of tomato juice, potato juice or orange juice were placed in small dishes. The enzyme preparation which was produced as in Example 1 was purified by means of an ion-exchange resin. 0.1 mg, 1 mg and 10 mg of the resulting purified enzyme preparation was added to the contents of the dishes, respectively. Some dishes containing the foodstuffs or drinks were not mixed with the enzyme preparation, as control. The dishes were maintained in air at 28°C for a period of time and then the occurrence of spoiling was checked by observing discoloration, appearance and odor of the juices. The test results are shown below. In this table, the symbol + denotes that spoiling occurred and the symbol – denotes no spoiling occurred.

Material	Addition of enzyme	Time lapse (days) 1	2	4	7
Tomato juice	Not added	–	+	+	+
	0.1 mg/10 ml	–	–	–	+
	1 mg/10 ml	–	–	–	–
	10 mg/10 ml	–	–	–	–
Potato juice	Not added	+	+	+	+
	0.1 mg/10 ml	–	–	+	+
	1 mg/10 ml	–	–	+	+
	10 mg/10 ml	–	–	–	–
Orange juice	Not added	–	–	+	+
	0.1 mg/10 ml	–	–	–	–
	1 mg/10 ml	–	–	–	–
	10 mg/10 ml	–	–	–	–

DIASTASE PROTECTIVE COVER FOR SILAGE

The loss of nutrients in silage is a result of the oxidation and mold growth which takes place during the normal fermentation process. As a result, part of the nutrient value contained in the silage, principally the sugars, is oxidized to carbon dioxide and water and thus is lost. This not only reduces the value of the silage as a nutrient but together with the mold also makes the silage unpalatable to animals. In addition to this loss of nutrients due to oxidation, loss of silage due to spoilage of the forage in contact with air occurs since the production of desirable acids such as propionic acid and the stimulation of beneficial bacteria is hindered.

Spoilage can also occur in the preparation of feed such as corn meal or grist which, due to the high moisture content contained in the feed material and the heat generated in the grinding of the feed materials, as for example, in the preparation of corn meal or grist, is conducive to molding, thus resulting in waste.

J.G. Forest and E.J. Czarnetzky; U.S. Patent 3,871,949; March 18, 1975 describe a fermentation control-containing laminate for retarding the spoilage of silage and like materials on storage. It comprises a polyvinylidene chloride layer and a layer of a kraft paper impregnated with sodium sulfate and a malt diastase.

The active ingredients, i.e., the sodium sulfate and the malt diastase, can be applied to the Kraft paper layer in any conventional manner. A solution of the sodium sulfate and the malt diastase, in a conventional solvent such as water or an aqueous solution, can be prepared, applied in a conventional manner such as brushing, dipping, spraying to the Kraft paper side of the laminate with the Kraft paper absorbing the solution of sodium sulfate and the malt diastase resulting in the impregnation in the Kraft paper layer of the active ingredients. The amount of the sodium sulfate is not critical and generally the amount of sodium sulfate/square foot of a 50 lb weight paper will range from 0.05 to 2.0 lb/ft^2.

The amount of the malt diastase will generally range from 1 to 5 parts of malt diastase to each 50 parts of sodium sulfate. In addition, the laminate may contain intermediate layers between the polyvinylidene chloride layer and the kraft paper layer to provide additional properties to the laminate. For example, an intermediate clay layer can be incorporated between the polyvinylidene chloride layer and the kraft paper layer to minimize absorption of the polyvinylidene chloride into the kraft paper layer in the preparation of the laminate and thereby provide maximum absorptive capability to the kraft paper for the impregnation of the sodium sulfate and the malt diastase. Where desired, a thermal insulation layer can also be included. Any of these additional layers are conventional and are not a critical characteristic of the laminate.

To illustrate the fermentation control-containing laminate, a laminate was prepared from a polyvinylidene chloride and a 40 lb kraft paper such that the laminate prepared resulted in a paper layer which was approximately 4½ mils thick and a polyvinylidene chloride layer which was approximately 1½ mils thick. A 20% solution of sodium sulfate containing malt diastase at a level of 1 part by weight to each 30 parts of sodium sulfate was applied to the kraft paper side of the laminate. Absorption by the paper of the solution of the sodium sulfate and the malt diastase is such that the kraft paper contained 6% by weight sodium sulfate, based on the weight of the paper.

This laminate containing the sodium sulfate and the malt diastase was then cut into sheets approximately 5" x 5" and inserted into quart containers such that the laminate formed a top-to-bottom divider. These containers were then filled with milo grain (having a moisture content of 18%) or with a high moisture content corn (having a 30% moisture content). In another run, quart containers filled with milo grain (having the moisture content of 18%) and high moisture content corn (having a 30% moisture content) was saturated with the sodium sulfate/malt diastase. A control of the laminate was prepared in exactly the same manner with the exception that the kraft paper layer was not impregnated with the sodium sulfate and malt diastase and was inserted into containers and subsequently filled with milo grain or high moisture content corn.

Microaerophyllic conditions were established for each of the samples and the containers were set at 79°F. After 5 days the containers containing the laminate and the containers containing the milo grain and high moisture content corn saturated with the sodium sulfate/malt diastase composition were examined with the following results:

Sample	Results
Milo with laminate impregnated with sodium sulfate and malt diastase	Mold had grown all over the sides of the container as well as over the top, except for a distance of about ¼ to ½" next to the kraft paper side of the laminate, both as observed from the outside of the container and as observed after emptying the container.
High moisture corn with laminate impregnated with sodium sulfate and malt diastase	The same results as were obtained above with the milo were observed, but the mold was much more luxuriant. However, mold was not observed on the

(continued)

Sample	Results
	kraft paper side of the laminate similar to the above results.
Milo impregnated with sodium sulfate and malt diastase per se	No mold growth was observed
High moisture corn impregnated with sodium sulfate and malt diastase	No mold growth was observed
Control (laminate without sodium sulfate and malt diastase)	Molds of several types grew luxuriantly. The milo sprouted after 10 days. Sprouting occurred in all cases with the milo. The high moisture corn did not sprout.

The above results show the preservative and protective qualities which can be obtained by using the preservative-containing laminate. Similar results are also obtained where a clay layer is placed between the polyvinylidene chloride layer and the kraft paper layer in the preparation of the laminate.

The fermentation control-containing laminate can be used in almost any application where protection from moisture and the inhibition of mold growth is desired, inhibiting the action of harmful bacteria causing silage to spoil is sought, and preservation of silage is desired. For example, the fermentation control-containing laminate can be used as a lining for trench silos to protect from air and moisture and as a covering material for the covering of silage stored in silos, since it not only provides preservatives but also provides a moisture and gas barrier. The preservative-containing laminate can also be used as bagging material, as for example, where normal moisture content or high moisture corn and other grains are to be bagged and subsequently stored for future use.

The fermentation control-containing laminate can be also used in other applications, as for example, as a liner for the holds of ships in which a fungible material is to be stored and transported. The laminate provides a physical barrier to molds and mold spores and a physical barrier to the exchange of gases necessary for the life processes of molds. In addition, the laminate provides physical bacteriological and other conditions permitting the formation of specific acids due to fermentation of the sugars contained in ensilage, the acids being specifically inhibitory to mold growth.

HYDROGEN PEROXIDE PRESERVATION METHODS

Milk Sterilization

Fluid food products such as milk often contain bacteria which is harmful or potentially harmful and undesirable with respect to its consumption as a food product and the product quality. The undesired bacteria in fluid food products are typically removed from the product in several alternative ways. One method of sterilization or destruction is commonly known as pasteurization wherein the fluid is raised to a pasteurization temperature and held there for a temperature-time relationship which assures killing of the bacteria. Another method is to add a chemical such as hydrogen peroxide to milk which kills the bacteria in the milk, followed by an additional treatment of the milk to remove the unused and residual hydrogen peroxide.

Still another way of sterilizing or removing undesired bacteria in fluid food products is by passing that fluid through a centrifugal purifying means wherein the bacteria being more dense are separated from the milk (and fat) or fluid serum by the appropriate application of gravitational forces.

When milk is being purified to be utilized in cheese making, the raising of the milk for sterilizing purposes to pasteurization temperatures for the required time period functions to adversely effect the flavor of the cheese. Similarly, the addition of hydrogen peroxide to sterilize milk prior to its use in cheese making results in a chemical flavoring of the cheese and an adverse effect on its texture or body. Moreover, the United States Federal Food & Drug Administration places severe limitations on how much hydrogen peroxide may be added to milk intended for cheese making.

With respect to the centrifugal method, apparatus has been developed which improves the functioning of the centrifugal method in removing bacteria from the milk being processed. In apparatus of this type the milk is passed through and subjected to centrifugal forces which separate the milk into a milk serum and fat component substantially free of harmful bacteria and a second sludge-like component containing substantially all of the harmful, deleterious, and undesired bacteria. The sludge-like component, moreover contains approximately 2.5% of the original milk input. Accordingly, 2.5% of the input milk is lost in the centrifugation operation.

However, even more serious is that the sludge-like component which is extracted from the usable milk contains much free protein, about 12%. Accordingly, although the centrifugal purifying or sterilizing method has potential in that it may be used to avoid raising the milk to a pasteurization level for the time period (or alternatively avoids the necessity of adding hydrogen peroxide to the milk), it has the detriments of reducing the milk by a percentage during each stage of centrifugation and, even more importantly, it subtracts from that milk a very desirable protein subcomponent.

In order to effectively use hydrogen peroxide, *F.V. Kosikowski; U.S. Patent 3,525,629; August 25, 1970; assigned to Cornell Research Foundation, Inc.* developed a process in which bacteria are removed by centrifugal separation methods and the sludge is sterilized and returned to the process. The sterilization is accomplished by hydrogen peroxide treatment followed by removal of excess peroxide with catalase.

Referring to Figure 11.1 there are shown two centrifugal purifying means **1** and **2** connected in series. Assuming a fluid food product such as milk to be processed, the milk is first passed through centrifugal means **1** and then through centrifugal means **2**. While two centrifugal means are shown, this number is merely illustrative and one or more than two may be used depending on the precise design requirements of a particular processing system. Each centrifugal means (assuming proper design) functions to separate the milk into two components on the basis of difference in density. Specifically, the potentially harmful, deleterious and unwanted bacteria of the input milk have a higher density than the milk serum and most of the fat of the input milk.

Accordingly, one component comprising a large part of the milk serum and fat passes out of separating means **1** in liquid form to the input of separating means **2**

Figure 11.1: Sterilization of Milk

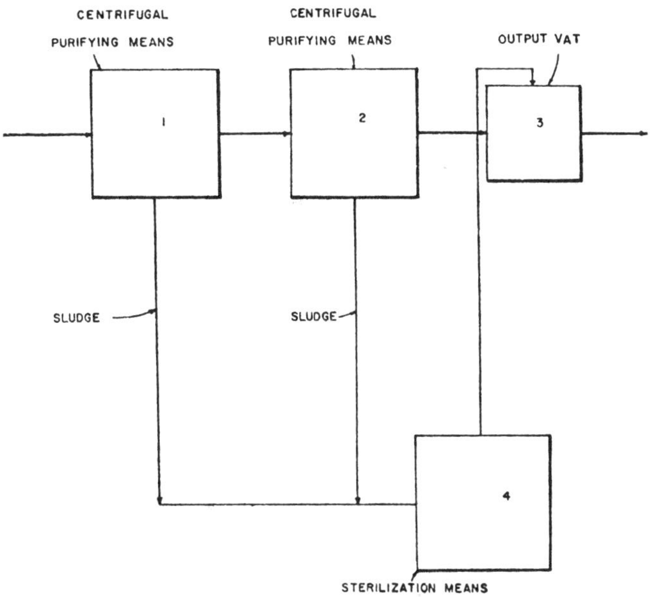

Source: U.S. Patent 3,525,629

whereas the other component leaves separating means **1** in sludge-like form by a separate route, as shown, and includes a high proportion of the bacteria and other microorganisms. The output from the separating means **1** is then passed through separating means **2** again separating the milk into two components. By the time that the milk has passed through two centrifugal means, as shown, substantially all of the bacteria is removed from the output milk in that excess of 90% of the bacteria is in the sludge-like output from means **1** and **2**. Milk serum and fat which passes through the centrifugal separating means (shown as means **1** and **2**) is passed on into an output container **3** where it is utilized in further processing a food product (cheese for example).

Block **4** illustrates the sterilization step and block **3** illustrates the return of the sterilized sludge into fluid mixture for further processing. Once the sterilized sludge is returned to fluid mixture with fluid component of milk serum and fat, the cheese making would follow a conventional process. Pasteurization of the sludge by raising its temperature to a sufficient level for a sufficient time period to kill substantially all of the bacteria in the sludge is not a detriment to the flavor of the ultimate product, such as cheese, because only a small proportion of the fluid mixture has been subjected to those temperatures. One sterilizing temperature and time period which has been used successfully for milk in cheese making is 165°F for one minute. The pasteurization of the sludge can be accomplished by conventional techniques.

The sludge may alternatively be sterilized by adding a quantity of hydrogen peroxide by following conventional techniques in this regard. For example, following the addition of the hydrogen peroxide (diluted with tap water on 1:1 basis), sludge containing hydrogen peroxide is heated to 130°F for about 16 seconds by conventional means. This is followed by the addition of an enzyme catalase for the purpose of causing any residual hydrogen peroxide to decompose into water and oxygen. The catalase is allowed to remain in contact with the milk for approximately 20 minutes or until a test such as a potassium iodide qualitative test indicates that no hydrogen peroxide remains.

The sterilized sludge is then returned to the centrifuge output product thereby returning to the fluid mixture the valuable subcomponent such as protein in milk and the normal loss herein attributed to the prior art technique is minimized. The hydrogen peroxide used for sterilization must be of the electrolytic type.

Because the chemical additive, hydrogen peroxide, as an example, is used to treat only the sludge and not the total output fluid, its effect upon the taste and texture or body of the final food product is minimal or nonexistent. For example, cheese made of a milk processed in this manner does not have a chemical flavoring nor is its texture or body adversely affected. Moreover, the various government regulations concerning the use of hydrogen peroxide become less stringent on a percentage basis. For example, less hydrogen peroxide is needed in milk processing than was heretofore possible.

Chocolate Syrup Sterilization

Conventional preparation of chocolate milk drinks includes a pasteurization step which kills bacteria, but the endospores survive the pasteurization and germinate in the chocolate milk drink preparation with the result that there is active bacterial growth during storage and the product is spoiled. This spoilage problem has been a major one for dairies for many years and many dairies consider chocolate milk spoilage to be their largest single ongoing problem.

Hydrogen peroxide is used by *B.T. Warkentin; U.S. Patent 3,860,730; Jan. 14, 1975; assigned to Chocolate International, Inc.* to destroy the endospores in chocolate syrup. Excess hydrogen peroxide is removed by a catalase.

A chocolate dry blend consists of cocoa powder, sugar, stabilizer, salt and vanillin. Tapioca flour and sweet whey powder can also be in the formulation. The dry blend is dissolved in water to yield a heavy syrup having a solids content in the range about 20 to 30% by weight. 30% by weight represents about the upper limit of solids content which can be handled in the process with facility and syrups having solids contents lower than 20% by weight have the disadvantage of imposing a heavier burden on the spray drying system in the final step of the process.

After the dry blend of solids has been dissolved in water to produce the heavy syrup, the syrup is heated to pasteurization temperature to kill bacteria. The pasteurization conditions are generally about the same as those for the pasteurization of milk, the syrup being heated to a temperature in the range of about 140° to 150°F and held at that temperature for at least half an hour.

After the pasteurization step is completed the syrup is cooled to ambient temperature (50° to 70°F) and allowed to stand at that temperature for at least 10 hours. During this period the endospores contained in the cocoa powder germinate and become vulnerable to thermal or chemical attack. After the time allowed for spore germination has elapsed concentrated hydrogen peroxide is added to the syrup. The quantity of hydrogen peroxide added is in the range of about 1 to 2% by weight of the syrup.

After addition of the hydrogen peroxide, which serves to sterilize the syrup by killing germinated endospores, the syrup is heated to moderately elevated temperature, usually in the range of 80° to 100°F and maintained at such temperature for 10 to 20 hours after which the syrup is heated to a higher temperature in the range of about 180° to 200°F and maintained at that higher temperature for at least 10 minutes as a finishing step in the killing of the germinated endospores. After the hydrogen peroxide treatment the syrup is cooled to ambient temperature (50° to 70°F) and from about 0.2 to 2.0% by weight based on the syrup, of catalase is added to the syrup. The syrup containing the added catalase is allowed to stand for about an hour during which time the catalase catalyzes the decomposition of residual hydrogen peroxide so that at the end of the hour residual hydrogen peroxide has been completely decomposed.

After the catalase treatment the syrup is spray dried in conventional spray drying equipment to yield a powder having a moisture content in the range of about 3 to 7% by weight. Spray drying temperature is about the decomposition temperature of catalase so the residual catalase is destroyed in the spray drying step. Catalase decomposes at 150°F and the spray drying step is carried out at temperatures well above 150°F, for example, at temperatures in the range of about 190° to 220°F.

BACTERIAL OXIDATION INHIBITOR

Antioxidants are normally used in foodstuffs to inhibit oxidation of oxidizable substances. "Oxidizable substances" refers to any substance liable to degradation by oxidation involving superoxide ion.

A.M. Michelson and J. Monod; U.S. Patent 3,920,521; November 18, 1975; assigned to Agence Nationale de Valorisation de la Recherche (ANVAR), France identified a superoxide dismutase as a foodstuff oxidation inhibitor as demonstrated by its antioxidant effect on substances such as mushrooms, apples and potatoes.

The superoxide dismutase enzyme can be extracted from any suitable source, such as: erythrocupreine, *Escherichia coli,* fungi such as *Pleurotus olearius,* marine bacterial strains such as *Photobacterium phosphoreum, Photobacterium leiognathi* and *Photobacterium sepia.* The activity of the enzyme is determined by inhibition of the chemiluminescence of luminol, a reaction produced by the enzyme system hypoxanthine/xanthine oxidase/oxygen. Each time the reaction was effected, 1 ml 0.33 x 10^{-4} M hypoxanthine was injected into a mixture of 0.3 ml 10^{-3} M luminol, 0.3 ml of 1 M, pH 9 NaOH-glycine buffer, 0.3 ml 10^{-3} M EDTA, 1.8 ml water and 0.005 ml of a 1 mg/ml xanthine oxidase solution. The experimental device used to determine light intensities (1) without, and (2) following inhibition by means of superoxide dismutase requires a silver-lined cuvette adapted to receive the mixture to be tested and positioned in front of a photomultiplier. The reaction was initiated by injecting the substrate into the cuvette, following which emission of a photon flux occurred which in response to the photomultiplier, gave rise to a current the intensity of which was measured with a picoamperemeter and recorded. If a suitable amount of the superoxide dismutase, to be determined, was introduced into the reaction mixture, prior to reaction initiation, the light emission was inhibited. The unit of activity selected corresponded arbitrarily to the amount of enzyme inducing a 50% reduction in the maximum light intensity in the absence of enzyme inhibition. The unit is therefore independent of the enzyme source and the superoxide dismutase enzyme used.

Example 1: 5 g of fresh mushrooms were used; these were thinly sliced and the colored gills removed as they might have affected the color of the medium and made it difficult to assess the results. The mushrooms so prepared were divided between several 100 ml beakers and 50 ml of a pH 7.8 (0.01 M phosphate buffer) buffered solution containing 0.02% ascorbic acid was added. 50 units of superoxide dismutase of *Photobacterium leiognathi,* strain No. ATCC 25521 was added to one beaker.

Each beaker was covered with a double sheet of paper and left to stand at laboratory temperature. 40 hours later, the optical density of the supernatant in each of the beakers was determined at 600 mμ. The optical density found made it possible to deduce an average increase of 0.173 in the optical density for the controls, whereas the increase in optical density was only 0.092 for the supernatant comprising the superoxide dismutase. Oxidation in the beaker so treated was therefore 47% lower than it was in the control beakers.

Example 2: Slices of apple were introduced into a solution of 3 units of superoxide dismutase of *Photobacterium leiognathi* strain No. ATCC 25521 per ml, in a pH 7.8 phosphate buffer. A few minutes later, the apple slices were removed from the solution, placed in Petri dishes and left to stand. Several days later, the slices so preserved were compared with slices prepared in the same manner but preserved without superoxide dismutase treatment. The latter were observed to be browner than those treated with the enzyme.

Example 3: Slices of potato were placed in beakers each containing 3 units of superoxide dismutase of *Photobacterium leiognathi* strain No. ATCC 25521 per ml in a pH 7.8 phosphate buffer. They were removed a few minutes later, placed in Petri dishes and left to stand. Four days later the treated slices appeared to be markedly less oxidized than identical slices subjected to the same treatment but without the addition of superoxide dismutase.

EGG YOLK COLOR RETENTION

Hard boiled eggs develop typical gray or green discoloration in the area of the interface of the yolk and white. This discoloration is particularly unattractive in commercially prepared egg products where the yolk material is surrounded by egg white. A combined treatment by a proteolytic enzyme and an oxidizing agent is used by *W. Ng; U.S. Patent 3,598,612; August 10, 1971; assigned to The Ralston Purina Company* to eliminate the color problem.

Liquid egg yolks are combined with a minor amount of a proteolytic enzyme. The enzyme serves to hydrolyze the protein that is present and provides for a smoother texture in the cooked yolk product. After a sufficient time for reaction between the enzyme and protein and lipoprotein of the yolk, the yolk is coagulated at elevated temperatures in a casing or mold of desirable shape. The yolk is removed from the mold or casing and subsequently passed through or immersed in a solution containing a minor amount of an oxidizing agent such as hydrogen peroxide.

In another method, the raw liquid egg yolks are combined with a minor proportion of proteolytic enzyme. After a sufficient time for reaction between the enzyme and the protein and lipoprotein of the yolk, a minor amount of an edible oxidizing agent such as hydrogen peroxide is added to the yolk mixture. The treated yolk is then placed into a mold or casing and heated at elevated temperatures sufficient to coagulate the yolk and also inactivate the enzyme. The yolk is cooled, removed from the casing or mold and subsequently processed in a manner such that coagulated egg whites surround the coagulated yolk to produce a finished egg product.

The proteolytic enzymes that may be used in the process are typically those types of enzymes that will substantially hydrolyze the protein and lipoprotein present in the yolk to which it has been added. Examples of such enzymes include papain, bromelin, pepsin and trypsin. The enzyme has a strength of about 6,000 hemoglobin units/g of the enzyme material. The liquid yolks and whites are heated during the process to elevated temperatures sufficient to cause the yolks and whites to coagulate and also to inactivate any proteolytic enzymes which may be used in the process. Such temperatures are found to be in the range of from about 180° to about 212°F.

Example: The whites and yolks from 72 fresh, cool eggs were separated and placed into individual containers. Into the yolk portion, weighing 1,000 g, was added 4 ml of a 5% solution of papain (0.02% by weight based on weight of yolk). The papain was reacted with the yolk for about 20 to 50 minutes. To the egg yolk was mixed 10 ml of a 5% hydrogen peroxide solution (0.05% by weight based on weight of the yolk). The treated yolk was placed into a standard sausage casing 20 mm in diameter and closed at both ends. The casing was placed into hot water at a temperature of about 200°F for about 25 minutes. The casing was then removed from the hot water and placed in cold water until the temperature of the yolk was about 70°F, the casing being subsequently removed from around the yolk.

About a 12¾" length (150 g) of the solid cooked yolk was placed into a second sausage casing about 45 mm in diameter. The casing contained about 350 g of raw liquid egg white and completely surrounded the solid egg yolk. The casing being closed at both ends was placed into hot water at a temperature of about 200°F for about 20 minutes. The resulting cooked egg roll product was removed from the hot water, placed in cold water until the temperature of the entire product was about 70°F. The cooked egg product was placed in a refrigerator at about 50°F for 72 hours to determine the amount of discoloration that would occur in the product. At the end of the period the product was removed, cut open and no discoloration was visible at the yolk-white interface.

BY-PRODUCT AND WASTE UTILIZATION AND ANIMAL FEED PRODUCTION

VEGETABLE AND FRUIT WASTE UTILIZATION

Oil Seed Fermentation

Enzyme action is effected via a fermentation in which vegetable oil seeds are digested to obtain proteins suitable for human or animal nutrition. According to *T.J. Staron; U.S. Patent 4,085,229; April 18, 1978; assigned to Institut National de la Recherche Agronomique, France* the most effective microorganisms for this fermentation are the fungi *Achrostalagmus aphidum* and *Geotrichum candidum*.

An aqueous suspension of a cake is formed, containing between 20 and 30% by weight of dry matter. The maceration conditions require a temperature slightly higher than ordinary ambient temperature, in the range of about 30° to 40°C, for example. The length of maceration should be sufficient to permit detoxification of the cake used and hydrolysis of the undigestible polysaccharides; it generally varies as a function of the temperature and the microorganism used; for example, the length of maceration is generally from about 24 to 45 hours, when the temperature is between about 30° to 40°C.

At such temperatures microorganisms do not develop, but their enzymes are then at a maximum of activity, resulting in optimal destruction of various carbohydrates, antinutritional substances, phenols and tannins. These multiple enzymatic activities produce about 20 to 30% solubilization of the hydrocarbon dry matter, resulting in a rapid and substantial decrease of the pH, which falls from about 6.2 to about 4 in 15 to 20 hours.

The microorganism used in the maceration step is added to the aqueous suspension of crude cake, assaying about 20 to 30% by weight of dry matter, at the rate of 0.1 to 0.5 liter of microbial culture per kilogram of cake. Maceration is carried out under stationary or stirred conditions; slow stirring in the order of 45 to 60 rpm is sufficient. It is best to ventilate the reaction medium to extract the gas produced during fermentation.

After maceration, the insoluble products are removed from the reaction medium by filtration or centrifugation. Protein solubilization is carried out by means of an alkali such as sodium hydroxide in the presence of an aldehyde. The presence of such an aldehyde makes it possible to inhibit the formation of lysinoalanine which is reputed to be toxic. The flowsheet below is a schematic representation of the steps of the process.

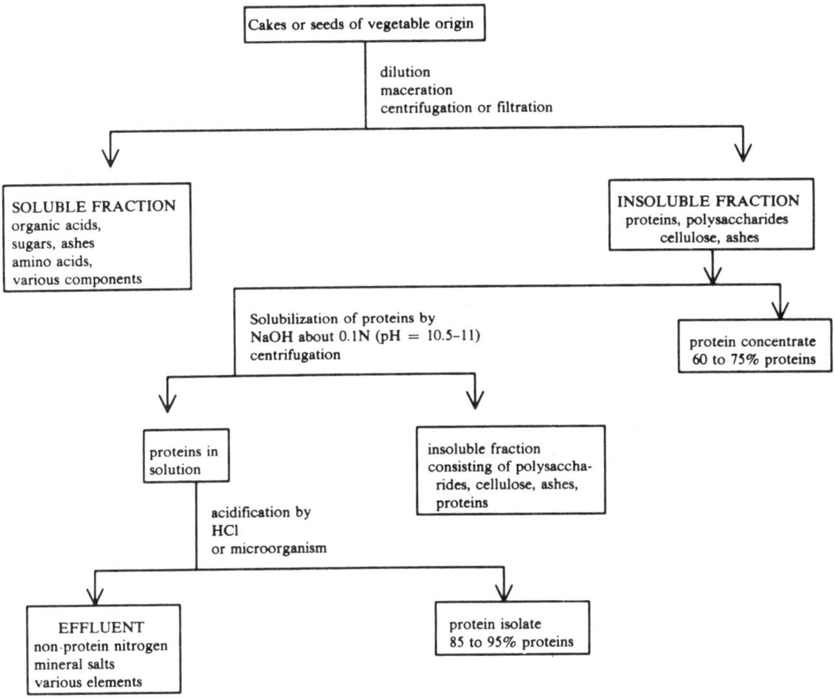

Example: Crude soy cake was mixed with water until a suspension was obtained assaying about 25% dry matter. 0.1 to 0.3 liter of *Geotrichum candidum* culture per kilogram of cake was then added. Maceration was carried out in a fermenter for 30 hours at a temperature of approximately 40°C. In order to maintain an even temperature in the fermenter tank, it was rotated at 45 to 60 rotations per minute and the reaction medium was ventilated to extract the gas produced during fermentation.

The insoluble products in the reaction medium were then recovered by filtration. The insoluble products recovered were then dried to obtain a product. The insoluble products were then treated with sodium hydroxide in order to make soluble the proteins contained in the insoluble products. For this purpose, 4 parts of 0.1 N sodium hydroxide containing 0.3% formaldehyde were added to the insoluble products to avoid the formation of lysino-alanine. The resulting reaction medium was heated for about 20 minutes at a temperature of about 40°C. The reaction medium was then centrifuged to remove the sodium-containing insoluble products. The proteins in the resulting liquid fraction were then precipitated by acidification with hydrochloric acid to pH 4.9 and a protein isolate was obtained.

Instead of using the yeast *Geotrichum candidum* in the maceration step, similar results were obtained by the use of one of the following microorganisms: *Rhizopus nigrieans; Penicillium roqueforti; Penicillium camemberti; Aspergillus oryzae; Verticilium alboatrum; Fusarium gramminaerum; Circospora beticola; Monilia laxa; Trichoderma viridae; Achrostalagmus aphidum; Saccharomyces cerevisiae* and *Candida lipolytica.*

Hemicellulose and Pectin Conversion

Plant by-products and plant wastes, high in carbohydrates and low in protein can be treated by an enzyme process to produce a uniform cohesive product of balanced nutritional value. In the process of *H.C. Silberman; U.S. Patent 3,615,721; October 26, 1971; assigned to Philip Morris Incorporated* raw material is converted to a slurry or paste, treated with enzymes and then dried by spraying, casting or extruding. The enzyme mixture contains cellulase, hemicellulase and pectinase activity.

Improvement in taste is realized as in the conversion of bitter tasting flavonoid glycosides into bland hydrolysis products and in the treatment of carrots that develop an apricot-like taste. During the conversion of such vegetable and fruit material, essentially the same chemical composition is maintained by adding nothing besides catalytically small amounts of enzymes and by taking away nothing while improving digestibility.

A requirement for the vegetable or fruit to be processed is the presence in them of hemicelluloses and pectin. Cellulose may be, but need not be, present. The enzyme systems should be such that they partially degrade the cellulose, the hemicelluloses and the pectin. Thus, suitable enzyme systems must exhibit hemicellulase activity and pectinase activity, i.e., they must include a hemicellulase and a pectinase. In order for the enzyme preparation to have some activity on the plant material, the plant material must contain pectin and/or such hemicelluloses which can be degraded into intermediate weight oligosaccharides.

Not all the polysaccharides, not all the hemicelluloses and not all the plant gums were being degraded in a similar manner by the enzyme preparations which were found suitable. This was found by the following experiment in which suitable enzyme preparations were exposed to pectin, xylan, guar gum, locust bean gum, algin, gum acacia and gum tragacanth. Five grams of these polysaccharides were mixed with 500 ml portions of an acetic acid-acetate buffer solution having a pH of 4.5 and a molarity of 0.04. Then one portion of 0.05 g of the enzyme system was added to each mixture, the mixture being maintained at 45°C for a period of 24 hours.

The resulting mixtures were then each placed in 1,000 ml of acetone, whereby precipitates formed which comprised undegraded or little-degraded polysaccharides. These precipitates were separately filtered, and dried in a forced air circulation oven at 45°C to constant weight. Controls were also run for each of the materials, under identical conditions, except that no enzyme was present.

All enzyme preparations, Cellulase 35 (a mixture of polysaccharases including cellulase, pectinase and hemicellulase), Pectinol R 10 (Pectinol is a mixture of pectin enzymes and includes other enzymes such as cellulase), Lipase B (an enzyme preparation standardized with salt exhibiting protease activity at pH 5 and hydrolysis of fats and fatty acid esters), all from Rohm and Haas Co., Pectinase (Nutritional Biochemical Corp.), Cellase 1000 (a fungal cellulase preparation derived from a selected strain of *Aspergillus niger*, and which contains trace amounts of protease, amylase, pectinase, beta-galactosidase and cellobiase, from Wallerstein Co.) and Cellulase 4000 (Miles Laboratories) are capable of producing good cohesive sheets, filaments, strands or bands. All the enzyme preparations contain enzymes which degrade pectin, hemicelluloses, cellulose and plant gums.

Certain fruits and vegetables give poor results because a viscous slurry and cohesive sheet are difficult to obtain. These unsuitable fruits or vegetables are those high in starch and to some extent those high in fatty oil and proteins or high in water content. In these instances, the end product, after drying, may be a powder or flake of mainly starch or syrup. Fruits like strawberries, seeds like nuts and vegetables like potatoes will probably be unsuitable for treatment with the enzyme systems.

However, so long as there are enough pectins and hemicelluloses present to be acted upon by the enzymes, the presence of starches, proteins and lipids does not interfere with the action of the sheet-forming enzymes. The starches, the proteins and the lipids can be considered as being more or less inert filling materials as far as sheet formation is concerned. One has, therefore, to concentrate on the presence of enough convertible pectic and hemicelluloses, rather than putting too much emphasis on the presence of starches, proteins and lipids.

Example: All sheets were prepared with 0.2 g Rohm and Haas Cellulase 35. All weights were dry weights based on water-free materials.

. . .Starting Material, g. . .		Evaluation of Cast Sheets
Carrots	18	Very good plastic sheet
Carrots and olive oil	18 16	Good plastic sheet; when sheet is squeezed, oil oozes out
Soybeans	38	Very weak, cracked, brittle sheet
Carrots and soybeans	9 28	Weak, brittle sheet
Carrots and soybeans	18 28	Good tough sheet

The soybeans, which contain 9% moisture, 40% proteins, 18% fats, 4.6 ash and 17% carbohydrates (mainly sucrose, raffinose, stachyose, and pentosans) cannot be made into a good sheet by the use of enzymes alone. However, through the mixture of some carrots (containing about 7% pectin on a dry substance basis) good sheets are obtained.

The treatment time is dependent upon the species of vegetable or fruit, the size of the plant material, the enzyme concentration, the degree of agitation, the temperature and the pH of the slurry. Generally, the treatment time will vary between 2 and 12 hours. Under certain circumstances the addition of a preservative to prevent spoilage may be desired. Such preservatives should not interfere with the hydrolytic action of the enzyme preparation. The pH during enzyme action may be between about 4.0 and 5.0. This pH is very often the natural pH of the plant material. Acid may have to be added if the plant material exhibits a pH of more than about 5.5. The treatment temperature can be between 25° and 60°C. Before casting, spraying or extruding, the plant-enzyme system must be agitated. This may be accomplished by continuous stirring or by periodic agitation. Stirring may be difficult and unneccessary at the beginning of the treatment when the plant material is of large size and relatively hard. Agitation will be easy and is necessary at the end of the enzyme action when the plant material has become mushy and forms a uniform slurry suitable for casting, spraying or extruding.

The uniform slurry can be applied on a moving endless belt. A dry sheet is applied which can be collected in big rolls, cut into large pieces, shredded or ground to a powder. The slurry can also be extruded into filaments, strands or bands. Drying of the sprayed, cast or extruded slurry is accomplished at elevated temperature in an air stream for rapid moisture removal. In certain instances, high temperatures may be desired to obtain a cooked product.

SILAGE TREATMENTS

Celloflour Production

Alfalfa and clover contain crude protein which may be used as food for ruminants and other animals. While these plants contain protein in amounts potentially useful as a food substance for humans, such plants have been infrequently used for this purpose. A major reason for this is that these plants are not readily digestible by humans. Another reason is the fact that these plants are unpalatable to humans in their unprocessed state.

A process is provided where a food product for humans high in protein, carbohydrates, vitamins, minerals and essential amino acids is produced from alfalfa or clover. When alfalfa meal is treated with an aqueous basic solution having a pH not less than 8, followed by treatment or digestion of the resulting slurry with pancreatin, a soluble fraction and an insoluble celloflour residue are obtained. Both fractions are useful as human food. The details of the process are given by *G.W. Edwards and A.W. Edwards; U.S. Patent 3,833,738; Sept. 3, 1974*. A particularly effective digestion system is a combination of pancreatin and bile materials.

Alfalfa or clover is treated with an aqueous basic solution; the resulting slurry is treated simultaneously or serially with pancreatin and a bile-containing material and the soluble and insoluble portions separated. The insoluble residue is dried, ground and resuspended in water and again digested with pancreatin and a bile-containing material, after which a second water-soluble fraction and a second water-insoluble residue are obtained. The second insoluble residue and second soluble fraction may be separated or they may be dried together and ground. The purpose of the second digestion is to extract and hydrolyze the saponins left after the first digestion.

The improved extraction and digestion increases both the nutritive value and the palatability of the product. The soluble fraction from the digestion slurry can, after separation from the residue, be separated into two fractions (the low and high density extracts). This can be done by centrifugation or by letting the soluble fraction settle. The fraction of low density is brown, while the high density fraction is dark grey to black, and the high density fraction contains more protein and fiber but less crude fat than the low density fraction. Analysis reveals that both the high and low density soluble fractions, after separation from the insoluble celloflour residue, contain an amount of protein which is about the same or slightly less than that of the unprocessed alfalfa or clover. The fat content of the soluble fraction is higher than that of the ether extract of the unprocessed plant material from which the soluble fraction is obtained. This indicates that the lipid in the alfalfa or clover is in the form of a lipoprotein which is released in the above digestive procedure. Also, the soluble fraction is higher in carbohydrate than is the alfalfa or clover from which it is prepared.

The insoluble residue is a celloflour which is slightly green in color. Its protein content is approximately equal to that of the unprocessed plant material from which it is prepared. The fat content of the celloflour is higher than that of the ether extract of the alfalfa or clover. The celloflour is higher in crude fiber, but lower in nitrogen-free extract (carbohydrate), than the alfalfa or clover. The celloflour is lower in ash than the alfalfa or clover.

One interesting characteristic of the celloflour residue is the fact that it is very similar in composition to oatmeal. Rolled oats, for example, has a carbohydrate content of 68.2%, as compared to 65.6% for the celloflour residue. (The carbohydrate content of the celloflour residue is found to be 30% crude fiber and 26.6% nitrogen-free extract.) The protein and fat content of the celloflour residue and rolled oats are approximately the same. The ash content of rolled oats is, however, much less than that of the celloflour residue.

These nutritional characteristics make celloflour a useful extender for addition to meats, such as sausage or hamburger, as well as other foods. When a portion of hamburger is replaced by celloflour, very little difference in flavor can be detected. When celloflour and/or the soluble fraction is added to yeast breads, biscuits or corn bread, the flavor of the bread is not appreciably altered. In fact, in this use, the celloflour and/or the soluble fraction serves as an excellent growth medium for yeast.

The pH of liver bile is 8 to 8.6, which is the pH at which the bile digestion step of this process is carried out. The main components of bile are bile salts, bile pigments, cholesterol and lecithin. The principal bile salts are glycocholic acid and taurocholic acid, which are related to the sterols.

Example 1: Add together and mix 100 g of alfalfa meal, 3.5 g of sodium hydroxide and sufficient water to form 750 to 800 ml of slurry. Heat 1 hour at the boiling point. Cool to 55°C and add 0.222 g of pancreatin and 0.02 g of gall sac powder. Allow to digest for 24 to 30 hours at 55°C. Separate the soluble fraction in the resulting digestion slurry from the insoluble residue. The insoluble residue may be colored with red food dye to the desired color, dried and ground to a celloflour. The soluble fraction extract may then be concentrated as desired.

Example 2: *Bread –*

Dry yeast, package	1
Water, cup	¼
Scalded milk, cooled to lukewarm, cup	¾

(continued)

Sugar, tablespoons	2
Salt, tablespoon	½
Vegetable oil, tablespoon	1
Soluble fraction or celloflour from Example 1, cup	¼
Whole wheat flour, cup	¼
White flour, cups	3*

*Approximate amount

Soften yeast in ¼ cup warm water, then add remaining water, sugar and salt; mix. Stir in oil. Add flour in two portions. Knead 5 minutes. Place in oiled bowl, cover with damp cloth and let rise at 85°F until mass doubles in volume (1½ to 2 hours). Punch down dough and proceed as with first rising. The dough should again double in volume in 30 minutes. Punch down, round up and mold into a loaf. Place in oiled loaf pan, cover with damp cloth and let rise at 85°F for 50 to 60 minutes until mass again doubles in volume. Bake 25 to 30 minutes at 425°F.

Pancreatin and Yeast Digestion

In another process *G.W. Edwards and A.W. Edwards; U.S. Patent 3,780,183; December 18, 1973* use a combination of pancreatin and yeast to obtain useful nutritional materials from alfalfa and clover. Alfalfa or clover is first brought into contact with an aqueous solution of a base which will not leave a residue which may adversely affect the suitability of the product or leave it unpalatable.

The aqueous basic solution digests the plant material and this yields an aqueous extract. The aqueous extract, with or without a prior separation of undissolved material, is then treated at least with pancreatin to convert starches into soluble carbohydrates, digest lipids and release protein hydrolysates. The resulting aqueous solution, after separation of lipids and lipoproteins, can be added to other foods without prior concentration, or it can be concentrated by removal of water before being so used.

The extraction can be readily effected by forming a slurry of the plant material in the aqueous basic solution. The aqueous basic solution can be preheated before it is combined with the plant material or a slurry of the plant material in the aqueous basic solution can be first formed and the slurry then heated. The extraction is considered completed after about 30 minutes to 2 hours, depending on the temperature of the extraction. When the slurry is boiled, an extraction time of 30 minutes to 1 hour is generally suitable.

After the digestion or extraction with the basic solution is terminated, the slurry is brought to a temperature of about 38° to 75°C, without separating the liquid phase from the dispersed solids, and pancreatin is added. When a strong basic solution is used for the digestion, the resulting extract will be more basic than suitable for pancreatin stability and effectiveness. An acid may be added to lower the extract pH to a level which favors pancreatin activity.

Following the pancreatin incubation, the slurry is separated into a solid phase and a liquid phase by any suitable means, such as filtration, centrifugation or decantation. When the product is filtered, the filtered extract may develop a greenish-black substance comprising decomposed chlorophyll and lipids which settles out. This substance can be removed by centrifugation or gravity settling. The liquid extract constitutes a human and animal food product rich in digestible protein, carbohydrates, essential amino acids, vitamins and minerals. The liquid extract, containing about 4 to 5% solids, can be added to other foods. It can be concentrated, however, by a number of means to raise the solids concentration up to about 50 to 100%. One method of concentration is to heat the liquid extract to boiling to evaporate water. However, the pH of the liquid extract advisably is adjusted to a pH of about 4 to 5 by the addition of a mineral acid before the evaporation. Phosphoric acid and hydrochloric acid or some other acid can be used for this pH adjustment.

One important supplementary procedure is to inoculate the slurry, obtained from the aqueous basic solution extraction, with a suitable yeast simultaneously with the addition of the pancreatin. The conditions used for the pancreatin treatment can be selected to be highly conducive to yeast growth in the slurry and the growth of yeast induces a highly beneficial conversion of products in the slurry into materials advantageously present in a human food.

The growth of yeast, in itself, provides a protein supplementation which is desirable in the product. Any suitable yeast can be employed, but for simplicity, common baker's yeast (*Saccharomyces cerevisiae*) is used. About 1 to 2 g of yeast per pound of plant-starting material is adequate although more or less yeast than this can be used. Adding yeast with the pancreatin tends to increase the protein content and to decrease the ash content, the latter of which is often desirable, of the product. When yeast is used, the product can be worked up in the same way as described for the use of pancreatin without yeast, using an incubation period of about 12 to 48 hours.

A second significant supplementary procedure which can be used constitutes a treatment of the slurry obtained from the pancreatin digestion, and before separation of undissolved solids, whether or not yeast has been used in conjunction with the pancreatin, with another enzyme or mixture of enzymes. Potato-corn starch amylase, pepsin, trypsin, papain, hog pancreas diastase, malt diastase, steapsin, cellulase and pectinase are specific enzymes which can be used. However, any enzyme which acts on protein, lipid or carbohydrate can be used. Potato-corn starch amylase is considered the best choice. The use of one or both of these enzymes in the process leads to an even more suitable product.

Example 1: 113 g (0.25 lb) of alfalfa meal is added to 850 ml of an aqueous 0.33% sodium hydroxide solution. The mixture is boiled for 1 hour. The solution is cooled to 45° to 50°C and then 0.125 g of pancreatin and 0.25 g of baker's yeast are added. The mixture is incubated for 48 hours. The greenish-black lipoprotein was separated by decantation and the liquid extract containing 5% solids was evaporated to a product containing 50% solids. The extract contained 28% of the 113 g of starting alfalfa meal. The extract solids analyzed: crude protein, 19.20%; fat, 0.231%; crude fiber, 0.037%; nitrogen-free extract carbohydrates, 80% and ash, 0.37%.

Example 2: Alfalfa Sloppy Joes –

Vegetable oil	3 tablespoons
Ground beef	2½ pounds
Onions, chopped	2

(continued)

Green pepper, chopped	1
Tomato soup, undiluted	2 10.5 oz cans
Catsup	1 cup
Alfalfa extract (10% solids, Example 1)	¾ cup
Flour	½ cup
Worcestershire sauce	2 teaspoons
Salt and pepper to taste	
Hamburger buns	

Brown the ground beef in vegetable oil; add onion and green pepper and cook about 5 more minutes. Stir in flour, soup, catsup, alfalfa extract and Worchestershire sauce and cook until thickened. Season with salt and pepper and serve on hamburger buns.

High Protein Recovery

Animals require generally the same nutrients as man. For maintenance, growth, reproduction and good health, animals require a diet of carbohydrates, protein, fat, minerals, and vitamins. Small amounts of protein are necessary for maintenance of the physical well-being of animals, in particular, for repair of muscles, internal organs and other body tissues. High-grade protein is usually supplied by eggs, milk, fish meal, meat by-products and soybean meal, which usually contain the correct amounts of essential amino acids. Low-quality protein, such as that in corn grain (maize), contains too little of one or more essential amino acids.

The quality of the protein in food is of little importance to animals, particularly to ruminants, including cattle, sheep, goats and other animals that have four stomachs, because bacteria in the digestive tract of these animals aid in the digestion of food and in using simple nitrogen compounds to build proteins.

Experiments dealing with digestion and balance have measured the degree to which the various components of animal feed are absorbed and retained by the animal body. Protein requirements may be expressed as the amount of digestable protein needed for growth or other body functions. The energy required by animals, is measured as digestable energy (DE), metabolizable energy (ME), net energy (NE), or total digestable nutrients (TDN).

J.P. Henry and J.D. Myers; U.S. Patent 3,968,259; July 6, 1976; assigned to M & M Livestock Products Co., Inc. show how to increase the digestible protein in silages to extend the usefulness of this material as a feedstock for animal production. Certain metal salts are used in combination with an enzyme mixture of protease, amylase and gumase. One source of the enzymes is the *Bacillus subtilis*.

The essential salts which must be present include salts of calcium, potassium, sodium and magnesium. Calcium should be present in amounts of not less than 2% by weight, and potassium in amounts of not less than 25%, magnesium in amounts of not less than 4%, and salt (sodium chloride) in amounts of not less than 8%. In addition, the composition contains trace amounts of iron, copper, cobalt, manganese and zinc, as well as phosphorus and iodine. The metals are present in the composition in the form of salts or oxides. The salts of the composition serve two purposes. The salts maintain the electrolyte level of the feedstocks, and are also necessary to the animal diet for maintaining pH balance and electrolyte content of the digestive tract, and more importantly aid and assist in the formation of new tissue and sustain growth of the animal. Further, inclusion of these salts maintains the electrolyte balance of the bood and reduces weight loss of animals under strain.

Example: A composition was prepared with the following components.

Component	Percent by Weight
Potassium chloride	50.00
Sodium chloride	10.00
Potassium sulfate	11.25
Magnesium sulfate	13.75
Calcium propionate	10.00
Fermentation products of *Bacillus subtilis* (source of enzymes)	1.50
Sodium tripolyphosphate	1.20
Iron carbonate	0.200
Iron sulfate	0.203
Iron oxide	0.100
Copper oxide	0.200
Ethylene diamine dihydroiodide	0.025
Cobalt carbonate	0.002
Manganese sulfate	0.32
Zinc oxide	0.25
Mineral oil	1.00
Total	100.00

The fermentation products of *Bacillus subtilis* (source of enzymes) contained protease, amylase and gumase in a ratio of 3,750, 6,240 and 10, respectively. Generally, the dosages require about ¼ to ¾ pound per ton of grain (corn, milo) silages, preferably ½ pound of the composition per ton of grain; for oatlage, the range of the composition is between ½ to 1½ pounds of the composition per ton of oatlage; approximately ½ pound to 1½ pounds of composition is then mixed with a ton of haylage.

The composition may be added directly to silage or haylage, or it may be diluted to assist distribution of the composition throughout the silage feed. When diluted, limestone, urea, salt, grains and millfeeds may be used as the diluent. Generally, 5 to 20 pounds of the composition in combination with the diluent is added to a ton of silage. The amount of enzymes in the composition (fermentation products of *Bacillus subtilis*) ranges from 3.50 to 4.00 parts per million in a ton of corn silage, when ½ pound of the composition is added to silage and up to 7 to 8 parts per million in a ton of silage, when one pound of the composition is added to a ton of silage.

Corn Hull Conversion

A process for cellulose conversion using corn hulls is described by *R.L. Antrim and D.W. Harris; U.S. Patent 4,089,745; May 16, 1978; assigned to Standard Brands Incorporated.* Corn hulls from a wet milling operation contain relatively large amounts of impurities with the fibrous, corn hull fraction. These impurities are in the form of fine material and contain the predominant amount of nonfibrous substances, such as a starch, protein, oil-containing material, lignified tip cap, etc. Removal of these materials may be accomplished, for instance, by screening through a screen of 6 U.S. Standard mesh. The particle size range of the corn hulls containing the predominant amount of impurities may vary, of course, depending upon the particular manner in which the corn hulls are treated and handled during the milling operation. Residual starch which may be present in the corn hulls after the screening operation can be removed by an enzymatic solubilization treatment, for instance, with α-amylase.

The relatively purified corn hulls may be considered to contain three predominant substances or materials: a cellulose fraction, a hemicellulose fraction, and a noncarbohydrate fraction. These fractions are obtained by contacting corn hulls with a sufficient amount of alkali to hydrolyze the hulls to effect liberation of the hemicellulose fraction so that it may be solubilized in water and to solubilize the noncarbohydrate fraction. Three fractions are recovered, i.e., a water solution of hemicellulose, an organic solvent extract of the noncarbohydrate fraction and an insoluble residue containing the cellulose fraction.

The two other fractions may be recovered by adding a sufficient amount of water-miscible organic solvent to the alkaline solution to precipitate the hemicellulose. Solvents which may be utilized are acetone, methanol, ethanol, propanol, isopropanol, secondary butyl alcohol, tertiary butyl alcohol and mixtures thereof. The solvent supernate contains the noncarbohydrate fraction and this fraction may be recovered by evaporating the solvent. Cellulase derived from *Aspergillus* species, *Trichoderma viride*, or other cellulase-producing organisms may be used for converting the cellulose fraction to glucose.

It was found that the cellulose derived from corn hulls is extremely susceptible to enzymatic saccharification. Because of this, lesser quantities of enzyme are necessary to achieve a relatively high degree of conversion compared to cellulose derived from a number of other sources. Also, the desired degree of conversion can be achieved in a shorter period. This may be due to the particular form or structure of the liberated corn hull cellulose.

It has been observed that the cellulose derived from corn hulls has a higher degree of water absorptivity than other types of cellulose which indicates that the former has a more open structure. This apparently permits easy access of the enzyme to the cellulose fibers where the enzyme can more readily catalyze the degradation of the cellulose to glucose.

Example: This example illustrates the treatment of cellulose derived from corn hulls with cellulase derived from *Aspergillus* sp. It also compares the amount of glucose formed by this treatment with the amount of glucose formed by treating another source of cellulose. Corn hulls from a wet milling operation were wet-screened through a 6 screen at about 50°C to remove the fine fiber, most of the starch and some of the protein and lipid material present. 9.5 kg, dry basis, of the screened material, having a moisture content of 65%, was mixed with 5.7 liters of water in a 190 liter jacketed tank.

To the mixture held at a temperature of 70°C was added 635 g of calcium oxide and the mixture stirred for 3 hours. Portions of the mixture were processed in a 3.785 liter Waring blender in a batchwise fashion for 30 seconds. Water was added during blending to promote mixing. A total of 84 liters of water was utilized. The material was dewatered in a centrifuge to separate the soluble hemicellulose and noncarbohydrate from the cellulose cake. The cellulose cake was then slurried in about 84 liters of water and again centrifuged. The cake was then reslurried, screened through a U.S. No. 20 mesh screen to remove tip caps and hull materials, and again dewatered by centrifugation.

As a control cellulose preparation, 5 g of bleached sulfite pulp was pretreated by suspension in 200 ml of deionized water. The pH of the suspension was adjusted to 12.4 with 50% NaOH solution. The suspension was then held at ambient temperature for 1 hour, filtered, and the resultant cake washed with water to a pH of 8.5. 5 g of corn hull cellulose having a moisture content of 92% prepared as described above, and 5 g of the pretreated sulfite pulp were individually suspended in 200 ml of deionized water. The suspensions were buffered with 0.7 ml of glacial acetic acid and the pH adjusted to 4.3 with NaOH solution.

560 mg of cellulase containing 30,000 units of *Aspergillus* sp. cellulase per gram was added to each suspension and the reaction held at a temperature of 50°C for 16 hours. The final reaction pH was 4.3. After 16 hours, the suspensions were filtered and the glucose content of the filtrates determined as reducing sugar by a Fehling's solution method. The results are shown below.

	. . . . .Glucose Yield.	
	Based on Approximate Cellulose Content (%, db)	**Based on Total Solids (%, db)**
Cellulose (sulfite pulp)	10	10
Corn hull cellulose	58	35

The data show that treatment of corn hull cellulose prepared with cellulase derived from *Aspergillus* sp. resulted in the production of almost 6 times as much glucose on an approximate cellulose content basis and 3.5 times as much glucose on a total solids basis as did similar treatment of bleached sulfite pulp.

INTERMEDIATE MOISTURE PET FOODS

Lipase and Protease Treatments

The flavor and palatability of animal foods are altered and improved by adding fat, protein and an enzyme combination of a lipase and protease. The process is given by *G.J. Haas and J.C. Lugay; U.S. Patent 3,857,968; December 31, 1974; and U.S. Patent 3,968,255; July 6, 1976; both assigned to General Foods Corporation.*

The process for conditioning the fat and protein involves emulsifying the fat and treating the mixture with an enzyme mixture of lipase and protease. The reaction between the emulsified fat-protein mixture and the enzyme mixture is believed to produce a complex array of reaction products. The fat reacts with the lipase to produce free fatty acids and mono- and diglycerides. The protein reacts with the protease to produce polypeptides and free amino acids. The exact reactions and reaction products responsible for the improvement in the palatability of animal foods are not presently identifiable.

It is considered necessary to emulsify the fat before treatment with the enzyme mixture. This is due to the fact that emulsification increases the fat-water interfacial area, thereby facilitating the heterogeneous reaction between the fat and the lipase. Emulsification can be obtained through the addition of an emulsifier; or, as in the case where the fat-protein mixture is a meat slurry, the meat slurry will impart a limited natural emulsifying effect.

The enzyme mixture should contain from about 20 to 250 lipase units per gram and from about 500 to 7,000 protease units per gram. A lipase unit is defined as that amount of the enzyme which will hydrolyze 0.885 g of olive oil calculated as triolein to diolein and oleic acid in 2 hours at 37°C. A protease unit is defined as that amount of the enzyme which will digest 1 mg of casein in 1 minute at 50°C at pH 7.5. Pancreatic lipase, which is an enzyme mixture containing about 220 lipase units per gram and about 6,000 protease units per gram, is particularly preferred.

It has been found that temperatures within the range of from about 35° to 50°C are effective for both heating the reactants and for maintaining the reaction. Any reaction temperature effective to sustain the enzymatic reaction resulting in the production of the palatability improving composition can be used. It is noted that lower temperatures can be used with somewhat diminished results due to incomplete emulsification and reduced reaction rates. Also, somewhat higher temperatures can be used, but are generally more costly than the increased reaction rates will justify. The aqueous emulsifier dispersion, which may also contain the promoter, is brought to a boil for about 10 minutes and then reduced to 35° to 50°C for mixture with the fat.

After emulsification, the enzyme mixture is mixed with the emulsion and constant efficient stirring is maintained for the reaction period to maintain a desirably high rate of reaction. The reaction will be continued for a period of time effective to sustain the enzymatic reaction resulting in the production of the palatability improving composition. After the desired period of reaction, the emulsion can be treated to inactivate the enzymes. Typically, it can be heated to an elevated temperature, on the order of about 70° to 95°C, for a period of time sufficient to inactivate the enzymes, on the order of about 5 to 15 minutes.

Example 1: This example illustrates a procedure for conditioning a fat-protein mixture. Promine D soy isolate (120 g), a proteinaceous emulsifier, was dispersed in 3.2 liters of an aqueous solution which was 0.0066 M with respect to $CaCl_2$ and 0.5 M with regard to NaCl. This mixture was brought to a boil with a steam coil and held at this temperature for about 10 minutes. The mixture was then cooled to 50°C and the pH was adjusted to 8.0 to 8.5 and 800 g of melted bleachable fancy tallow was added.

The mixture was then homogenized by mixing in a Waring blender at high speed for 3 minutes. The pH was adjusted to the indicated range by the addition of sodium hydroxide when necessary. To the homogenized mixture was added 8 g of pancreatic lipase 250 (Miles). The reaction mixture was incubated for about 2 hours while adjusting the pH to 8.0 to 8.5 every 30 minutes. At the conclusion of the conditioning reaction, the material was frozen.

Example 2: The frozen emulsion prepared in Example 1 was thawed in warm water for application to an animal food. In those instances where a quantity of the emulsion broke upon thawing, it was reemulsified by homogenizing in a Waring blender at high speed for about 3 minutes. This emulsion was applied to a commercially available, granular, dry animal food having a coating amounting to 3% by weight of nonconditioned bleachable fancy tallow, a moisture content of 9% and having the following dry basis formulation: 40.4% corn; 23.1% soy meal; 20.8% wheat middlings; 11.6 to 50% meat meal; and 4.1% vitamin, mineral, dye and flavoring mix.

The emulsion contained about 20% of the enzyme-conditioned fat-protein mixture and was sprayed onto the animal food by means of a spray gun into a cement mixer containing the dry food to give an enzyme-conditioned fat-protein mixture content of 1%. The animal food was then dried by heating for 5 to 10 minutes at 90° to 110°C to reduce the final moisture content of the animal food to about 8 to 12%.

Example 3: A mixture containing about 53.6% tripe, 15.8% trimmings, 3.9% defatted beef tissue, 1.7% lungs, 1.1% emulsifier, 12.9% propylene glycol, 6.4% corn syrup, and 4.6% water is pulverized in a homogenizer to obtain a meat slurry. 400 lb of this meat slurry is brought to 50°C with efficient stirring. The pH is adjusted to 8.0 to 8.5 by the addition of 2 N NaOH.

Then, 0.36 lb of pancreatic lipase 250, equivalent to 1% of the fat in the meat slurry, is added to the slurry. Constant efficient stirring is maintained for the 37-minute reaction period, during which the pH is monitored continuously. Each time the pH drops to 7.8, sufficient 2 N NaOH is added to raise the pH to about 8.2. The pH is adjusted to 8.2 at the conclusion of the reaction, and the mixture is heated to 90° to 95°C to inactivate all enzymes.

Example 4: Three separate portions of enzyme-treated meat slurry prepared in Example 3 are blended with additional amounts of the untreated meat slurry to obtain combined slurries containing 11.2, 33.6 and 67.2% of the enzyme-treated slurry. Then, 44.6 parts of each of these combined slurries is mixed with 1.9 parts of soy hulls, 5.9 parts of a vitamin, mineral, flavoring and dye mix, 15.7 parts sucrose, and 31.9 parts of soy flakes. These mixtures are then extruded through a cooled extruder to form 3 animal foods: the first containing 5% of the enzyme-treated meat slurry; the second, 15%; and the third, 30%.

Starch Conversions

Conversion of starch to glucose is possible by an enzyme process in a solid component system where moisture is present only within the solid piece or absorbed on the surface. The process according to *E.H. Borochoff; H.N. Dunning and T.W. Craig; U.S. Patent 3,617,300; December 28, 1967; assigned to General Mills Inc.* is particularly suitable for soft, moist pet foods.

There are several advantages to the process. For example, a solid pet food product sweetened by dextrose may be produced in a continuous process by starting with the cheaper and more easily handled raw starch (gelatinized starch may also be used). The pet food produced by this process has high acceptance from dogs, from a palatability standpoint. While processes for the conversion of starch to dextrose by the sequential addition of α-amylase and amyloglucosidase or an acid in a fluid system are known, there has been no successful process for the production of dextrose in a food product which is essentially a solid system. The two enzymes used in this process are α-amylase and amyloglucosidase. The α-amylase randomly attacks the alpha 1-4 bonds in a starch or dextrin molecule resulting in the fragmentation of both linear and branched fragments of starch, i.e., dextrinization. If sufficient time is allotted, under certain conditions, α-amylase alone can be made to convert linear glucose chains mostly to a mixture of maltose and dextrose, and branched chains to a mixture of maltose, dextrose and panose, the latter being a trisaccharide containing an alpha 1-6 linkage.

α-Amylase is known as a liquefying enzyme because as commonly used for dextrose conversion, it converts a starch slurry to dextrins, thereby thinning it out considerably. The second enzyme used in the process is an amyloglucosidase which works directly from the nonreducing ends of starch chains, but splits off single glucose units. It acts on alpha 1-4 linkages and alpha 1-6 linkages, the rate of reaction being about 20 times faster on the former than on the latter.

Contacting may be done by a variety of means. The enzymes in their powdered form or in a liquid form may be mixed with a dough if the product itself is in dough form. The enzymes may be dissolved or suspended in a small amount of water and this may be sprayed or coated on the surface of the solid amylaceous material. If the latter method is used, the minimum amount of water needed is used so that the system is still solid and no slurry or suspension is formed. In order to accomplish the desired enzyme conversion, the temperature after the enzymes have been included should not exceed the inactivation temperature of the enzymes, between about 80° and 90°C.

However, by elevating the temperature beyond that point after the desired amount of dextrose has been produced, an excellent method for controlling the amount of conversion is available. It is also possible to add the enzymes after whatever heat treatment normally used in the processing of a particular food product is concluded, allowing the conversion to dextrose to occur during storage. It was found that a solid amylaceous system converted by the process seems to be affected by moisture. Water is needed at a level of about 25% by weight of the product to obtain a reaction. Maximum conversion is obtained near about 40% moisture.

Example 1: A typical grain-based pet food was made using the ingredients listed below.

Ingredient	Percent by Weight
Soy bean meal (solvent extracted)	19.34
Flour (70% starch)	40.87
Bone meal (steamed)	3.18
Whey	1.82
Propylene glycol	3.18
Soy bean oil (crude, degummed)	1.18
NaCl	0.63
Vitamin and mineral mix	0.69
Red color	0.02
Beef tallow mixed with formula	2.86
Beef tallow added to surface	3.36
Meat by-products (dehydrated)	2.07
Water	20.84

(continued)

Ingredient	Percent by Weight
α-Amylase	0.1
Amyloglucosidase	0.1

With the exception of the meat by-products, surface tallow, water and enzymes, the ingredients were mixed together in a ribbon blender. Water and enzymes were blended with the mixture and the mixture was then formed. The formed mass was cut and enrobed with tallow and dehydrated meat by-products. The product was split into 2 lots and held at room temperature. The first lot was held for 3 days and the second for 6 days. After holding, total sugar was measured with the following results. In the fat held for 3 days, the total sugar obtained was 7.95% by wt of the final product. In the fat held for 6 days, the total sugar was 12.0% by wt of the final product.

ANIMAL FEED

Digested Defatted Bones

The bones used in the process of *E. Levin and J.D. Mosser; U.S. Patent 3,944,655; March 16, 1976; assigned to VioBin Corporation* are those which contain marrow in their centers. The marrow contains a valuable edible fat and also contains hemoglobin and other protein. The bone itself consists of an organic basis of albuminous nature in which a great amount of inorganic matter is deposited.

The organic substance is mainly collagen with a minor amount of osseomucoid and very small amounts of osseoalbuminoid. The inorganic material makes up about 40% of the dry residue of the bone and is chiefly calcium phosphate and calcium carbonate, but there are also present a little magnesium and a trace of fluoride and chloride. Whole bones, marrow and all, are generally used to make gelatin, the inorganic material being discarded and the constituents of the bone marrow, including bone marrow protein, not being separated from the other material of the bone during the process of making gelatin.

The bone marrow contains a nutritious protein which has never been separated either wholly or partially from the other protein of the bone, and which never has been prepared and made suitable for use as a food. Defatted and dehydrated particles containing bone and bone marrow are digested in an aqueous mixture of trypsin, amylase and lipase in an amount of each enzyme of at least 0.001% of the dry bone marrow by weight at a temperature of 15° to 23°C and a pH of about 7 to 9, and for a time sufficient to dissolve a substantial amount of the protein of the bone marrow while dissolving not more than a portion of the protein of the bone.

The solution obtained by the digestion process is separated from the residue by decanting, filtering or centrifuging. The solution may be used as a liquid protein food or a solid protein may be recovered from the solution, such as by low temperature evaporation, or the solution may be spray-dried to obtain a water-soluble protein powder. The solid residue may be used to make a soluble collagenous product and a mineral product suitable for use in a pet food. It has been found that trypsin, amylase and lipase must all be used to obtain satisfactory results in the digestion process, but that other enzymes present in pancreatin may also be present. It is particularly desirable to use a pancreatin product which contains approximately all the enzymes present in raw pancreas. It has been found that a suitable and efficient source of the enzymes trypsin, amylase and lipase is present in the pancreatin product sold as Viokase. This Viokase has an activity of such enzymes in defatted pancreas tissue such that when it is mixed with powdered defatted bone and water, only 0.1% of the Viokase, on the basis of the dry bone, is required.

The amount of water appears to be unimportant so long as contact of the Viokase with the bone is maintained, but the total solids of bone and Viokase can be 5 to 10% by wt of the total mixture without substantially affecting the solubilizing reaction. Viokase is a dehydrated and defatted powdered pancreas in which the pancreas is defatted with a fat solvent such as ethylene dichloride or other fat solvent which forms an azeotrope with water and dehydrated by azeotropic distillation at moderate temperatures, such as below 60°C.

Viokase or any other pancreatin having the following activity is particularly suitable for this purpose: (a) The activity of trypsin of at least 4 times the activity of the USP pancreatin powder made by comminuting fresh animal pancreas in the presence of water and thereafter adding absolute alcohol to precipitate a composition containing enzymes. The United States Pharmacopoeia sets forth standards of activity for this pancreatin. An activity of trypsin sufficient to render at least 25 times its own weight of casein nonprecipitable by acid-alcohol after incubation for 1 hour at 40°C is desirable.

(b) Amylase activity at least 4 times the activity of the USP standard for the precipitated type pancreatin, and able to hydrolyze at least 25 times its own weight of potato starch past the erythrodextrin stage in 5 min at 40°C and (c) a lipolytic activity at least 4 times that of the USP pancreatin and for at least 2,000 U/g as determined by the method of Desnuelle et al, *Bull Soc Chi Biol,* 37, 285, 1955.

Example 1: Production of Edible Fat – Bones which contain bone marrow were cleaned from extraneous meat and fat, and the clean bones with their marrow were pulverized to a powder. The powdered bones were mixed in a distillation vessel with approximately 10 times their volume of ethylene dichloride. At a temperature of 40°C and a pressure of approximately 215 mm of mercury, an azeotrope of ethylene dichloride and water was distilled and this distillation was continued until about 98% of the water contained in the bone was removed.

At no time was the temperature allowed to exceed approximately 50°C. When the bones had been dehydrated to the desired point, the solvent was drained out through a screened opening in the vessel and fresh solvent was added to the residue. The mixture was stirred and finally the solvent was withdrawn to the full extent that it would drain out. The pressure within the vessel was then reduced to about 50 to 100 mm and the vessel warmed until all the ethylene dichloride had been removed from the dehydrated bones.

The dehydrated bones, which contained the nonfat constituents of the bone marrow, including the protein, were used to produce solubilized protein food. The solvent contained the fat from the bone, principally from the bone marrow. The solvent of this fat solution was evaporated at below about 80°C, viz, 60°C leaving a residual fat. The azeotropic distillation at low temperature to remove water makes the removal of fat from the bones by extraction much more efficient. The removal of the fat can be done in less time and to a greater degree.

The fat is a high-grade soup stock. The dry defatted bones are in better condition for digestion with enzymes than bones defatted without an azeotropic distillation of the water.

Example 2: Solubilization of Protein of the Bone Marrow – Powdered defatted dry bone made as in Example 1 was mixed in water and then the powdered defatted pancreas was added. This pancreatin material had the trypsin, amylase and lipase activity as follows: trypsin, approximately 1.00 hemoglobin unit per gram of dried pancreatin; amylase, an ability to hydrolyze 25 times its own weight of potato starch past the erythrodextrin stage in 5 minutes at 40°C; and lipase, activity of 4,000 units per gram as determined by the method of Desnuelle et al, supra.

The pancreatin material of the above activity was used in an amount of 0.15 grams per 100 grams of the bone on a dry basis. The powdered bone and pancreatin was present in the water in about 10% by weight. The water was made slightly alkaline with sodium hydroxide and the mixture maintained at a pH of approximately 8.0. The temperature of the mixture was maintained at approximately 20°C. The bone and bone marrow were kept in contact with the pancreatin and its enzymes for 5 hours, the digestion of marrow protein and conversion of this protein to a soluble form being substantially complete in about 4 hours.

At the end of the 5 hours, the aqueous solution of protein was separated from the bone residue. This solution was a mild and pleasant tasting liquid protein food suitable for animal and human consumption. A breakdown of the materials used and obtained by the above process is given below, with components specified in percent by weight.

	Starting Material*	Enzymatic Digestion**	Product from First Digestion***
Protein	38.4	85.2	27.9
Ash	58.5	10.6	69.0
Moisture	4.0	5.5	2.8
Fat	trace	trace	trace

*895 lb (44.7%) defatted, dehydrated beef bone obtained from 2,000 lb bone. 435 lb (21.7%) crude fat of excellent quality and 670 lb (33%) water were also obtained.
**165 lb soluble, edible, marrow protein.
***730 lb bone residue.

Mucin Viscosity Reduction

For many years, the poultry industry has been interested in feed additives and other methods of increasing the feed efficiency of chickens, turkeys and other forms of poultry. Because of the large amounts of feed consumed by poultry and the small margin of return on capital invested in raising chickens, turkeys, etc., even increases as small as 1% are meaningful economically. The term feed efficiency generally is based upon measuring the amount of weight gain in relation to the amount of feed consumed.

In the process of *F.J. Hartdegen and K.M. Abdo; U.S. Patent 4,073,884; Feb. 14, 1978; assigned to W.R. Grace & Co.* feed efficiency in animals, e.g., cattle, swine, poultry, etc., is increased by feeding controlled amounts of an enzyme fraction obtained by culturing *Streptomyces griseus* in a fermentation medium. The fraction is characterized as having a relatively small proteolytic activity, approximately ½ the Hooreman enzyme (U.S. Patent 3,823,072) on an equal weight basis. The enzyme fraction possesses a high capacity to reduce mucous viscosity, approximately 3 times that of the Hooreman enzyme. The "mucinase" activity of the enzyme fraction is also inhibited by trypsin inhibitor.

The fraction possesses enzymatic activity which is measured in relation to the effect of the fraction on gastric mucin. More precisely stated, 1 unit of enzyme activity is defined as that amount of enzyme fraction which when added to 1 ml of water, and that solution added to 1 ml of prepared solution of gastric mucin, reduces its viscosity by 50% in 2 hours at 39°C. The solution of gastric mucin is a 7.5% (w/w) solution of gastric mucin, bacteriological in 0.2M phosphate buffer, pH 7.0.

The enzyme fraction obtained from fermentation of *Streptomyces griseus* is characterized as follows: 1 g of enzyme fraction represents about 500 units of enzyme activity as defined above; the extract is insoluble in acidic acetone; centrifuged broth from the fermentation medium contains 0.3 to 1.0 unit of enzyme fraction per milliliter broth; and storage of the dry enzyme fraction at 109°F for 28 weeks or at 122°F for 8 weeks results in less than 10% activity decrease. The culture used to produce the enzyme fraction is grown on potato dextrose agar.

The fraction is recovered by utilizing the fact that it is insoluble in slightly acidic acetone. Thus, the following steps are undertaken. First, the fermentation broth is centrifuged to remove all insoluble material including cells. This centrifuged broth is a solution of the acetone-insoluble enzyme fraction and may be used to improve feed efficiency in poultry and other animals. However, the broth is dilute and contains only 0.3 to 1.0 unit of enzyme fraction per milliliter of broth.

Therefore, after centrifugation, the pH of the supernatant solution is then adjusted to 5.5 by the addition of glacial acetic acid. The fraction containing enzymatic activity is precipitated by adding 3 liters of acetone per 1 liter of solution. Optimum yield is obtained by using cold acetone and cooling the suspension to -20° for 18 hours. The acetone-insoluble fraction is separated by filtration. The resultant solid fraction is then washed with dry acetone to remove residual water. Finally, the fraction is dried in vacuo at 25°C to remove acetone. It is stable at room temperature. One gram of the dried product so prepared generally represents about 500 units of enzyme activity.

Example: Chicken Feeding Trial – A chicken trial of 2 groups of 60 birds each (4 pens of 15 birds for the first 4 weeks, and 2 pens of 30 birds for the second 4 weeks) was run for 56 days using a commercial diet consisting of 20% crude protein. For one group, the enzyme fraction in the form of a dry powder was added at a level of 0.04% of the feed. Approximately 0.2 unit of enzyme activity was used for each gram of feed. For the other group (control), no additives were added to the diet. Feed and water were offered ad libitum. Weight gain and feed efficiency were determined. After 56 days, the results were:

Group 1	
Average daily gain, g	35.1
Feed efficiency, %	2.26

(continued)

Group 2, control	
Average daily gain, g	33.7
Feed efficiency, %	2.34

As can be seen from the above figures, the first group of birds showed an improvement over the control group of 4.3% in average daily gain and 3.42% in feed efficiency.

Administration of Enzyme Fraction: The following table sets forth daily amounts of enzyme fraction for administration to various types of animals to obtain increased feed efficiencies.

	Average Daily Amount, (units)
Swine	40–400
Chickens	2–20
Turkeys	4–40
Cattle	160–1,600

The daily feeding level for other animals is increased over the level used for chickens to reflect the larger size of turkeys, pigs, cattle, etc. Here again, the above ranges are average levels and reflect the fact that lower levels will be fed initially with more fraction being fed as the animal grows and eats more. For example, for turkeys it is preferred to feed about 6 to 15 units per day for the first 3 weeks following hatching. For pigs, it is preferred to feed about 80 to 300 units per day for the first 4 weeks after the animal is weaned. For cattle, the corresponding level is about 320 to 1,200 units per day for the first 7 weeks after weaning.

Another factor contributing greatly to the success of the feeding program is to begin feeding the enzyme fraction as soon as possible. For chickens, it has been found that delays of a week, for example, greatly decreased the improvement in weight gain. Therefore, for birds such as chickens and turkeys, feeding of the enzyme fraction should begin 1 to 2 days after hatching. For pigs, cattle and similar animals, the fraction should be administered within 1 to 7 days after the animal is weaned, and continuously thereafter until the desired weight level is obtained.

Acid Curd Process

A combination of autolysis and proteolysis is used to digest animal tissue protein followed by an acid-producing fermentation. In the process of *I.E. Burrows and P.A. Cheney; U.S. Patent 4,016,295; April 5, 1977; assigned to Pedigree Petfoods Limited* the fermented material is added to a colloidal solution of a binder protein such that the binder protein is precipitated by the acid in the form of a curd that entraps and binds the fermented material, and the curd and entrapped material are separated off and pressed into a compact mass.

The proteinaceous material is also subjected to autolysis or to proteolysis by added proteolytic enzymes or microorganisms where the fermented material has a liquid or semiliquid consistency when it is added to the colloidal solution of the binder protein. Autolysis should accompany fermentation, but where proteolysis is conducted through the use of proteolytic enzymes from an exogenous source, other conditions may apply. Thus, in the case of alkaline or neutral proteases the preferred system allows proteolysis to proceed for up to 4 hours prior to initiating fermentation. When raw meats, fish or other protein material are used then obviously the autolytic enzymes would contribute some activity during this period. Acid proteases will normally work during the fermentation as for optimum activity they depend upon the lower pH which develops during fermentation. In certain cases it may be desirable to heat-sterilize the materials before fermentation. In such circumstances, autolysis would be unable to proceed due to denaturation of the enzyme systems and exogenous enzymes must be added if proteolysis is to be brought about.

The proteinaceous material to be fermented is animal tissue protein, for example, meat, fish or offal. The product may also contain coloring matters, added vitamins, mineral supplements, antioxidants and binding systems such as gelable hydrophilic colloids or coagulable proteins, to confer stiffness or gel properties. It may also contain an antimycotic such as a sorbic acid compound, for example, potassium sorbate. Since certain microorganisms can metabolize sorbate, it may be necessary to adjust the concentration of sorbate in the fermenting material to ensure that the desired final level of sorbate is achieved in the product.

The proteinaceous starting materials are comminuted, usually by mincing through a 1.0 to 2.0 cm plate. Desirable fermentation additives such as fermentable carbohydrate and organic acid may be added, and antimycotic may also be added at this stage. For example, 10% of a fermentable carbohydrate such as glucose, 0.5% of potassium sorbate and 0.2% citric acid based upon the weight of proteinaceous starting materials may be added and mixed in well. The mixture is inoculated with a starter culture of lactic acid producing bacteria, e.g., *L. casei* or *S. lactis*, and the mixture incubated at 30°C until the pH of the system reaches 4.0.

The temperature may be varied depending upon the strain or species of microorganism used. When protease enzymes of exogenous origin are used, they should be added directly after comminution. The enzyme or mixture of enzymes and the proteinaceous materials are then incubated for the optimum period of time, i.e., one to two hours at the optimum temperature for enzyme activity.

When proteolysis has proceeded to a sufficient degree, the carbohydrate, potassium sorbate and citric acid may be mixed in. The mixture is cooled to 30°C and inoculated with viable lactic-acid-producing bacteria. Further incubation then takes place and the mixture is allowed to ferment until the pH of the slurry falls to about 4.0. After incubation, the material is added slowly with agitation to a colloidal solution of milk solids, for example, pasteurized milk or a solution of spray-dried skim milk.

After complete addition of the fermented material, the curd is allowed to separate and the whey and residual liquor from the proteinaceous material ferment is removed. The curd is then pressed and drained and can then be stored in a conventional manner.

Example: A fermented autolyzed spray slurry was prepared by first comminuting 1,000 parts by weight of raw deep frozen sprats by passage through a mincer fitted with a 1.0 cm plate. 100 parts by weight of lactose, 6 parts by weight of potassium sorbate and 0.2 part by weight of citric acid were added and mixed well in. 110 parts by weight of a 24 log hour old culture of *Lactobacillus lactis* were added and mixed well. The whole mixture was incubated at 37°C until the pH of the mixture reached 4.0. During this time, autolysis took place and a fluid slurry was produced. 30 parts by weight of this fermented and autolyzed sprat slurry were added slowly with stirring to 70 parts by weight of 20% spray-dried skim milk solution held at 45°C. During the course of the addition, the pH of the milk was reduced by the lactic acid in the fermented fish and as the milk proteins precipitated, a curd was formed which contained the solid components and some of the liquors of the added fish.

The fish contained 0.5% potassium sorbate, but if necessary, the concentration should be adjusted prior to addition of the acidified or low pH component. Gentle mixing by a slowly revolving magnetic follower was provided during the addition to ensure adequate mixing of the components without homogenization of the curd. The mixture was allowed to cool to ambient temperature and the liquors drained from the curd by decantation. The very moist and soft curd was then subjected to slow vacuum filtration until as much water and liquors as possible had been removed in this way.

The curd was then removed from the vessel, and packaged or wrapped in a conventional way prior to storing at subambient temperature. The product had a texture similar to cheddar cheese, a pH of 4.5 to 5.0 and a pleasant fish aroma.

Excreted Waste Conversion

A major source of environmental pollution is the excreted waste products of farm animals in feedlots, particularly beef cattle. The daily discharge from a feedlot with 10,000 head of average 800 lb cattle is a slurry containing about 45 tons (dry weight) of solids in about 70,000 gallons of water, plus additional water from overflow of watering troughs and from rainstorm runoff. Each such 10,000 head feedlot module creates the equivalent of the sewage pollution potential of a city of 100,000 persons. Feedlot installations for 50,000 head or more are not unusual.

A process for production of high-quality single cell bioprotein animal feed from excreted waste was developed by *L.G. Erickson and H.E. Worne; U.S. Patent 4,041,182; August 9, 1977.* The process requires five specific steps as follows.

(1) Grinding or shredding the input raw materials to coarse particulate size.

(2) Biolytic decomposition in a liquid solution containing a broad spectrum enzyme complex capable of hydrolyzing the insoluble high molecular weight proteins, starches, and fats into soluble low molecular weight intermediates.

(3) Separation of cellulose-lignin particulate solids as a by-product, dewatered to less than 50% moisture content to inhibit biochemical decomposition during long-term dry storage. Dewatering to this level is possible due to biochemical preconditioning during the biolytic decomposition step. Thus, an increased proportion of nutrients in liquid form is released for subsequent processing.

(4) Cellular synthesis of the organic intermediate liquid nutrient substrate materials in aerated solution, inoculated with selected

microorganisms to metabolize and resynthesize the intermediate nutrient substrate materials, plus oxygen and nitrogen from air, into cellular material. These cells are single cell bacteria which proliferate exponentially by growth and cell division into a biomass of cells limited only by availability of nutrients and air in solution.

(5) Separation of bioprotein cellular biomass, harvested for use as the primary animal feed product. This material in a dehydrated state is biologically stable containing 40 to 50% high-grade protein and may be directly incorporated in either a dehydrated or hydrated state as a feed supplement for cattle or other farm or domestic animals.

The liquid residual by-product remaining after the five-step process is biologically stabilized, deodorized, and has a low biological oxygen demand. It may be stored indefinitely in a pond for interim use as a hydraulic fluid for flush cleaning of animal feedlots and when needed may be distributed to agricultural lands as irrigation water. This water residual contains a residue of nonvolatile organic materials in solution of value as a fertilizer soil conditioning material. The cellulose-lignin dry solids by-product may be used for industrial purposes, or burned as a fuel, or may be recombined in solution during irrigation operations and recycled back to the land for its mulch-fertilizer soil-building value.

Referring to Figure 12.1, there is shown schematically a flushable cattle confinement floor **11** providing a discharge flow of liquid manure slurry via pipeline **12** as the source of raw material to be processed. The first step involves grinding and shredding of the raw material, if necessary, into coarse particulate size by means of grinder **13**. The purpose of this operation is to reduce the raw material solids to reasonably sized particles to facilitate attack by enzymes and to facilitate suspension in solution. Water may be added to adjust solids content to about 10 to 15%.

Figure 12.1: Excreted Waste Conversion

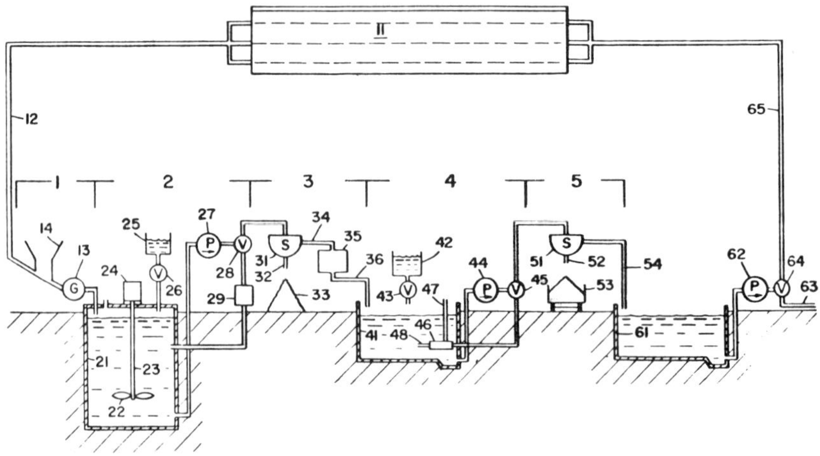

Source: U.S. Patent 4,041,182

The second step involves use of a vented biolytic tank **21** to receive the input raw material in solution in batch quantities to substantially fill the tank. Recirculation as necessary to maintain the particulate matter in suspension may be accomplished by means of a subsurface impeller **22** driven by shaft **23**. The biolytic decomposition process is initiated by addition from batch tank **25** of a solution of broad spectrum hydrolyzing enzymes into tank **21**. The enzymes used are composed of proteolytic, amylolytic, lipolytic and cellulolytic enzymes capable of hydrolyzing the high molecular weight substrates into low molecular weight intermediates.

The enzymes may either be prepared by growing selected organisms in media especially prepared for enzyme production, after which the organisms are separated and the enzyme broth is used to treat the substrate; or powdered, broad spectrum hydrolyzing enzymes may be dissolved in water and added to the biolytic tank **21**. The amount of enzyme concentrate required for effective hydrolysis is about 0.1% by weight relative to the dry weight of volatile organic solids in solution.

In addition, a biodegradable surface active nonionic wetting agent suited to reduction of surface and interfacial tension of water is added to a concentration of one part in 2,500 to 5,000 to aid in emulsification and subsequent biodegradation of fats and oils. The enzyme concentration is sufficient to hydrolyze and liquify the insoluble volatile organic material fraction within 1 to 4 hours. This material is mainly fine particulate organic matter as compared to the relatively nonvolatile fractions of cellulose-lignin components which decompose much more slowly.

The third step involves mechanical separation and dewatering of the nonvolatile solids fraction of the raw material remaining in biolytic tank **21**. This consists mainly of cellulose and lignin materials partially decomposed by action of the hydrolytic enzymes of step two upon the hemicellulose and other volatile nutrients bound within the cellular structure of such cellulose-lignin materials in the raw state. As a result of such biolytic preconditioning, the contained water will be more completely released, for instance, by means of centrifugal separator **31** into which the fluidized contents of tank **21** are discharged.

The solids by-product is discharged to storage pile **33** with a moisture content less than 50%, low enough to substantially inhibit biochemical decomposition. If necessary, a small amount of lactic acid or other edible acid may be added to more completely reduce biochemical decomposition. This material may be held indefinitely in dry covered storage without appreciable deterioration. Without such biolytic preconditioning, a moisture content of about 60% would remain after conventional dewatering operations and the material would be subject to continuing biochemical decomposition and deterioration if held in storage.

The liquid discharged from primary separator device **31** may be passed through a secondary separator **35** to remove fine particulate matter larger than about 10 μ in diameter. The liquid discharged into synthesis tank **41** contains practically all of the metabolite products of the biolytic process of step two in the form of soluble low molecular weight intermediates diffused in water. This nutrient solution provides the nutrient substrate materials for feeding the endogenous synthesizing microorganisms used in the fourth step of the cellular synthesis process.

The fourth step, cellular synthesis process, is initiated by addition from batch tank **42** of an inoculating solution of synthesizing microorganisms into tank **41**. If a liquid culture inoculum of heterotrophic aerobic cell synthesizing microorganisms is used, the culture is grown to a concentration of ten million microorganisms per milliliter and added, in the proportion of about 5% by volume into the nutrient substrate solution in tank **41**. If a dried culture is used, it is dispersed in water in the proportion of 1 lb/25 gal two hours prior to inoculum addition into cell synthesis tank **41** in the proportion of 1 lb of crude, dried microorganisms per 1,000 lb of intermediate nutrient substrate.

The fifth step involves mechanical separation and dewatering of the biomass of cells grown and synthesized in synthesis tank **41**. A bowl-type centrifugal separator **51** may be used into which the fluidized contents of tank **41** are discharged. The bioprotein cellular material separated is discharged to container **53** with a moisture content of about 45%, most of which is intracellular protoplasm fluid. A relatively clean final product is obtained in this fifth step separation process for the reason that most particulate solids remaining after the second step biolytic process have previously been removed in the third step separation process.

The intermediate nutrient solution supplied to synthesis tank **41** is practically free of particulate matter and the function of separation device **51** can be specifically adapted to separation of the biomass of cells grown and synthesized out of the inoculated solution of the fourth step cellular bioprotein production operation.

Keratinase and Disulfide Conversions

Conversion of nonedible protein-containing wastes such as feathers, nails, and offal to edible forms of animal feeds was developed by *L.E. Weeks and B.S. Wildi; U.S. Patent 3,578,460; May 11, 1971; assigned to Monsanto Company.* The liquid digestive medium used in the process comprises water, keratinase and a reductive type disulfide splitting agent, referred to as reducing agent. Other materials such as buffering agents and surfactants may be included in the liquid digestive medium if desired.

The liquid digestive medium may be prepared, for example, by mixing keratinase and the reducing agent in water. Typically, the medium is a dilute aqueous solution containing from about 0.1 to 5.0% by weight each of the reducing agent and keratinase. Keratinase enzyme is obtained by cultivating a keratinase-producing microorganism in a nutrient fermentation medium. Examples of microorganisms used in the preparation of keratinase include keratinase-producing microorganism of the *Streptomyces* and *Anixiopis* genera, such as, for example, *Streptomyces fradiae* and *Anixiopis recticulispora*.

A particular strain of microorganism used in the preparation of keratinase is a strain of *Anixiopis recticulispora* designated as *Anixiopis recticulispora* ATCC 14362. Any suitable reductive type disulfide splitting agent (reducing agent) may be used. Such reducing agents include mercaptoethanol, sodium sulfite, sodium sulfide, potassium sulfite, sodium hydrosulfite, potassium hydrosulfite, sodium metabisulfite, zinc chlorobenzoate, hydrogen sulfide, cystine, and thioglycolic acid. Excellent results are obtained with mercaptoethanol and sodium sulfite.

Any suitable nonedible protein material obtained from domestic animals may be used in the process. Such materials typically include keratinaceous materials such as hair, nails, hoofs and feathers from chickens, turkeys, hogs, and cattle, and other nonedible protein materials obtained, for example, from the butchering of domestic animals, such as blood, entrails, lungs, etc. In carrying out the process, the nonedible protein material is subjected to the action of the liquid digestive medium at a temperature and for a period of time sufficient to digest the protein material, but insufficient to cause substantial degradation of amino acids. Temperatures above about 70°C tend to cause degradation of valuable amino acids and therefore should be avoided especially for extended periods of time.

Example: A solution was prepared by mixing 0.24 g of mercaptoethanol, 0.10 g sodium sulfite and 0.25 g keratinase (as an aqueous solution) in 10 ml of water. To the resulting solution was added 0.1 g of ground chicken feathers of particle size sufficient to pass a 30 mesh screen. The resulting mixture was gently heated at 37°C with stirring for 1 hour. At the completion of the 1 hour period, digestion of the feathers was substantially complete.

For comparative purposes, the above procedure was repeated exactly, with the exception that keratinase was not included in the solution. At the completion of 1 hour, no digestion of feathers was observed. Heating at 37°C was continued for an additional 15 hour period. At the completion of the additional heating period, no digestion of feathers was observed.

A comparison of the results from the above tests effectively shows the improvement obtained in the digestion of feathers with an aqueous digestive medium containing reducing agent and keratinase, over an aqueous digestive medium containing reducing agent alone. Additional examples of the digestion of nonedible protein materials of the process are given below. In each experiment, a solution of keratinase and reducing agent in water was prepared. Ground feathers were added to the solution and the resulting mixture was stirred at the temperature and for the period of time indicated below.

Reducing Agents	Temperature, °C	Time, hours	Result
Mercaptoethanol	37	16	Complete digestion
Sodium sulfite	37	16	Partial digestion
Mercaptoethanol	25	21	Partial digestion
Mercaptoethanol and sodium sulfite	25	21	Complete digestion

Protease and Disulfide Conversions

In a similar approach *L.E. Weeks and B.S. Wildi; U.S. Patent 3,578,461; May 11, 1971; assigned to Monsanto Company* use a mixture of proteases and a reductive type of disulfide to convert nonedible protein to animal feed. The enzyme used is a protease selected from the group consisting of neutral protease and alkaline protease. Protease enzymes are obtained by cultivating a protease-enzyme-producing microorganism in a nutrient fermentation medium. Examples of microorganisms used in the preparation of protease enzymes include protease-enzyme-producing microorganisms of the *Bacillus, Aspergillus* and *Streptomyces* genera, such as, for example, *Bacillus subtilis, Aspergillus oryzae* and *Streptomyces griseus.*

A particular strain of microorganism used in the preparation of mixtures of neutral protease and alkaline protease is a mutant strain of *Bacillus subtilis*. A culture of this organism is deposited in the culture collection of the U.S. Department of Agriculture and is available to the public. Most protease-enzyme-producing microorganisms yield a mixture of neutral protease and alkaline protease.

The digested proteinaceous material can be recovered from the liquid medium, for example, by spray-drying, oven-drying or by simple evaporation. The proteinaceous material may also be recovered by precipitation, as for example, with a salt, such as magnesium sulfate. The recovered material may be incorporated into an animal feed composition and serves as a source of protein in the feed. If desired, the liquid medium containing the digested protein may be absorbed by spraying onto an absorbent edible carrier having nutritional value, and then fed directly to the animal. Examples of suitable carriers include soy grits, wheat middlings, cornmeal, fishmeal, and ground corncobs. The liquid medium may also be dried by evaporation and then fed directly to the animal.

Example 1: A solution was prepared by mixing 0.2 g of mercaptoethanol, 0.1 g sodium sulfite and 0.0125 g of a mixture of alkaline protease and neutral protease in 10 ml of water. The enzyme mixture was shown by assay to contain 1.327×10^6 protease units per gram and 0.219×10^6 alkaline protease units per gram. To the resulting solution was added 0.1 g of ground chicken feathers of particle size sufficient to pass a 30 mesh screen. The resulting mixture was gently heated at approximately 37°C with agitation. Digestion as evidenced by solubilization of the feathers was complete after two hours. A proteinaceous material rich in nutritionally valuable amino acids and useful as a source of protein in animal feeds is obtained.

Example 2: A solution was prepared by mixing 0.0125 g of a mixture of alkaline protease and neutral protease in 10 ml of water. To the solution was added 0.1 g of ground chicken feathers of particle size sufficient to pass a 30 mesh screen. The resulting mixture was gently heated at approximately 37°C with agitation. Digestion of the feathers was effected in approximately 16 hours.

Fish Waste Conversion

When marine life and livestock are processed, a waste liquor is produced which contains a high amount of water-soluble proteins and other food products. This waste liquor has been normally discharged into water bodies such as rivers and oceans. However, with the advent of stringent anti-water-pollution legislation, the direct discharge of the waste liquor is no longer possible. Accordingly, efficient processes for treating such waste liquors are now required.

During the treatment of marine products and livestock, the products are normally soaked in water to remove blood, fats and extraneous masses from the torn meat. Thus, water-soluble proteins cause an increase of the BOD of the waste liquor. The increased BOD is a major contributing factor to water pollution, and therefore the proteins have to be separated and recovered. A process has been known for recovering proteins from such waste liquors by adding polyacrylates to the waste liquor to form a coagulum of the proteins after adjusting the pH of the waste liquor, and forming a froth by means of microfine air bubbles generated by introducing water containing air dissolved under pressure into the waste liquor. However, in this process it is extremely difficult to dehydrate and dry the protein coagulum in the froth and it is also hard to separate and recover the fats and oils. An improved process for treating the waste liquor froth for recovery of fats and proteins was developed by *T. Sato and K. Ishida; U.S. Patent 3,891,770; June 24, 1975; assigned to Kyowa Hakko Kogyo Co., Ltd., Japan.*

It was found that the proteins in the froth formed by adding, for example, polyacrylates to the waste liquor can be solubilized either by adjusting the pH of the froth to alkalinity and then treating with protease, or by subjecting the froth to thermal pretreatment to denature the froth and then treating the denatured froth with protease. In such manner, the fats and oils contained in the treated froth are readily separated. Furthermore, it was found that the residual protein solution can be processed into a paste by concentration.

Polyacrylate is added to a waste liquor containing water-soluble proteins, fats and oils, at a concentration of 10 to 320 ppm based on the weight of the waste liquor. The polyacrylate used is a salt of a polyacrylic acid having a repeating structure of:

$$-CH_2-\underset{\displaystyle COOH}{\underset{|}{CH}}-$$

A froth is formed using applied pressure such as by microfine aeration, and the resulting froth is separated. The froth is then denatured and mixed with protease to carry out enzymatic decomposition. The following two methods are available for the pretreatment, either of which is applicable depending upon the apparatus used: the pH of the froth is adjusted to alkalinity by adding an alkali such as sodium hydroxide; or the froth is heated to a temperature of 90°C and then immediately cooled to room temperature.

About 0.1 to 0.5% of protease, based on the weight of the dry froth is added to the pretreated froth and enzymatic decomposition is carried out for several hours, while stirring the froth with heating at 30° to 35°C. As a result, the peptide bonds of the proteins constituting the froth are ruptured by enzymatic decomposition, and the proteins are solubilized. At the same time, the fats and oils are caused to be more readily separable from the liquor.

Endopeptidase having a strong ability to solubilize proteins is used. Exopeptidase may also be present. Alkali and neutral proteases obtained from microbial, animal and vegetable origins can be used and a crude enzyme preparation or enzyme culture broth are also applicable. The froth, which has been subjected to enzymatic decomposition is subjected to centrifugal separation to separate and recover the fats and oils.

The residue obtained by separating the fats and oils can, after adjustment to a desired pH, be concentrated under reduced pressure or atmospheric pressure to a paste containing water in a ratio of about 30 to 15%. The resulting paste can be preserved at room temperature for several months without any putrefaction. On the other hand, froth, which has not been subjected to the enzymatic treatment, has no fluidity and cannot be concentrated under reduced pressure. The paste or its mixture with brans may be commercially used as an animal feed. Otherwise, the paste may be subjected to further enzymatic or acid decomposition to produce amino acids and utilized as a seasoning agent.

Example: 2,000 kg of a waste liquor resulting from soaking torn meat of pollacks in water was adjusted to a pH of 5.5 with HCl and 500 g of sodium polyacrylate was added. A froth layer was formed by introducing water maintained under applied pressure into the waste liquor, where 72.5 kg of a froth containing 83% water was obtained.

The froth was adjusted to a pH of 8.5 with sodium hydroxide, and 55 g of a protease having an activity of 100 proteolytic units per milligram was added. Enzymatic decomposition was carried out at 50°C for 3 hours with stirring. The resulting solution was sterilized by heating at 85°C for 20 minutes, and then subjected to centrifugal separation with a Scharples-type centrifuge. As a result, 2.1 kg of fats and oils was separated from the fine particles and soluble fraction and recovered. The later was concentrated under reduced pressure, whereby 14 kg of the paste (total solid content, 70%) was obtained.

COMPANY INDEX

The company names listed below are given exactly as they appear in the patents, despite name changes, mergers and acquisitions which have, at times, resulted in the revision of a company name.

INVENTOR INDEX

U.S. PATENT NUMBER INDEX

NOTICE

Nothing contained in this Review shall be construed to constitute a permission or recommendation to practice any invention covered by any patent without a license from the patent owners. Further, neither the author nor the publisher assumes any liability with respect to the use of, or for damages resulting from the use of, any information, apparatus, method or process described in this Review.